Soil Microbiology and Sustainable Crop Production

Geoffrey R. Dixon • Emma L. Tilston
Editors

Soil Microbiology and Sustainable Crop Production

Springer

Editors
Geoffrey R. Dixon
Centre for Horticulture & Landscape
School of Biological Sciences
University of Reading, Whiteknights
Reading, Berkshire, RG6 6AS, UK
geoffrdixon@btinternet.com

Emma L. Tilston
Scottish Universities Environmental
Research Centre
Scottish Enterprise Technology Park,
East Kilbride, G75 0QF, UK
e.tilston@suerc.gla.ac.uk

ISBN 978-90-481-9478-0 e-ISBN 978-90-481-9479-7
DOI 10.1007/978-90-481-9479-7
Springer Dordrecht Heidelberg London New York

Library of Congress Control Number: 2010934733

Cover illustration:
Background: Sustainable crop production as represented by bed production of salad leaves with adjacent wildlife refuge and trees (copyright: Geoff Dixon)
Small left photo: Three roots of swede showing galling caused by *Plasmodiophora brassicae* (Clubroot); (copyright: Geoff Dixon)
Small middle photo: Roots of pea plant showing nodules caused by nitrifying bacteria (copyright: Oksana Shtark)
Small right photo: Three roots of winter wheat showing moderate to severe damage caused by *Gaeumannomyces graminis* var. tritici (Take-All) (copyright: Emma Tilston).

Printed on acid-free paper

Springer is part of Springer Science+Business Media (www.springer.com)

Preface

Nothing is so fatal to the progress of the human mind as to suppose .... that there are no new worlds to conquer. (Humphrey Davy, English engineer and physicist, public lecture 1810)

Soil and in particular its microbial diversity remains largely an unexplored world. A few researchers have provided insights into the outer edges of this world but it remains mostly unknown and inhabited by a huge diversity of organisms whose biology is open to speculation. Yet it is this world and its inhabitants which arguably hold many of the properties which will enable mankind to surmount the huge problems resulting from a burgeoning population and diminishing land supply. Increasing food production in parallel with conserving and protecting our environment while allowing producers adequate financial returns are the primary challenges facing agricultural science research in the twenty-first century. These factors of food production, environmental protection and producers' profit form a triangle which defines agrarian sustainability. Sustainably raising crop production will only be achieved by gaining far greater understanding of the physics and chemistry of the soil environment in which roots grow and the impact of benign and pathogenic microbes on them. It is at least as important that we understand the world beneath our feet as we do the Earth's atmosphere and oceans and those of neighbouring planets. Increasing the benefits obtained from soil microbes must be linked with cautious care for the world which they inhabit. Ill-judged and ignorant exploitation has lead to, and continues leading to, devastated land where the soil is degraded into lifeless dust-bowls where structure, texture and biological activity are lost and salinity rises. The very nature of soils has slowed their study until recently. Now the tools of molecular biology are offering powerful new ways of unravelling complex relationships and simplifying interactions.

This Book provides an insight into the developing knowledge of soil microbes and points to ways by which they can be utilised in support of agronomically and environmentally sustainable crop production. This context is introduced in the first chapter which sets out the parameters of sustainable production. It is succeeded by analyses of microbiology in natural, unfarmed soil defining the baselines from which agriculture has modified soil resources. Chapters describing nutrient cycling and the development of soil organic matter clearly demonstrate the impact of mankind's activities and means by which these may be tailored to achieve sustainable objectives. Detailed studies of beneficial and pathogenic soil- and root-borne microbes follow identifying the continuous interactions between plants and the organisms with which they co-exist. Outcomes of this co-existence can be either immensely valuable in terms of raising crop health and productivity or totally disastrous leading to disease and death. Husbandry practices affect the balance between these outcomes. Gradually the significance of the way in which land-use interacts with crop production and the potential for its manipulation to raise sustainable yields is being extending into mainstream crop production. This philosophy is not new but had been ignored and side-lined for at least the latter part of the twentieth century by crop production methods based almost solely on approaches which have targeted the plant as opposed to the soil. Crop agronomy using a soil-based approach demands as a priority the revision of plant breeders' targets. A result should be new cultivars tailored to attain maximum yields and quality in harmony with beneficial soil-borne microbes. This approach will enable crop producers to make far greater use of integrated systems for the control of pests and pathogens. Biological control used with cultivars fitted for increased productivity growing in soil managed by husbandry systems which enhance beneficial microbial populations could deliver the sustainable yield enhancements needed by population growth. Greater understanding and manipulation of the soil environment must include knowledge of the aerial environment and the manner by which it is changing. Both environments interact and influence each other consequently consideration must be given to the impact of global climate change both directly and indirectly on soil microbial populations. Changes to air temperature, precipitation and wind will have direct and substantial effects on soil borne microbes. Evidence for this is already apparent in the movements of aggressive pathogenic species in to previously un-colonised regions. Ultimately, increasing crop productivity and caring for the soil environment can only be considered to be fully sustainable if farmers and growers are able to maintain viable and successful businesses. Soil is the first resource which suffers from lack of care where economic sustainability is absent. Ensuring adequate incomes for soil users is integral and essential for the achievement of sustainability in both crop production and environmental care and conservation. Science and its practitioners can, given the necessary resources, open up and more fully explore soils for the greater benefit of mankind and the Earth's environment and its natural biodiversity. It is the task of politicians to understand the opportunities which this offers and ensure that the general public, the tax-payers, recognise why financial resources should be applied to the crucial task of acquiring knowledge of the microbes in soil. Regrettably, for at least the past generation there has been a worldwide failure by politicians and their advisors to provide adequately for studies of the agricultural and soil sciences. Unless this situation is reversed in the very short-term none of the opportunities outlined in this Book will be realised. That will exacerbate the famines which afflict mankind and continue the destruction of the Earth's soil, its environment and its inhabitants.

This Book was conceived as a contribution towards the international debate on population growth, food insecurity and the conservation of biodiversity. Gradually there is a recognition that soils demand as much attention as the atmosphere as an integral part of the biosphere. It has been written by an international team of researchers recruited from around the world. Substantial effort, expert knowledge and energy have brought this project to fruition for which the Editors Geoffrey R. Dixon and Emma L. Tilston sincerely thank each member of the team of authors. All have been prepared to devote considerable time to fulfilling this task and have shared with the Editors the joys and frustrations inherent in writing such a Book. We thank all of them for the care with which they have submitted excellently authoritative manuscripts. The Editors have been supported by many family and friends. Particular thanks go to Mrs Kathy Dixon who with much patience and good humour aids and accepts her husband's literary activities. Both Editors have made much use of the facilities in the Library and our respective Departments (Horticulture and Landscape and Soil Science) at the University of Reading for which we offer very grateful thanks and acknowledgement. Authors have offered acknowledgements at the ends of their Chapters.

Professor Geoffrey R. Dixon and Dr. Emma L. Tilston

March 2010 Sherborne, Dorset and Glasgow

Contents

Contributors

Teri C. Balser
Department of Soil Science, University of Wisconsin-Madison,
1525 Observatory Drive, Madison, WI 53706, USA
tcbalser@wisc.edu

Alexey Y. Borisov
All Russia Research Institute for Agricultural Microbiology (ARRIAM),
Laboratory of Genetics of Plant-Microbe Interactions, Department of Biotechnology,
Podbelsky chausee 3, St. Petersburg, Pushkin 8, 196608, Russia
ayborisov@yandex.ru

Michael Brownbridge
Vineland Research and Innovation Centre,
P.O. Box 8000, Vineland Station, ON LOR 2EO, Canada
michael.brownbridge@vinelandresearch.com

Peter Clinton
Scion, P.O. Box 29-237, Fendalton, Christchurch 8041, New Zealand
peter.clinton@scionresearch.com

Leo Condron
Agriculture and Life Sciences, Lincoln University,
P.O. Box 84, Lincoln 7647, Christchurch, New Zealand
leo.condron@lincoln.ac.nz

Geoffrey R. Dixon
Centre for Horticulture & Landscape, School of Biological Sciences,
Whiteknights, University of Reading, Berkshire, RG6 6AS, UK
geoffrdixon@btinternet.com

Jennifer A.J. Dungait
Rothamsted Research – North Wyke, Okehampton, Devon, EX20 2SB, UK
jennifer.dungait@bbsrc.ac.uk

Vadakattu V.S.R. Gupta
CSIRO Entomology, PMB No. 2, Glen Osmond South Australia 5064, Australia
gupta.vadakattu@csiro.au

Jessica L.M. Gutknecht
Department of Soil Ecology, Helmholtz-Centre for Environmental Research-UFZ, Theodor-Lieser-Str, 4, D06120 Halle (Saale), Germany
Jessica.gutknecht@ufz.de

Jim A. Harris
Natural Resources Department, School of Applied Sciences, Cranfield University, Cranfield, Bedfordshire, MK43 0AL, UK
j.a.harris@cranfield.ac.uk

Robert A. Hill
Bio-protection Research Centre, Lincoln University, PO Box 84, Lincoln University, Lincoln 7647, Christchurch, New Zealand
robert.hill@lincoln.ac.nz

David W. Hopkins
Scottish Crop Research Institute, Invergowrie, Dundee, DD2 5DA, UK
david.hopkins@scri.ac.uk

Zhiqun Huang
Scion, PO Box 29-237, Fendalton, Christchurch 8041, New Zealand
zhiqun.huang@scionresearch.com

Trevor A. Jackson
AgResearch, Private Bag 4749, Christchurch 8140, New Zealand
trevor.jackson@agresearch.co.nz

Chao Liang
Department of Soil Science, University of Wisconsin-Madison, 1525 Observatory Drive, Madison, WI 53706, USA
chaoliang@wisc.edu

Laura Lipps
Gaylord Nelson Institute for Environmental Studies, University of Wisconsin-Madison, Madison, WI 53706, USA
laalipps@yahoo.com

Petra Marschner
Soils, School of Agriculture, Food & Wine, The University of Adelaide, Adelaide, South Australia 5005, Australia
petra.marschner@adelaide.edu.au

Lindsey K. Moritz
Department of Soil Science, University of Wisconsin-Madison, 1525 Observatory Drive, Madison WI, 53706, USA
lkmoritz@gmail.com

Andrew D. Noble
International Water Management Institute,
c/o National Agriculture and Forestry Research Institute (NAFRI),
Ban Nongviengkham, Xaythany District, Vientiane, Lao PDR
a.noble@cgiar.org

Maureen O'Callaghan
AgResearch, Private Bag 4749, Christchurch 8140, New Zealand
Maureen.o'callaghan@agresearch.co.nz

Wilfred Otten
SIMBIOS Centre, University of Abertay, Bell Street, Dundee, DD1 1HG, UK
w.otten@abertay.ac.uk

Nikolay A. Provorov
All Russia Research Institute for Agricultural Microbiology (ARRIAM),
Podbelsky chausee 3, St. Petersburg, Pushkin 8, 196608, Russia
provorov@newmail.ru

Zed Rengel
School of Earth and Environment, The University of Western Australia,
Crawley, Western Australia 6009, Australia
zed.rengel@uwa.edu.au

Sawaeng Ruaysoongnern
Faculty of Agriculture, Khon Kaen University, Khon Kaen, Thailand
sawaeng@kku.ac.th

Harpinder S. Sandhu
CSIRO Sustainable Ecosystems, PMB No. 2, Glen Osmond,
South Australia 5064, Australia
Harpinder.Sandhu@csiro.au

Oksana Y. Shtark
All Russia Research Institute for Agricultural Microbiology (ARRIAM),
Laboratory of Genetics of Plant-Microbe Interactions, Department of
Biotechnology, Podbelsky chausee 3, St. Petersburg, Pushkin 8, 196608, Russia
oshtark@yandex.ru

Tom Sizmur
Department of Soil Science, University of Reading, Whiteknights,
Berkshire, RG6 6DW, UK
t.p.sizmur@reading.ac.uk

Christine Stark
Environmental and Geographical Sciences,
Manchester Metropolitan University, Manchester, UK
c.stark@mmu.ac.uk

Alison Stewart
Bio-protection Research Centre, Lincoln University,
PO Box 84, Lincoln University, Lincoln 7647, New Zealand
alison.stewart@lincoln.ac.nz

Igor A. Tikhonovich
All Russia Research Institute for Agricultural Microbiology (ARRIAM),
Podbelsky chausee 3, St. Petersburg,
Pushkin 8, 196608, Russia
contact@arriam.spb.ru

Emma L. Tilston
Scottish Universities Environmental Research Centre,
Scottish Enterprise Technology Park, East Kilbride, G75 0QF, UK
e.tilston@suerc.gla.ac.uk

Devin Wixon
Department of Botany, University of Wisconsin-Madison,
430 Lincoln Drive, Madison, WI 53706, USA
dwixon@gmail.com

Stephen D. Wratten
Bio-Protection Research Centre, Lincoln University,
P.O. Box 84, Lincoln 7647, New Zealand
steve.wratten@lincoln.ac.nz

Vladimir A. Zhukov
All Russia Research Institute for Agricultural Microbiology (ARRIAM),
Laboratory of Genetics of Plant-Microbe Interactions,
Department of Biotechnology, Podbelsky chausee 3,
St. Petersburg, Pushkin 8, 196608, Russia
vladimir.zhukoff@gmail.com

Chapter 1
The Nature of Sustainable Agriculture

Andrew D. Noble and Sawaeng Ruaysoongnern

Introduction

By 2050 the world's population will have increased by a third and demand for agricultural products will rise by 70% with meat consumption doubling. Lateral expansion of the agricultural sector through the clearing of land is untenable without significant negative implications on already stressed natural ecosystems and the range of drivers, including climate variability, that farmers will have to cope with, will require changes in the way we undertake agriculture. In meeting future food production demands without consuming more land and water will require technological innovation and changes in the way agriculture is undertaken. The chapter discusses the future global demands for food and highlights the importance of addressing soil chemical and physical constraints in increasing the productivity of degraded production systems. The role of clays in permanently changing the surface charge characteristics of soils and the potential for selected grass species to remediate compacted soil layers are presented as possible options in addressing the sustainability of degraded production systems.

The fragility of our current agriculture systems was brought into sharp focus during the recent global food crisis of 2007–2008. Not since the 1970s when the tumultuous famines of south Asia and the Horn of Africa rocked the world, have we seen such turmoil in global food supply and prices. National governments imposed restrictions on grain exports and people came out on the streets protesting dramatic increases in the prices of basic commodities, with the poorest and most marginalized sectors of the global community being most vulnerable. Governments

A.D. Noble (✉)
International Water Management Institute, c/o National Agriculture and Forestry Research Institute (NAFRI), Ban Nongviengkham, Xaythany District, Vientiane, Lao PDR
e-mail: a.noble@cgiar.org

S. Ruaysoongnern
Faculty of Agriculture, Khon Kaen University, Khon Kaen, Thailand
e-mail: sawaeng@kku.ac.th

G.R. Dixon and E.L. Tilston (eds.), *Soil Microbiology and Sustainable Crop Production*,
DOI 10.1007/978-90-481-9479-7_1, © Springer Science+Business Media B.V. 2010

around the world were caught off guard and scrambled to address the problem. How could this happen in the twenty-first century?

It has been argued that the recent food crisis was the confluence of a range of drivers that included but was not limited to, the effects of competition for cropland from the growth in biofuels; low global cereal stocks; high oil prices; speculation in food markets; restrictions on grain exports in key countries; and extreme weather events (Nellemann et al. 2009). The impacts of reduced food availability, higher food prices and thus lower access to food by many people have been dramatic. It is estimated that the crisis resulted in a dramatic increase in several central commodity prices, drove an estimated 110 million people into poverty and added 44 million more to the already undernourished (World Bank 2008).

The underlying supply and demand tensions are still present today as evidenced by the elevated food price index and price indices of major commodities when compared to the recent past (Fig. 1) (FAO Statistics 2009). These recent failures may reflect our inability to adapt to transformative changes as we push our agro-ecosystems towards a threshold. Coping with short-term food price volatility is daunting enough, particularly for low and middle income countries, but the long-term challenge of avoiding a perpetual food crisis under conditions of climate change and global warming is far more serious (Battisti and Naylor 2009) and will possibly challenge our notion of civilization. We are living in times of great uncertainty with respect to food supply and hence the sustainable management of our agricultural production systems is an imperative and one that will challenge farmers, scientist and policy makers over the coming decades.

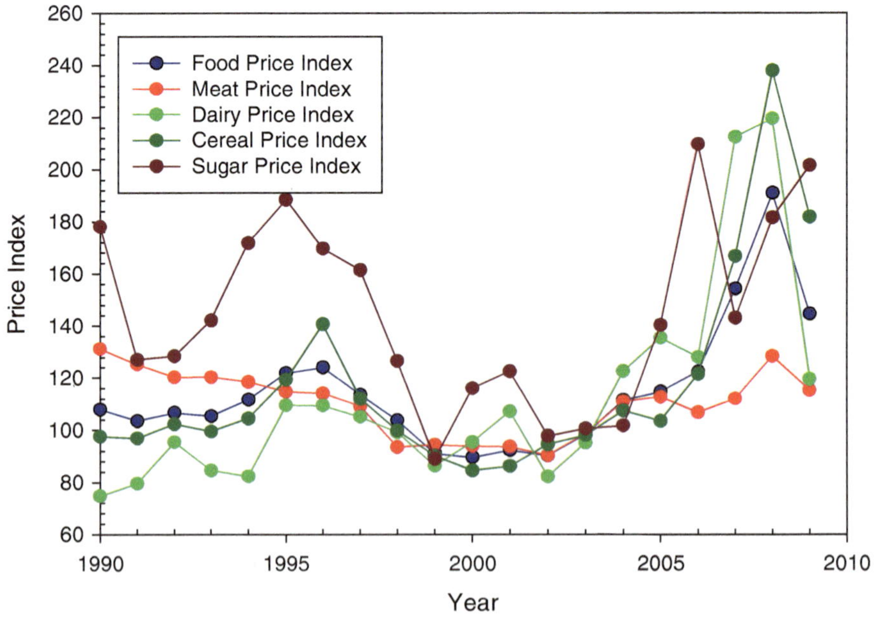

Fig. 1 Food price indices for a range of commodities (FAO Statistics 2009)

Recent Challenges in the Global Foodscape

Food and feed crop demand will nearly double in the coming 50 years. The two main factors driving how much food we will need are population growth and dietary change (Comprehensive Assessment of Water Management and Agriculture 2007). Over a 24 h period the global population grows by 200,000 with a concomitant impact on overall food demand. Projections in global population growth indicate that by 2050, a further 2.5 billion people will be added to the current 6.7 billion people that inhabit the planet (UN Population Division 2007). These new additions to the global community will require adequate housing and food placing greater demands on already stressed and compromised agro-ecosystems and natural resources. This growth in populations is geographically specific with the largest projected increases occurring in China, India and Southeast Asia. On a relative basis, Africa will experience the most rapid growth, over 70% faster than that in Asia (annual growth of 2.4% versus 1.4% in Asia) and in sub-Saharan Africa alone, the population is projected to increase from about 770 million to nearly 1.7 billion by 2050 (UN Population Division 2007).

The implications of global population growth on poverty are highlighted in a report released in 2008 by the World Bank (2008). It concluded that the number of people living in extreme poverty may be higher than previously thought. With a threshold of extreme poverty set at US$1.25 a day (2005 prices), the number of people living in extreme poverty in 2005 was estimated 1.4 billion. Reflecting on the consequences of extreme poverty, it is estimated that approximately ten million die of hunger and hunger-related diseases annually (World Bank 2008). This dire state of affairs is compounded by the fact that the number of children that are underweight exceeds 140 million in developing countries. The legacy of poor nutrition at critical stages in a child's development is evident throughout an entire generation with significant implications on GDP, productivity and health services. There are large regional disparities in the aforementioned trends. For example, China has been extremely successful in more than halving the proportion of underweight children between 1990 and 2006. Contrasting this and despite improvements since 1990, it is estimated that 50% of children are underweight in South Asia, a region that alone accounts for more than half the worlds malnourished children (Nellemann et al. 2009).

Concomitant with the rising demand for food associated with population growth there is an observable and significant shift in the dietary intake of communities. With rising incomes and continuing urbanization, food habits change toward more nutritious and more varied diets – not only toward increasing consumption of staple cereals but also to a shift in consumption patterns among cereal crops and away from cereals towards livestock and fish products and high-value crops (Comprehensive Assessment of Water Management and Agriculture 2007).

The global production of cereals is the cornerstone of world food supply accounting for approximately 50% of the calorie intake of humans (FAO 2003). Any changes in the production of or in the use of cereals for non-human consumption has a direct effect on the calorie intake of a large fraction of the world's population. As nearly half of the world's cereal production is used to produce animal feed, the dietary proportion of meat has a major influence on global food demand (Keyzer et al. 2005). With meat consumption projected to increase from 37.4 kg person^{-1} year^{-1} in 2000 to over 52 kg person^{-1} year^{-1} by 2050 (FAO 2006), cereal requirements for more intensive meat production may increase substantially to more than 50% of total cereal production (Keyzer et al. 2005). They conclude that compared to other factors that are generally expected to affect the future world food situation, the quantitative impact of the increased cereal feed demand will greatly exceed that of genetically modified organisms (GMOs) and climate change in the coming 3 decades (Keyzer et al. 2005).

With urbanization, industrialization, energy demand and population growth there has been a growing trend in converting cropland to other uses. This is best exemplified in China where 14.5 million hectares of arable land has gone out of production between 1979 and 1995 (ICIMOD 2008). It is estimated that 120 million hectares will be needed to support the traditional growth in food production by 2030, mainly in developing countries (FAO 2003). These estimates of cropland and associated yield increases to meet future demand for food will inevitably contribute to environmental degradation and losses in the provision of ecosystem services.

The production of biofuels has grown exponentially over the past decade fueled by higher global oil prices and an initial perception of their role in reducing CO_2 emissions (FAO 2008). It is estimated that biofuels, including biodiesel from palm oil and ethanol from sugarcane, corn and soybean, accounted for about 1% of the total road transport in 2005, and could potentially reach 25% by 2050, with the EU having set targets as high as 10% by 2020 (World Bank 2008; FAO 2008). For several countries including Indonesia and Malaysia, biofuels are seen as an opportunity to improve rural livelihoods and contribute to economic growth through exports (Fitzherbert et al. 2008).

While biofuels are a potential low-carbon energy source, the conversion of rainforests, peatlands, savannas, or grasslands to produce biofuels in the US, Brazil and Southeast Asia may create a "biofuel carbon debt" by releasing 17 to 420 times more carbon dioxide than the annual greenhouse gas reductions that these biofuels would provide by displacing fossil fuels (Fargione et al. 2008; Searchinger et al. 2008). In contrast, biofuels made from waste biomass or from biomass grown on degraded and abandoned agricultural lands planted with perennials incur little or no carbon debt and can offer immediate and sustained green house gas advantages (Fargione et al. 2008).

There are a range of drivers that will influence global food supply in this millennium. However, the greatest challenge for the global community will undoubtedly be adapting to climate change. The impacts of a changing climate are already being experienced with increased frequencies in the occurrence of droughts, floods, and heat waves bringing greater risks to our agricultural production systems. Before discussing the notion of sustainable agriculture it is pertinent to briefly reflect upon the dominant drivers that will influence the nature of agricultural systems and the challenges that the sector faces in the coming decades.

Land Degradation and It Impact on Sustainability

Central to the discussion of the notion of sustainable agriculture is the impact of a range of drivers on soil degradation and quality. Changes in land use associated deforestation and inappropriate land management has had a negative impact on approximately 2 billion hectares of agricultural (Pinstrup-Andersen and Pandya-Lorch 1998). However, this estimate hides large differences between zones and the particular vulnerabilities of soils in the tropics (Stocking 2003). It is estimated that unsustainable land use practices result in net losses of cropland productivity that averages approximately 0.2%$year^{-1}$. The combined effects of competition for land from growing populations, reduced opportunity for migration and rotation along with higher livestock densities, result in frequent overgrazing and, hence, loss of long-term productivity (Nellemann et al. 2009). Satellite measurements show that between 1981 and 2003, there was an absolute decline in the productive land area (as Net Primary Productivity) across 12% of the global land area. The areas affected are home to about 1–1.5 billion people, some 15–20% of the global population (Bai et al. 2008).

Land degradation effectively results in a decline in the productive capacity of soils that can be attributed to changes in physical, chemical, and biological attributes from some ideal state brought about by natural or anthropogenic influences (Latham 1994; Lal 1990). These all contribute to what in general terms are referred to as soil quality. Land degradation is often assessed as the amount of soil material that is removed from a landscape by water and wind erosion, since such physical changes are obvious and quantifiable. Effects of degradation on chemical properties, soil nutrient cycling, and biological activity of soils may be less visible, but are functionally as important as the former, particularly in developing countries (Drechsel et al. 2004; Eswaran et al. 2000).

At its most fundamental level, soil degradation is caused when natural factors such as wind, rain, evaporation and gravity combine with human management practices to remove, or add, soil or its nutrients at a given site leading to reduced productivity of land, water and/or other ecosystem services. Whilst soil degradation could be considered a 'natural' process associated with pedological development within a geological timeframe, it is human interventions which often results in accelerated rates of these processes that are of social concern at the human time scale. Of the range of human actions associated with soil degradation, deforestation and the removal of natural vegetation, overgrazing, and poor agricultural practices have been cited as the most common causes (Olderman et al. 1991). As Lal (Lal et al. 2004) shows, 'ploughing' is a particularly destructive human intervention in tropical agricultural systems. Inappropriate water and waste management and use of heavy equipment can also lead to degradation in the form of salinization, pollution and compaction. The human role in soil degradation is now relatively well understood. Globally, degradation of land resources, and as a consequence water resources, threaten food security, particularly in the hundreds of millions of households on marginal lands, and they reduce the level of environmental services that societies derive from the natural resources, including the provision of clean water (Penning de Vries et al. 2002).

The resilience of an ecosystem to buffer change, including resistance to a shock, is a fundamental property of ecosystems and provides a measure of vulnerability to degradation (Stocking and Murnaghan 2001). The concept suggests that a system, such as an agro-ecosystem, responds more or less gradually to changes in its management or environment (climate) but only up to a certain threshold. Once this threshold is surpassed, adjustments cannot take place rapidly enough and the system collapses. Realization of the existence of a 'resilience threshold' for natural resources has important consequences for their management (Sayer and Campbell 2003). Resilient soils (e.g. Vertisols) may be degraded to a significant degree and still be restored to some equilibrium state (e.g. in traditional slash and burn systems). Other soil types (e.g. Ultisols) are much less resilient. Once a threshold is reached, these systems collapse and cannot easily be brought back to their former state. This has been referred to as retrogressive succession as species diversity and functionality changes due to pedogenesis and landscape age (Walker et al. 2000; Jenny 1941). Soils that degrade beyond some threshold level are often damaged beyond repair at time scales of human relevance and may never recover

Fluxes between soil organic carbon and the atmosphere are of importance when considering overall changes in global carbon budgets. Changes in land use and associated conversion to agriculture have over the last 50 years contributed an estimated 50,000,000 Mg C to the atmosphere which represents approximately one third of the total loss from soil and vegetation (IPCC 2000). The rate of soil organic matter (SOM) mineralization depends predominantly on temperature and oxygen availability, land use, cropping system, soil and crop management (Lal 1990). Clearly inappropriate management of soil resources that result in a loss in SOM represent resource degradation and have contributed significantly to overall carbon emissions. Through the adoption of regenerative sustainable production systems that focus on the conservation of organic matter and its enhancement in soils, it is estimated that globally 1,302 Mt C year^{-1} could be achieved by 2010 (IPCC 2000).

Whilst it is customary to focus on the solid soil component when discussing land and soil degradation such approaches ignore an important dimension in the land degradation equation, since the loss of functionality of the agro-ecosystem refers as much to 'water' as to 'soil'. Degradation of soil has direct impacts on soil water holding capacity, erosion, increased sediment loads and siltation of water storage facilities and a decline in water quality, and often has significant landscape impacts. These impacts on water productivity have long been significant, especially in arid regions, but are increasingly important globally as levels of water stress increase.

To place soil degradation in perspective, selective examples are explored that focus on soil chemical and physical degradation and its remediation. Whilst biological components associated with the functionality of soils is equally important in soil quality, this aspect will be covered in several chapters that follow in this Book. What is important in discussing the sustainability of our agricultural soils is that all three soil attributes, namely chemical, physical and biological, need to be considered holistically and not as distinct components.

Chemical Degradation

In their natural state, fertile soils provide nutrients and water to sustain plant biomass production. Both water and nutrients are held by the inorganic soil substrate (ISS) and soil organic matter (SOM). The water and nutrient holding capacities of both of these components can be reduced both naturally and through anthropogenic activities. With low holding capacities, soil water drains easily down below the rooting zone and nutrients are leached. At a pedological level, a reduction in the water and nutrient holding capacities of a profile constitutes soil degradation.

The nutrient holding capacity of the ISS and SOM is related to the electrical charge characteristics of these two substrates. As a consequence of chemical degradation, the surface charge properties of soils are reduced, thereby affecting their ability to retain base cations. This has significant implications for the retention and supply of essential plant nutrients, and their subsequent leaching. For quantitative assessment of this decline in base cation retention, a method termed 'surface charge fingerprinting' (SCF) has been developed. A SCF is the graphical response curves of the soil's cation and anion exchange capacity (CEC, AEC) to changes in pH. The full concept and methodology for determining these fundamental properties are comprehensively described elsewhere (Gillman and Sumpter 1986a, b; Menzies and Gillman 1997). The CEC can be separated into two components, namely, the basic cation exchange capacity (CEC_B) which is the total amount of basic cations (normally measured using Ca^{2+} as the exchanging ion) that can be retained in an exchangeable form at any particular solution pH and ionic strength; and total cation exchange capacity (CEC_T) which is the total amount of basic and acidic cations (i.e. Al^{3+}) that can be retained in an exchangeable form at any particular solution pH and ionic strength (Gillman and Sumpter 1986a, b). By comparing the CEC of degraded and non-degraded soils, a quantifiable measure of the degree of chemical degradation a soil has undergone can be established. Using the concept of SCF, the degree of chemical degradation that occurs both naturally as well as through human intervention can be demonstrated.

In general, in their natural state soils maintain productive and diverse ecosystems that are dependent on efficient resource utilization. A characteristic of these systems is their reliance on SOM to cycle nutrients from the soil through the plant and back to the soil through plant debris. Soil organic matter effectively acts as a slow release nutrient delivery system that mediates the cycling of nutrients between soil and vegetation. However, vegetation structure and functionality of natural ecosystems are a function of pedogenesis and landscape age (Walker et al. 2000). In northern Queensland Australia, soils developed from recent quaternary basalt flows that are comparatively young having undergone limited pedogenesis are nutrient-rich, have high levels of organic carbon in their pristine state, high CEC_T (Table 1) and support complex notophyll vine forests (Walker et al. 2000). Contrasting this, soils that are formed from highly weathered metamorphic schist are inherently infertile when compared to younger soils, have lower levels of organic carbon, limited CEC_T and support simple notopyll vine forests (Table 1). These soils are naturally low in basic cation content associated with advanced weathering, with the

Table 1 Differences in selected soil chemical properties of contrasting Notophyll Vine Forests on soils derived from metamorphic schist and basalt in Northern Queensland, Australia (Gillman and Abel 1986; Noble et al. 2001)

	Basalt derived soils		Metamorphic schist derived soils	
Soil attribute	Complex notophyll forest	Tea plantation	Simple notophyll forest	Sugarcane
Soil pH	5.2	4.9	4.3	4.4
Organic carbon (g kg^{-1})	66.0	35.0	19.0	10.0
Ca^{2+} ($cmol_c$ kg^{-1})	1.30	0.24	0.07	0.07
Mg^{2+} ($cmol_c$ kg^{-1})	1.02	0.17	0.26	0.05
K^+ ($cmol_c$ kg^{-1})	0.23	0.26	0.09	0.07
Na^+ ($cmol_c$ kg^{-1})	0.09	0.04	0.06	0.02
Al^{3+} ($cmol_c$ kg^{-1})	0.25	0.15	1.40	1.00
CEC_T at pH 5.5 ($cmol_c$ kg^{-1})	4.94	1.85	2.68	1.52

CEC largely being satisfied by exchangeable aluminium (Table 1). The fragile nature of these relatively infertile schist derived soils can be masked by climax rainforest vegetation in their undisturbed state, where a delicate balance is maintained between nutrients in the biomass, organic matter in surface soil horizons and the efficient cycling of nutrients between these components and the soil. If a pedologically young system is disturbed, it will tend to return to something similar to its previous state relatively quickly, a process/concept termed progressive succession (Walker et al. 2000). However, in an old system (highly weathered and in an advanced state of soil formation) such returns to a pre-disturbance state or a variant thereof, is highly unlikely, this being termed retrogressive succession.

The importance of soil organic carbon under these two contrasting vegetation types is clearly demonstrated when these systems of converted to agriculture (Table 1). Declines in soil organic carbon, exchangeable cations and CEC_T are clearly evident. The decline in CEC_T for every percent decline in organic matter is 1.28 and 0.99 $cmol_c$ kg^{-1} for the highly weathered metamorphic schist and basalt derived soils respectively, reflecting differences in the quality of the organic matter produced under these contrasting forest systems. Hence conversion of soils that have undergone a greater degree of pedogenesis requires high levels of inputs in order to maintain productivity, an issue that is discussed below. When climax ecosystems are disturbed through clearing and continuous cultivation, the productivity of strongly weathered soils often declines rapidly due to a loss in SOM, accelerated soil acidification, crusting and erosion (Gillman and Abel 1986; Aweto et al. 1992; Kang and Juo 1986; Kooistra et al. 1990; Willett 1995; Gillman et al. 1985; Plamondon et al. 1991; Van der Watt and Valentin 1992). Consequences of such a catastrophic decline in fertility can best be illustrated by comparing the abundance of vegetation in undisturbed systems (pristine systems) with the meager and unpalatable vegetation of those that have undergone some form of changed land use. The influence of changed land management induced by man on soil chemical degradation can be found in two representative sites from Northeast Thailand both of which

Table 2 Selected soil chemical properties of the surface 0–10 cm depth interval from paired sites in Northeast Thailand indicating differences in attributes associated with conversion to agricultural production (Noble et al. 2003)

Attribute	Undisturbed	Disturbed	Undisturbed	Disturbed
Location	RS4		CS1	
Cropping/plant system	*Dipterocarp* forest	Rice	*Dipterocarp* forest	Cassava
Parent material	Alluvium		Alluvium	
Soil pH	4.87	5.18	5.18	5.00
Organic carbon (g kg^{-1})	8.5	2.1	6.7	3.3
Ca^{2+} ($cmol_c$ kg^{-1})	0.34	0.14	0.74	0.25
Mg^{2+} ($cmol_c$ kg^{-1})	0.20	0.03	0.40	0.11
K^+ ($cmol_c$ kg^{-1})	0.07	0.04	0.09	0.03
Na^+ ($cmol_c$ kg^{-1})	0.03	0.01	0.00	0.00
Exch. acidity ($cmol_c$ kg^{-1})	0.57	0.19	0.12	0.35
CEC_B pH 5.5 ($cmol_c$ kg^{-1})	1.33	0.48	1.57	0.83
Sum bases	0.65	0.22	1.24	0.39
Clay content (%)	4.9	2.9	4.9	4.7
pH buffer capacity (cmol H^+ kg^{-1}.unit pH^{-1})	1.07	0.41	0.92	0.62

are on light textured sandy soils (Table 2) (Noble et al. 2003). Over the past 3 decades significant land clearing has occurred throughout Northeast Thailand associated with the expansion of upland cropping systems. The original climax *Dipterocarp* forests have been replaced by intensive cropping systems that are dominated by rice, cassava and sugarcane. Remnant *Dipterocarp* forests can be found in isolated pockets that allow direct comparisons to be made between an undisturbed state and the current agronomic enterprise that is in close proximity. Changes in soil chemical properties for two of these sites under climax *Dipterocarp* forest and adjacent farmer fields under production for the past 40 years are presented in Table 2 for the 0–10 cm depth interval. Assuming a soil bulk density of 1,300 kg m^{-3}, the loss in soil organic carbon (OC) is equivalent to 8.3 and 4.4 t C ha^{-1} in the top 0–10 cm depth interval for sites RS4 and CS1 respectively. Accompanied with this decline in OC are a loss of exchangeable cations and an increase in exchangeable acidity in the case of site CS1. By comparing the difference in CEC between the undisturbed site and the adjacent agronomic production system at pH 5.5, the degree of permanent change in the charge characteristics associated with changed land management becomes clear. There has been a 64% and 47% decline in the CEC_B with conversion to agriculture in the previous examples. The loss of plant nutrients in the same layer has been even more dramatic: as much as 70% of the nutrients are removed, due to enhanced leaching and to extraction by crops without full replenishment by inorganic fertilization or other approaches.

Soil degradation associated with changed land use is not confined to light textured soils. Highly weathered clay based soils of the humid tropics converted to intensive agricultural systems exhibit significant degradation of both physical and chemical attributes. The properties of an Oxisol cleared of climax rain forest 53 years previously and currently under tea production in north Queensland were compared with an adjacent undisturbed forest (Noble et al. 2001). Soil pH declined by approximately 0.6 units under the tea production system. The loss of soil organic carbon from the tea site amounted to an equivalent of 58.1 t ha^{-1} over the 0–20 cm depth. This dramatic decline in soil organic carbon has had a significant impact on the soils ability to retain cations and hence the inherent fertility of the disturbed site as evident in the charge fingerprint for the 0–10 cm depth interval (Fig. 2). However, with depth, differences between the disturbed and undisturbed systems declined so that at 50 cm there were no differences between systems (Fig. 2).

Contrasting the previous examples, which represent strongly weathered soils (Ultisols and Oxisols) of the tropics whose mineralogy is dominated by either kaolinite or sesquoxides, an example of a soil whose ISS is dominated by smectite-vermiculite (Vertisol) is presented from the sub-tropics. The soils are from a long-term trash management trial established in 1939 at Mt Edgecombe, South Africa and are part of the oldest sugarcane soil management trials in the world. For brevity, selected results of SCF from soil samples collected in 1996 are presented. Figure 3 represents the charge fingerprints of composite samples from the control treatment (under grass for the duration of the trial) and an extreme treatment that had been burnt, harvest residues removed, and an annual fertilizer dressing of 140 kg N ha^{-1}, 20 kg P ha^{-1} and 140 kg K ha^{-1} applied. Over the duration of the study there has been a significant decline in soil pH and associated base stripping in the 0–5 cm depth interval (Van Antwerpen and Meyer 1998). It is of note that the inherent charge characteristics of these soils are an order of magnitude larger than the previously discussed examples but clearly show a decline associated with changed management which is predominantly driven by a decline in soil organic matter (Graham et al. 2002). Intuitively these soils have a greater capacity to resist charge diminution due to contrasting management systems as they are dominated by permanently charged smectitic clays. This is clearly demonstrated when one compares the decline in percentage charge between the grass ley managed system compared to the burnt cane managed system at a comparable pH of 5.5. The burnt system has undergone a mere 14% reduction in charge over a 57 year period whilst under continuous cane production. This demonstrates the resilience of these soils to buffer changed management and contrasts dramatically with the previous two examples.

The examples presented above demonstrate the contrasting impact of changed land management on soil chemical properties. The degree of impact is in part contingent upon the clay mineralogy of the soils. If soils are dominated by low activity kaolinitic and sesquioxidic clays as in the first two examples, the amount of charge associated with this inorganic phase is small. Consequently these soils have a small 'permanent' charge component. In permanent surface-charge minerals, the charge is due to imperfections in the lattice structure resulting from isomorphous substitution. As is evidenced by the change in surface charge with an increase in pH, there is a second form of charge generation namely 'variable' charge. In variable charge soil systems, the surface charge is created by the adsorption of the potential-determining ions (p.d.i.) or other specifically adsorbed ions onto the surface. The charging process

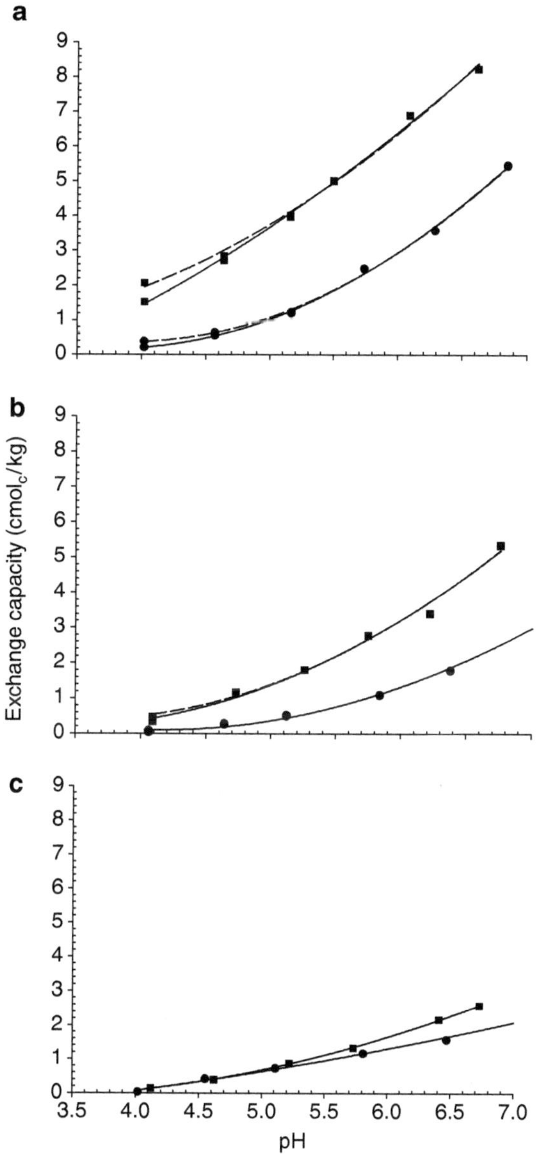

Fig. 2 Cation exchange characteristics for the 0–10 cm (**a**), 20–30 cm (**b**) and 50–70 cm (**c**) of soils collected from a tea (•) and adjacent climax tropical rain forest (▪) collected from north Queensland. CEC_T is represented by the dotted lines and CEC_B by the solid lines (Noble et al. 2001)

requires the presence of these ions in the soil solution in sufficient quantities for adsorption to occur. Variable charge or amphoteric constituents in soils may have organic (i.e. organic matter) and inorganic origin (i.e. sesquioxidic surfaces).

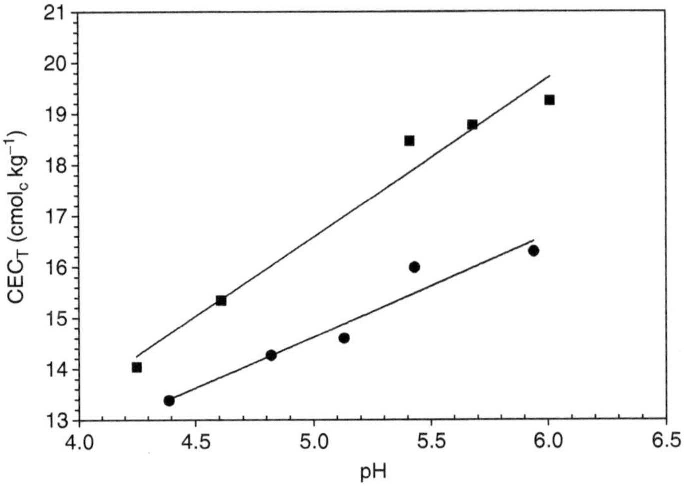

Fig. 3 Surface charge fingerprints of soil collected from the control (▪) and trash burnt/fertilized (•) treatments of a long-term trial established at Mt Edgecombe, South Africa on a Vertisol in 1939

With the loss of soil organic matter through mineralization and accelerated acidification due to crop removal and the application of acidifying nitrogenous fertilizers, the variable charge component declines dramatically upon conversion to agriculture. The first three cases discussed above would typify soils that are dependent on this form of charge. Contrasting this, soils that are dominated by permanently charged clay minerals, as demonstrated in the previous example, show significantly higher inherent charge capacity and relatively small declines in this attribute over a significant period of crop production. The major difference in this case is that these soils are dominated by high-activity 2:1 layer smectitic clays that have 'permanent' charge. This charge is not affected by changes in pH and is little influenced by soil organic matter. Clearly, soils that are dominated by these clay minerals are more resilient to changed management without a significant decline in productive capacity.

Amelioration of Charge Degradation

Declining productivity on degraded soils can in part be ascribed to a reduction in the water holding capacity of the profile, a decline in the physical characteristics (i.e. reduced infiltration, greater runoff, or increased drainage) and a reduction in the fertility status of the soil largely driven by a decline in the cation exchange capacity. In degraded soils, in order to sustain agronomic production, larger inputs of water and inorganic fertilizers are required to sustain yield levels, leading to greater losses associated with leaching and potentially to lower nutrient efficiencies. Organic amendments including manures and composted sewage-sludge have been used to add nutrients in a slow release form as well as increase the available water holding capacity of soils (Tester 1990; Bauer and Black 1992). However, in the case of organic amendments the positive effects on soil chemical and physical attributes are often transient and require ongoing additions to sustain these effects.

Alternative strategies such as the application of crushed basalt rock, zeolites and other geological materials have been tested with varying degrees of success (Gillman 1980; Chesworth et al. 1987). Using these techniques the CEC is increased with a concomitant increase in basic cations. Studies have shown that remediation of chemical degradation and its associated declining productivity can be achieved through the addition of bentonite that has been shown to be highly economic (Noble et al. 2001; Saleth et al. 2009).

Changing the production system to one that incorporates soil organic carbon conservation can lead to significant improvements in the fertility and nutrient holding capacity of soils. Changing land use so as to conserve SOM, fundamental surface charge characteristics can be manipulated to one's advantage. Following a change in harvesting practice in the sugar industry in north Queensland, Australia, from burnt cane harvesting to green cane trash blanketing, there have been recorded instances of SOM enhancement (Wood 1991; Prove et al. 1986; Spain and Hodgen 1994). For instance, on an Inceptisol under a high rainfall regime in tropical Queensland, changed management practice through green cane trash blanketing caused organic carbon to increase from 1.33% to 2.26% in the top 0–5 cm interval over a relatively short period of 7 years (Noble et al. 2003). This in turn favorably altered surface charge characteristics as evidenced in the charge fingerprints (Fig. 4). The degree of increase in charge characteristics of the surface soil under the green cane trash blanketing system was estimated to be 150%. It is important to note that improvements in the overall CEC were confined to the surface layer and that below 5 cm there was no evidence of improvements in charge. This generation of increased surface charge can only be viewed as beneficial. However, it should be realized that the increase in charge characteristics associated with elevated soil organic matter is transient under these tropical conditions. When the system is perturbed through cultivation a significant proportion of this increased charge will be lost through organic matter mineralization.

Alternative strategies such as a grass ley in rotation with crops would also increase the soil organic carbon content and have a direct benefit on the charge characteristics of the soil (Noble et al. 1998, 2001). However, to resource poor farmers in developing countries whose primary objective is house-hold food security, the implementation of a long term grass ley into their farming systems is remote. There is mounting empirical evidence to suggest that 'conservation agriculture' systems are an effective alternative to conventional farming (tillage based) systems that effectively exploit the natural resources upon which they are based without degrading them, and in some cases allowing their restoration (Bot and Benites 2001). The two essential features of 'conservation agriculture' are no-tillage and the maintenance of a cover on the soil surface. Significant increases in productivity have been achieved through the adoption of 'conservation agriculture' (Bot and Benites 2001). However, these responses are contingent on the inherent fertility status of the soil. Furthermore on highly degraded production systems, the positive productivity benefits associated with the adoption of organic matter

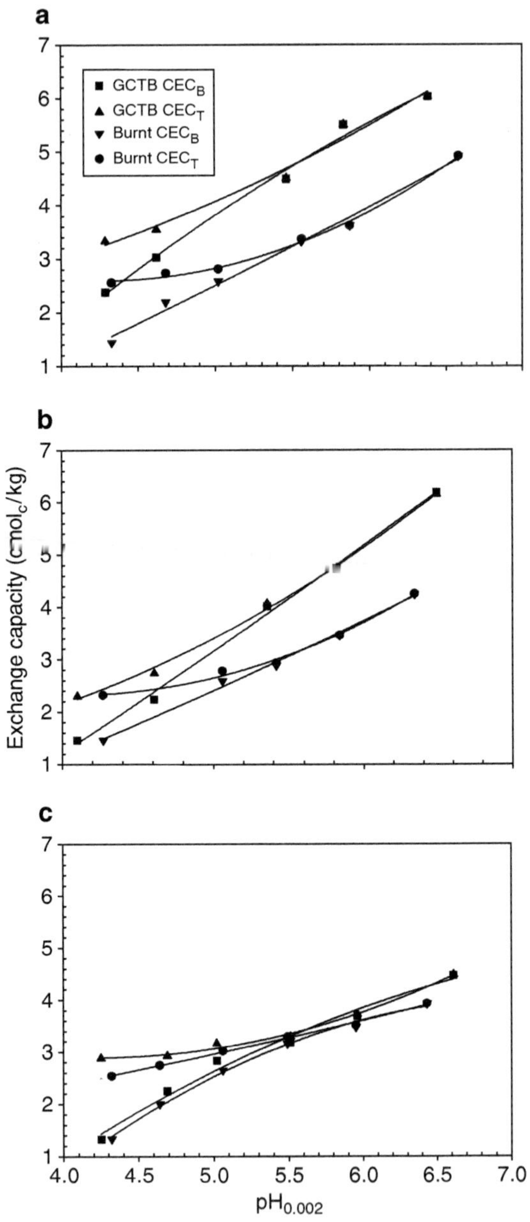

Fig. 4 Surface charge fingerprints of the 0–2 cm (**a**), 2–5 cm (**b**) and 5–10 cm (**c**) Depth intervals under a long-term Green Cane Trash Blanketing and Burnt Cane System (Noble et al. 2003)

conservation systems may have a lag phase associated with a buildup of SOM to some critical level. This may influence adoption of such practices under particular socio-economic circumstances.

As alluded to in the previous discussion, one constraint in the adoption of sustainable agronomic practices is the lag time between implementation of a practice(s) and associated increases in productivity. In some cases there is often a slight reduction in productivity before significant increases are achieved. In an effort to mitigate this lag effect, the concept of directly addressing the key chemical constraints of degraded soils (namely low nutrient and water holding capacity), has been effected through the incorporation of permanently charged high activity clays to degraded soils (Noble et al. 2001). Through the addition of cation beneficiated bentonite to degraded soils, Noble (Noble et al. 2001) have shown significant responses in yield and permanent changes in the charge characteristics of soils. Surface charge characteristics from a subsequent study where varying rates of beneficiated bentonite (0, 20, 40 and 60 t ha^{-1} containing a ratio 4:2:1 of Ca/Mg/K) were applied to two contrasting soil types are presented in Fig. 5. The cation exchange capacity at soil

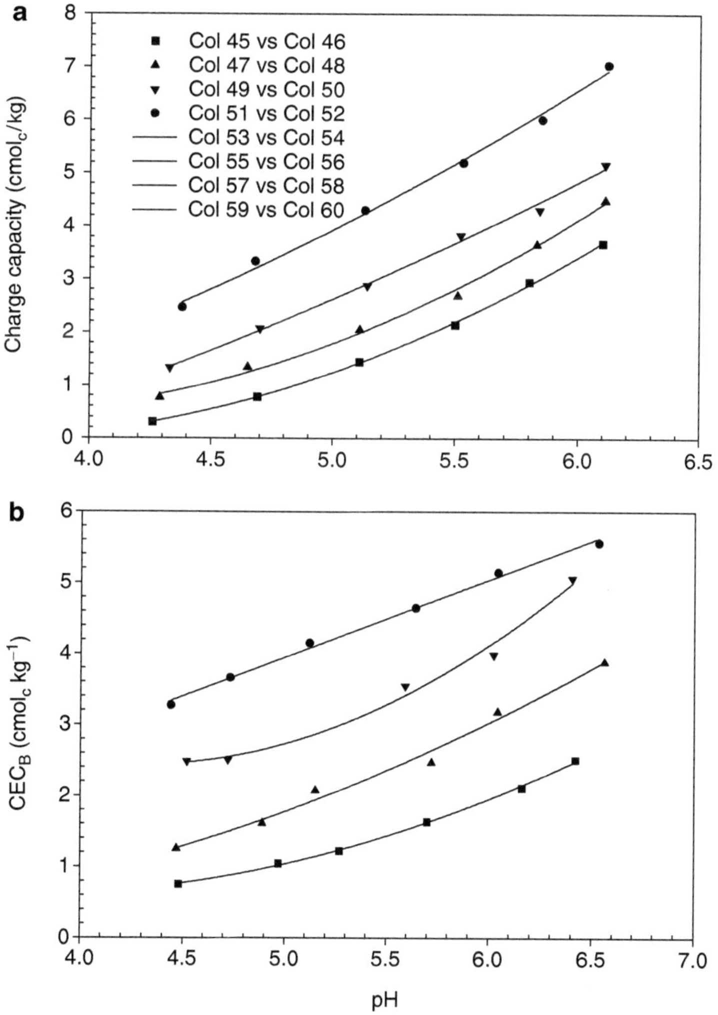

Fig. 5 Surface charge characteristics of (**a**) an Oxisol and (**b**) Ultisol after the addition of varying rates of Bentonite (Noble et al. 2001)

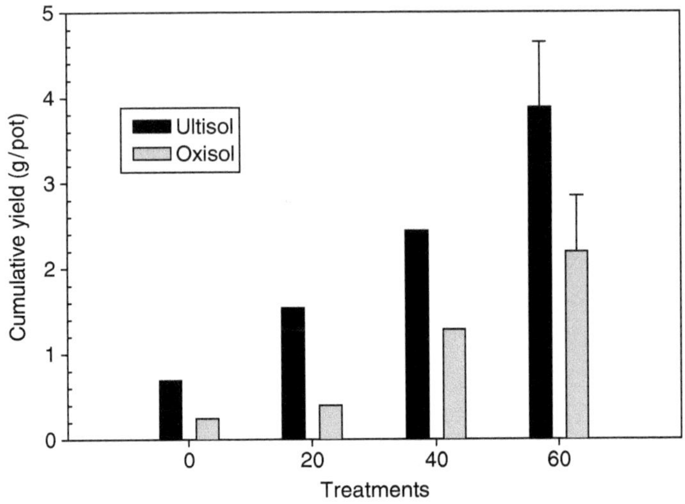

Fig. 6 Total above ground yield response of forage sorghum to varying additions of beneficiated bentonite (t/ha) on two contrasting soil types (Noble et al. 2001). Vertical bar represents the LSD at the 5% level

pH was increased from 0.97 to 4.08 $cmol_c\ kg^{-1}$ on a light textured Ultisol and from 0.58 to 2.95 $cmol_c\ kg^{-1}$ on an Oxisol through the addition of an equivalent of 60 t ha^{-1} of beneficiated bentonites. Since the surface charge on bentonite clay is permanently negative, i.e. not affected by changes in soil pH, the additional CEC now present in these soils will be a permanent feature. Concomitant with the improved charge characteristics was a significant and sustained increase in forage sorghum biomass production with increasing additions of these materials on both soil types (Fig. 6). The cumulative increase in yield between the control and 60 t ha^{-1} bentonite was 5.6 and 9.0 fold for the Ultisol and Oxisol soil types respectively.

In studies undertaken in Northeast Thailand to assess the efficacy of a range of soil amendment techniques in rejuvenating a degraded light textured sandy soils, highly significant increases in biomass production were observed in those treatments receiving the traditional termite mound material, bentonite and bentonite + compost when compared to the control, dredged material and composted leaf litter (Fig. 7) (Noble et al. 2004). In the case of the 2003 growing season it was not possible to establish a forage sorghum crop even after repeated plantings on the control, dredged and compost treatments this being ascribed to the variable precipitation that was experienced. The responses with respect to the termite mound materials and bentonite treatments can in part be attributed to the enhanced nutrient supplying capacity and increased CEC associated with these materials (Table 3). Total above ground dry matter (DM) production over the 2 years ranged from 0.22 t DM ha^{-1} in the control treatments where plant production completely failed in the second year of cropping to 10.8 and 22.4 t DM ha^{-1} for the locally sourced termite mound material (120 t ha^{-1}) and local bentonite (50 t ha^{-1}) + leaf

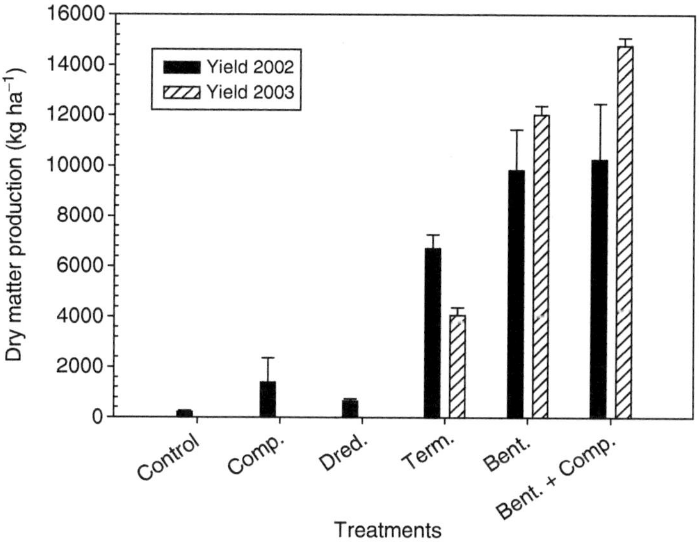

Fig. 7 Yield response of forage sorghum to selected treatments applied to a degraded light textured soil in Northeast Thailand over two consecutive years. Treatments include Control = Control (current farmer practice); Dred. = Dredged Material at 240 t ha^{-1}; Comp. = Leaf Litter Compost at 10 t ha^{-1}; Term. = Termite Mound Soil at 120 t ha^{-1}; Bent. = Local Bentonite at 50 t ha^{-1}; and Bent. + Comp. = Local Bentonite at 50 t ha^{-1} with Leaf Litter Compost of 10 t ha^{-1}. Vertical bars are the standard errors of treatment means (Noble et al. 2004)

Table 3 Selected chemical characteristics of Bentonite clay and Termite mound material used in a field study in Northeast Thailand. Values in brackets are the standard error of the means of four replicates

	Exchangeable cations (cmol$_c$ kg^{-1})				Sum bases	CEC
Source	Ca^{2+}	Mg^{2+}	K^{+}	Na^{+}	(cmol$_c$ kg^{-1})	(cmol$_c$ kg^{-1})
Termite mound	17.04 (±0.48)	5.92 (±0.11)	0.41 (±0.03)	1.23 (±0.17)	24.62 (±0.30)	13.68 (±0.66)
Bentonite	25.59 (±0.36)	12.78 (±0.09)	0.55 (±0.09)	6.44 (±0.10)	45.37 (±0.44)	27.68 (±0.53)

litter compost (10 t ha^{-1}), respectively. The responses to the treatments persisted into the second season and continued to increase under the bentonite treatments. It is of note that the CEC of the bentonite was approximately double that of the termite mound material, and similarly, the sum of exchangeable bases in each case was approximately double that of the CEC indicating a significant proportion of the bases were not being held on the exchange complex (Table 3). A further significant benefit associated with the application of bentonite to these light textured soils is the potential to increase the water holding capacity of these soils (Suzuki et al. 2007).

Addressing Soil Physical Degradation

In addition to the unfavorable chemical properties of soils, many soils are prone to sub-surface compaction which is often observed as a high resistance to penetration that inhibits the development of crop root systems below 20–40 cm (Bruand et al. 2004; Hartmann et al. 2001). This has a significant impact on the crops ability to take up nutrients and stored subsurface moisture leaving the crop vulnerable to drought stress. Deep ploughing and sub-soiling have been shown to be ineffective in overcoming this problem on coarse textured soils due to their low structural stability (Ball-Coelho et al. 2000) and hence re-establishment of compacted layers occurs within the following growing season. The vast majority of soil compaction in agricultural production systems is attributed to traffic load associated with heavy machinery.

Comparisons of penetrometer resistance between remnant *Dipterocarp* forest in an upland landscape position in Northeast Thailand and adjacent crop land show the effect of changed land use on soil physical properties (Fig. 8). A layer of extreme resistance is present over the 20–40 cm depth interval that would significantly reduce a crops ability to penetrate this layer in order to accesses stored water. The depth of this compacted region in the profile corresponds with the working depth of most tillage implements and therefore is attributed to both the increased loads associated with wheel traffic and smearing effect of tillage implements (i.e. disc plough). The consequences of these areas of high soil strength are reduced root proliferation and enhanced risk of the crop being predisposed to drought stress that is common to the region.

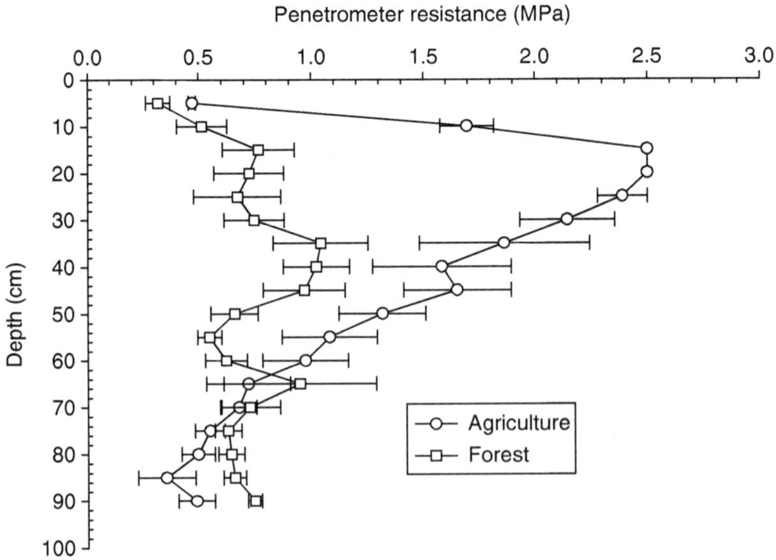

Fig. 8 Effect of changed land use on penetrometer resistance for an upland farming systems in Northeast Thailand. Horizontal bars represented the standard deviation of the mean

The impact of soil structure on root growth and proliferation has received considerable attention over the past decade (Passioura 1991). It has been shown that root proliferation in the soil is closely dependent on the presence of macropores (Hatano and Sakuma 1990; Hatano et al. 1988; Stewart et al. 1999). Mechanical modification of the soil profile, through deep-ploughing or subsoiling will significantly increase the porosity of soils. However, such profile modifications require high energy inputs that is often beyond the means of resource-poor farmers and invariably, the benefits diminish rapidly after the first heavy rains due to the inherently unstable nature of these soils (Hartmann et al. 2002). Actively growing plant root systems have the potential to ameliorate subsoils in poor physical condition by biological drilling (Cresswell and Kirkegaard 1995). Decaying roots leave a continuous network of vertically-oriented macropores that subsequent plants can use. In studies conducted with the grass species *Andropogen gayanus* cv Kent (Gamba) and the legume *Stylosanthes hamata* (Stylo) on a light textured sandy soil with a distinct compacted layer between 20 and 40 cm have clearly demonstrated the contrasting potential of these species in penetrating these compacted layers (Fig. 9). Despite the existence of the dense compacted layer, development and proliferation of Gamba roots to depth were significantly higher when compared to Stylo with a rooting depth reaching down to 1.0 m. Taking advantage of the ability of Gamba root to penetrate compacted layer, Gamba is a highly effective biological soil remediation agent on these compacted

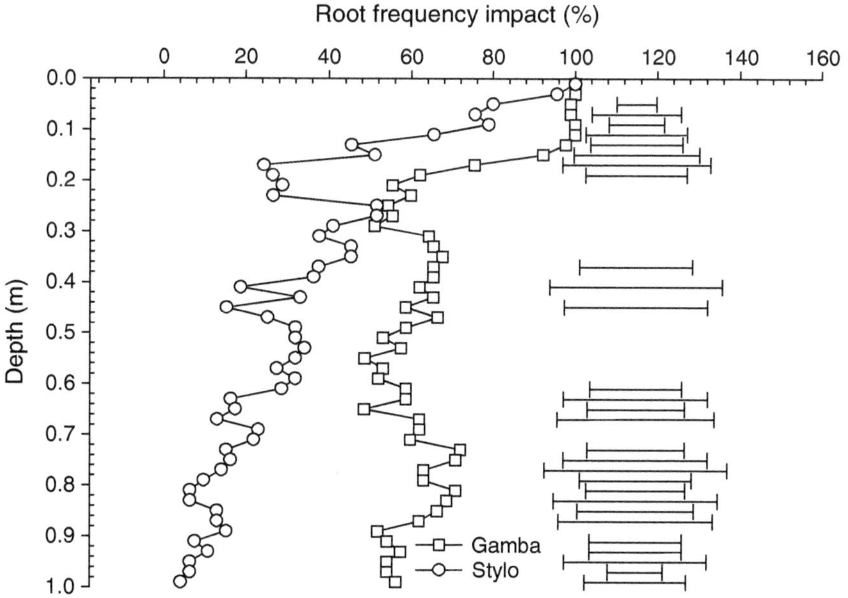

Fig. 9 Mean root frequency impacts for *Andropogen gayanus* cv Kent (Gamba) and *Stylosanthes hamata* (Stylo) treatments. Horizontal bars are the least significant difference between treatment means ($n = 3$)

soils and would fit into a grass ley system. Moreover, the establishment of stable macropores through these compacted layers could potentially provide subsequent crops with biological channels through which roots could grow and access stored sub-surface moisture thereby enhancing water productivity under these rainfed production system.

In a similar series of studies with the same physical impediments using the aforementioned species, soil organic carbon (OC) levels were increased to depth with the establishment of pasture species within these profiles (Noble et al. 2008). Of importance in assessing the amount of carbon stored in soil profiles is the net carbon contents that fall below the plough layer (0–30 cm) that is assumed to be fixed with a much greater residence time. Figure 10 represents the net storage of carbon over the 0–30 cm and 30–110 cm depth intervals in studies undertaken in Northeast Thailand. The mean increase in soil organic carbon in the surface 0–30 cm was 776 kg ha^{-1} from that of the 2001 with all treatments having significantly higher levels of OC than the initial values. What is important is the over sixfold increase in soil organic carbon stored at >30 cm from that of the 2001 (Fig. 10). Fisher (Fisher et al. 1994) observed substantial increases in organic carbon under Gamba pastures in South American savannas.

The role of permanent leys, whether legume or grass, in cropping systems clearly demonstrates the ability of these species to penetrate compacted layers

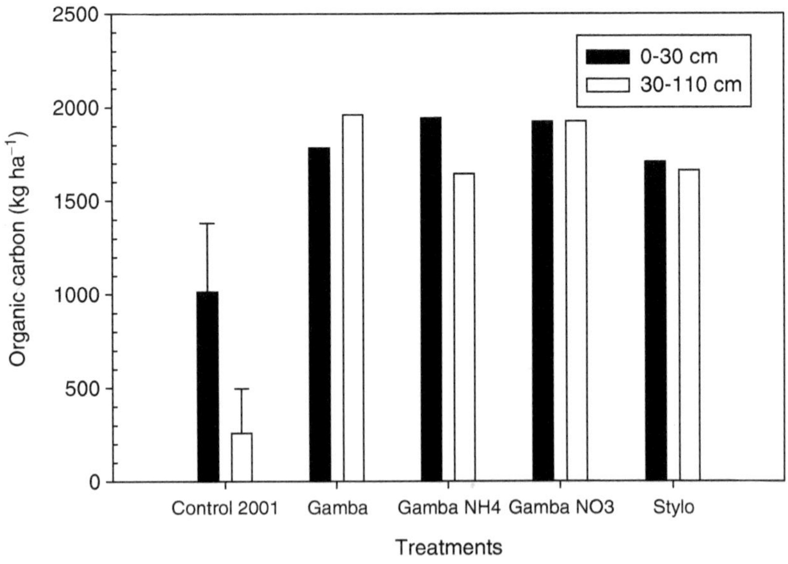

Fig. 10 Total organic carbon stored in the 0–30 cm and 30–110 cm depth intervals 3 years after the implementation of a range of treatments that included a control (bare fallow); Gamba (no fertilizer); Gamba NH_4 (Gamba fertilized 278 kg N as NH_4 over 3 years); and Gamba NO_3 (Gamba fertilized 278 kg N as NO_3 over 3 years). Vertical bars represent the LSD0.05 between treatment means over the same depth interval (Noble et al. 2008)

resulting in the development of permanent stable macropores and the storage of organic carbon to depth. This has positive benefits under rainfed production systems since stored subsoil moisture can be accessed reducing the risk of drought stress as well as increasing the productivity of these systems.

Transforming the Agricultural Landscape

In 1974 Henry Kissinger, the then USA secretary of state, announced at a global food conference that no child would go hungry within 10 years. Just over 35 years later a United Nations food summit in Rome stated one billion people will go to bed hungry (The Economist 2009). By 2050 the world's population will have increased by a third and demand for agricultural produce will rise by 70% with meat consumption doubling. Lateral expansion of the agricultural sector through the clearing of land is untenable without significant negative implications on already stressed natural ecosystems and the range of drivers, including climate variability, that farmers will have to cope with, will require changes in the way we undertake agriculture.

In meeting future food production demands without consuming more land and water will require technological innovation. This may include changes in the way we use water resources through more efficient irrigation; the adoption of no-till farming systems; better use of fertilizers and pesticides; and raising yield potential through genetically modified (GM) crops that are adapted to a range of environmental stresses.

From a soils perspective in the introduction of conservation agriculture (CA) that is founded upon three guiding principles: (1) minimum or no mechanical soil disturbance; (2) permanent organic soil cover (consisting of a growing crop or a dead mulch of crop residues); and (3) diversified crop rotations has been successfully implemented in North and South America with positive impacts on soil conservation (Bolliger et al. 2006; Triplett and Warren 2008). However, recent issues over claims of increased yields under CA and its role in the context of Sub-Saharan African (SSA) agriculture highlight the necessity for further review of where this approach to soil resource management is applicable (Giller et al. 2009). Concerns raised include decreased yields often observed with CA, increased labour requirements when herbicides are not used, an important gender shift of the labour burden to women and a lack of mulch due to poor productivity and due to the priority given to feeding of livestock with crop residues (Giller et al. 2009). Despite the publicity claiming widespread adoption of CA, the available evidence suggests virtually no uptake of CA in most SSA countries, with only small groups of adopters in South Africa, Ghana and Zambia. We conclude that there is an urgent need for critical assessment under which ecological and socio-economic conditions CA is best suited for smallholder farming in SSA. Critical constraints to adoption appear to be competing uses for crop residues, increased labour demand for weeding, and lack of access to, and use of external inputs. These are significant challenges that are faced by the majority of global farmers, namely smallholder poor farmers.

While there is considerable uncertainty about the impacts of climate change, there is little doubt that its effects will be felt most strongly by the poor. Climate change impacts on the habitats that marginal producers depend on will further exacerbate their livelihood insecurity stemming from historical neglect and discrimination. This underscores the gravity of their vulnerability to the cumulative impacts of climate and other anthropogenic activities that degrade the natural resources and result in a decline in ecosystem services (ES).

Agriculture does not of itself compromise ecosystem services. On the contrary, agriculture is essentially a means of concentrating and enhancing provisioning services – but this can often come at the cost of other functions, particularly regulating services which are intimately related to land cover. For example: deforestation reduces the ability to regulate erosion; loss of wetlands reduces the ability of streams to regulate floods and water quality; and clearance of mangroves reduces the ability to regulate storm impacts.

Landscapes can be modified to protect or enhance a specific ES; for example: afforestation in watersheds to prevent erosion and improve water quality; man-made wetlands for sewage treatment; restoring environmental flows in rivers to preserve aquatic ecosystems; and construction of man made reefs and replanting of mangroves to mitigate storm damage in coastal zones. An important aspect of protecting ES is identifying critical thresholds, beyond which the resilience of the ecosystem is compromised, and managing agricultural systems within these limits.

The challenge for the future is to create productive agro-ecosystems that maintain and enhance a range of ecosystem services that are delivered. In many cases, this can best be achieved by mimicking aspects of natural systems (Lefroy et al. 1999); for example:

- Paddy fields mimic the water retention of natural wetlands to provide multiple benefits including rice and fish production, flood mitigation, ground water recharge, soil erosion control, water purification.
- Plantations can mimic natural forests, with well developed understorey to prevent soil erosion, provide habitat and promote biodiversity.

It is important that actions to preserve and improve the quality of ecosystem services enhance, rather than compete with, livelihoods either directly, through improvements in local environments and production, or indirectly through financial compensation or payments for ecosystem services.

References

Aweto AO, Obe O, Ayanniyi OO (1992) Effects of shifting and continuous cultivation of cassava (*Manihot esculenta*) intercropped with maize (*Zea mays*) on a forest alfisol in south-western Nigeria. J Agric Sci Cambridge 118:195–198

Bai ZG, Dent DL, Olsson L et al (2008) Proxy global assessment of land degradation. Soil Use Manage 24:223–234

Ball-Coelho BR, Roy RC, Swanton C (2000) Tillage and cover crop impacts on aggregation of a sandy soil. Can J Soil Sci 80:363–366

Battisti DS, Naylor RL (2009) Historical warnings of future food insecurity with unprecedented seasonal heat. Science 323:240–244

Bauer A, Black AL (1992) Organic carbon effects on available water capacity of three soil textural groups. Soil Sci Soc Am J 56:248–254

Bolliger A, Magid J, Amado TJC et al (2006) Taking stock of the Brazilian "zero-till revolution": a review of landmark research and farmers' practice. Adv Agron 91:47–110

Bot A, Benites J (2001) Conservation agriculture. Case studies in Latin America and Africa. In: FAO Soils Bulletin 78. FAO, Rome

Bruand A, Hartmann C, Ratana-Anupap S et al (2004) Composition, fabric and porosity of an Arenic Haplustalf in Northeast Thailand: relation to penetration resistance. Soil Sci Soc Am J 68:185–193

Chesworth W, Van Straaten P, Smith P et al (1987) Solubility of apatite in clay and zeolite bearing systems: application to agriculture. Appl Clay Sci 2:291–297

Comprehensive Assessment of Water Management and Agriculture (2007) Water for food, water for life: a comprehensive assessment of water management in agriculture. Earthscan/International Water Management Institute, London/Colombo

Cresswell HP, Kirkegaard JA (1995) Subsoil amelioration by plant roots – The processes and the evidence. Aust J Soil Res 33:221–239

Drechsel P, Giordano M, Gyiele LA (2004) Valuing nutrients in soil and water: concepts and techniques with examples from IWMI studies. Research Report 82. International Water Management Institute, Colombo, Sri Lanka

Eswaran H, Lal R, Reich PF (2000) Land degradation: an overview. In: Bridges EM, Hannam ID, Oldeman LR, Penning de Vries FWT, Scherr SJ, Sombatpanit S (eds) Response to land degradation. Oxford/IBH Publishing, New Delhi

FAO (2003) World agriculture: towards 2015/2030. FAO, Rome. http://ftp.fao.org/docrep/fao/004/y3557e/y3557e.pdf

FAO (2006) World Agriculture, towards 2030/2050. FAO, Rome. http://www.fao.org/es/ESD/AT2050web.pdf

FAO (2008) The state of food and agriculture 2008. FAO, Rome. http://www.fao.org/docrep/011/i0100e/i0100e00.htm

FAO Statistics (2009)) FAOSTAT. http://faostat.fao.org/default.aspx

Fargione J, Hill J, Tilman D et al (2008) Land clearing and the biofuel carbon debt. Science 319:1235–1238

Fisher MJ, Rao IM, Ayarza MA et al (1994) Carbon storage by introduced deep-rooted grasses in the South American savannas. Nature 371:236–238

Fitzherbert EB, Struebig MJ, Morel A et al (2008) How will oil palm expansion affect biodiversity? Trends Ecol Evol 23:528–545

Giller KE, Witter E, Corbeels M et al (2009) Conservation agriculture and smallholder farming: the heretic's view. Field Crops Res 114:23–34

Gillman GP (1980) The effect of crushed basalt scoria on the cation exchange properties of a highly weathered soil. Soil Sci Soc Am J 44:465–468

Gillman GP, Abel DJ (1986) A summary of surface charge characteristics of the major soils of the Tully-Innisfail area, north Queensland. CSIRO Australia Division of Soils Divisional Report No 85. CSIRO, Melbourne, Australia

Gillman GP, Sumpter EA (1986a) Surface charge characteristics and lime requirements of soils derived from basaltic, granitic, and metamorphic rocks in high rainfall tropical Queensland. Aust J Soil Res 24:173–192

Gillman GP, Sumpter EA (1986b) Modification to the compulsive exchange method for measuring exchange characteristics of soil. Aust J Soil Res 24:61–66

Gillman GP, Sinclair DF, Knowlton R et al (1985) The effect on some soil chemical properties of the selective logging of a north Queensland rainforest. For Ecol Manage 12:195–214

Graham MH, Haynes RJ, Meyer JH (2002) Changes in soil chemistry and aggregate stability induced by fertilizer applications, burning and trash retention on a long-term sugarcane experiment in South Africa. Eur J Soil Sci 53:589–598
Hartmann C, Poss R, Singhatat V (2001) Soil compaction and plant growth in northeast Thailand: the case of the Nam Phong soil series. In: Proceedings of the planning workshop on ecoregional approaches to natural resources management in the Korat Basin, Northeast Thailand. International Rice Research Institute, Los Baños, Philippines
Hartmann C, Poss R, Janeau JL et al (2002) Use of granular material theories to interpret structural changes in a sandy soil. In: Proceedings of the 17th world congress of soil science, Bangkok, Thailand
Hatano R, Sakuma T (1990) The role of macropores on rooting pattern and movement of water and solutes in various field soils. In: 14th congress of international society of soil science, Kyoto, Japan
Hatano R, Iwanaga K, Okajima H et al (1988) Relationship between the distribution of soil macropores and root elongation. Soil Sci Plant Nutr 34:535–546
ICIMOD (2008) Food security in the Hindu Kush-Himalayan region. ICIMOD, Chengdu, China
IPCC (2000) Land use, land-use change, and forestry special report. Cambridge University Press, Cambridge
Jenny H (1941) Factors of soil formation. McGraw-Hill, New York
Kang BT, Juo ASR (1986) Effect of forest clearing on soil chemical properties and crop performance. In: Lal R, Sanchez PA, Cummings RW (eds) Land clearing and development in the tropics. AA Balkema, Rotterdam, The Netherlands
Keyzer MA, Merbis MB, Pavel IFPW et al (2005) Diet shifts towards meat and the effects on cereal use: can we feed the animals in 2030? Ecol Econ 55:187–202
Kooistra MJ, Juo ASR, Schoonderbeek D (1990) Soil degradation in cultivated Alfisols under different management systems in southwestern Nigeria. In: Douglas LA (ed) Soil micromorphology: a basic and applied science. Elsevier, Amsterdam, The Netherlands
Lal R (1990) Soil erosion and land degradation: the global risk. Adv Soil Sci 11:129–172
Lal R, Griffin M, Apt J et al (2004) Managing soil carbon. Science 304:393
Latham M (1994) Evaluation of the soil and land resource. In: Syers JK, Rimmer DL (eds) Soil science and sustainable land management in the tropics. CAB International, Wallingford, UK
Lefroy EC, Hobbs RJ, O'Connor MH et al (1999) What can agriculture learn from natural ecosystems. Agroforest Syst 45:425–438
Menzies NW, Gillman GP (1997) Chemical characterization of soils of a tropical humid forest zone: a methodology. Soil Sci Soc Am J 61:1355–1363
Nellemann C, MacDevette M, Manders T et al (2009) The environmental food crisis – The environment's role in averting future food crises. A UNEP rapid response assessment. United Nations Environment Programme, GRID-Arendal
Noble AD, Thompson CH, Jones RJ et al (1998) The long-term impact of two pasture production systems on soil acidification in southern Queensland. Aust J Exp Agric 38:335–343
Noble AD, Gillman GP, Nath S et al (2001) Changes in the surface charge characteristics of degraded soils in the tropics through the addition of beneficiated bentonite. Aust J Soil Res 39:991–1001
Noble AD, Moody P, Guodao L et al (2003) Quantification of soil chemical degradation and its remediation in tropical Australia, China and Thailand. Pedosphere 13:31–39
Noble AD, Ruaysoongern S, Penning de Vries FWT et al (2004) Enhancing the agronomic productivity of degraded soils in Northeast Thailand through clay-based interventions. In: Seng V, Craswell E, Fukai S, Fischer K (eds) Water and agriculture, vol 116, ACIAR Proceedings 2004. ACIAR, Canberra, pp 147–160
Noble AD, Suzuki S, Soda W et al (2008) Soil acidification and carbon storage in fertilized pastures of Northeast Thailand. Geoderma 144:248–255
Olderman LR, Hakkeling RTA, Sombroek WG (1991) World map of the status of human-induced soil degradation: an explanatory note, 2nd edn. United Nation Environment Programme, Nairobi
Passioura JB (1991) Soil structure and plant growth. Aust J Soil Res 29:717–728

Penning de Vries FWT, Acquay H, Molden D et al (2002) Integrated land and water management for food and environmental security. Comprehensive Assessment Research Report Paper 1. Comprehensive Assessment Secretariat, IWMI, Colombo, Sri Lanka
Pinstrup-Andersen P, Pandya-Lorch R (1998) Food security and sustainable use of natural resources: a 2020 Vision. Ecol Econ 26:1–10
Plamondon AP, Ruuiz RA, Morales CF et al (1991) Influence of forest protection on soil and water conservation (Oxapampa, Peru). For Ecol Manage 38:227–238
Prove BG, Truong PN, Evans DS (1986) Strategies for controlling caneland erosion in the wet tropical coast of Queensland. In: Proceedings Australian Society of Sugar Cane Technologists, pp 77–84
Saleth RM, Inocencio A, Noble A et al (2009) Economic gains of improving soil fertility and water holding capacity with clay application: the impact of soil remediation research in Northeast Thailand. J Dev Eff 1:336–352
Sayer JA, Campbell BM (2003) Research to integrate productivity enhancement, environmental protection and human development. In: Campbell BM, Sayer JA (eds) Integrated natural resources management: linking productivity, the environment and development. CABI, Wallingford, UK
Searchinger T, Heiimlich R, Houghton RA et al (2008) Use of US croplands for biofuels increases greenhouse gases through emissions from land-use change. Science 319:1238–1240
Spain AV, Hodgen MJ (1994) Changes in the composition of sugarcane harvest residues during decomposition as a surface mulch. Biol Fertil Soils 17:225–231
Stewart JB, Moran CJ, Wood JT (1999) Macropore sheath: quantification of plant root and soil macropore association. Plant Soil 211:59–67
Stocking MA (2003) Tropical soils and food security: the next 50 years. Science 302:1356–1359
Stocking MA, Murnaghan N (2001) Handbook for the field assessment of land degradation. Earthscan, London
Suzuki S, Noble AD, Ruaysoongnern S et al (2007) Improvement in water-holding capacity and structural stability of a sandy soil in Northeast Thailand. Arid Land Res Manage 21:37–49
Tester CF (1990) Organic amendment effects on physical and chemical properties of a sandy soil. Soil Sci Soc Am J 54:827–831
The Economist (2009) How to feed the world. 19th November issue
Triplett GB, Warren AD (2008) No-tillage crop production: a revolution in agriculture! Agron J 100:153165
UN Population Division (2007) UN 2006 population revision. UN, New York. http://esa.un.org/unpp/;2007
Van Antwerpen R, Meyer JH (1998) Soil degradation – II: effect of trash and inorganic fertilizer application on soil strength. Proc S Afr Sugar Technol Assoc 72:152–158
Van der Watt H, Valentin C (1992) Soil crusting: the African view. In: Stewart BA (ed) Soil crusting: chemical and physical processes. Lewis Publishing, Boca Raton, FL
Walker J, Thompson CH, Olley J et al (2000) Retrogressive succession in old landscapes. In: Proceedings of 42nd international association of vegetation science. Opulus Press, Uppsala
Willett IR (1995) Role of organic matter in controlling chemical properties and fertility of sandy soils used for lowland rice in Northeast Thailand. In: Lefroy RDB, Blair GJ, Craswell ET (eds) Soil organic matter management for sustainable agriculture Australian Centre for international agricultural research proceedings, vol 56. Brown Prior Anderson, Melbourne
Wood AW (1991) Management of crop residues following green harvesting of sugar cane in north Queensland. Soil Tillage Res 20:69–85
World Bank (2008) Rising food and fuel prices: addressing the risks to future generations. World Bank, Washington, DC. http://siteresources.worldbank.org/DEVCOMMEXT/Resources/Food-Fuel

Chapter 2
The Microbiology of Natural Soils

Teri C. Balser, Devin Wixon, Lindsey K. Moritz, and Laura Lipps

Introduction

Soil microorganisms, such as bacteria and fungi, control ecosystem functioning through decomposition and nutrient cycling and may serve as indicators of land-use change and ecosystem health (Doran and Zeiss 2000; Waldrop et al. 2000; Yao et al. 2000). However, the study of soil microorganisms is difficult and our current understanding limited. The sheer number, astonishing diversity and small size of these communities become more apparent as our technologies to explore them have improved in recent years (Cardon and Gage 2006). With the rapid rise of molecular techniques, microbial ecologists are now able to walk through the world with the equivalent of the naturalist's "field notebook," cataloguing and classifying species. However, we often do not know what they do functionally or ecologically, or why they are found in some soils and not others (Balser et al. 2006). What we do know is that soil microbial communities are dynamic and diverse (Sylvia et al. 2005) almost beyond measure (Schloss and Handelsman 2006), and that some patterns seem to hold on a global scale.

In this chapter we survey a variety of ecosystems and summarize what we know about their underground microscopic inhabitants. Our knowledge is limited by the

T.C. Balser (✉) and L.K. Moritz
Department of Soil Science, University of Wisconsin-Madison, 1525 Observatory Drive, Madison, WI 53706, USA
e-mail: tcbalser@wisc.edu

D. Wixon
Department of Botany, University of Wisconsin-Madison, 430 Lincoln Drive, Madison, WI 53706, USA
e-mail: dwixon@gmail.com

L. Lipps
Gaylord Nelson Institute for Environmental Studies, University of Wisconsin-Madison, Madison, WI 53706, USA
e-mail: laalipps@yahoo.com

G.R. Dixon and E.L. Tilston (eds.), *Soil Microbiology and Sustainable Crop Production*,
DOI 10.1007/978-90-481-9479-7_2, © Springer Science+Business Media B.V. 2010

relative lack of studies and the particular challenge of linking microbial function with identity. However, trends such as astounding belowground diversity in the tropics, and the importance of fungi in forest soils, are gaining wider attention.

Methods of Study

Any treatment of microbial communities or ecology must include consideration of how we study them. More than most areas of endeavour, the study of soil microbiology is methodologically constrained: the size and fantastic diversity of microorganisms have long been a challenge to researchers. Historically, most techniques have relied on growing the organisms in a laboratory (culturing), but we now know that less than 1% of bacteria and an estimated 17% of fungi are culturable (Torsvik et al. 1990; Amann et al. 1995; Bridge and Spooner 2001; Horner-Devine et al. 2004). Currently we have many new tools for biochemical and molecular analysis of microorganisms, including phospholipid fatty acid (PLFA) analysis and genetic sequencing. These tools allow the targeting of the entire microbial community, rather than just those that grow in culture (Zelles 1999; Nannipieri et al. 2003; Kirk et al. 2004; Leckie 2005). In PLFA (lipid biomarker) analysis, soil microbial community members are identified based on the structure of fatty acids in the microbial cell membrane (Vestal and White 1989). Cell membranes of microorganisms degrade rapidly upon cell death and thus the fatty acids extracted from a soil sample provide an estimate of viable microbial community biomass and composition (Balser 2000; White et al. 1979). Amino sugar analysis instead assesses compounds from cell walls that persist for a very long time after cell death in the soil (Liang et al. 2008). In contrast to lipid biomarker analysis, molecular methods use DNA or RNA sequencing to identify genes unique to microorganism groups, potentially providing a higher level of resolution of the soil microbial community (Nannipieri et al. 2003). New molecular methods attempt to consider the entire community genome rather than that of particular inhabitants (metagenomics, Handelsman 2005).

Regardless of what biochemical or molecular method is used to examine the structure of a microbial community, they usually have the same shortcoming: they primarily provide information about what groups of microorganisms are present, rather than allowing insight into the functionality, or what the microorganisms are *doing* in the soil (Zelles 1999; Nannipieri et al. 2003; Kirk et al. 2004; Leckie 2005). Other methods do allow determination of functionality, i.e., we know what they consume based on how rapidly they consume food (substrate-induced respiration, or SIR), community-level physiological profiling (CLPP), or the enzymes microbes produce (extracellular enzyme assays). However, while these methods can assess the potential and actual activities of the soil biota, they don't tell us exactly who is doing what. In other words – using a given method we can determine *who* is there, but not what they are doing, or we can determine what they are doing, but not who it is. Currently, key studies often combine identification-based methods with activity-based methods. Future research will depend on finding better ways to do both simultaneously.

Soil as a Habitat

Soil microbial community structure and activity depend to a large extent on the status of their soil habitat. Within this habitat, soil organisms are eating, respiring, competing, cooperating, and responding to changes in their immediate environment. Indeed, the majority of the microbial community may be dormant at any given time in most soils, ready to respond as conditions for a particular group become favorable (Stenström et al. 2001). The soil habitat is perhaps best envisioned as a complex matrix with pores and soil aggregates of differing sizes (Sylvia et al. 2005). Certain bacteria and fungi tend to congregate in the soil immediately adjacent to plant roots (the rhizosphere), where they may feed off the sugars that plant roots exude or actually physically associate with the plant root system and exchange sugars and nutrients in a (usually) mutualistic relationship (mycorrhizas). The soil community and its habitat are influenced by an interconnected web of variables that differ among ecosystems, making each ecosystem somewhat unique in its microbial community (Wixon and Balser 2009). Across the globe, as with vegetation, community structure is perhaps most influenced by soil temperature and moisture (Sylvia et al. 2005), though it changes with the seasons (e.g. Lipson 2007), and is strongly affected by soil acidity or alkalinity (pH).

Within a given ecosystem, depth in soil is a primary consideration for microbial habitat, and many key habitat characteristics (e.g. oxygen levels, availability of food and nutrients) change through the soil profile. Carbon availability (and often quality) declines, as does overall microbial biomass (Fig. 1). Soil structure, such as particle size fractions and stable aggregates also change with depth and impact the soil biological habitat (Van Gestel et al. 1996; Ranjard et al. 2000; Sessitsch et al. 2001; Poll et al. 2003). Few studies have considered, however, the importance of both soil parent material and resultant soil texture (Ulrich and Becker 2006; Rasmussen et al. 2007).

Soil organic carbon (SOC) is the largest terrestrial component of the global carbon budget (Jobbágy and Jackson 2000). Worldwide, the top 1 m of soil contains two to three times more carbon than the amount stored in all aboveground vegetation (Brady and Weil 2002). Studies of soil carbon and microbial communities often concentrate on the upper 20–30 cm of soil, as this is considered to be the most biologically active portion of the soil profile (Fierer et al. 2007; Jobbágy and Jackson 2000; Veldkamp et al. 2003; Baisden and Parfitt 2007; Goberna et al. 2006). However, the majority of carbon in soil occurs below 20 cm, and thus by discounting lower depths we are missing up to 50–65% of the carbon (Jobbágy and Jackson 2000). As a result, many current ecosystem models of land-use and climate change inadequately model carbon turnover and microbial communities because they disregard the carbon stocks and biological activities of deeper soil horizons (Baisden and Parfitt 2007). Generally, most microbial community studies down the soil profile have occurred in grasslands and agricultural lands (Fierer et al. 2007; Lavahun et al. 1996; Blume et al. 2002; Taylor et al. 2002; Allison et al. 2007), with fewer studies in boreal and temperate forests (Goberna et al. 2005; 2006;

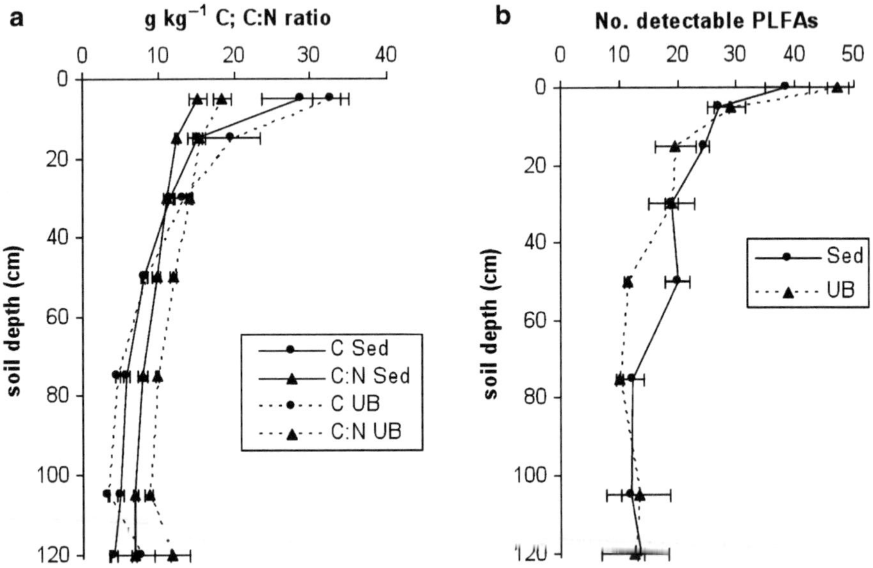

Fig. 1 Microbial and soil properties vary with depth. Depth profile graphs from Sedimentary and Ultrabasic sites in Borneo (See Moritz, 2008). (**a**) Total soil C and carbon:nitrogen (C/N) ratios with depth. Soil C/N differed significantly ($p < 0.05$) between sites under the conditions of two-way ANOVA. (**b**) The number of identified lipid peaks (detectable PLFAs) at each soil sampling depth, representing PLFA richness. Error bars are ± 1 SEM and $N = 3$. Error bars are ± 1 standard error of the mean (SEM) and $N = 3$

Ekelund et al. 2001), and even fewer in tropical forests (Veldkamp et al. 2003). In general, studies demonstrate that fungal-to-bacterial ratios decline with depth, as does overall microbial biomass. Deeper soils become more dominated by bacteria, particularly the slower growing Gram-positive types. In exceptionally deep soils bacterial diversity declines, though organisms are still capable of living at great depths.

Survey of Ecosystems

Soil formation is generally considered to be driven by five factors: parent material, climate, biota, topography and time (Brady and Weil 2002). As we saw briefly above, parent material provides the framework for soil 'architecture', which provides habitable space for microorganisms. Parent material and climate influence weathering rates, which influence soil chemical properties. Over time soils evolve and change and develop in accordance with their environmental and biological factors. The resulting global mix of soil-ecosystems differs in many ways that impact microbial communities. Tropical, temperate, agricultural and wetland ecosystems each have distinct soils, and pose strikingly different challenges for soil microbial life. For example, tropical systems often have low levels of nutrients in the soil, and generally constant warm, moist temperatures, while high-latitude systems store massive amounts of carbon and can experience dramatic temperature shifts (Brady and Weil 2002). Finally, human land use can also greatly alter ecosystems and thus soil life and its functioning.

Although some generalization by ecosystem is possible, the diversity of factors impacting microbial community habitats implies some caution is needed. Even in adjacent areas, the microbial community of soils of differing types can be strikingly different. For example, two studies in the Pacific Northwest of the United States have compared microbial communities between adjacent soils of differing types (serpentine and non-serpentine) and found that microbial communities in the former were more similar to each other than non-serpentine communities (Oline 2006). The serpentine soils had reduced microbial biomass (DeGrood et al. 2005), suggesting that these offer an inferior environment for microbial growth.

Among the least well understood areas in terms of soils and their microbial community structure and function are the tropical and wetland ecosystems. The vast majority of studies are focused on temperate grasslands and forests, with some emphasis on taiga or boreal forest communities. However, tropical forest and wetland soils are some of the most endangered in the world. They are critical in their potential response to current and future climate change, and as potential sites for agriculture in a world with ever increasing human population size. Below we review some of what is known about the microbiology of these soils.

Tropical Forests

Tropical forests are complex natural ecosystems endowed with unrivaled biodiversity. Located between the Tropics of Cancer and Capricorn, and comprising only 7–12% of the earth's total landmass, tropical forests contain approximately 50% of all known species (Laurance 1999; FAO 2001). In addition to supporting these high levels of biodiversity, tropical forests offer a multitude of ecosystem services including: human habitat, pharmaceuticals, food, and other natural products; watershed stability, flood amelioration, and soil conservation; control of regional climate patterns through evapotranspiration; and regulation of global carbon and nutrient budgets (Laurance 1999; Reiners et al. 1994; Serrão et al. 1996; Bawa et al. 2004; Davidson and Artaxo 2004; Fearnside 2005).

The importance of tropical forests has recently taken on greater importance as the scale of global climate change becomes appreciated. The tropics are at the forefront of concerns for the risks posed by global change because of the high levels of de-forestation and rapid rates of land conversion to cropping, with subsequent loss of biodiversity and capacities for carbon sequestration. From 1981 to 2000, 21% of the area occupied by tropical forests was de-forested (FAO 2001; Bawa et al. 2004) and these rates have not slowed (FAO 2005); if anything they have accelerated. Efforts are being made to catalogue the enormous species diversity in tropical forests and assess what ecosystem characteristics are altered by land-use changes in order to assess the implications of their loss. In 2001, tropical biologists convened in

Bangalore, India, to establish research priorities for the tropics and refined these objectives and directions over subsequent years by additional meetings, workshops, and retreats (Bawa et al. 2004). A primary broad ecological research goal that emerged from these efforts aimed to achieve a better understanding of the structure and function of tropical ecosystems. Emphasis was placed on studies of soil microbial communities and belowground ecosystem structure and function including specifically: describing genetic and species diversity in poorly known regions, groups, and habitats of the tropics. This included defining the fungi and other microorganisms in soil; exploring relationships between components of soil biodiversity, nutrient cycling, and productivity; and predicting the functional response of tropical ecosystems to naturally and anthropogenically induced change.

The tropics are an especially understudied microbial ecosystem. Those studies which do exist have focused on tropical soil microbial communities in the Neotropics, primarily in the Amazon, Costa Rica, and Hawaii, with infrequent scattered studies in other locations (Veldkamp et al. 2003; Borneman and Triplett 1997; Nüsslein and Tiedje 1999; Burke et al. 2003; Carney and Matson 2006; Gomez-Alvarez et al. 2007, Kim et al. 2007; Cleveland et al. 2003). There has been little research on soil microorganisms in the Tropics of Africa, Southeast Asia, and Oceania (Waldrop et al. 2000; Amir and Pineau 1998; Krave et al. 2002; Bossio et al. 2005; Venkatesan and Senthurpandian 2006), with no studies of microbial communities yet published for tropical Australia. The majority of these studies have examined shifts in microbial community composition and function when tropical forest is converted to other land uses, such as pasture, plantations, and agriculture (e.g. Waldrop et al. 2000; Borneman and Triplett 1997; Burke et al. 2003). Other studies have examined seasonal influences of tropical wet and dry seasons (Carney and Matson 2006; Krave et al. 2002), and the effects of plant species and aboveground diversity on soil microbial communities (Carney and Matson 2005; 2006). Below we consider several studies from around the world that have investigated soil microbial communities in the tropics using culture-independent methods (Table 1).

Neotropics: Amazonia, Costa Rica, Hawaii

Amazonia. The Amazon Basin is the largest area of contiguous primary rainforest in the world (Davidson and Artaxo 2004; FAO 2005; Rudel 2005). Amazonia is four times larger than the world's second major expanse of rainforest in the African Congo Basin, and spans nine countries across the northern portion of South America (FAO 2005). Brazil contains nearly two thirds of Amazonian tropical forests and is home to an estimated 7,780 native tree species (FAO 2001; 2005; Rudel 2005). From 2000 to 2005, Brazil had the largest annual net loss of forested area worldwide with 3.1 million hectares lost per year (FAO 2005). The dominant land-use change is conversion to pasture, accounting for approximately 50–70% of the de-forestation, with broadacre cropping and logging being the other major forms of land-use change (Serrão et al. 1996; Fearnside 2005).

Table 1 Summary of tropical soil microbiology studies included in review

Authors	Location	Study scheme	Sampling depth	Microbiological methods
Borneman and Triplett 1997	Eastern Amazon (Para, Brazil)	Mature forest; Pasture	0–10 cm	SSU rDNA
Kim et al. 2007	Western Amazon (Rondonia, Brazil)	Pristine forest; Terra preta soils	0–10 cm	16S rRNA
Cleveland et al. 2003	Costa Rica	Oxisol forest and pasture; Mollisol forest and pasture	0–10 cm	Chloroform fumigation-extraction; phosphatase enzyme assay; substrate-induced growth response
Veldkamp et al. 2003	Costa Rica (La Selva)	Alluvial forest and pasture; residual forest and pasture	0–3 m (alluvial) 0–4 m (residual)	Basal respiration; substrate-induced respiration
Carney and Matson 2005; 2006	Costa Rica (La Selva)	Plant diversity gradient of 1, 3, 5, or >25 species	0–10 cm	Phospholipid fatty acid analysis (PLFA)
Nüsslein and Tiedje 1998	Hawaii (Big island of Hawaii)	200 year old volcanic ash deposit with low plant diversity	0–7.5 cm	Guanine-plus-Cytosine (G+C)of DNA; SSU rDNA; denaturing gradient gel electrophoresis (DGGE)
Nüsslein and Tiedje 1999	Hawaii (Big island of Hawaii)	Pristine forest; Pasture	0–10 cm	G+C of DNA; SSU rDNA
Gomez-Alvarez et al. 2007	Hawaii (Big island of Hawaii)	Volcanic deposits of four different ages	Litter + 1 cm mineral soil	16S rRNA
Burke et al. 2003	Hawaii	Pasture; sugarcane	0–20 cm	PLFA
	Brazil	Forest; sugarcane	0–20 cm	
	Ecuador	Forest; pasture; SC	0–10 cm	
Bossio et al. 2005	Western Kenya	Range of soil textures; forest, woodlot, or agriculture land use; improved fallow, traditional ag., or tea	0–5 cm	Chloroform fumigation-extraction; five enzyme assays; BIOLOG; PLFA; 16S rRNA DGGE

(continued)

Table 1 (continued)

Authors	Location	Study scheme	Sampling depth	Microbiological methods
Venkatesan and Senthurpandian 2006	Western Ghats, India	Pristine forest; tea plantations	0–200 cm	Five enzyme assays
Amir and Pineau 1998	New Caledonia	Five ultrabasic soils of varying plant cover	0–20 cm	Dilute plate count; fluorescein diacetate hydrolysis
Waldrop et al. 2000	Tahiti, French Polynesia	Forest; pineapple plantations of varying ages	0–12.5 cm	Chloroform fumigation-extraction; seven enzyme assays; BIOLOG; PLFA
Krave et al. 2002	Java, Indonesia	Seasonal effects (wet and dry season)	Litter + 0–10 cm mineral soil	16S rRNA DGGE
Moritz 2008	Borneo (Mt. Kinabalu, Sabah, Malaysia)	Pristine forest on ultrabasic or meta-sedimentary bedrock	0–120 cm	Two lignocellulose degrading enzyme assays; PLFA; amino sugar analysis

The traditional first step in conversion of primary forest to some other land-use is to slash and burn, which may or may not be preceded by selective logging (Hölscher et al. 1997; Fearnside and Imbrozio Barbosa 1998; Desjardins et al. 2004). The slash and burn process results in a rapid redistribution of nutrients because of the substantial removal of aboveground biomass. Large amounts of carbon are removed from the system during burning while other nutrients, including phosphorus, potassium, calcium, magnesium, and sometimes ammonium-nitrogen, are concentrated in the ash layer. These nutrients are deposited on top of the soil and incorporated throughout the surface layer, resulting in a temporary influx of basic nutrients and an increase in alkalinity (Reiners et al. 1994; de Moraes et al. 1996; Müller et al. 2004). This temporary enrichment of soil nutrients allows alternate land uses on the otherwise nutrient-poor soils of the tropics (Hölscher et al. 1997; Müller et al. 2004). Over time, the conversion from primary tropical forest to pasture decreases the bulk density and porosity of the soil (Reiners et al. 1994; Eden et al. 1991), and has varying effects on carbon content (Fearnside and Imbrozio Barbosa 1998). Some studies in Amazonia have quantified increases in soil carbon under pasture management (Desjardins et al. 2004; de Moraes et al. 1996; Trumbore et al. 1995; Neill et al. 1996); others report decreases in soil carbon (Desjardins et al. 2004; Eden et al. 1991). Some studies document the absence of changes to soil carbon after conversion (Müller et al. 2004; Buschbacher et al. 1988). These conflicting results appear to be related to forms of pasture management and age of pasture, though there may be an as-yet unspecified role for microorganisms in carbon cycling in such changes to land-use (Balser and Firestone 2005).

While the effects of conversion of primary tropical forest to pasture on soil properties have been relatively well documented in Amazonia, comparatively very little is known about the effects of conversion and anthropogenic influences on soil microbial communities in the Region. Few enough studies have examined soil microorganisms in the tropics at all, while even fewer have been conducted using culture-independent techniques (Borneman and Triplett 1997; Kim et al. 2007). In the first published study using molecular methods which surveyed Amazonian soils, Borneman and Triplett (1997) analyzed 100 small-subunit rRNA gene sequences (SSU rDNA) from northeastern Pará, Brazil, from active pasture and mature forest (50 clones from each land-use). They found 98 bacterial and two archaeal sequences that were unique (with no duplicate sequences), and all were novel sequences that had not been previously documented. Representative major bacterial phyla included *Planctomyces*, *Clostridium*, *Fibrobacterium*, and *Proteobacteria*. Phylogenetic analysis of the sequences indicated that in general, most clones were distinct from other organisms that had been previously described in sequence libraries and 18% of the sequences could not be placed into a known bacterial kingdom. Accordingly, Borneman and Triplett (1997) suggested that these tropical soils contained high levels of microbial diversity composed of some unusual microorganisms. The authors then used rRNA intergenic spacer analysis (RISA) to investigate the effects of vegetative cover and land-use changes on microbial communities and found different banding patterns for the mature forest and pasture soils, indicating that distinct microbial communities were present under each form of land-use. This study made a major impact in soil and tropical microbiology by providing evidence of novel soil microorganisms and vast microbial diversity. Further, this study showed that de-forestation in the tropics and subsequent conversion to pasture resulted in changes in the composition of the bacterial community, which may also indicate or result in altered nutrient cycling and ecosystem functioning. Unfortunately, the authors did not assay microbial activity together with the community composition. The influence of de-forestation on the role of microbial communities in ecosystem processes (such as nutrient cycling) under each land-use remains unknown.

The next molecular survey of Amazonian soils was published a decade later by Kim et al. (2007). In this work, the authors used oligonucleotide fingerprint groupings (OFRG) to sort through 1,500 16S rRNA clone libraries and DNA sequencing to compare bacterial communities of pristine tropical forest soils and 'terra preta' forest soils in Rondonia, Brazil, in Western Amazonia. Terra preta soils were anthropogenically created by disturbance during pre-Colombian times by the indigenous practice of "slash and char" agriculture. This alternative to slash and burn involves converting the biomass to charcoal, or charring, rather than burning, and results in a lower impact on the environment. Today these soils contain thick epipedons (e.g. the 'A' horizon was 1 m thick at the study site of Kim et al.) with high organic matter contents (Lima et al. 2002; Mann 2002). There is interest in terra preta soils for their ability to stabilize large amounts of organic matter and maintain high fertility in spite of the presence of favorable conditions for rapid organic matter decomposition (Amundson 2001; Lehmann et al. 2003). Kim et al. (2007) found that terra preta soils had greater bacterial diversity and were significantly different from pristine forest soils, with approximately 25% greater species richness. This is particularly interesting given that both soils had similar aboveground species composition and structure, and indicates that the legacy effect from past land-use and the alteration of soil properties remained apparent in the extant microbial community, regardless of subsequent vegetational cover. Phylogenetic analysis showed 14 major bacterial groups were present on terra preta, compared with nine in the pristine forest soil. Phyla common to both soils included the predominant *Acidobacterium*, *Proteobacteria*, *Actinobacteria*, *Planctomycetes*, and *Verrucomicrobia.* Similarly, Borneman and Triplett (1997) found *Planctomycetes* and *Proteobacteria* dominated in eastern Amazonian soils. An important difference between this study by Kim et al. (2007) and Borneman and Triplett (1997) was that the methodology and technology for clone analysis had greatly improved in a decade, thereby facilitating a more comprehensive survey using 1,500 clones in 2007 versus the 100 sequenced by Borneman and Triplet in 1997. In the same way, the sequence databases were much larger following 10 years of additional research, and consequently, fewer novel sequences were found in the study by Kim et al. (2007), although they did identify three potential new subgroups of *Acidobacterium.* A similarity between the two studies was that neither employed a functional measure of the microbial community alongside their molecular analysis and therefore the role of each microbial community remained unknown. The taxonomic information generated using molecular methods only allows us to quantify the diversity of soil microorganisms and facilitates comparison, but cannot currently give information about ecological functions (Fierer et al. 2007).

Nevertheless, both studies showed that soil microbial communities in Amazonia changed alongside anthropogenic disturbance, be it a soil legacy effect from hundreds of years ago or recent de-forestation and pasture conversion. Further, both studies indicated that Amazonian tropical soils harbour distinct microbial communities belowground, paralleling the distinctness found in aboveground biodiversity.

Costa Rica. Central America contains approximately 4.5% of the world's tropical forests and houses an estimated 7% of the world's biodiversity (FAO 2001). From 1990 to 2005, Central America experienced one of the highest rates of de-forestation in the world (FAO 2001; 2005). As in Amazonia, cattle ranching and broadacre cropping have generally been the forces driving land use change, with urban expansion playing a lesser but still important role (FAO 2001; Rudel 2005). Costa Rica contains just over 2.5% of Central American tropical forests and has the lowest rate of de-forestation in the region (FAO 2001). Costa Rica is one of the pre-eminent locations for tropical forest research in Central America and the world. In the 1960s, pristine tropical forest was set aside in northeastern Costa Rica for ecological research and named La Selva Biological Station of the Organization for Tropical Studies (Veldkamp et al. 2003). Three studies have focused on soil microbial activities and communities at La Selva (Veldkamp et al. 2003; Carney and Matson 2005; 2006), with one additional study of soil microbial dynamics located elsewhere in Costa Rica (Cleveland et al. 2003).

At La Selva Biological Station in 1991, an abandoned cacao plantation was cleared, the merchantable trees extracted, and then the area was slash burned (Haggar and Ewel 1995). Subsequently, plant communities with gradients of diversity were established, including plots planted with one, three and five species, and >25 species (natural regeneration). Carney and Matson (2005; 2006) utilized these experimental plots to investigate the influence of aboveground diversity on belowground soil microbial communities using PLFA (Carney and Matson 2005; 2006), catabolic potential and litter decomposition assays (Carney and Matson 2005). They found that microbial communities shifted in composition with each level of plant diversity and bacterial communities (not fungal) were primarily responsible for these differences (Carney and Matson 2005). Catabolic and litter decomposition assays suggested that each microbial community under different levels of plant diversity harboured distinct functional characteristics. The authors suggested that these shifts were related to the diversity of secondary compounds available for microorganisms to use in their metabolism; for example, a monoculture that produces a carbon source of uniform composition results in a microbial community with lower enzyme diversity. Thus, a reduction in aboveground diversity may affect the catabolic capacity of a belowground microbial community, thereby altering carbon and nutrient cycling and ecosystem functioning. This could indicate an important linkage between aboveground and belowground biodiversity. This finding is particularly relevant to de-forestation issues in the tropics, where primary forest is often cut down and replaced with monocultural plantations (e.g. Waldrop et al. 2000; Krave et al. 2002; Venkatesan and Senthurpandian 2006).

In 2006, Carney and Matson published a follow-up to their 2005 study, where they presented comparisons among different monocultures and focal members in the three species combinations and ranges of sampling dates (representative of wet and dry seasons). They found that microbial community composition differed significantly among different monocultures and focal species, further supporting the suggested influence of individual plant species in shaping microbial communities. Examination of the microbial communities from different sampling dates showed no significant effect of season relating to community composition. Taking the results of both studies together, it may be concluded that land-use change that alters plant species diversity and composition affects soil microbial communities and their functional roles in ecosystems. These are notable studies because Carney and Matson (2005; 2006) successfully utilized functional and compositional measures of the microbial community to provide an insight into altered ecosystem functioning under land-use change (an important goal of tropical ecology [Bawa et al. 2004]).

At La Selva, in addition to the experimental set-up for studies of aboveground and belowground diversity used by Carney and Matson (2005; 2006), there is primary forest that lies adjacent to de-forested areas cleared for pasture during the 1970s (Veldkamp et al. 2003). Thus, an existing system is in place to study the ecological effects of pasture conversion on native tropical forests. Veldkamp et al. (2003) used the pristine forest and adjacent pasture to look at microbial activity throughout the profile (0–3 m) using basal respiration (BR) and substrate-induced respiration (SIR). Both BR and SIR measure the carbon dioxide production of a soil sample over a period of time, but SIR adds a labile substrate (e.g. glucose) to maximize the respiration rate (representative of the "potential" of the soil microbial community), while BR represents the capacity of the microbial community in situ. Soil profiles in the tropics are highly weathered and often many metres deep, containing substantial stores of soil carbon (Veldkamp et al. 2003; Trumbore et al. 1995; Nepstad et al. 1994). Thus, an energy source is available for microorganisms at depth, but very little is known about their activities in the subsoil. Veldkamp et al. (2003) detected both BR and SIR down to 3 m depth in both forest and pasture. The pasture had 30–50% higher SIR than the forest above 2 m, below which there was no significant difference between land-use cover types. By contrast, BR was higher in the forest than the pasture in surface sampling depths (10–50 cm), with no significant difference below 50 cm. It is interesting that the potential community activity under pasture was greater than under forest cover, but the actual activity of the community was greater under forest cover than pasture. However, the focus of the Veldkamp et al. (2003) study was primarily to assess the existence of substantial carbon pools and microbial activity at depth in tropical soils and therefore soil microbial communities and the ecology of microorganisms were not generally addressed.

A final study in Costa Rica was located outside of La Selva on the Osa Peninsula, and also examined soil microorganisms under forest and pasture (Cleveland et al. 2003). In contrast to the depth study by Veldkamp et al. (2003), the focus of Cleveland et al. (2003) was to investigate how land conversion to pasture affects soil microbial communities using two contrasting soil types, Oxisols and Mollisols. (Oxisols are highly weathered, low-nutrient tropical soils, while Mollisols are generally higher-nutrient grassland soils) (Brady and Weil 2002). Cleveland et al. (2003) used chloroform fumigation-extraction to quantify microbial biomass, and phosphatase enzyme assays and substrate-induced growth response (SIGR) to measure microbial activities. They found that both soil carbon and microbial biomass decreased upon conversion to pasture, with a more pronounced effect on Oxisols. The authors suggested that low fertility soils (e.g. Oxisols and Ultisols; also the most common in the tropics) are more susceptible to reductions in microbial biomass than soils of higher fertility (e.g. Mollisols). Analysis of microbial activities between pasture and forest showed different physiological capacities, with forest communities being more responsive to changes in resource availability. Further, forest communities appeared to be better adapted to decompose recalcitrant carbon compounds (as evidenced by SIGR using salicylate) than pasture soils. This indicates alterations in microbial community composition after land-use conversion from high biodiversity forest vegetation to less diverse pasture vegetation, and supports the conclusions of Carney and Matson (2005) who suggested shifts in microbial communities under different levels of plant diversity were related to the diversity of carbon sources. However, Carney and Matson (2005) used both compositional and functional measures of the microbial community, while Cleveland et al. (2003) did not use a compositional measure of the microbial community, but rather inferred it from shifts in the microbial activity with different types of cover.

Hawaii. The Hawaiian Islands are the most secluded archipelago in the world (Gomez-Alvarez et al. 2007), formed in the Pacific Ocean over millions of years from basaltic magma that originated from a "hotspot" in the earth's crust, creating a chain of volcanic islands that migrate outward from the centre, activity that continues today (Crews et al. 1995). The geographic isolation and continuing volcanic activity of the Hawaiian Islands make the archipelago an attractive system for ecological studies (Gomez-Alvarez et al. 2007; Crews et al. 1995; Nüsslein and Tiedje 1998). Several studies of soil microbial communities have taken place in the Hawaiian Islands, including examination of soil bacteria in volcanic deposits (Gomez-Alvarez et al. 2007; Nüsslein and Tiedje 1998) and the effects of land-use change on soil microorganisms (Nüsslein and Tiedje 1999; Burke et al. 2003).

Nüsslein and Tiedje (1998) were the first to study soil microbial communities in Hawaii using molecular methods. For their study site, they selected a 200 year-old volcanic ash deposit with low plant diversity and used genetic nucletotide guanine-plus-cytosine (G+C) content to distinguish members of the community. An advantage of G+C content is that it is comprehensive for all DNA and not subject to methodological bias (Kirk et al. 2004; Nüsslein and Tiedje 1999). Prior to the study, they hypothesized there would also be a low level of bacterial diversity on the parent material because of its youth and lack of aboveground diversity. By contrast, they found such high bacterial diversity that they could not determine the community structure using G+C content, as this measure of microbial diversity has relatively coarse resolution. To attempt to reduce the diversity to more manageable levels, they fractionated the G+C content into 63% and 35% (indicative of certain bacterial groups), but still found few duplicates and high diversity. The 63% fraction represented the dominant bacterial biomass and consisted of fewer bacterial taxa, including *Pseudomonas*, *Rhizobium-Agrobacterium*, and *Rhodospirillum*, while the 35% fraction was not dominant but had higher bacterial diversity. The authors suggested that this reflects the ecological paradigm (Levine 1976) where the most successful competitors consist of fewer species, while the less competitive minority consists of more diverse organisms that occupy smaller, more defined niches. This initial study of microbial communities in Hawaii identified high levels of bacterial diversity, even on a young substrate.

The next study of bacterial diversity in Hawaii, by Gomez-Alvarez et al. (2007) was conducted nearly a decade later. The authors used molecular methods with higher resolution and compared soil bacterial communities on three volcanic deposits ranging in age and included an established old-growth tropical forest (Gomez-Alvarez et al. 2007). The study sites ranged in vegetative cover from a complete absence of vegetation to mature forest. The sequencing and phylogenetic analysis of 16S rRNA showed 56% of the sequences were unclassified and largely distinct. The three volcanic deposits had 60–80% of sequences unclassified, while the mature forest had only 6% unclassified. The oldest volcanic deposit had the greatest bacterial diversity, while the intermediate-age volcanic deposit with an absence of vegetative cover showed the least. Only *Acidobacteria* and *Actinobacteria* were common to all four study sites and no phylotype showed >97% homology among sites. The results of Gomez-Alvarez et al. (2007) indicated a high diversity of soil bacteria within a relatively small area of the Kilauea volcano caldera and found a large number of previously un-described clone sequences in Hawaiian Andisols. Similar to the study of Nüsslein and Tiedje (1998), the work of Gomez-Alvarez et al. (2007) emphasized the diverse and ubiquitous abilities of bacteria to colonize young, even un-vegetated soil, while providing more detailed genetic sequencing of the microbial community.

Another study in Hawaii by Nüsslein and Tiedje (1999) used G+C content fractionation to compare soil bacterial communities under land conversion from native tropical forest to pasture. Similar to other studies in other tropical areas, they found a significant G+C shift (49%) in microbial community structure with the change in vegetative cover (Veldkamp et al. 2003; Borneman and Triplett 1997; Cleveland et al. 2003). *Fibrobacter*, an organism that specializes in cellulose degradation, dominated the 63% G+C content in the forest, with a shift to *Proteobacteria* (general Gram-negative type bacteria) dominance in the pasture. This shift in dominance may indicate that the physiological capacity of the soil bacterial community also changed upon land-use conversion, but there was no quantitative or qualitative measure of microbial function in this study. Thus, while Nüsslein and Tiedje (1999) described genetic diversity of the soil bacterial community, the inter-relationships between bacterial community structure and ecosystem functioning were not addressed.

The final study we consider in the Hawaii and Neotropics section used of phospholipid fatty acid (PLFA) analysis to examine shifts in microbial community composition in forest, pasture, and sugarcane soils in Hawaii, Brazil, and Ecuador (Burke et al. 2003). Lipid analysis provides a more general perspective of microbial community structure; rather than genetic information, it provides information on ecological functional groups, such as Gram-positive bacteria, Gram-negative bacteria, actinomycetes, and saprophytic fungi (Vestal and White 1989). Generally, Burke et al. (2003) found that soil type (e.g. Hawaiian Andisols versus Brazilian Oxisols) was most closely related to the relative abundances of Gram-positive and Gram-negative bacteria.

Conversely, land-use was the major determinant of fungal, actinomycete, and protozoal abundance. Overall, biomass and more specifically, Gram-negative biomarkers, declined upon conversion from forest to sugarcane in Hawaii and Brazil. Gram-positive bacteria were more abundant in agricultural soils, most likely reflecting the less labile carbon substrate that is available on cropland without continuous additions of fresh litter. The study by Burke et al. (2003) reaffirmed that land-use conversion affects the structure of soil microbial communities and introduced the concept that these changes show some regional variability (i.e. Hawaii versus Ecuador versus Brazil) but can be broadly applied to a range of tropical soils.

Conclusions: Neotropics

The majority of tropical soil microbiology studies have had sites in the Neotropics and have concentrated on the effects of land-use conversion on soil microbial communities. These studies have utilized a wide range of methods for the study of soil microorganisms, including PLFA, DNA-based methods, and measures of microbial function. However, we can collectively use the results from these studies to conclude that the effects of land-use change on vegetative cover type and plant species diversity changes the structure and function of the soil microbial communities (Veldkamp et al. 2003; Borneman and Triplett 1997; Nüsslein and Tiedje 1999; Burke et al. 2003; Cleveland et al. 2003; Carney and Matson 2005; 2006). There is indication that the conversion to pasture reduces microbial activity and biomass (Veldkamp et al. 2003; Burke et al. 2003; Cleveland et al. 2003) and that microbial communities have a less diverse physiological response as aboveground biodiversity decreases (Cleveland et al. 2003; Carney and Matson 2005). Other studies have indicated there is an immense diversity of soil microorganisms in the tropics, with many previously unclassified organisms (Borneman and Triplett 1997; Kim et al. 2007; Nüsslein and Tiedje 1998). The majority of studies restricted soil sampling to the upper 10 cm of soil, and given the substantial carbon pools at depth in tropical soils, further studies of microbial activity in tropical subsoils are much needed (Veldkamp et al. 2003). In addition, more studies that use both compositional and functional measures of tropical soil microbial communities are needed to help understand what changes in ecosystem functioning occur under land-use changes.

Tropics of Africa, Southeast Asia, and Oceania

Africa. In 2000, Africa was estimated to contain approximately 33% of global tropical forests (FAO 2001). The Congo Basin, located in the heart of Africa, is the second largest expanse of primary rainforest in the world (FAO 2001; 2005). From 2000 to 2005, six out of the top ten countries with the highest de-forestation rates in the world were located in Africa (Congo, Nigeria, Sudan, Tanzania, Zambia and Zimbabwe), making tropical African forests a rapidly diminishing resource. Although it is difficult to generalize for the whole of Africa, the major force behind de-forestation is often conversion to small-scale subsistence agriculture (FAO 2001; Rudel 2005). While much is unknown about the tropics, the tropical forests of Africa are perhaps the most understudied of all tropical areas, with little known about the structure and function of these systems and the effects of de-forestation and land-use conversion.

A single study by Bossio et al. (2005) investigated tropical soil microbial communities in Western Kenya under a variety of land uses and employed a consortium of microbiological analyses. This study included comparisons among forest, woodlot, and agriculture land-use types; agriculture practices of traditional maize cropping (two crops per year), management with improved fallow (alternating maize with a species of nitrogen-fixing tree, *Tephrosia candida*), and perennial tea cultivation; and a range in soil textures from sandy (5% clay) to clay (60% clay). To assess microbial community activity, Bossio et al. (2005) used BIOLOG and enzyme assays, and used PLFA and genetic analysis (DGGE analysis of 16S rRNA) to determine microbial community structure. This is the only study that used such a comprehensive assemblage of microbial characterization techniques (Table 1). Molecular methods showed that soil type (such as sandy, clay) appeared to induce the primary differentiation among soil microbial communities, while land-use (wooded or agriculture) was the secondary determinant. The DGGE analyses showed there was no significant difference in bacterial diversity between wooded and agricultural soils. While PLFA data showed greater relative abundance of Gram-negative bacterial indicators in wooded soils than agricultural soils, and agricultural soils had higher proportions of actinomycetes and Gram-positive bacteria. These same shifts in lipid biomarkers under land-use change were reported in the Neotropics by Cleveland et al. (2003) and are likely to be linked with varying availabilities of carbon sources under different land-uses. Bossio et al. (2005) found that soil microbial communities from wooded sites had the most diverse abilities for substrate utilization, a finding also reported by Cleveland et al. (2003). Finally, the authors found substantially lower soil carbon, microbial biomass, and total enzyme activities under traditional maize cultivation, all of which generally increased under improved fallow management. Using a complex suite of methods and study sites, Bossio et al. (2005) found that the microbial community shifted with different soil types and land-uses, but overall levels of microbial diversity were similar among land uses. Shifts in activity or functions of the microbial community were closely linked to management practices, but showed less specificity than measures of microbial community structure. This study reiterates the important influences of land-use change and management practices on soil microbial communities in the tropics. This study by Bossio et al. (2005) is the only study we are aware of that focused on soil microbial communities in tropical Africa. Clearly, there is a need for more studies on this continent with its vast tropical forest reserves and rapid de-forestation.

India. India is unique among the other tropical regions examined in this literature review in that it has had a high population density exerting pressure on the land since at least the nineteenth century (Rudel 2005). While pristine tropical forests are being newly encroached upon and exploited in Amazonia, Africa, and parts of Southeast Asia, tropical forests in India have been de-aforested and fragmented in response to poverty and anthropogenic pressures for over a century. As a result, in the 1980s, only 2% of India's land area contained primary forest, with much of the remaining land consisting of cutover and degraded forests. The government responded by establishing forestry management plans and intensive re-forestation efforts, resulting in a slight increase (38 ha $year^{-1}$) in forested area from 1990 to 2000 with 72% of India's forests under management plans. Not surprisingly, the majority of soil microbiology studies in the tropical forested regions of India have examined the effects of land conversion and management practices on soil microorganism populations. Changes in microbial biomass under different land uses have been researched intensively in India and studies have generally shown that microbial biomass declines upon conversion from primary forest (Srivastava and Singh 1991; Basu and Behera 1993; Singh and Singh 1995; Behera and Sahani 2003). However, no studies in India were found that explored shifts in microbial community structure upon land-use changes using culture-independent methods (PLFA or molecular methods), while one study was found that investigated differences in microbial enzyme activities between pristine forest and land converted to tea plantations (Venkatesan and Senthurpandian 2006).

In the middle of the nineteenth century, the Western Ghats, a region in Southern India, was covered with native tropical wet evergreen forests (FAO 2001; Kumara et al. 2004). In the 1880s, private landowners cleared large tracts of this rainforest and established tea plantations on the deeply weathered tropical soils, creating a mosaic of plantations and primary forest (Venkatesan and Senthurpandian 2006; Kumara et al. 2004). A study by Venkatesan and Senthurpandian (2006) examined differences in microbial enzyme activities between these two land uses in the Western Ghats. They sampled every 25 cm down the soil profile to 200 cm depth in an effort to quantify microbial activities throughout the typical rooting zone of tea. All enzyme activities (with the exception of protease) were detectable in the deepest sampling depths, indicating an active microbial community in the subsoil. Veldkamp et al. (2003) reported a similar result with measurable microbial respiration at 3 m depth in tropical pasture and forest soils in Costa Rica. Venkatesan and Senthurpandian (2006) found that acid phosphatase activities were highest in both tea and forest soils, followed by alkaline phosphatase activities. At any given depth, urease activity was higher under tea plantations than under forest, and showed no significant decline with depth under tea crops. The authors attributed this result to the continuous application of urea used in plantation management. Aryl sulfatase activity was considerably higher in the top 50 cm of the tea plantation than the forest, below which the forest had consistently higher aryl sulfatase activity. Generally, Venkatesan and Senthurpandian (2006) summarized their findings of enzyme activities and attributed observed patterns to organic matter content and beyond this provided little insight into their results. Thus, data are presented, but lack an ecological context, which is a common shortcoming in soil microbiology. The enzymatic patterns observed by Venkatesan and Senthurpandian (2006) were generally irregular with no distinct conclusion to be drawn between types of cover. However, they overlooked that all enzyme activities were greater below 150 cm in the forest than the tea plantations (with the exception of urease). This may be attributed to the deeper rooting patterns of tropical wet evergreen forests that extend well below the 2 m zone of tea plantations. This study is important because of its focus on microbial activity with depth in the tropics, a relatively unknown realm of study (Veldkamp et al. 2003). However, further studies of depth that include determinations of microbial community structure along with measures of activity are needed in order to arrive at ecologically meaningful interpretations of ecosystem functioning.

Pacific Southeast Asia and Oceania. The tropical forests of the Southeast Asian Pacific and Oceania have some of the highest reported levels of biodiversity in the world (FAO 2001; 2005; Myers and Mittermeier 2000). However, recent estimates have shown that Pacific Southeast Asia retains only 3–15% of its primary vegetation, while Oceania, including New Caledonia and Polynesia/Micronesia, preserves 22–28% of its original vegetation (Myers and Mittermeier 2000). Indonesia lost almost 1.9 million hectares per year from 2000 to 2005, the second largest annual net loss of forest behind Brazil (FAO 2005). Conversely, from 1990 to 2000, New Caledonia and French Polynesia, two countries considered in this review, had no net loss of tropical forests. De-forestation in Southeast Asia results primarily from logging and the timber trade (Laurance 1999; FAO 2001). Similar to the tropical forests in Africa, the forests of Southeast Asia and Oceania have few studies of soil microorganisms and ecosystem functioning (Waldrop et al. 2000; Amir and Pineau 1998; Krave et al. 2002).

On the island of Java, Indonesia, nearly all primary tropical forests have been harvested and today approximately 50% of the forest is has been converted to pine plantations (Krave et al. 2002). A study by Krave et al. (2002) investigated seasonal influences (wet and dry season) on soil bacterial communities in Javanese *Pinus merkusii* plantations. Their objective was to achieve a better understanding of nutrient cycling and so improve plantation management. They sampled litter, duff,[1] and mineral soil samples during successive dry and rainy seasons and then used DGGE analysis of 16S rRNA to assess bacterial community structure. They found that each layer of forest floor and mineral soil had significantly different physical and chemical properties, including organic matter content, and the bacterial communities were stratified accordingly. Moisture content, pH, and nitrogen levels of the three forest layers changed significantly with season, but bacterial communities in the duff and mineral soil were not influenced. Only the litter microbial communities varied with season. Similarly, Carney and Matson (2006) found that season did not influence the composition of soil microbial communities in Costa Rica. Krave et al. (2002) constructed a clone library for one litter sample taken in the wet season and found that it contained high levels of bacterial diversity. *Proteobacteria* (specifically *Rhizobium-Agrobacterium*) and Gram-positive bacteria with high G+C content dominated the profiles. *Rhizobium-Agrobacterium* was reported in the study of Hawaiian soils by Nüsslein and Tiedje (1998) and Gomez-Alvarez et al. (2007),

---

[1]Duff: the layer of partially and fully decomposed organic materials lying below the forest flar litter and immediately above the mineral soil.

but was not documented in phylogenetic analyses of tropical soils by Borneman and Triplett (1997) and Nüsslein and Tiedje (1999). Overall, this study suggests that microbial communities in duff and mineral soil are better buffered from seasonal changes than litter microbial communities, and each layer harbours a distinct community of considerable diversity.

Another study of tropical soil microbial communities in the Southeast Asian Pacific was conducted by Amir and Pineau (1998) in New Caledonia using culture-dependent methods to identify microbial community structure and culture-independent methods to assay microbial activity. Approximately one third of New Caledonia is underlain by ultramafic outcrops. Soils weathered from ultramafic rock, often also referred to as ultrabasic or serpentine soils, pose special challenges for plant growth and survival. These rocks and their resulting soils are characterized by high levels of metals (e.g. nickel, cobalt); low levels of nitrogen, phosphorus, and potassium; high levels of magnesium with low calcium; and low soil moisture (Walker 1954; Proctor 1999; Brady et al. 2005). Ultrabasic outcrops often have poor productivity and contain many endemic species that are specially adapted to the potentially toxic levels of magnesium and other metals (Walker 1954). Most ultramafic outcrops are exposed in orogeny (mountain-building) zones and are rare, occupying approximately 1% of the Earth's surface (Proctor 1999). In New Caledonia, many ultramafic outcrops have been mined for nickel and revegetation efforts are underway to reclaim land degraded by mining activities (Amir and Pineau 1998). Amir and Pineau (1998) examined soil microorganisms in a series of ultrabasic soils that varied in vegetative cover (e.g. bare ground, cluster of vegetation and crop cover) to investigate methods for improving soil fertility on reclaimed lands. Soil microbial populations were estimated using dilution plate counts (a culture dependent method) and microbial activity was estimated using the fluorescein diacetate (FDA) hydrolysis test (a culture-independent method). The authors found that microbial populations and activity in New Caledonian ultrabasic soils were lower than other estimates reported in studies of nonserpentine soils. These low numbers were attributed to minimal organic matter contents in the ultrabasic soils. Actinomycetes were the predominant bacteria, accounting for 70–90% of the total culturable microorganisms, and two very common culturable fungal families, *Tuberculariaceae* and *Dematiaceae*, were absent. The authors suggest these results may be related to the metal tolerance of these microbial groups. The ultrabasic soils varied widely in microbial characteristics depending on the level of root colonization of soil, with the highest biomass and activity under vegetation and the lowest values on bare uncolonised ground. Cropping appeared to affect microbial communities positively, as bacterial and fungal diversity increased under cultivation. The authors caution that their results are difficult to compare with other soils because of the uniquely extreme conditions of ultramafic outcrops. Amir and Pineau (1998) provided preliminary information about soil microbial communities in tropical ultrabasic soils, but more comprehensive studies using culture-independent methods are needed along with further studies of microbial function.

A further study in Oceanic tropical soils used enzyme activity assays and BIOLOG along with PLFA to investigate responses of the soil microbial community to land-use conversion from primary forest to pineapple plantations in Tahiti (Waldrop et al. 2000). Researchers found that conversion to plantation agriculture increased soil acidity and decreased carbon content, and accordingly soils had one quarter of the microbial biomass of forest soils. Soil microbial communities were significantly different between primary forest and plantations, with greater relative abundances of fungi and actinomycetes in plantation soils. BIOLOG did not show differences between forest and plantation treatments. This contrasts with the hypothesis presented by Carney and Matson (2005) in their study of the effects of plant diversity on soil microbial communities in Costa Rica where they suggested that under less diverse vegetative cover (e.g. plantation) soils have lower diversity in their degradative abilities. Waldrop et al. (2000) further found that enzyme activities were similar between forest and plantation soils, with the exception of β-glucosidase and sulphatase enzymes. However, the pineapple plantations are managed distinctly from other crops and every 5 years the entire pineapple crop is cut down and incorporated into the soil, 'resetting' the system with organic matter input. Thus, pineapple plantations may have carbon inputs comparable to natural forests, resulting in similar microbial community enzyme activities between forest and plantation soils. Similarly, Venkatesan and Senthurpandian (2006) found no discernible differences in enzymatic patterns (with the exception of urease) between tea plantations and tropical forest soils in India. Relating microbial community composition to function, Waldrop et al. (2000) were able to correlate specific enzyme activities (total activity divided by microbial biomass) and relative abundance of specific indicator lipids. They found β-glucosidase and β-xylosidase, enzymes used to degrade simple carbon compounds, were correlated with Gm–, fast-growing bacteria that metabolize simple sugars. Thus, the authors successfully deduced the function of a member of the soil microbial community, lending insight into its role in ecosystem functioning. This study by Waldrop et al. (2000) confirmed that land-use change in the tropics alters soil properties and soil microbial communities. Further, it provided a successful interpretation of microbial structure and function, achieving a major goal of tropical ecology (Bawa et al. 2004).

Conclusions: Old World Tropics

These studies of soil microbial communities were as varied as their locations. They examined the effects of soil type and land-use in Africa (Bossio et al. 2005); microbial activity through depth in tea plantations and tropical forests in India; wet and dry seasons on litter and mineral soil in Java (Krave et al. 2002); vegetative cover in New Caledonia (Amir and Pineau 1998); and conversion from primary forest to pineapple plantations in Tahiti (Waldrop et al. 2000). Perhaps the best collective conclusion from these five studies of the tropics is that little research has been done in this vast region of the world concerning soil microbial communities and ecosystem functioning. Clearly, there is a huge range of diversity of soils and land uses in the tropics, and many questions concerning ecosystem functioning remain to be answered.

Wetlands

Wetlands are transient or permanently saturated terrestrial habitats. The term wetland has been applied to freshwater marshes, bogs, peatlands and swamps (EPA 2004). It also encompasses coastal areas such as salt marshes and mangroves (Shaw and Fredine 1956). This excessively broad definition has long hampered our understanding of these ecosystems and its inhabitants. Overall wetland ecosystems comprise a small proportion of the earth surface (at least 9% by some estimates [Ramsar Convention Secretariat 2007]), yet their contribution to biochemical cycles, water cycle (as source and filter), and the maintenance of species richness overshadow that of other larger terrestrial environments (Basu and Behera 1993). Additionally, wetlands are used by humans for water, flood control, fiber, food, and recreation. The services delivered by wetlands have been arguably valued at US$14 trillion annually (Ramsar Convention Secretariat 2007). Worldwide, wetlands have long been underappreciated and drained. Wetland loss and degradation has primarily been driven by land conversion for agriculture, infrastructure development, water extraction for irrigation or drinking water, eutrophication and pollution and over-exploitation (Ramsar Convention Secretariat 2007). Climate change is expected to exacerbate the loss and degradation of wetlands, both through drying out and resulting shifts in precipitation (Ramsar Convention Secretariat 2007).

Wetland ecosystems link terrestrial, aquatic, and biosphere level processes through their contribution to biogeochemical cycles and global climate change (Fig. 2). This includes interactions between environmental, hydrological and biotic processes. Among these the physical processes, e.g. hydrology, have been studied in more detail than the biotic processes. A recent review of wetland microbial communities by Gutknecht et al. (2006) notes that research in wetland biogeochemistry has most often been focused on microbially-mediated processes (e.g., methanogenesis, denitrification and methanotrophy), and less often on microbial communities or on populations of specific microorganisms of interest (Fig. 3). Work to date indicates an important process level role for hydrology and soil nutrient status. The impact of plant species composition on processes is potentially critical, but is as yet poorly understood.

Research on microbial communities in wetland soils has primarily focused on anaerobic bacteria responsible for methanogenesis (production of methane gas), denitrification and sulfate reduction (Gutknecht et al. 2006). There has been less work on taxonomic groups such as those responsible for nitrogen fixation, or aerobic processes such as nitrification. While less well known, it is also possible that aerobic organisms such as nitrifiers and methanotrophs may play an important role in wetland ecosystem functioning.

Work on general community composition and on wetland mycorrhizal fungi is particularly sparse. Despite longstanding conventional wisdom that says fungi are not abundant or important in wet systems, it has now been demonstrated that mycorrhizal fungi are often abundant in wetlands (Mentzer et al. 2006) and may play a significant role, and mycorrhizal association was the focus of a recent review (Anupam 2003).

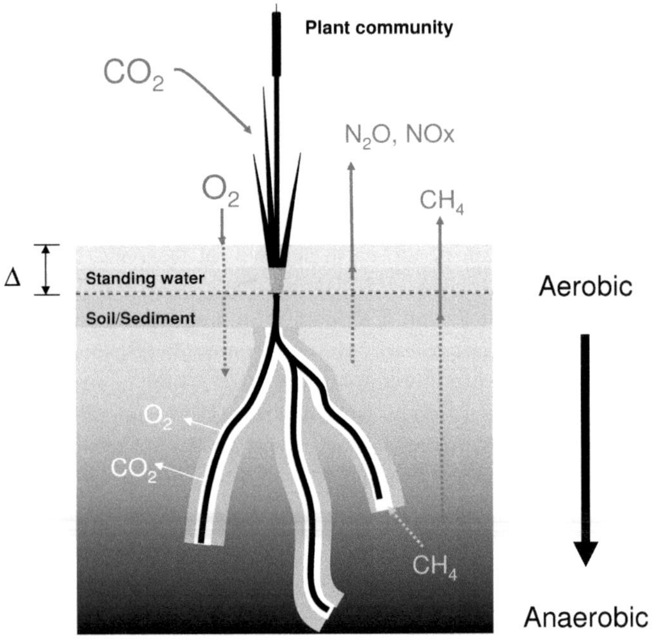

Fig. 2 Wetland structure. Water table height, depth from surface, and distance from plant roots create oxic to anoxic gradients. The result is a complex interplay between anaerobic and aerobic conditions that allows for a wide range in processes to occur in wetland soils

While arbuscular mycorrhizas (AMF) can be found at all levels along moisture gradients from dry to excessive moisture, abundance may decrease under very high water content (Miller and Bever 1999). In particular, AMF species accustomed to dry environments may be very sensitive to high water levels, indicating that water level is an important factor in AMF colonization. Mycorhizzal community composition is also likely sensitive to plant community structure (Gutknecht et al. 2006).

The most conspicuous effect of alterations to a wetland habitat is perhaps the drastic change in vegetation composition and structure. Wetlands habitats receive high levels of pollution (i.e. nitrogen, phosphorus and pesticides) as runoff and ground water from agricultural activities and human settlements, which has contributed to their high susceptibility to invasive species (Zedler and Kercher 2004), and a high proportion of local extinctions. Invasions alone could bring in drastic changes in vegetation composition and litter chemistry, which ultimately influence soil processes. Over a span of decade to centuries, changes at plant and microbial levels would influence ecosystem properties such as geomorphology, hydrology, biochemistry and disturbance regime (Gordon 1998). Although invasive species are spreading rapidly in wetland habitats at local and global levels (Zedler and Kercher 2004), their current impact in nutrient cycling and soils processes are not well

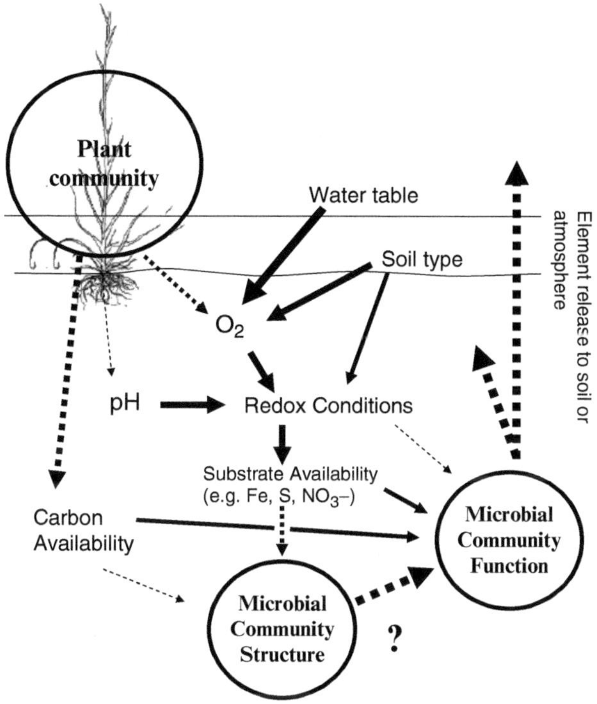

Fig. 3 Relationships among controls over wetland ecosystem microbial communities and element cycling. Arrows indicate relationships, and width of arrows indicates relative importance of relationship for ecosystem functioning. Dashed arrows represent interactions that are poorly understood, even though they may be important

understood (Windham and Ehrenfeld 2003). One study has shown that different levels of an aggressive wetland invader (reed canary grass, *Phalaris arundinacea*) do differentially impact the microbial community, affecting both biomass and the structure of the biota (Kao-Kniffin and Balser 2007).

In recent years the value of wetlands has been increasingly recognized. The cost associated with constructing and restoring wetlands contributes to this appreciation.

Soil Microbial Communities and Changing Agricultural Management Regimes

Humans have converted one-quarter of the Earth's surface to cultivated systems, largely by changing native ecosystems to arable lands within the past 50 years (Millennium Ecosystem Assessment 2005). Agricultural expansion and intensification is likely to accelerate as the world human population grows, and soil functions at all levels will be affected, including the structure and functions of soil microbial communities. An important area of study is community-level responses to land-use changes rather than effects particular species or groups such as arbuscular mycorrhizal fungi or nitrogen-fixing bacteria, which are well known to play key roles in agroecosystems.

It is well established that the conversion of native ecosystems to agricultural uses can strongly affect microbial community structure, composition and diversity. For example, conversions of tropical forest to plantations (Waldrop et al. 2000) have been found to engender distinct soil microbial community structures, and agricultural intensification has been reported to decrease microbial diversity (Steenwerth et al. 2006). Additionally, the type of land management practices used in agroecosystems also affects microbial community structure and function through a variety of different mechanisms. Numerous studies have documented changes in microbial community structure resulting from physical disturbance, especially tillage (Frey et al. 1999; Guggenberger et al. 1999). Tillage represents a severe disturbance to fungi by severing hyphal connections. However, no-till systems favour the development of fungi as compared with bacterial community components (Minoshima et al. 2007; Kennedy and Schillinger 2006). Conversions to agriculture and attendant cultivation practices also alter microbial communities through changes to temperature, soil moisture (through irrigation and alteration of soil structure), and other physical parameters.

Land-use changes alter soil microbial community structure through changes in carbon availability and quality, pH (Cookson et al. 2007), nutrient availability, or other chemical parameters. For example, studies comparing agroecosystems and natural systems report that adding nitrogen decreases the relative abundance of fungi by comparison to bacteria (Bradley et al. 2006; Bardgett et al. 1999). Seghers et al. (2004) found that nitrogen fertilizers decreased populations of methanotrophs and root endophytes in the bulk soil microbial community. They also found differential effects of organic fertilizers versus inorganic, which was consistent with other studies. For example, Wander et al. (1995) (Cookson et al. 2007) reported that manure-amended plots showed less diverse populations of microorganisms than cover cropped soil, but that microbial biomass was more metabolically active (Wander et al. 1995). Ulrich et al. (2008) found that manure applications led to an increase in the population densities of cellulytic bacteria within the soil microbial community (see chapters elsewhere in this book). Fungal-to-bacterial ratios are commonly measured as indicators of microbial community structure, and the relative proportions of fungi are increased by no-till practices, crop rotations, and use of cover crops (Six et al. 2006).

In addition to physical disturbance effects, alterations in vegetation, plant diversity and species-specific plant traits can cause changes in aboveground litter quantity and quality, and belowground root dynamics. However, alterations in vegetation tend to cause idiosyncratic effects of particular plant species or particular functional traits, and it is difficult to draw more generalized patterns (Porazinska et al. 2003). The effects on plant litter quality and quantity, in particular, become limiting factors to microorganisms, and thus species-specific differences in plant litter can affect microbial community structure and function (Wardle et al. 2004).

Effects may persist for many years and decades after a given land-use has stopped (Steenwerth et al. 2006). For example, Fraterrigo et al. (2006) (Ulrich et al. 2008) found long-term microbial alteration in forest stands that had been cultivated but not logged. Fungal markers, especially, were lower in previously cultivated sites, suggesting that fungi may need extensive periods of time to recover from agriculture (Fraterrigo et al. 2006). Spiegelberger et al. (2006) found changes in the microbial community due to lasting pH changes resulting from liming 70 years after agricultural abandonment (Spiegelberger et al. 2006). Evidence is beginning to accumulate that the history of land-use can leave significant legacies in the soil microbial community, ultimately influencing successional dynamics of future plant communities (Kardol et al. 2007) and hence, providing a mechanism by which changes due to agricultural management practices may persist.

Conclusion

This chapter has discussed studies of soil microbial communities in the tropics, highlighted the importance of wetland communities, and reviewed potential agricultural impacts on soil microbial communities. We have explored general patterns in microbial communities, and within communities in these systems in particular. Overall, the complexity of soil microbial communities pose a challenge that has limited the number and power of studies. The recent emergence of new methods is beginning to supply much-needed information, but connecting the function and identity of microbes remains a key research area. Major human impacts have altered both wetland and tropical ecosystems, which are rapidly diminishing. Studies have focused on pasture conversion effects in the tropics and reveal a diversity of microbial life underground, and the functional and structural changes brought with such conversion. Wetland systems experience strong fluctuations in environmental and biotic conditions. Recent research points to the sensitivity of the microbial community to both hydrology and plant structure. Finally, agricultural regimes strongly impact the soil microbial community and its functions, in particular decreasing fungal biomass and overall microbial diversity. These changes in belowground community can be persist on the order of decades, even when the aboveground community shifts.

Acknowledgments The authors would like to acknowledge the members of the Balser laboratory at the University of Wisconsin-Madison, especially Dr. Harry Read and Kevin Budsberg, for their untiring support and assistance. We would also like to thank Balser laboratory alumna Drs. Jessica Gutknecht and Jenny Kao-Kniffin for valuable discussions about these ideas over the years. Finally, we thank the editors for their valuable comments in strengthening this chapter.

References

Allison V, Yermakov Z, Miller R, Jastrow J, Matamala R (2007) Using landscape and depth gradients to decouple the impact of correlated environmental variables on soil microbial community composition. Soil Biol Biochem 39(2):505–516

Amann R, Ludwig W, Schleifer K (1995) Phylogenetic identification and in situ detection of individual microbial cells without cultivation. Microbiol Rev 59(1):143–169

Amir H, Pineau R (1998) Influence of plants and cropping on microbiological characteristics of some New Caledonian ultramafic soils. Aust J Soil Res 36(3):457

Amundson R (2001) The carbon budget in soils. Annu Rev Earth Planet Sci 29(1):535–562

Anupam B (2003) Mycorrhizae in wetlands: a review. Int J Forest Manage 4:34–40

Baisden W, Parfitt R (2007) Bomb ^{14}C enrichment indicates decadal C pool in deep soil? Biogeochemistry 85(1):59–68

Balser TC (2000) Linking soil microbial communities and ecosystem functioning. PhD dissertation, University of California-Berkeley, Berkeley, CA

Balser TC, Firestone MK (2005) Linking microbial community composition and soil processes in a California annual grassland and mixed-conifer forest. Biogeochemistry 73(2):395–415

Balser T, McMahon K, Bart D, Bronson D, Coyle D, Craig N et al (2006) Bridging the gap between micro – and macro-scale perspectives on the role of microbial communities in global change ecology. Plant Soil 289(1):59–70

Bardgett RD, Mawdsley JL, Edwards S, Hobbs PJ, Rodwell JS, Davies WJ (1999) Plant species and nitrogen effects on soil biological properties of temperate upland grasslands. Funct Ecol 13(5):650–660

Basu S, Behera N (1993) The effect of tropical forest conversion on soil microbial biomass. Biol Fertil Soils 16(4):302–304

Bawa KS, Kress WJ, Nadkarni NM, Lele S (2004) Beyond paradise – meeting the challenges in tropical biology in the 21st century. Biotropica 36(4):437–446

Behera N, Sahani U (2003) Soil microbial biomass and activity in response to Eucalyptus plantation and natural regeneration on tropical soil. For Ecol Manage 174(1–3):1–11

Blume E, Bischoff M, Reichert JM, Moorman T, Konopka A, Turco RF (2002) Surface and subsurface microbial biomass, community structure and metabolic activity as a function of soil depth and season. Appl Soil Ecol 20(3):171–181

Borneman J, Triplett E (1997) Molecular microbial diversity in soils from eastern Amazonia: evidence for unusual microorganisms and microbial population shifts associated with deforestation. Appl Environ Microbiol 63(7):2647–2653

Bossio D, Girvan M, Verchot L, Bullimore J, Borelli T, Albrecht A et al (2005) Soil microbial community response to land use change in an agricultural landscape of western Kenya. Microb Ecol 49(1):50–62

Bradley K, Drijber R, Knops J (2006) Increased N availability in grassland soils modifies their microbial communities and decreases the abundance of arbuscular mycorrhizal fungi. Soil Biol Biochem 38(7):1583–1595

Brady NC, Weil RR (2002) The nature and properties of soils, 13th edn. Prentice Hall, Upper Saddle River, NJ

Brady KU, Kruckeberg AR, Bradshaw H Jr (2005) Evolutionary ecology of plant adaptation to serpentine soils. Annu Rev Ecol Evol Syst 36(1):243–266

Bridge P, Spooner B (2001) Soil fungi: diversity and detection. Plant Soil 232(1):147–154

Burke RA, Molina M, Cox JE, Osher LJ, Piccolo MC (2003) Stable carbon isotope ratio and composition of microbial fatty acids in tropical soils. J Environ Qual 32(1):198–206

Buschbacher R, Uhl C, Serrao EAS (1988) Abandoned pastures in Eastern Amazonia. II. Nutrient stocks in the soil and vegetation. J Ecol 76(3):682–699

Cardon ZG, Gage DJ (2006) Resource exchange in the rhizosphere: molecular tools and the microbial perspective. Annu Rev Ecol Evol Syst 37(1):459–488

Carney K, Matson P (2005) Plant communities, soil microorganisms, and soil carbon cycling: does altering the world belowground matter to ecosystem functioning? Ecosystems 8(8):928–940
Carney K, Matson P (2006) The influence of tropical plant diversity and composition on soil microbial communities. Microb Ecol 52(2):226–238
Cleveland CC, Townsend AR, Schmidt SK, Constance BC (2003) Soil microbial dynamics and biogeochemistry in tropical forests and pastures, southwestern Costa Rica. Ecol Appl 13(2):314–326
Cookson W, Osman M, Marschner P, Abaye D, Clark I, Murphy D et al (2007) Controls on soil nitrogen cycling and microbial community composition across land use and incubation temperature. Soil Biol Biochem 39(3):744–756
Crews TE, Kitayama K, Fownes JH, Riley RH, Herbert DA, Mueller-Dombois D et al (1995) Changes in soil-phosphorus fractions and ecosystem dynamics across a long chronosequence in Hawaii. Ecology 76(5):1407
Davidson EA, Artaxo P (2004) Globally significant changes in biological processes of the Amazon basin: results of the large-scale biosphere-atmosphere experiment. Glob Change Biol 10(5):519–529
de Moraes JFL, Volkoff B, Cerri CC, Bernoux M (1996) Soil properties under Amazon forest and changes due to pasture installation in Rondônia, Brazil. Geoderma 70(1):63–81
DeGrood SH, Claassen VP, Scow KM (2005) Microbial community composition on native and drastically disturbed serpentine soils. Soil Biol Biochem 37(8):1427–1435
Desjardins T, Barros E, Sarrazin M, Girardin C, Mariotti A (2004) Effects of forest conversion to pasture on soil carbon content and dynamics in Brazilian Amazonia. Agric Ecosyst Environ 103(2):365–373
Doran JW, Zeiss MR (2000) Soil health and sustainability: managing the biotic component of soil quality. Appl Soil Ecol 15(1):3–11
Eden MJ, Furley PA, Mcgregor DFM, Milliken W, Ratter JA (1991) Effect of forest clearance and burning on soil properties in northern Roraima, Brazil. For Ecol Manage 38:283–290
Ekelund F, Rønn R, Christensen S (2001) Distribution with depth of protozoa, bacteria and fungi in soil profiles from three Danish forest sites. Soil Biol Biochem 33(4–5):475–481
EPA (2004) Wetlands overview, EPA 843-F-04-011a [Internet]. http://www.epa.gov/owow/wetlands/pdf/overview.pdf
FAO (2001) Global forest resources assessment 2000 – main report. Forestry Paper 140. Rep. 140
FAO (2005) Global forest resources assessment 2005. Forestry paper 147. Rep. 147
Fearnside PM (2005) Deforestation in Brazilian Amazonia: history, rates, and consequences. Conserv Biol 19(3):680–688
Fearnside PM, Imbrozio Barbosa R (1998) Soil carbon changes from conversion of forest to pasture in Brazilian Amazonia. For Ecol Manage 108(1–2):147–166
Fierer N, Bradford MA, Jackson RB (2007) Toward an ecological classification of soil bacteria. Ecology 88(6):1354–1364
Fraterrigo JM, Balser TC, Turner MG (2006) Microbial community variation and its relationship with nitrogen mineralization in historically altered forests. Ecology 87(3):570–579
Frey S, Elliott E, Paustian K (1999) Bacterial and fungal abundance and biomass in conventional and no-tillage agroecosystems along two climatic gradients. Soil Biol Biochem 31(4):573–585
Goberna M, Insam H, Pascual J, Sánchez J (2005) Storage effects on the community level physiological profiles of Mediterranean forest soils. Soil Biol Biochem 37(1):173–178
Goberna M, Sánchez J, Pascual J, García C (2006) Surface and subsurface organic carbon, microbial biomass and activity in a forest soil sequence. Soil Biol Biochem 38(8):2233–2243
Gomez-Alvarez V, King GM, Nüsslein K (2007) Comparative bacterial diversity in recent Hawaiian volcanic deposits of different ages. FEMS Microbiol Ecol 60(1):60–73
Gordon D (1998) Effects of invasive, non-indigenous plant species on ecosystem processes: lessons from Florida. Ecol Appl 8(4):975–989
Guggenberger G, Frey SD, Six J, Paustian K, Elliott ET (1999) Bacterial and fungal cell-wall residues in conventional and no-tillage agroecosystems. Soil Sci Soc Am J 63(5):1188–1198

Gutknecht J, Goodman R, Balser T (2006) Linking soil process and microbial ecology in freshwater wetland ecosystems. Plant Soil 289(1–2):17–34

Haggar JP, Ewel JJ (1995) Establishment, resource acquisition, and early productivity as determined by biomass allocation patterns of three tropical tree species. Forest Sci 41:689–708

Handelsman J (2005) Metagenomics or megagenomics? Nat Rev Microbiol 3(6):457–458

Hölscher D, Ludwig B, Möller RF, Fölster H (1997) Dynamic of soil chemical parameters in shifting agriculture in the Eastern Amazon. Agric Ecosyst Environ 66(2):153–163

Horner-Devine MC, Carney KM, Bohannan BJM (2004) An ecological perspective on bacterial biodiversity. Proc R Soc Lond B Biol Sci 271(1535):113–122

Jobbágy EG, Jackson RB (2000) The vertical distribution of soil organic carbon and its relation to climate and vegetation. Ecol Appl 10(2):423–436

Kao-Kniffin J, Balser T (2007) Elevated CO_2 differentially alters belowground plant and soil microbial community structure in reed canary grass-invaded experimental wetlands. Soil Biol Biochem 39(2):517–525

Kardol P, Cornips N, van Kempen M, Bakx-Schotman J, van der Putten W (2007) Microbe-mediated plant-soil feedback causes historical contingency effects in plant community assembly. Ecol Monogr 77(2):147–162

Kennedy A, Schillinger W (2006) Soil quality and water intake in traditional-till vs. no-till paired farms in Washington's Palouse region. Soil Sci Soc Am J 70(3):940–949

Kim J, Sparovek G, Longo RM, De Melo WJ, Crowley D (2007) Bacterial diversity of terra preta and pristine forest soil from the Western Amazon. Soil Biol Biochem 39(2):684–690

Kirk JL, Beaudette LA, Hart M, Moutoglis P, Klironomos JN, Lee H et al (2004) Methods of studying soil microbial diversity. J Microbiol Meth 58(2):169–188

Krave AS, Lin B, Braster M, Laverman AM, Straalen NMV, Röling WFM et al (2002) Stratification and seasonal stability of diverse bacterial communities in a *Pinus merkusii* (pine) forest soil in central Java, Indonesia. Environ Microbiol 4(6):361–373

Kumara HN, Kumar MA, Sharma AK, Sushma HS, Singh M, Singh M (2004) Diversity and management of wild mammals in tea gardens in the rainforest regions of the Western Ghats, India: a case study from a tea estate in the Anaimalai Hills. Curr Sci 87:1282–1287

Laurance WF (1999) Reflections on the tropical deforestation crisis. Biol Conserv 91(2–3): 109–117

Lavahun MFE, Joergensen RG, Meyer B (1996) Activity and biomass of soil microorganisms at different depths. Biol Fertil Soils 23(1):38–42

Leckie SE (2005) Methods of microbial community profiling and their application to forest soils. For Ecol Manage 220(1–3):88–106

Lehmann J, Pereira da Silva J, Steiner C, Nehls T, Zech W, Glaser B (2003) Nutrient availability and leaching in an archaeological Anthrosol and a Ferralsol of the Central Amazon basin: fertilizer, manure and charcoal amendments. Plant Soil 249(2):343–357

Levine SH (1976) Competitive interactions in ecosystems. Am Nat 110(976):903

Liang C, Fujinuma R, Balser T (2008) Comparing PLFA and amino sugars for microbial analysis in an Upper Michigan old growth forest. Soil Biol Biochem 40(8):2063–2065

Lima HN, Schaefer CER, Mello JWV, Gilkes RJ, Ker JC (2002) Pedogenesis and pre-Colombian land use of "Terra Preta Anthrosols" ("Indian black earth") of Western Amazonia. Geoderma 110(1–2):1–17

Lipson DA (2007) Relationships between temperature responses and bacterial community structure along seasonal and altitudinal gradients. FEMS Microbiol Ecol 59(2):418–427

Mann CC (2002) Agriculture: the real dirt on rainforest fertility. Science 297(5583):920–923

Mentzer J, Goodman R, Balser T (2006 June) Microbial response over time to hydrologic and fertilization treatments in a simulated wet prairie. Plant Soil 284(1–2):85–100

Millennium Ecosystem Assessment (2005) Ecosystems and human well-being: synthesis [Internet]. http://www.millenniumassessment.org/documents/document.356.aspx.pdf

Miller SP, Bever JD (1999) Distribution of arbuscular mycorrhizal fungi in stands of the wetland grass Panicum hemitomon along a wide hydrologic gradient. Oecologia 119(4): 586–592

Minoshima H, Jackson L, Cavagnaro T, Sanchez-Moreno S, Ferris H, Temple S et al (2007) Soil food webs and carbon dynamics in response to conservation tillage in California. Soil Sci Soc Am J 71(3):952–963
Moritz LK (2008) Soil habitat and microbial communities on Mount Kinabalu, Borneo. MS thesis, University of Wisconsin-Madison, Madison, WI
Müller MML, Guimarães MF, Desjardins T, Mitja D (2004) The relationship between pasture degradation and soil properties in the Brazilian amazon: a case study. Agric Ecosyst Environ 103(2):279–288
Myers N, Mittermeier RA (2000) Biodiversity hotspots for conservation priorities. (Cover story). Nature 403(6772):853
Nannipieri P, Ascher J, Ceccherini MT, Lan Dl L, Pietramellar G, Renell G (2003) Microbial diversity and soil functions. Eur J Soil Sci 54(4):655
Neill C, Fry B, Melillo JM, Steudler PA, Moraes JFL, Cerri CC (1996) Forest- and pasture-derived carbon contributions to carbon stocks and microbial respiration of tropical pasture soils. Oecologia 107(1):113–119
Nepstad DC, de Carvalho CR, Davidson EA, Jipp PH, Lefebvre PA, Negreiros GH et al (1994) The role of deep roots in the hydrological and carbon cycles of Amazonian forests and pastures. Nature 372(6507):666–669
Nüsslein K, Tiedje JM (1998) Characterization of the dominant and rare members of a young Hawaiian soil bacterial community with small-subunit ribosomal DNA amplified from DNA fractionated on the basis of its guanine and cytosine composition. Appl Environ Microbiol 64(4):1283–1289
Nüsslein K, Tiedje JM (1999) Soil bacterial community shift correlated with change from forest to pasture vegetation in a tropical soil. Appl Environ Microbiol 65(8):3622–3626
Oline D (2006) Phylogenetic comparisons of bacterial communities from serpentine and nonserpentine soils. Appl Environ Microbiol 72(11):6965–6971
Poll C, Thiede A, Wermbter N, Sessitsch A, Kandeler E (2003) Micro-scale distribution of microorganisms and microbial enzyme activities in a soil with long-term organic amendment. Eur J Soil Sci 54(4):715–724
Porazinska DL, Bardgett RD, Blaauw MB, Hunt HW, Parsons AN, Seastedt TR et al (2003) Relationships at the aboveground-belowground interface: plants, soil biota and soil processes. Ecol Monogr 73(3):377–395
Proctor J (1999) Toxins, nutrient shortages and droughts: the serpentine challenge. Trends Ecol Evol 14(9):334–335
Ramsar Convention Secretariat (2007) Wise use of wetlands: a conceptual framework for the wise use of wetlands. Ramsar handbooks for the wise use of wetlands, 3rd edn, vol 1 [Internet]. http://www.ramsar.org/pdf/lib/lib_handbooks2006_e01.pdf
Ranjard L, Poly F, Nazaret S (2000) Monitoring complex bacterial communities using culture-independent molecular techniques: application to soil environment. Res Microbiol 151(3): 167–177
Rasmussen C, Southard RJ, Horwath WR (2007) Soil mineralogy affects conifer forest soil carbon source utilization and microbial priming. Soil Sci Soc Am J 71(4):1141–1150
Reiners WA, Bouwman AF, Parsons WFJ, Keller M (1994) Tropical rain-forest conversion to pasture – changes in vegetation and soil properties. Ecol Appl 4(2):363
Rudel TK (2005) Tropical forests regional paths of destruction and regeneration in the late twentieth century. Columbia University Press, New York
Schloss PD, Handelsman J (2006) Toward a census of bacteria in soil. PLoS Comput Biol 2(7):e92
Seghers D, Wittebolle L, Top E, Verstraete W, Siciliano S (2004) Impact of agricultural practices on the *Zea mays* L. endophytic community. Appl Environ Microbiol 70(3):1475–1482
Serrão EAS, Nepstad D, Walker R (1996) Upland agricultural and forestry development in the Amazon: sustainability, criticality and resilience. Ecol Econ 18(1):3–13
Sessitsch A, Weilharter A, Gerzabek MH, Kirchmann H, Kandeler E (2001) Microbial population structures in soil particle size fractions of a long-term fertilizer field experiment. Appl Environ Microbiol 67(9):4215–4224

Shaw SP, Fredine CG (1956) Wetlands of the United States – their extent and their value to waterfowl and other wildlife. Version 05JAN99 [Internet]. http://www.npwrc.usgs.gov/resource/wetlands/uswetlan/index.htm

Singh S, Singh JS (1995) Microbial biomass associated with water-stable aggregates in forest, savanna and cropland soils of a seasonally dry tropical region, India. Soil Biol Biochem 27(8):1027–1033

Six J, Frey SD, Thiet RK, Batten KM (2006) Bacterial and fungal contributions to carbon sequestration in agroecosystems. Soil Sci Soc Am J 70(2):555–569

Spiegelberger T, Hegg O, Matthies D, Hedlund K, Schaffner U (2006) Long-term effects of short-term perturbation in a subalpine grassland. Ecology 87(8):1939–1944

Srivastava S, Singh J (1991) Microbial C, N and P in dry tropical forest soils: effects of alternate land-uses and nutrient flux. Soil Biol Biochem 23(2):117–124

Steenwerth K, Jackson L, Carlisle E, Scow K (2006) Microbial communities of a native perennial bunchgrass do not respond consistently across a gradient of land-use intensification. Soil Biol Biochem 38(7):1797–1811

Stenström J, Svensson K, Johansson M (2001) Reversible transition between active and dormant microbial states in soil. FEMS Microbiol Ecol 36(2–3):93–104

Sylvia DM, Fuhrmann JJ, Hartel PG, Zuberer DA (2005) Principles and applications of soil microbiology, 2nd edn. Prentice Hall, Upper Saddle River, NJ

Taylor JP, Wilson B, Mills MS, Burns RG (2002) Comparison of microbial numbers and enzymatic activities in surface soils and subsoils using various techniques. Soil Biol Biochem 34(3):387–401

Torsvik V, Goksoyr J, Daae FL (1990) High diversity in DNA of soil bacteria. Appl Environ Microbiol 56(3):782–787

Trumbore SE, Davidson EA, Camargo PBD, Nepstad DC, Martinelli LA (1995) Belowground cycling of carbon in forests and pastures of eastern Amazonia. Glob Biogeochem Cycles 9(4):515–528

Ulrich A, Becker R (2006) Soil parent material is a key determinant of the bacterial community structure in arable soils. FEMS Microbiol Ecol 56(3):430–443

Ulrich A, Klimke G, Wirth S (2008) Diversity and activity of cellulose-decomposing bacteria, isolated from a sandy and a loamy soil after long-term manure application. Microb Ecol 55(3):512–522

Van Gestel M, Merckx R, Vlassak K (1996) Spatial distribution of microbial biomass in microaggregates of a silty-loam soil and the relation with the resistance of microorganisms to soil drying. Soil Biol Biochem 28(4–5):503–510

Veldkamp E, Becker A, Schwendenmann L, Clark DA, Schulte-Bisping H (2003) Substantial labile carbon stocks and microbial activity in deeply weathered soils below a tropical wet forest. Glob Change Biol 9(8):1171–1184

Venkatesan S, Senthurpandian VK (2006) Comparison of enzyme activity with depth under tea plantations and forested sites in south India. Geoderma 137(1–2):212–216

Vestal JR, White DC (1989) Lipid analysis in microbial ecology. Bioscience 39(8):535–541

Waldrop MP, Balser TC, Firestone MK (2000) Linking microbial community composition to function in a tropical soil. Soil Biol Biochem 32(13):1837–1846

Walker RB (1954) The ecology of serpentine soils: a symposium. II. factors affecting plant growth on serpentine soils. Ecology 35:259–266

Wander M, Hedrick D, Kaufman D, Traina S, Stinner B, Kehrmeyer S et al (1995) The functional significance of the microbial biomass in organic and conventionally managed soils. Plant Soil 170(1):87–97

Wardle DA, Bardgett RD, Klironomos JN, Setala H, van der Putten WH, Wall DH (2004) Ecological linkages between aboveground and belowground biota. Science 304(5677):1629–1633

White DC, Davis WM, Nickels JS, King JD, Bobbie RJ (1979) Determination of the sedimentary microbial biomass by extractible lipid phosphate. Oecologia 40(1):51–62

Windham L, Ehrenfeld J (2003) Net impact of a plant invasion on nitrogen-cycling processes within a brackish tidal marsh. Ecol Appl 13(4):883–896

Wixon D, Balser T (2009) Complexity, climate change and soil carbon: a systems approach to microbial temperature response. Syst Res Behav Sci 26(5):601–620

Yao H, He Z, Wilson MJ, Campbell CD (2000) Microbial biomass and community structure in a sequence of soils with increasing fertility and changing land use. Microb Ecol 40(3): 223–237

Zedler J, Kercher S (2004) Causes and consequences of invasive plants in wetlands: opportunities, opportunists, and outcomes. Crit Rev Plant Sci 23(5):431–452

Zelles L (1999) Fatty acid patterns of phospholipids and lipopolysaccharides in the characterisation of microbial communities in soil: a review. Biol Fertil Soils 29(2):111–129[1]Duff: the layer of partially and fully decomposed organic materials lying below the forest flar litter and immediately above the mineral soil

Chapter 3
Soil Microbiology and Nutrient Cycling

David W. Hopkins and Jennifer A.J. Dungait

Introduction

Soil organisms play a central role in the recycling of nutrients in soils, making them available to plants, transforming some nutrient elements to gaseous forms which can be lost from soil, and other transformations which predispose nutrients to loss. In this chapter we will illustrate these processes in the context of sustainable crop production. To do so requires some consideration of what is meant by 'sustainable crop production'. To be truly sustainable nutrient supply should correspond to demand, with the rate of nutrient removal from the fields by crops matched by replacement into the plant-soil system. There is no escape from the thermodynamic principles that underlie the law of conservation of mass matter. This implies that for any process occurring in a closed system, the mass of the reactants equals the mass of the products. When viewed from this absolute standpoint, no cropping system would be truly sustainable unless all the residues of the consumers were returned to the fields. The use of composts, manures and slurries on fields used for food production is widely practiced, and in some cases even the collection of human excreta ("night soil"; Fig. 1) and its application to fields is not unknown. With the increasing urbanization of the human population and the necessary development of waste water and sewerage systems, the return loop for, or connection between, animal (including human) and food wastes and land used for food production, is broken in many parts of the world. This effectively leads to the net transfer of vast quantities of nutrients

D.W. Hopkins (✉)
Scottish Crop Research Institute, Invergowrie, Dundee, DD2 5DA, UK
e-mail: david.hopkins@scri.ac.uk

J.A.J. Dungait
Rothamsted Research – North Wyke, Okehampton, Devon, EX20 2SB, UK
e-mail: jennifer.dungait@bbsrc.ac.uk

G.R. Dixon and E.L. Tilston (eds.), *Soil Microbiology and Sustainable Crop Production*, DOI 10.1007/978-90-481-9479-7_3, © Springer Science+Business Media B.V. 2010

Fig. 1 Human excreta ("night soil") that has been collected and applied to horticultural plots in south eastern China

from sites of food production, which are predominantly rural, to cities, usually to be dissipated to the atmosphere or water courses and the ocean, or locked-up (at least for the short-term) in waste disposal sites, such as land-fill dumps.

Soil microorganisms almost certainly have a role in the transformation of all the bioelements (Table 1), but the quantities required by crop plants are usually relatively small and can be supplied from the trace quantities present in soil atmospheric deposits and recycled alongside those elements required in larger quantities. The focus of this chapter is on the major bioelements, carbon and nitrogen, with brief mention also of phosphorus. We have concentrated on the roles of soil microorganisms in the supply of these elements. In order to improve the sustainability of nutrient cycling in an agricultural system, organic residues and resources need to be both exploited efficiently and conserved as far as possible in the system. For this reason, it is essential that the cycling of plant nutrients, such as nitrogen, is considered alongside the turnover of carbon and the processing of energy in the soil.

Table 1 The major bioelements, their main functions in organisms and key contributions made by soil microorganisms to the availability. The elements are present in approximate decreasing order of typical requirement by crop plants

Element	Principal sources in cropping systems	Biological roles	Contribution made to nutrient supply made by soil microorganisms
Carbon	Atmospheric CO_2	Main cellular constituent	Decomposition to release C from organic molecules to CO_2
Oxygen	Atmospheric O_2 H_2O organic molecules CO_2	Main cellular constituent Respiratory electron acceptor in aerobic respiring organisms	
Hydrogen	H_2O organic molecules NH_4^+	Main cellular constituent	
Nitrogen	NH_4^+ NO_3^- N_2 organic molecules Atmospheric deposition (pollution)	Main cellular constituent Respiratory electron acceptor by some bacteria under anoxic conditions	Conversion to plant available (often) inorganic forms (mineralization) Conversion from NH_4^+ to NO_3^- (nitrification) which can predispose N to loss from soils Conversion from NO_3^- to gaseous forms N_2O and N_2 (denitrification) which is a direct loss mechanism Conversion from atmospheric N_2 to NH_3 (N fixation) which is a gain mechanism
Sulphur	SO_4^{2-} organic molecules Atmospheric deposition (pollution)	Component of several biomolecules Structural role in proteins and co-enzyme Respiratory electron acceptor by some bacteria under anoxic conditions	Conversion to inorganic forms (mineralization) Dissimilatory reduction to H_2S under anoxic conditions which can lead to gaseous loss
Phosphorus	HPO_3^{2-} Organic molecules	Component of several biomolecules Constituent of nucleic acids, ATP/ADP, and phospholipids	Conversion from organic to inorganic forms (mineralization) Solubilization of inorganic P minerals
Potassium	K^+	Component of several biomolecules Co-factor in many enzymes	

(continued)

Table 1 (continued)

Element	Principal sources in cropping systems	Biological roles	Contribution made to nutrient supply made by soil microorganisms
Calcium	Ca^{2+}	Component of several biomolecules Co-factor in many enzymes Important regulatory roles	
Iron	Fe^{2+} and Fe^{3+}	Component of several biomolecules Co-factor in many enzymes	
Zinc	Zn^{2+}	Component of several biomolecules Co-factor in many enzymes	

The Global Biogeochemical Cycle of Carbon

Globally, the biogeochemical carbon cycle is quantitatively dominated by massive pools of carbon in geological deposits as sedimentary carbonates and organic deposits, which contribute in part to fossil fuel stocks, and marine sediments. However, these are relatively static compartments compared to the smaller components of carbon represented by soil organic matter, atmosphere and biomass. Globally, the atmosphere contains about 780 Pg C predominantly as carbon dioxide, although trace quantities of methane and other hydrocarbons and chlorofluorocarbons also have important biogeochemical roles; the global biomass represents a similar amount, approximately 800 Pg C. The amount of carbon in soil organic matter (SOM) is approximately the same as the atmosphere and biomass added together, at around 1,500 Pg (Fig. 2). This pool of carbon is in a permanent state of flux, undergoing decomposition and subsequent release, returning once again to the atmosphere. Atmospheric carbon enters the biomass (plants and some soil bacteria) principally by photosynthesis (110 Pg C year^{-1}), with small contributions from chemoautotrophic carbon fixation. This primary input is approximately balanced by carbon dioxide released to the atmosphere as a result of the collective respiration by consumers and decomposers.

Organic Matter in Soils and Its Turnover

Decomposition is the progressive breakdown of organic materials, ultimately into inorganic constituents, and is mediated mainly by soil microorganisms, which derive energy and nutrients from the diverse range of molecules in SOM. Microorganisms are the major direct contributors to the flux of carbon from soil to the atmosphere (Hassink et al. 1994; Hopkins and Gregorich 2005). The contribution made by decomposers, i.e. those organisms which utilize dead remains and waste organic materials from organisms as their main or sole source of energy and carbon for biosynthesis, to the return of carbon dioxide from the biosphere to atmosphere is between 50 and 60 Pg C year^{-1}, i.e. approximately 50% of the total respiratory return of carbon dioxide. The process of decomposition is central to the efficient use and recycling of nutrients in managed, agricultural, semi-natural and natural ecosystems because it is the route whereby complex biomolecules are degraded and bioelements, particularly nitrogen, are released in plant-available forms.

The substrates for decomposition include a wide range of materials, forming a continuum from recently added plant litter and carbon transferred into the soil by physical processes such as rainfall, or biological processes via root exudation, the activity of fungi and bioturbation, to very stable, highly altered organic matter. Many components of fresh plant litter decompose very quickly, because they are rich in energy, readily accessible to organisms and, particularly in the case of

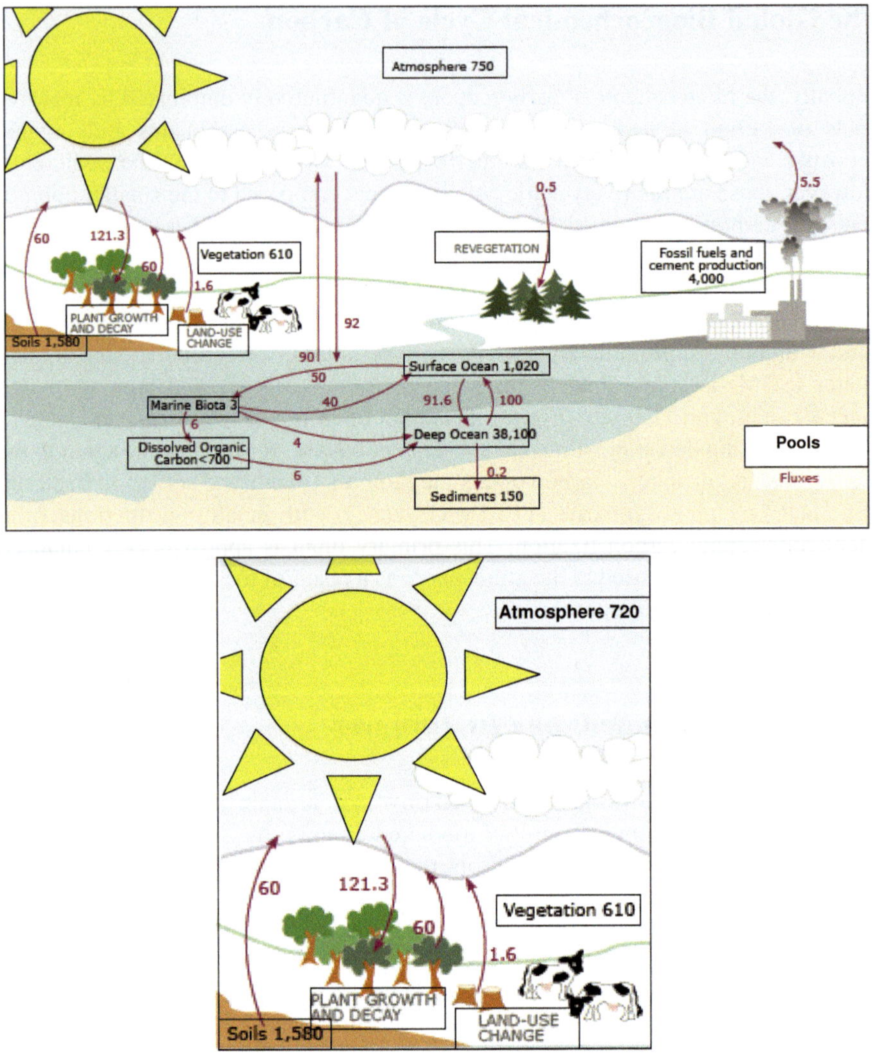

Fig. 2 Global biogeochemical cycle of carbon. The units for pools are Pg (= 10^{15} g = 10^{12} kg), and those for fluxes are Pg year^{-1})

simple sugars and peptides, rapidly assimilated. Consequently, though it represents only a small proportion of carbon in soil, about half of the carbon dioxide output from soil on a global basis comes from decomposition of this active pool of biologically available carbon from annual litter inputs to the soil (Coûteaux et al. 1995) – a fraction of organic matter that is not typically included in SOM models. The stable component of organic matter, at the other extreme, decomposes very slowly over centuries or millennia (Campbell et al. 1967); the size of this pool is comparatively large (Table 2).

Table 2 Pool sizes and turnover times of carbon in temperate arable soil at equilibrium simulated using the Rothamsted carbon model with an annual carbon input of 1 t C ha^{-1} year^{-1} (Jenkinson and Rayner 1977). See also Fig. 3

Pool	Equilibrium content (t C ha^{-1})	Turnover time (years)
Decomposable plant residues	0.01	0.2
Resistant plant residues	0.47	3.3
Microbial biomass	0.28	2.4
Physically protected organic matter	11.3	71
Chemically protected organic matter	12.2	2,900
Total	24.2	1,450

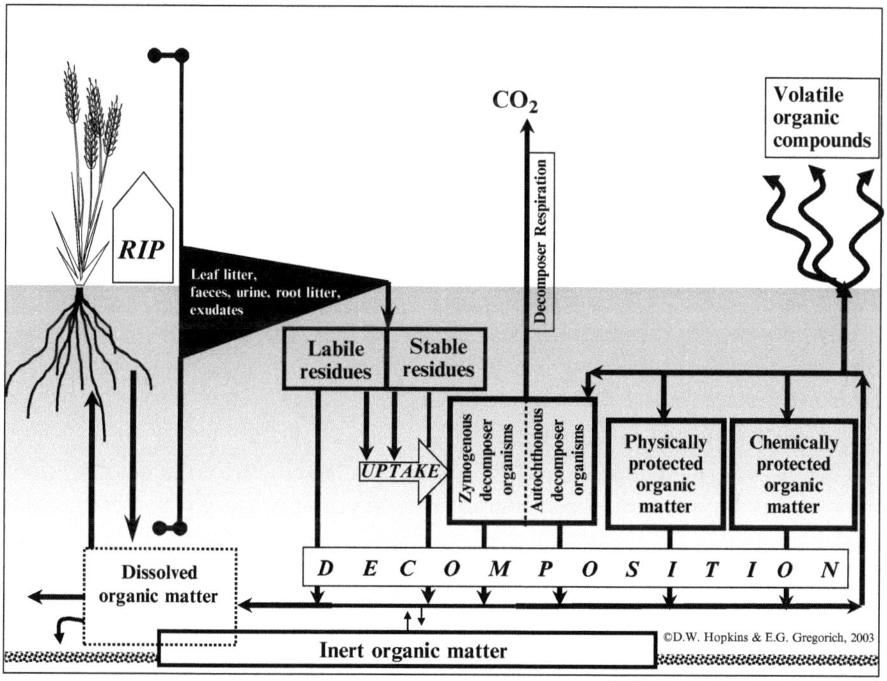

Fig. 3 Decomposition and carbon turnover in soil. A conceptual diagram summarising the main elements of the initial Rothamsted carbon model (Jenkinson and Rayner 1977). To this we have added other small, but potentially functionally important compartments: the volatile organic carbon and the dissolved organic carbon derived during both decomposition of litter and exudation from plants. An inert organic matter pool is added as this appears in latter versions of the Rothamsted model (Reproduced from Hopkins and Gregorich [2005]). See also Table 2

The closest we have come to a unified concept for summarising the interaction of carbon and organisms in soils is embodied in models of soil carbon turnover. The conceptual model summarised in Fig. 3 owes much to the Rothamsted carbon model (Jenkinson and Rayner 1977). Unlike some other models, which perform well in their individual ways, this model is based on compartments with functional relevance. In later versions an inert carbon compartment has been added to improve the performance of the model. This is not a unique feature of soil carbon models; for example, the "Century" model has a "passive organic matter" compartment (Parton et al. 1987) that also functions to slow down the model. Clearly, there cannot be a completely inert soil carbon fraction in soil (how could it have arisen?), and modelling simply points to the presence of a very stable organic matter fraction. Making the link between the presence of this "inert" soil carbon and its biological role remains elusive, but even soils that have been deprived of fresh organic matter inputs for prolonged periods (decades) are still biologically active, and activity must be being sustained by degraded components of the recalcitrant SOM remaining in these 'carbon exhausted' soils (Lawson et al. 2000).

In addition to the compartments represented in the Rothamsted carbon model, functionally significant compartments have been added such as dissolved and volatile organic carbon (Mackie and Wheatley 1999; Gregorich et al. 2003). Dissolved organic carbon is important for two main reasons. Firstly, in some high latitude and altitude ecosystems, organic nitrogenous compounds make an important contribution to the nitrogen economy of plants (Näsholm et al. 1998). Secondly, because it is a largely uncharacterised but biologically active component of the soil carbon which contributes significantly to its export from soils (Grieve 1984), and acts both as a driver of rhizosphere function and diversity, and of the initial exploitation of fresh plant litter (Hopkins and Gregorich 2005; Webster et al. 2000).

Factors Controlling Decomposition

The factors controlling decomposition of organic materials are determined by three sets of interacting factors – substrate quality, organisms and environment (Swift et al. 1979); thus, decomposition does not occur at uniform rate in either time or space. In the initial phase of decomposition, which typically lasts less than 1 year, the majority of readily metabolised components, such as unprotected sugars and oligosaccharides, amino acids and proteins, are rapidly exploited by decomposer organisms leaving more recalcitrant components. Recalcitrant components may resist breakdown because of their biochemistry, or by physical protection within the soil matrix. The application of modern spectroscopic and stable isotopic approaches to investigate soil organic matter turnover is revealing the incredible complexity of organic matter breakdown (Hopkins and Gregorich 2005). In particular, the long-held view that lignin components and lignin derivatives are inherently more stable than other components of plant materials is being questioned (Dungait et al. 2008a, 2009). Even simple components of organic substrate may become stabilized for millenia in the 'passive' organic matter pool (Paustian et al. 1992) by interaction with soil mineral colloids, encapsulation in soil aggregates where the local environmental conditions preclude rapid decomposition (e.g. because of limited oxygen diffusion), and interaction between organic compounds leading to increased chemical complexity that is resistant to microbial decomposition.

Substrate Quality

The ability of a soil to supply nutrients, store water, release greenhouse gases, modify pollutants, resist physical degradation and produce crops is strongly affected by the quality and quantity of organic matter it contains (Carter 2001). The physiochemical characteristics of organic material, for example its solubility in water or hydrophobicity, water content, nitrogen and other nutrient content, biochemical recalcitrance and toxicity, and physical protection strongly influence decomposition. Thus, the substrate quality of SOM can be regarded as a suite of combined properties that influence the supply of carbon and energy to heterotrophic soil organisms. Although this is a simple concept, the ability to assess substrate quality is not easy. Early studies recognised that different components of plant litter were decomposed at different rates, which was considered to be a reflection of their resource value to decomposer organisms (Tenney and Waksman 1929; Minderman 1968). More recently theoretical approaches that regard detritus as a continuum of biochemical components from the recalcitrant to the highly labile have been developed (Ågren and Bossatta 1996). Despite substantial progress in the development and application of increasingly sensistive analytical technques, we are still a long way from being able to characterise all potential biological substrates in SOM, and for key groups of organic compounds, most notably those containing nitrogen, biochemical characterisation remains remarkably sparse. This is because the analytical chemistry involved in properly characterising and quantifying organic matter at the molecular level is sometimes difficult and challenging. Indeed, there is a massive and increasing body of literature on the chemical characterisation of mysterious 'humic substances' that are obtained from soil by simple extraction with strong alkalis and acids. The ecological relevance of these substances has been questioned by soil science "dissidents" (Tate 2001), which include the authors of this chapter, because it is well documented that such preparations contain artefacts from sample preparation and are entirely non-selective with respect to biological entities in soils (Baldock et al. 1991). Nevertheless, deeply cherished terms such as "humic acid" have, like some of the molecules it purports to represent, proved remarkably resistant to decay.

The primary source of SOM is plant litter are various stages of degradation. Plant residues usually enter the soil as litter, including leaves, stems, roots and root exudates. The typical composition of plant materials is approximately 50% cellulose, 20% hemicellulose (polymers of hexoses, pentoses and uronic acids), 15% lignin, 5% proteins, 5% amino acids and sugars, 1% pectin and 1% waxes and pigments (Swift et al. 1979; Dungait et al. 2008a; Killham 1994). Excreta from livestock is another source of organic matter in soils (Fig. 4). Faeces contain not only the undigested diet (10–40%), but intestinal-dwelling bacteria (50–85%) and endogenous wastes from animal metabolism (10–15%) (Van Soest 1994). In agriculture, manures and slurries derived from ruminant grazing animals, such as cattle, are commonly added to soil as a source of nitrogen to promote plant growth. The organic matter in the manure and slurries is largely composed of biochemically transformed plant residues from dietary fodder and forage. Simple substrates, such

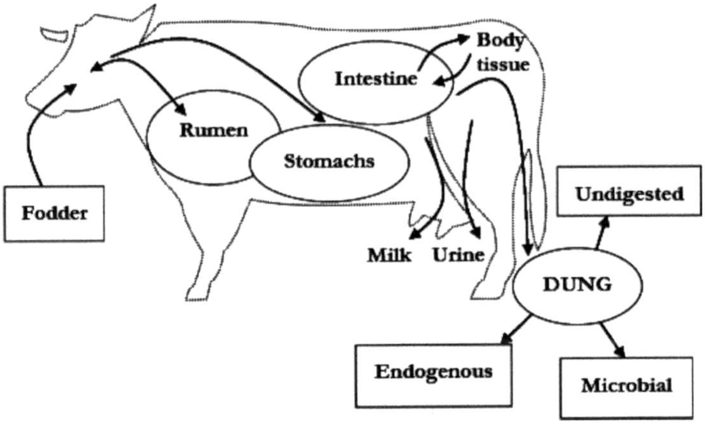

Fig. 4 Ruminant digestion and the origin of faecal matter, showing major components of dung (Figure reproduced from Dungait et al. [2010] and adapted from Van Soest [1994])

as proteins, amino acids and sugars are wholly digested, whilst more complex polymers such as cellulose and hemicellulose are partially degraded. Other components, such as chemically resistant lignin and waxes pass almost unchanged through the gut. In faecal structures, such as cow pats, decomposition that started in the animal gut may continue and is later overtaken by the activity of coprophilous flora and fauna that further modify the organic matter before it is incorporated into the soil. The modification of plant biomass by passage through the animal gut may contribute to slower turnover of carbon pools in soil following manure addition due to the proportional increase in polymers that are resistant to decay (Dungait and Bol 2005) recognised as an associated increase in SOM content in manured soils (Haynes and Naidu 1998).

Organisms

One to 5% of SOM is the soil microbial biomass, about 90% of which is fungi. These microorganisms are capable of decomposing the majority of organic material. Although probably not strictly correct, it is often assumed that all the microorganisms necessary to complete the decomposition of any natural compounds (and many anthropogenic compounds) are present in soil and that collectively they are infallible. This assumption holds for many decomposition processes because of the large functional redundancy in decomposer microbial communities, but it is not safe to assume that there is redundancy amongst all organisms performing specific functions in soils. For example, the high degree of specificity in the interaction between plant hosts and their mycorrhizal symbionts means ubiquity cannot always be assumed.

The Rothamsted model (Fig. 3) distinguishes between organisms which notionally respond to addition of fresh substrate (equivalent to the zymogenous biomass; *sensu* Winogradsky [1924]) and those which notionally eek out an existence on the older, more stable organic matter (equivalent to the autochthonous biomass; *sensu* Winogradsky [1924]). The soil community is probably not as sharply divided as Winogradsky's definitions would imply. Similarly, the distinction between *r*-selected and *K*-selected organisms, i.e. either those showing rapid proliferation following a pulse of substrate availability, or maintaining near constant population by efficient use of low concentrations of organic substrate, is probably not rigid. Therefore so-called "zymogenous" and "autochthonous" categories of microorganisms are approximately analogous to *r*-selected and *K*-selected organisms, respectively. Both concepts are useful for understanding soil carbon dynamics, but cannot be related directly to particular taxa, which may switch strategies (Chapman and Gray 1981), or applied unreservedly in a complex environment such as soil where many factors other than carbon supply may affect biological activity and biomass.

As mentioned above, microorganisms are the major drivers of decomposition in soils, but soil animals also play a major role in facilitating the decomposition process by physically fragmenting and mixing organic residues into the soil, thereby increasing the surface area of substrates and their exposure to microbial activity. Soil invertebrates, which include mesofauna such as Collembola, consume a wide range of organic materials which differ in their physiochemical accessibility, spatiotemporal distribution and nutritional value. Macroinvertebrates, such as earthworms, are important agents of organic matter decomposition in soils, accelerating rates of communition and dispersal through feeding activities (Dungait et al. 2008b), which has led them to be described as 'soil engineers' (Wolters 2000). They also influence decomposition by inoculating substrates with microbes, altering microbial activity (e.g. by deposition of mucus or wastes), altering the composition of the decomposer community, and the maintaining soil structure. The direct faunal contribution to soil to atmosphere carbon flux is, typically between 5% and 15% of the total flux (Hassink et al. 1994; Alphei et al. 1996).

The diversity of biochemical reactions involved in decomposition is vast. Briefly, the main processes involved are predigestion in animals (in some cases), communition by soil invertebrates, depolymerization of macromolecules, and uptake and use of the organic molecules by decomposer organisms. These processes do not occur in sequence because the bodily remains and waste materials from the decomposer organisms contribute to the substrate pool (Fig. 3).

Environmental Conditions

The combined physical and chemical conditions in which decomposition occurs both regulates the rate of decomposition and, in conjunction with the characteristics of the resources, imposes a selection pressure on the decomposer community thereby influencing its composition. Soil temperature, water content, acidity or alkalinity, oxygen supply and texture are amongst the most significant factors controlling the environment.

Like all other biological processes, the rate of decomposition of organic matter and the other relevant biogeochemical transformations increases with temperature in line broadly with the van t'Hoff rule that the rate of reaction doubles for a 10°C rise in the range 10–40°C (Jenkinson 1981; Hartley and Ineson 2008). In temperate regions, the temperature of the upper surface of the soil may occasionally rise to about 30°C, particularly if there is no plant cover, and in tropical regions soil temperatures consistently in the range 20–35°C are common. At greater depths in the soil temperature is typically in the range 5–20°C, with the size of the daily and seasonal fluctuations declining with increasing depth (Payne and Gregory 1988). Consequently biological processes in the sub-soil are typically slower. Recent concern about increasing atmospheric temperature globally has focussed attention on the effect on soil processes, particularly carbon cycling and the depletion of soil carbon reserves as a result of increased decomposition. There are a number of feedbacks which affect the net effect of increased temperature on the soil carbon reserves which may direct extrapolation difficult (Hartley et al. 2008).

Biological processes require a supply of liquid water, without which they ultimately cease. Even under the driest conditions in soils, there is rarely absolutely no water, but it is not uniformly distributed and forms a film over soil particles and colloids leading to a mixture of air- and water-filled pores and voids. Decomposition slowly proceeds even in naturally dry soil because of the activity of those organisms able to access this water. Griffin (1972) reported that fungi are more active in dry soils than bacteria because of the ability to translocate water across air-filled pores and voids. Because of the different range of water contents over which different components of the soil microbial community are active, there is a rather flat optimum water content for both decomposition and mineralization (Clement and Williams 1962). Fluctuations in microbial activity accompany rapid changes in moisture content, for example Birch (1957) showed that rewetting a dry soil enhances the short-term rate of organic matter decomposition and nitrogen mineralization, and that this effect increases with repeated drying and rewetting. These observations are consistent with the increased turnover of microbial biomass due to wet/dry cycles in the soil (Jenkinson and Ladd 1981).

Under hot or drying conditions, decomposition and other biological nutrient transformations may become restricted because of desiccation. This affects both plant residues and animal wastes including faeces, manure and slurry, and in the case of animal waste on the soil surface, "capping" may result. Capping occurs as a hydrophobic crust forms during surface drying and can limit subsequent gas flow (Dickinson et al. 1981). The presence of faeces, particularly cow pats can have significant effects on the underlying soil moisture and temperature (Dickinson et al. 1981), as well as acting as point sources of inorganic nitrogen that influence nitrate leaching (Ryden et al. 1984).

At the wet extreme, biological activity in soil can be affected by lack of or limited oxygen because of the low solubility and slow diffusion in water. This has important implication for the nitrification and denitrification discussed below. The decomposition of organic materials in soils is qualitatively different under long-term anoxic conditions. This is related to the fact the breakdown of lignin in plant materials is carried out by a restricted number of organisms which require oxygen (Kirk 1984).

There are many examples of decomposition being slower under acidic soil conditions compared with near neutral soils (Jenkinson 1977). This is in part because of the reduced overall microbial activity in acidic soils and the smaller abundances and activity of earthworms in acidic soils.

The Biogeochemical Cycle of Nitrogen

The main components of the nitrogen cycle are shown in Fig. 4. The individual transformations mediated by microorganisms that are relevant to nutrient cycling in agricultural systems are nitrogen mineralization which is intimately linked to decomposition, nitrogen fixation, nitrification and denitrification.

Nitrogen Mineralization

Nitrogen mineralization is the result of a set of processes allied to decomposition which lead to the conversion of nitrogen in organic compounds to inorganic nitrogen, usually ammonium (NH_4^+), carried out by a wide range of microorganisms (Harmsen and van Schreven 1955). This is significant because for many plants inorganic nitrogen is the main form in which nitrogen in taken up from the soil. This is particularly the case for crop plants.

Nitrogen mineralization operates in effective opposition to nitrogen immobilization because, depending on the nitrogen supply in the substrate being utilized by decomposer organisms, there may be a surplus of nitrogen in relation to carbon, or a deficit. Jansson and Persson (1982) describe these processes as the mineralization-immobilization turnover process in which inorganic nitrogen is released into the soil solution and assimilated from the soil solution. There is more recent evidence that soil microorganisms assimilate nitrogen from organic sources in the apparently more efficient (from the microorganisms' perspective), "direct" immobilization route with only the nitrogen surplus to the microorganisms' requirements being released in to the soil solution (Barraclough 1997):

MINERALIZATION-IMMOBILIZATION TURNOVER

Mineralization
→
Organic nitrogen in organic matter and microbial biomass | Inorganic nitrogen in the soil
←
Immobilization

DIRECT ASSIMILATION

Organic nitrogen in organic matter → Organic nitrogen in micro organisms → Inorganic N in soil solution

If the substrate offers a nitrogen surplus to the decomposer organisms, inorganic nitrogen will be released and, if the substrate offers a nitrogen deficit to the decomposer organisms, they will uptake nitrogen from the external (soil) environment, i.e. immobilization. Thus, the carbon-to-nitrogen ratio of the substrate strongly influences whether inorganic nitrogen is released during decomposition or whether the decomposer organisms make a demand on the soil nitrogen pool, whereby nitrogen is assimilated into microbial tissues. Typically a carbon-to-nitrogen ratio of about 20 is the threshold for nitrogen mineralization above which it is immobilized, because the substrate offers a nitrogen deficit to the decomposers, and below which nitrogen mineralized, because there is a surplus of nitrogen relative to carbon in the substrate (Harmsen and van Schreven 1955). However, the substrate quality and the composition of the decomposer community influence the relationship between carbon-to-nitrogen ratio of the decomposing substrate and mineralization and immobilization (Fig. 5)

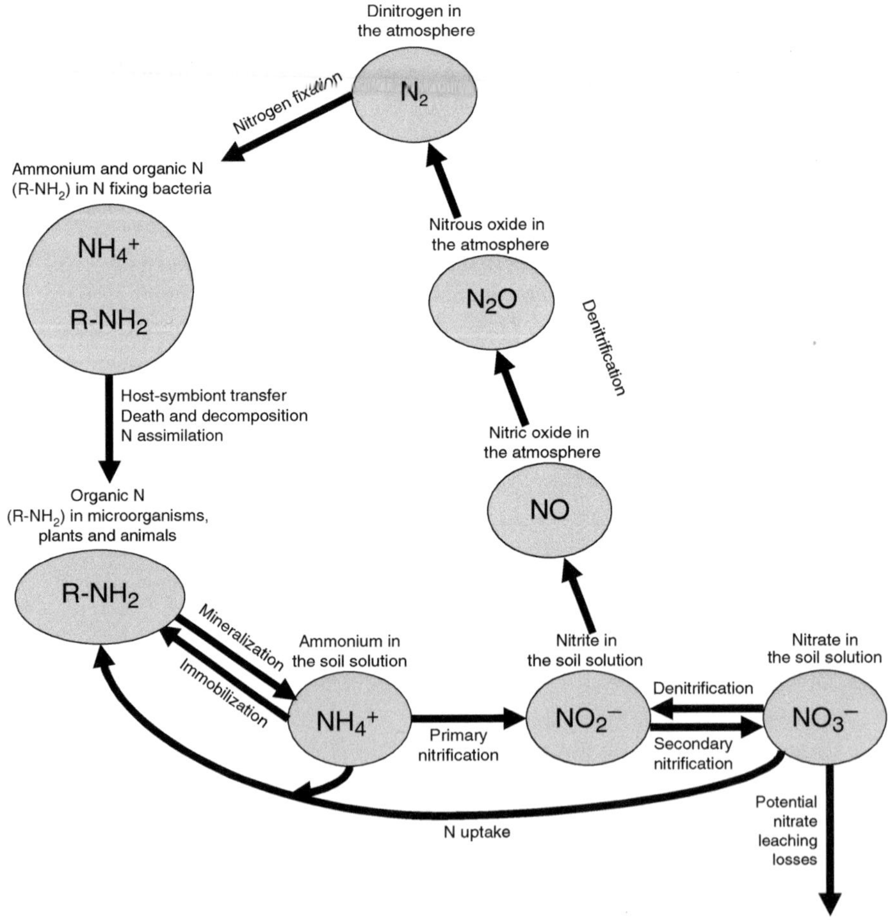

Fig. 5 The biochemical transformations and major pools of soil nitrogen

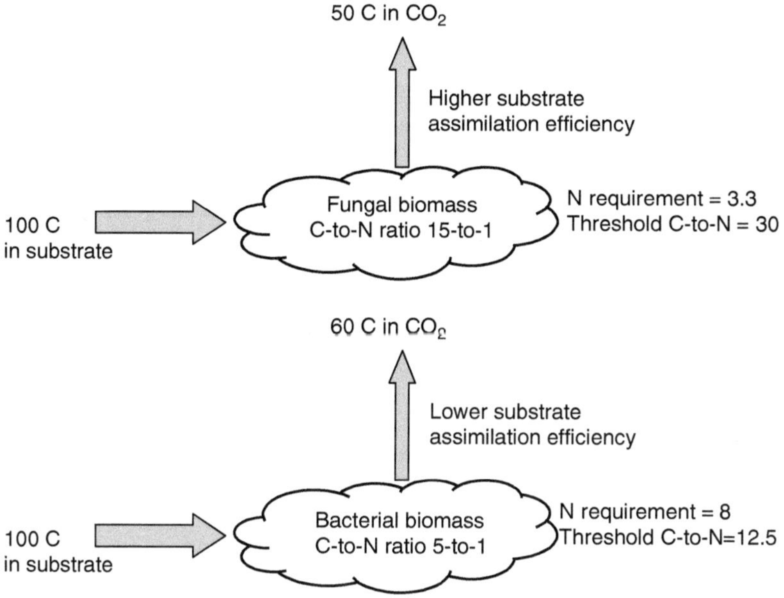

Fig. 6 Relationships between C-to-N ratio and the threshold values for net N mineralization. Fungi tend to have a larger C-to-N ratio than bacterial because bacteria contain a larger proportion of structural peptides than bacteria so the threshold for net N mineralization for the substrate is greater than for bacteria. It should also be noted that the assimilation efficiency of C by fungi is greater than that for bacteria, so the proportion of C lost as CO_2 is less for fungi for bacteria and this affects amount of C the organisms will assimilate and therefore the amount of N that they will require from the substrate (Redrawn from Killham 1994)

(Swift et al. 1979; Killham 1994). Wheat straw has a carbon-to-nitrogen ratio typically in the range 80 to 100:1 and will lead to net nitrogen immobilization, whereas leaf and stem residues from a nitrogen-fixing legume with a carbon-to-nitrogen ratio in the range 12 to 15:1 will lead to net mineralization. This provides an opportunity to manage the synchrony of nitrogen supply to plants, or to remove inorganic nitrogen from the soil pool and thus reduce the chance of nitrogen loss by denitrification and leaching (Figs. 6 and 7).

Nitrogen Fixation

Biological nitrogen fixation is the process whereby atmospheric nitrogen or dinitrogen (N_2) is fixed by organisms. This accounts for approximately 140 Mt nitrogen per year and is the main route by which nitrogen fixed by natural processes enters the biosphere, far exceeding nitrogen fixation by oxidation during lightning discharges (Killham 1994; Sylvia et al. 1995). Biological nitrogen fixation is carried out exclusively by bacteria with the nitrogenase enzyme complex. In developed countries the nitrogen input to agricultural systems from biological

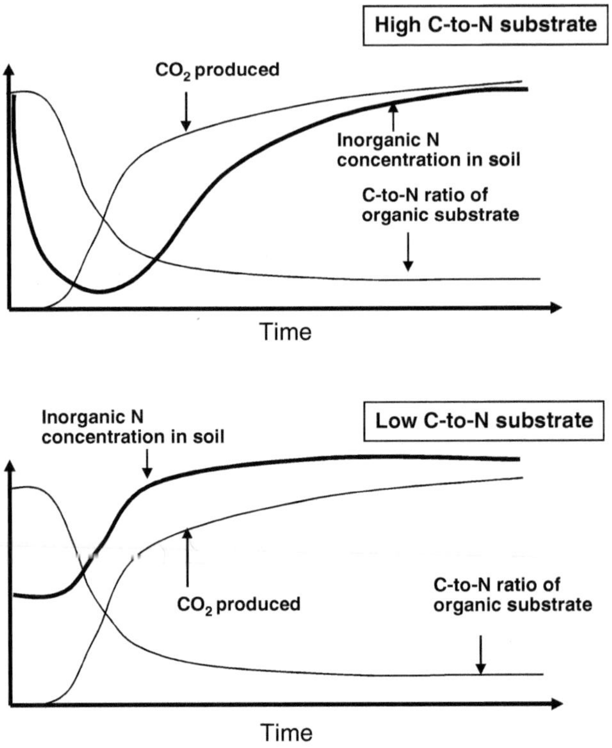

Fig. 7 Dynamics of CO_2 production, inorganic N concentration in the soil and the C-to-N ratio of the organic substrate as the substrate decomposes for two contrasting substrates: one with a high C-to-N ratio (above the threshold for net N mineralization) and one with a low C-to-N ratio (below the threshold for net N mineralization)

nitrogen fixation is relatively small by comparison with fertilizer nitrogen produced by the energy-demanding Haber-Bosch and other processes, whereas biological nitrogen fixation is the main source of additional nitrogen in developing regions where nitrogen fertilizers are expensive relative to the local economies (Sylvia et al. 1995).

Biological nitrogen fixation is carried out by: (i) free-living bacteria such as members of the genera *Bacillus*, *Klebsiella* and *Clostridium*, (ii) bacteria of the genus *Rhizobium* and close relatives which live symbiotically in nodules on the roots of leguminous plants such as beans, alfalfa (lucerne) and clover, (iii) actinobacteria (actinomycetes) in the genus *Frankia* which fix nitrogen in root nodules of some non-leguminous plants such as alder, free-living cyanobacteria such as *Nostoc* which are photosynthetic and live on or close to the soil surface, and (iv) nitrogen fixation by bacteria such as members of the genera *Azotobacter* and *Azospirillium* living in association with plant roots (though not in symbiotic association), described as 'associative nitrogen fixation'. The N_2 molecule is very stable and the reaction to reduce it to biologically available NH_4^+ requires a source of energy. In the case of nitrogen fixation by cyanobacteria this is provided by organic carbon from photosynthesis, and in the case of other bacteria from organic substrates in the soil-plant system. The symbiotic nitrogen fixers receive organic carbon from plants and the associative nitrogen fixers receive some organic carbon from rhizodeposits.

Chemical nitrogen fixation:	$N_2 + 3H_2 \rightarrow 2NH_3$ $\Delta G = 53$ kJ
(Haber-Bosch):	20 MPa (200 atmospheres) pressure and 400-500°C
Biological nitrogen fixation:	$N_2 + 8H^+ = 8e^- + 16ATP \rightarrow 2NH_3 + H_2 + 16ADP + 16Pi$
	Atmospheric pressure and environmental temperature

Denitrification

Denitrification is the dissimilatory reduction of nitrate (NO_3^-), to N_2 that occurs under anoxic conditions by a wide range of facultatively anaerobic bacteria which use nitrogen species as an alternative to oxygen as a respiratory electron acceptor (Sylvia et al. 1995). The reaction proceeds via a series of intermediates catalyses by separate enzymes, not all of which are necessarily expressed in all denitrifying bacteria or induced simultaneously. In many cases the enzymes are induced in a cascade so that as anoxic conditions develop the further stages of the process occur (Cooper and Smith 1963). However, as all steps are not induced simultaneously, and because not all denitrifiers are capable of the full reduction pathway from NO_3^- to N_2, intermediates can accumulate. The main product, gaseous nitrogen diffuses from the soil in to the atmosphere and is, thus, a route by which nitrogen is lost from the soil. Smaller quantities of nitrous oxide (N_2O) and nitric oxide (NO) are also produced because the coupling between the different steps in the overall pathway is complete (Firestone and Davidson 1989). The production of nitrous oxide is of particular interest because it is both a greenhouse gas and involved in the depletion of stratospheric ozone.

NO_3^- $\rightarrow$	NO_2^- $\rightarrow$	NO $\rightarrow$	N_2O $\rightarrow$	N_2
Nitrate	Nitrite	Nitric oxide	Nitrous oxide	Dinitrogen

Denitrification is promoted by any factor that reduces oxygen concentration in the soil, such as the water-logging or the supply of labile organic substrate, producing conditions where microorganisms have the potential to consume oxygen faster than it can diffuse to the site of decomposition in the soil. Managing denitrification losses from soil is an important objective in a sustainable agricultural system, but because it is driven in part at least anoxic conditions, the weather is main controlling factor.

Management by ensuring that the nitrate concentration in the soil is minimised during wet season is, in principle, possible using high C-to-N ratio organic amendments.

Nitrification

Nitrification is the oxidation of NH_4^+ to NO_3^- performed mainly by chemoautotrophic bacteria under oxic conditions which couples the reducing power of oxidation to the reduction of carbon dioxide from the atmosphere into organic carbon (Sylvia et al. 1995). The reaction typically occurs in two stages carried out by different genera of nitrifying bacteria, but because of the high affinity of the organisms carrying out the second stage for nitrite (NO_2^-), the intermediates rarely accumulate. Indeed, because the energetic yield from either NH_4^+ or NO_2^- oxidation is very small, provided environmental conditions permit (oxic and near neutral pH) and the pertinent organisms are present, free ammonium ions rarely persist in soils and are rapidly oxidized to nitrate ions by chemoautotrophic nitrifiers for which ammonium is the only source of energy. Chemoautrophic nitrification is restricted to a few genera of bacteria, termed the "nitroso" genera, *Nitrosomonas*, *Nitrosovibrio*, *Nitrosolobus* and *Nitrosococcus*, that carry out oxidation of ammonium to nitrite and the "nitro" bacteria, *Nitrobacter*, *Nitrospira*, *Nitrococcus* and *Nitrospina*, that carry out the oxidation of nitrite to nitrate. There is increasing evidence that fungi may also oxidize ammonium, but this is believed to be restricted to acidic forest soils.

$$NH_4^+ \xrightarrow[\text{Nitroso-bacteria}]{\text{Primary nitrification}} NO_2^- \xrightarrow[\text{Nitro-bacteria}]{\text{Secondary nitrification}} NO_3^-$$

Nitrification does not directly influence the availability of nitrogen to plants, as most crop plants are able to take up both NH_4^+ or NO_3^-, but it does indirectly influence nitrogen supply to plants because it predisposes the nitrogen to loss from the soil. Nitrate is an anion so it is poorly held at exchange sites in soils by comparison with the cationic NH_4^+; nitrification therefore predisposes to nitrogen loss by leaching. As mentioned above, nitrate is the starting point for denitrification, so nitrification can promote denitrification even though the two processes are promoted by contrasting environmental conditions (denitrification is an anoxic process whereas nitrification is an oxic process). This can occur because nitrate may be produced at oxic microsites in the soil and, being mobile, diffuse to anoxic microsites. In addition, it may be produced under oxic conditions by nitrification and be reduced by denitrification when environmental conditions become anoxic.

The potential to manage nitrification in soils exists through the manipulation of ammonium as outlined above or by using nitrification inhibitors. However, nitrification inhibitors have not been widely adopted because of expense, and the fact that they are often unreliable because they cannot be always be applied at the precise sites in the soil where nitrification is occurring.

Microbial Contributions to Phosphorus Cycling in Soil

In developed countries, there has been considerable attention to the cycling of phosphorus though soil because losses to surface water contribute to aquatic pollution. However, in terms of sustainable agriculture, attention is turning increasingly to sources of phosphorus to sustain agriculture because the exploitable phosphorus rock sources are now seriously depleted and at the current rate of consumption are likely to be exhausted within decades. This has focussed attention on exploiting sources of phosphorus in the soil. This includes manipulating the soil-plant system to solubilize phosphorus from inorganic sources and promoting the mineralization of phosphorus from organic sources.

Solubilization of inorganic phosphorus by soil microorganisms is the result of acidification, chelation and exchange reactions. However, there is relatively little understanding of how these processes actually operate and interact, and which specific free-living organisms are involved, although both bacteria and fungi are implicated. Mycorrhizal fungi are however, known to promote the solubilisation and uptake of phosphorus. These are considered in a separate chapter in this book.

Concluding Remarks

At the beginning of this chapter we presented the argument that cropping systems are inherently unsustainable if one uses an absolute definition because they result in the export of nutrients from the plant-soil system. However, rather than considering sustainable nutrient supply in cropping systems in impossibly absolute terms, it is more helpful to adopt a more practical stand point and incorporate the following processes in to nutrient management:

1. Recycling of nutrients in crop residues
2. Using of green manuring crops which fix atmospheric nitrogen
3. Minimising nutrient losses
4. Returning of animal wastes (manures and slurries) to fields
5. Relying on nutrient release by weathering
6. Adding nutrients as atmospheric "pollutants"

Soil microorganisms play important roles in these processes either directly or indirectly.

Acknowledgements SCRI (Scottish Crop Research Institute, Mylnfield, Dundee) is funded in part by the Scottish Government's Rural Environment Research and Analysis Directorate (RERAD). Rothamsted Research – North Wyke, Devon is funded in part by the UK Biotechnology and Biological Sciences Research Council (BBSRC). This article is an output of the RERAD programme on Sustainable and Profitable Agriculture Plants and the BBSRC Cross-Institute programme on Sustainable Soil Function. Both funders are gratefully acknowledged.

References

Ågren GI, Bossatta E (1996) Quality: a bridge between theory and experiment in soil organic matter experiments. Oikos 76:522–528

Alphei J, Bonkowski M, Scheu S (1996) Protozoa, nematode and lumbricidae in the rhizosphere of *Hordelumus europaeus* (Poaceace): faunal interactions, response of microorganisms and effects of plant growth. Oecologia 106:111–126

Baldock JA, Currie GJ, Oades JM (1991) Organic matter as seen by solid state ^{13}C NMR and pyrolysis tandem mass spectrometry. In: Wilson WS (ed) Advances in soil organic matter research: the impact on agriculture and the environment. Royal Society of Chemistry, Cambridge, pp 45–60

Barraclough D (1997) The direct or MIT route for nitrogen immobilization: a ^{15}N mirror image study with leucine and glycine. Soil Biol Biochem 29:101–108

Birch HF (1957) The effect of drying on humus decomposition and nitrogen availability. Plant Soil 10:9–31

Campbell CA, Paul EA, Rennie DA, McCallum KJ (1967) Applicability of the carbon-dating method of analysis to soil humus studies. Soil Sci 104:217–224

Carter MR (2001) Organic matter and sustainability. In: Rees RM, Ball BC, Campbell CD, Watson CA (eds) Sustainable management of soil organic matter. CABI, Oxford

Chapman SJ, Gray TRG (1981) Endogenous metabolism and macromolecular composition of *Arthrobacter globisformis*. Soil Biol Biochem 13:11–18

Clement CR, Williams TE (1962) An incubation technique for assaying the nitrogen status of soils newly ploughed from leys. J Soil Sci 13:82–91

Cooper GS, Smith R (1963) Sequence and products formed during denitrification. Soil Sci Soc Am Proc 27:659–662

Coûteaux M-M, Bottner P, Berg B (1995) Litter decomposition, climate and litter quality. Trend Ecol Evol 10:63–66

Dickinson CH, Underhay VXH, Ross V (1981) Effect of season, soil fauna anmd water content on the decomposition of cattle dung pats. New Phytol 81:129–141

Dungait JAJ, Bol R (2005) Evershed RP Quantification of dung carbon incorporation in a temperate grassland soil following spring application using bulk stable carbon isotope determinations. Isot Environ Health Stud 41:3–11

Dungait JAJ, Stear NA, van Dongen BE, Bol R, Evershed RP (2008a) Off-line pyrolysis and compound-specific stable carbon isotope analysis of lignin moieties: a new method for determining the fate of lignin residues in soil. Rapid Commun Mass Spectrom 22:1631–1639

Dungait JAJ, Briones MJI, Bol R, Evershed RP (2008b) Enhancing the understanding of earthworm feeding behaviour via the use of fatty acid $\delta^{13}C$ values determined by gas chromatography-combustion-isotope ratio mass spectrometry. Rapid Commun Mass Spectrom 22: 1643–1652

Dungait JAJ, Bol R, Bull ID, Evershed RP (2009) Tracking the fate of dung-derived carbohydrates in a temperate grassland soil using compound-specific stable isotope analysis. Org Geochem 40:1210–1218

Dungait JAJ, Bol R, Lopez-Capel E, Bull ID, Chadwick D, Amelung W, Granger S, Manning DC, Evershed RP (2010) Applications of stable isotope ratio mass spectrometry in cattle dung carbon cycling studies. Rapid Commun Mass Spectrom 24:495–500

Firestone MK, Davidson EA (1989) Microbiological basis of NO and N_2O production and consumption in soil. In: Andreae MO, Schimel DS (eds) Exchange of trace gases between terrestrial ecosystems and the atmosphere. Wiley, Chichester, pp 7–21

Gregorich EG, Beare MH, Stoklas U, St-Georges P (2003) Biodegradability of soluble organic matter in maize-cropped soils. Geoderma 113:237–252

Grieve IC (1984) Concentrations and annual loadings of dissolved organic matter in a small moorland stream. Freshw Biol 14:533–537

Griffin DM (1972) Ecology of soil fungi. Chapman & Hall, London

Harmsen GW, van Schreven DA (1955) Mineralization of organic nitrogen in soil. Adv Agron 7:299–398

Hartley IH, Ineson P (2008) Substrate quality and the temperature sensitivity of soil organic matter decomposition. Soil Biol Biochem 40:1567–1574

Hartley IH, Hopkins DW, Garnett MH, Sommerkorn M, Wookey PA (2008) Soil microbial respiration in arctic soil does not acclimate to temperature. Ecol Lett 11:1092–1100

Hassink J, Neutel AM, de Ruiter P (1994) C and N mineralization in sandy and loamy grassland soils: the role of microbes and microfauna. Soil Biol Biochem 26:1565–1571

Haynes RJ, Naidu R (1998) Influence of lime, fertilizer and manure applications on soil organic matter content and soil physical conditions: a review. Nutr Cycl Agroecosyst 51:123–137

Hopkins DW, Gregorich EG (2005) Carbon as a substrate for soil organisms. In: Bardgett RD, Usher MB, Hopkins DW (eds) Biodiversity and function in soils, British ecological society ecological reviews. Cambridge University Press, Cambridge, pp 57–79

Jansson SL, Persson J (1982) Mineralization and immobilization of soil nitrogen. In: Stevenson FJ (ed) Nitrogen in agricultural soils. America Society of Agronomy, Crop Science Society of America, Soil Science Society of America, Madison, WI, pp 229–252

Jenkinson DS (1977) Studies on the decomposition of plant material in soil. V. The effects of plant cover and soil type on the loss of carbon from ^{14}C labelled ryegrass decomposing under field conditions. J Soil Science 28:424–434

Jenkinson DS (1981) The fate of plant and animal residues in soils. In: Greenland DJ, Hayes MHB (eds) The chemistry of soil processes. Wiley, Chichester, pp 505–561

Jenkinson DS, Ladd JN (1981) Microbial biomass in soil – measurement and turnover. In: Paul EA, Ladd N (eds) Soil biochemistry, vol 5. Marcel Dekker, New York, pp 415–471

Jenkinson DS, Rayner JH (1977) The turnover of soil organic matter in some of the Rothamsted classical experiments. Soil Sci 123:298–305

Killham K (1994) Soil ecology. Cambridge University Press, Cambridge

Kirk TK (1984) Degradation of lignin. In: Gibson DT (ed) Lignin biodegradation: microbiology, chemistry and potential applications. Marcel Dekker, New York, pp 399–437

Lawson T, Hopkins DW, Chudek JA, Janaway RC, Bell MG (2000) Interactions of the soil organisms with materials buried for up to 33 years in the Wareham archaeological experimental earthwork. J Arch Sci 27:273–285

Mackie AE, Wheatley RE (1999) Effects and incidence of volatile organic compound interactions between fungal and bacterial isolates. Soil Biol Biochem 31:375–385

Minderman G (1968) Addition, decomposition, and accumulation of organic matter in forests. J Ecol 56:355–362

Näsholm T, Ekblad A, Nordin A, Giesler R, Högberg M, Högberg P (1998) Boreal forest plants take up organic nitrogen. Nature 392:914–916

Parton WJ, Schimel DS, Cole CV, Ojima DS (1987) Analysis of factors controlling soil organic matter in Great Plains grasslands. Soil Sci Soc Am Proc 53:1173–1179

Paustian K, Parton WJ, Persson J (1992) Modelling soil organic matter in organic-amended and nitrogen-fertilized long term plots. Soil Sci Soc Am J 56:476–488

Payne D, Gregory PJ (1988) The temperature of the soil. In: Wild A (ed) Russell's soil conditions and plant growth, 11th edn. Longman, Harlow, UK, pp 282–297

Ryden JC, Ball PR, Garwood EA (1984) Nitrate leaching from grassland. Nature 311:50–53

Swift MJ, Heal OW, Anderson JM (1979) Decomposition in terrestrial ecosystems. Blackwell, Oxford

Sylvia DM, Fuhrmann JJ, Hartel PG, Zuberer DA (1995) Principles and applications of soil microbiology. Prentice Hall, Upper Saddle River, NJ

Tate RL (2001) Soil organic matter: evolving concepts. Soil Sci 166:721–722

Tenney F, Waksman SA (1929) Composition of natural organic compounds and their decomposition in soil. IV. The nature and rapidity of decomposition of various organic complexes in different plants under aerobic conditions. Soil Sci 28:55–84

Van Soest PJ (1994) Nutritional ecology of the ruminant, 2nd edn. Cornell University Press, Ithaca, NY, Chapter 2, p 8

Webster EA, Chudek JA, Hopkins DW (2000) Carbon transformations during decomposition of different components of plant leaves in soil. Soil Biol Biochem 32:301–314

Winogradsky S (1924) Sur la microflore autochthone de la terre arable. C R Acad Sci (Paris) 178:1236

Wolters V (2000) Invertebrate control of soil organic matter stability. Biol Fertil Soils 31:1–19

Chapter 4
The Role of Microbial Communities in the Formation and Decomposition of Soil Organic Matter

Leo Condron, Christine Stark, Maureen O'Callaghan, Peter Clinton, and Zhiqun Huang

Introduction

Organic matter is mainly present in the top 20–30 cm of most soil profiles and is essentially an array of organic macromolecules consisting principally of combinations of carbon, oxygen, hydrogen, nitrogen, phosphorus and sulphur. Soil organic matter is commonly measured as the quantity of organic carbon. The global pool of organic carbon in soil to a depth of 1 m has been estimated at 1,200–1,550 Pg (2 m: 2,370–2,450 Pg), and as such is significantly greater than either the biological-biota (560 Pg) or atmospheric (760 Pg) carbon pools (Baldock 2007). Almost all organic matter in soil is directly and indirectly derived from plants via photosynthesis. Thus atmospheric carbon dioxide is transformed by reduction into simple and complex organic carbon compounds, which in combination with key nutrients enable the plant to function and grow. Carbon dioxide is released directly from plants by respiration, but most of the fixed carbon is retained and ultimately transferred to the soil ecosystem via a combination of spatially distinct pathways over a variety of timescales. The most important pathways are the direct addition of senes-cent material as

L. Condron (✉)
Agriculture and Life Sciences, Lincoln University, PO Box 84, Lincoln 7647, Christchurch, New Zealand
e-mail: leo.condron@lincoln.ac.nz

C. Stark
Environmental and Geographical Sciences, Manchester Metropolitan University, Manchester, UK
e-mail: c.stark@mmu.ac.uk

M. O'Callaghan
AgResearch, Private Bag 4749, Christchurch 8140, New Zealand
e-mail: Maureen.o'callaghan@agresearch.co.nz

P. Clinton and Z. Huang
Scion, PO Box 29-237, Fendalton, Christchurch 8041, New Zealand
e-mail: peter.clinton@scionresearch.com; zhiqun.huang@scionresearch.com

G.R. Dixon and E.L. Tilston (eds.), *Soil Microbiology and Sustainable Crop Production*,
DOI 10.1007/978-90-481-9479-7_4, © Springer Science+Business Media B.V. 2010

above-ground and below-ground detritus, return of ingested plant matter in animal faeces, and exudation of soluble organic compounds from roots (Howarth 2007).

Plant and animal detritus and root exudates represent essential sources of energy and nutrients for soil microbial and faunal communities. Bacteria and fungi represent 95%+ of the biomass present in most soils, where they interact with a combination of micro-fauna (nematodes, protozoa), meso-fauna (acari, Collembola, mites) and macro-fauna (earthworms, termites, molluscs) in complex soil food-web systems that determine the turnover of organic matter and associated nutrients in the soil environment (Wardle 2002; Coleman and Wall 2007). Decomposition of organic carbon in soil is driven primarily by the activities of bacteria and fungi, while only 10–15% of soil carbon flux can be directly attributed to the actions of fauna (Hopkins and Gregorich 2005). The vast majority of soil microorganisms are heterotrophs that rely on organic matter for energy and nutrients. These can be divided into microorganisms that respond primarily to the addition of fresh carbon substrates (zymogenous or *r*-selected biomass) and those that derive their energy mainly from the decomposition of older, more recalcitrant forms of organic carbon (autochthonous or *K*-selected biomass) (Hopkins and Gregorich 2005).

Soluble plant root exudates account for 10–40% of the total carbon fixed by photosynthesis and are composed mainly of mixtures of sugars, amino acids, sugar alcohols, organic acids and secondary metabolites (Bais et al. 2006). Root exudates are particularly important drivers of microbial and faunal activity in soil due to a combination of their relatively high bioavailability compared with senescent plant detritus, their role in controlling the bioavailability of nutrients (e.g., phosphorus) and phytotoxic elements (e.g., aluminium), together with the fact that they are added to soil on a regular/semi-continuous basis (Singh and Mukerji 2006; Neumann 2007). Thus microbial numbers and activity in the immediate vicinity of growing roots (1–3 mm – rhizosphere) are commonly orders of magnitude greater than in non-rhizosphere soil. Ratios of microbial communities in rhizosphere and adjacent 'bulk' soil vary with plant species and environmental conditions, although 12–25-fold differences in bacterial and fungal populations have been observed (Kennedy 2005). Plants benefit directly from exudate enhanced biological activity in the rhizosphere, mainly via improved acquisition of sparingly soluble and organic soil nutrients mobilised by microorganisms in response to the provision of energy-rich carbon substrate. This includes the specific symbiotic relationship between plant roots and mycorrhizal fungi, whereby fungi living in close association with plant root cells obtain a supply of soluble carbon from the plant (up to 20% of assimilated carbon) in exchange for improved access to and mobilisation of sparingly-soluble mineral and organic forms of soil nutrients (Sylvia 2005). This is mainly attributed to the proliferation of fungal hyphae and associated bacteria into surrounding soil which markedly increases overall root surface area and the potential depletion zone for key nutrients (principally phosphorus and nitrogen) compared with non-mycorrhizal plant roots (Timonen and Marschner 2006; Powell and Klironomos 2007).

In most natural and managed ecosystems up to half of the organic carbon added to soil on an annual basis in plant detritus and root exudates is rapidly consumed by microbial and faunal activity and released as carbon dioxide (Hopkins and Gregorich 2005).

It has been shown that 64–86% of root exudates are rapidly respired by rhizosphere microorganisms (Hutsch et al. 2002). However, the remainder of the added organic matter, together with organic compounds synthesised by soil microorganisms and fauna during decomposition and released mainly as detritus, persist in the soil for an extended period. It is important to note that all the organic carbon added to soil is eventually mineralised and released as carbon dioxide by the combined actions of microorganisms and fauna (Hopkins and Gregorich 2005; Wolf and Wagner 2005). The process of plant residue decomposition is comprised of a number of stages, which occur over periods ranging from days to centuries (Fig. 1).

The quantity, form and spatial and temporal distribution of organic substrate addition to soil and its subsequent decomposition are primarily influenced by the nature and productivity of different ecosystems. Ecosystem productivity is determined by a combination of factors such as climate and soil type, which in turn are closely linked to human disturbance, including changes in land-use and management, together with inputs of nutrients in fertilisers (Baldock 2007; Janzen 2004; Six et al. 2002). For example, it has been found that biological activity, moisture content and aggregate stability are greater in soil managed under no-till compared with conventional till, which in turn results in increased levels of organic carbon in no-till soils (Frey et al. 1999; Six et al. 2000; White and Rice 2009).

The study of organic matter dynamics in soil is widely acknowledged as being extremely challenging given the diversity of biological, chemical and physical properties and processes involved. An often-bewildering array of methods has been developed to define different pools of soil organic matter (Baldock 2007; Hopkins and Gregorich 2005). These include separating living from non-living organic matter components, which can then be divided into more specific fractions based on chemical (alkali-acid solubility), physical (size fractions), and kinetic-functional (susceptibility to decomposition) variables (Fig. 2). The non-living soil organic matter can be divided into a number of fractions based on a combination of physical size and chemical form. Identifiable plant and faunal detritus ranging in size from over 2 mm down to 50 μm can be physically separated from soil by sieving and density floatation, although these fractions are often not considered part of soil organic matter (Hopkins and Gregorich 2005). The remainder of the non-living soil organic matter is comprised of organic macromolecules collectively known as 'humus'. Humus can be further divided into two broad categories of compound, namely 'non-humic substances' (ca. 30%) and 'humic substances' (ca. 70%). Non-humic substances are defined as being chemically identifiable plant, microbial, and faunal constituents, which includes nucleic acids, peptides and amino acids, sugars and polysaccharides, lipids and lignin. On the other hand, humic substance is a collective term for unidentified polydispersed organic macromolecules that are produced during decomposition of organic substrates in the soil environment. Humic substances in soil are characterised according to extractability and subsequent solubility as fulvic acid (alkali extractable – acid soluble), humic acid (alkali extractable – acid insoluble), and humin (non-extractable) (Stevenson 1994). Non-humic substances are generally regarded as being more readily decomposed in the soil environment compared with humic substances.

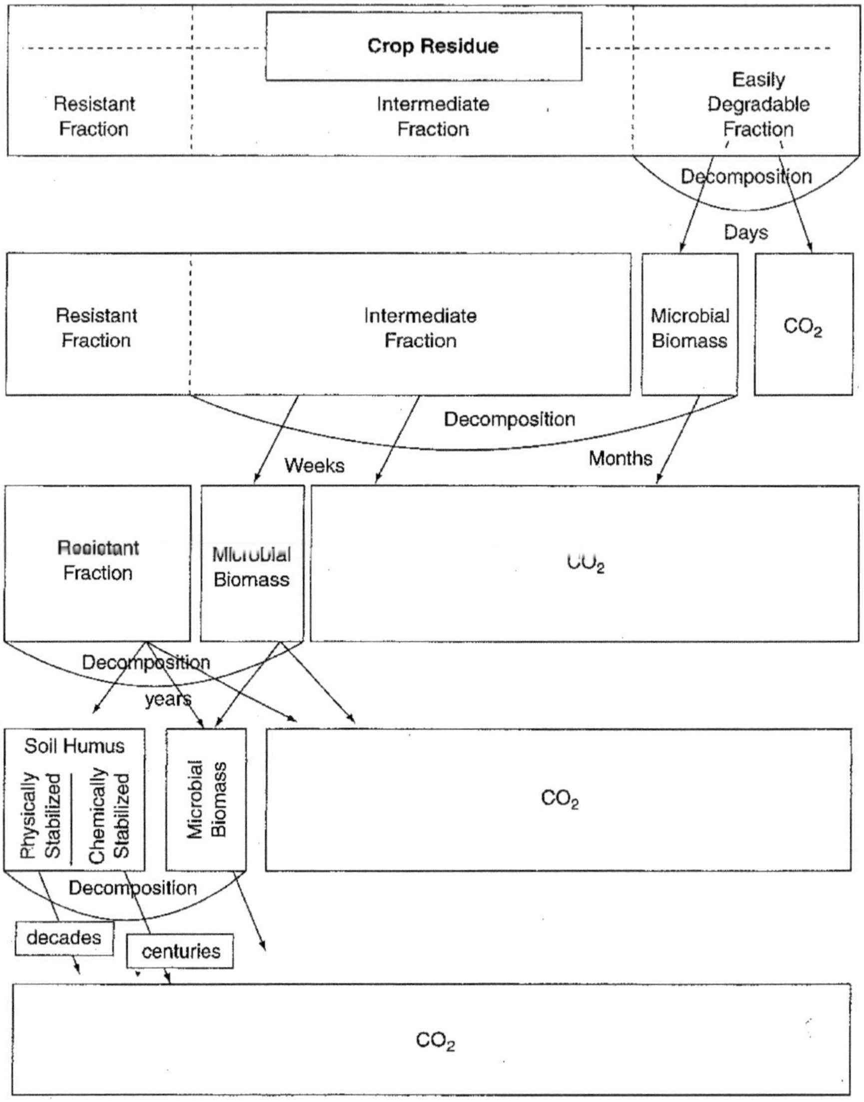

Fig. 1 The stages of crop residue decomposition in soil (changing box sizes indicate the relative quantities of carbon in the various fractions as decomposition proceeds to completion) (Wolf and Wagner 2005) (Reproduced with permission of Pearson Prentice Hall, USA)

The techniques described above have been used extensively over many years to investigate the processes involved in the formation and decomposition of organic matter in soil, with an emphasis on elucidating the chemical nature and synthesis of non-humic and humic substances (Baldock 2007; Howarth 2007; Hopkins and Gregorich 2005; Six et al. 2002; Stevenson 1985, 1994). There have been many mechanisms and pathways proposed which encompass the biological-biochemical and

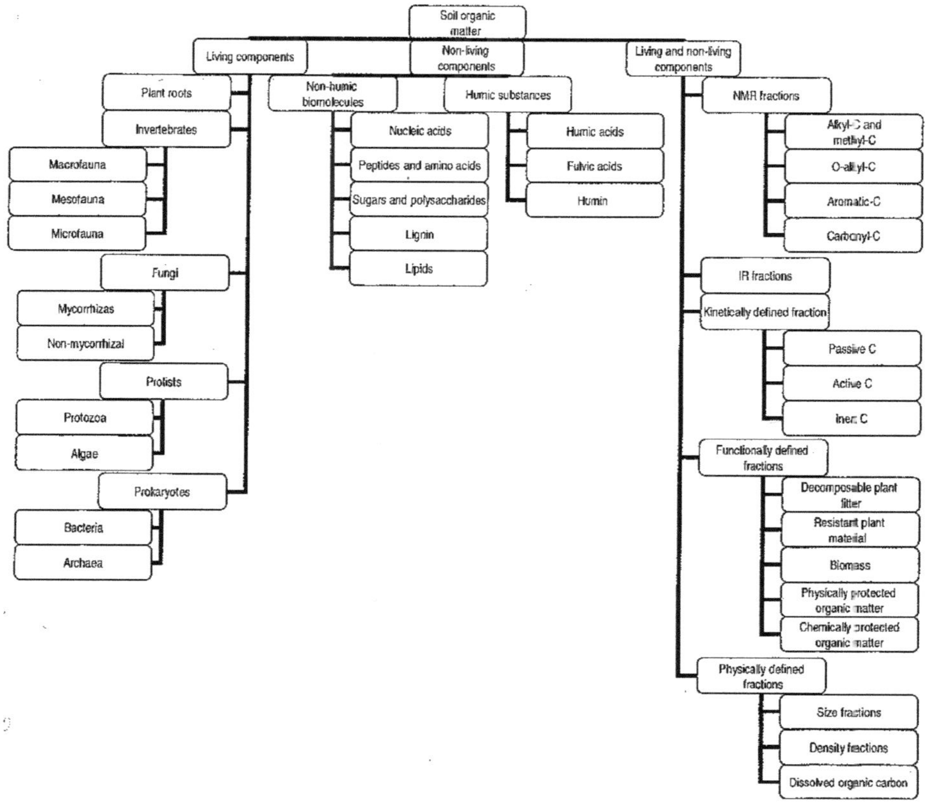

Fig. 2 Summary of soil organic matter fractions (This figure was published in Biological Diversity and Function in Soils (Editors RD Bardgett, MB Usher, DW Hopkins), 2005, Chapter 4 – Carbon as substrate for soil organisms, DW Hopkins and EG Gregorich, p. 60, copyright Cambridge University Press – reprinted by permission of the publisher)

physicochemical processes responsible for organic matter dynamics in soil. The most commonly hypothesized mechanisms for the formation of humic substances in soil are based around condensation of cellular metabolites, organic nitrogen compounds (amino acids, amino sugars), and lignin and phenolic substrates of microbial origin, which in turn are stabilised by a combination of chemical interaction with mineral colloids and physical incorporation into soil aggregates (Fig. 3). The presence and/or addition of relatively inert charcoal carbon may also influence the stability and turnover of organic carbon in soil due to its effect on microbial activity and nutrient availability (Hopkins and Gregorich 2005; Lehmann 2007; Wardle et al. 2008).

The ongoing formation and decomposition of organic matter in soil plays a major role in determining plant and ecosystem productivity, via a combination of its role in the storage and provision of nutrients and water, together with the development and maintenance of physical structure. In most soils over 90% of total nitrogen and sulphur, together with over 50% of total phosphorus, are associated with the microbial biomass and organic matter, and so the cycling and bioavailability of these key nutrients in soil are primarily controlled by organic matter

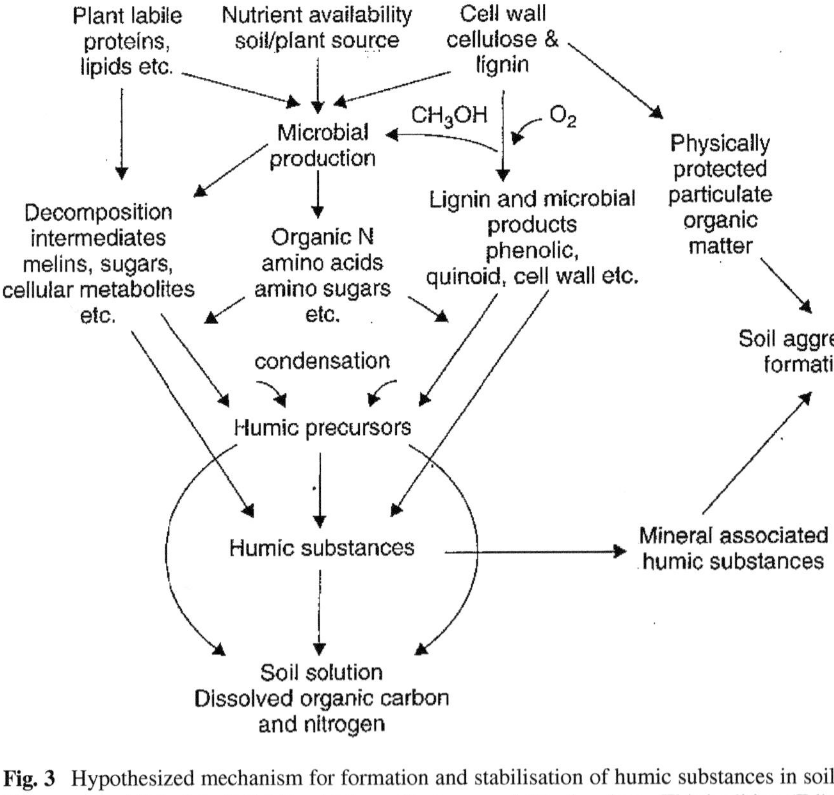

Fig. 3 Hypothesized mechanism for formation and stabilisation of humic substances in soil (This figure was published in Soil Microbiology, Ecology, and Biochemistry, Third edition (Editor EA Paul), 2007, Chapter 12 – Carbon cycling and formation of soil organic matter, W Horwath, p. 330, copyright Elsevier – reprinted by permission of the publisher)

transformations linked to microbial and faunal activity (McNeill and Unkovich 2007; Bünemann and Condron 2007). Soil organic matter is the major source of negative charge in most soils and therefore determines the retention and bioavailability of nutrient cations such as calcium, potassium, magnesium and ammonium. Organic matter also plays a vital role in the retention and availability of water in soil, and the association of soil organic matter with secondary minerals such as clay to form aggregates underpins soil structure, promoting aeration and water infiltration necessary for plant growth. In addition, the formation and decomposition of soil organic matter regulates production of carbon dioxide and nitrous oxide and consumption of methane (Howarth 2007). As soil organic matter may act as both a sink and source of carbon during global environmental change (Johnston et al. 2009), effective management of soil organic matter has become a key objective of research aimed at protecting the environment.

The remainder of this chapter will focus on description and discussion of selected key topics that relate to the role and function of microbial communities in the dynamics of organic matter in the soil environment, including microbial diversity, microbial-faunal interactions, the influence on nutrients on organic matter mineralization, and the microbial origin of organic matter.

Importance of Soil Microbial Community Composition and Diversity

The ability of soils to provide long-term carbon sequestration and the nutrients necessary for plant productivity are largely dependent on soil organic matter dynamics (Billings and Ziegler 2005). Thus, in the view of increasing sustainability of agricultural production systems, soil organic matter, which is at once part and driver of ecosystem processes such as primary productivity (= the total sum of organic matter production through assimilation of atmospheric carbon dioxide primarily by photosynthesis) and decomposition (= breakdown of soil inherent or added organic matter) demands to be continuously maintained and improved. Carbon accumulates in soils when productivity, i.e., the input of carbon substrates, exceeds decomposition, which leads to the build up of organic matter. In turn, decomposition entails a reduction in the soil carbon stock. Typically, the decomposition of inherent soil or added organic materials releases essential nutrients for uptake by plant and soil communities, whereas decomposer organisms, mainly bacteria and fungi, use carbon sources that are supplied mainly by primary producers, like plants. Soil organic matter cycling is therefore closely linked to the decomposer activity, which mineralise organic compounds and make essential nutrients plant available. It is therefore reasonable to assume that the abundance, composition and/or diversity of microorganisms in soils matter to ecosystem functioning.

Although changes in microbial community composition are increasingly seen as the driving factor behind functional capabilities in soils, there is only limited direct and specific evidence with regard to the relevance of microbial diversity in soil processes (Schimel and Gulledge 1998; Griffiths et al. 2000, 2001; Nannipieri et al. 2003).

This is especially true with regard to ecosystem processes related to the carbon cycle, such as decomposition and carbon sequestration. It is widely believed that these processes are too universal to be affected by changes in microbial diversity and that in functionally very diverse communities these processes are taken over by other members of the community if a microbial group is lost (cf. functional redundancy hypothesis) (Nannipieri et al. 2003). Moreover, microbial community composition, diversity and abundance are often deemed entirely unimportant to ecosystem processes and are generally not considered in ecosystem models, esp. in terrestrial environments. Most models, including CENTURY (Parton et al. 1987) and TEM (terrestrial ecosystem model) (McGuire et al. 1993), work on the assumption that changes within the microbial community only have limited effects on soil processes. This idea only holds true if the microbial community in question is resistant[1], resilient[2] and/or functionally redundant[3] (Allison and Martiny 2008). In their review, Allison and Martiny (2008) demonstrate the increasing recognition of the importance of microbial community composition to ecosystem processes. The authors note that the composition of soil microbial communities is generally not resistant to change as it is often sensitive to perturbations, including climate change scenarios such as increases in temperature and carbon dioxide, or agricultural management, such as mineral fertilisation or organic matter amendments. They also suggest that over the medium term time-scale of a few years disturbed communities still differ in composition from undisturbed ones. While the degree of functional redundancy is not easily quantified due to methodological and experimental difficulties, the authors conclude that disturbance induced changes to microbial composition often persist over time and have an effect on ecosystem process rates (also cf. Griffiths et al. 2000; Ekschmitt and Griffiths 1998; Balser and Firestone 2005; Wertz et al. 2006). This suggests that at least some groups within soil microbial communities are functionally dissimilar and not functionally redundant.

Links and interactions of above-ground (i.e., plant) diversity with primary productivity and their impact on below-ground communities have been studied extensively and are firmly established (Schulze and Mooney 1993; Loreau et al. 2001; Carney and Matson 2005; Catovsky et al. 2002; McGrady-Steed et al. 1997). Carbon availability is a key determinant of microbial growth and activity in soils, which links them closely to primary production, rhizosphere activity and litter substrate quality (Smith and Paul 1990). As root exudation and rhizodeposition vary in amount, quality and quantity with plant species type and abundance as well as in space and time, plant identity and plant community composition have significant bearing on the composition, diversity and activity of soil microbial communities, which affects the rate of

[1]Resistance, the degree to which microbial composition remains unchanged following disturbance.

[2]Resilience, the rate at which microbial composition returns to its original composition after the disturbance.

[3]Functional redundancy, the ability of one microbial group to carry out a process at the same rate as another under the same environmental conditions (according to Allison and Martiny 2008).

decomposition of the added carbon compounds (Catovsky et al. 2002; Marschner et al. 2001; Johnson et al. 2003; Brant et al. 2006; Hamilton and Frank 2001; Grayston et al. 1996). There is evidence that through rhizodeposition plants actively select for microbial communities to promote the supply of limited soil resources (Hamilton and Frank 2001). Similarly, a microbial community with a specific composition and functional diversity might select for a plant community that ensures the supply of favoured carbon compounds (Rillig 2004; Wardle 2005). Generally, more diverse plant communities exhibit greater primary productivity, resulting in higher carbon assimilation from the atmosphere. This adds to soil carbon storage, nutrient retention and energy turnover in soils, resulting from higher microbial diversity, and leading, yet again, to changes in the quality and quantity of rhizodeposition and stimulates decomposition (Dang et al. 2005; Broughton and Gross 2000; Ekschmitt et al. 2001). This clearly shows how primary productivity is not only directly affected by plant-related factors, but equally stimulated from below-ground, by decomposition rates, root processes, microbial community composition and soil organic matter content (Bausenwein et al. 2008; Tiunov and Scheu 2005).

Although no less crucial in ecosystem functioning, the complementary process in the carbon cycle – the decomposition or mineralisation of organic matter – and its interactions with soil microbial diversity have received significantly less attention. Decomposition is a mostly microbially mediated process, although its actual rate and extend are influenced by environmental variables, including soil temperature, moisture, oxygen, nitrogen content, the quality and quantity of available carbon substrates as well as soil management (physical disturbance, for example, might expose previously protected organic matter) and presence/absence of microbivorous soil fauna (Janzen 2004; Johnston et al. 2009; Swift et al. 1979). The implications soil microbial species composition and diversity have on carbon cycle related ecosystem processes are not entirely clear, and determining microbial diversity effects on soil processes while controlling for other environmental parameters is a huge challenge, not least as a result of the interactions between microbial communities, environment variables and ecosystem processes (Reed and Martiny 2007) (Fig. 4).

It is necessary to consider various levels of diversity, which might be differently affected by, or have different effects on, decomposition rates. We differentiate between species or microbial abundance (total number of individuals of a species or total size of the microbial biomass, respectively), species richness (total number of species), species evenness (relative number of species), species identity or community composition (types of species/taxa present) and species or ecosystem diversity, which accounts for number as well as frequency of species. For example, changes in microbial abundance have been identified to be more important to soil carbon mineralisation than differences in microbial community composition (Carney and Matson 2005). A basic, positive relationship can be observed between organic matter quantity and microbial biomass size. Higher soil organic matter content is generally associated with a larger microbial biomass, while organic matter amendments lead to increases in biomass (e.g., Nannipieri et al. 2003; Plassart et al. 2008; Bastida et al. 2008; Gunapala and Scow 1998). Similarly, the differential effects of substrate quantity and quality on the size and composition of soil microbial communities need

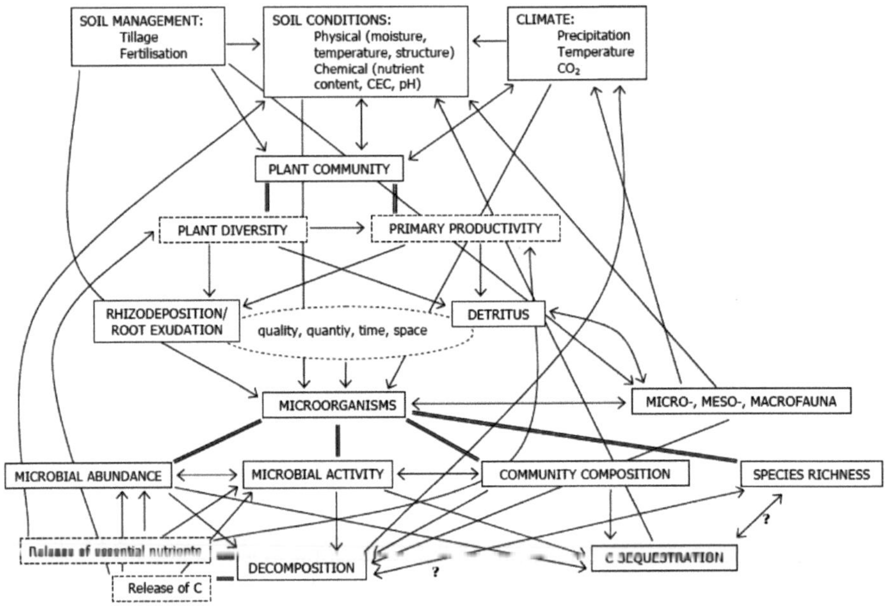

Fig. 4 Interactions and feedback effects of below- and above-ground communities, environmental conditions and soil carbon cycling

to be emphasised. Generally, carbon compounds of higher quality are characterised by a lower carbon-to-nitrogen ratio, which makes the substrate more readily available to decomposer organisms, in particular bacteria, while low quality substrates with high carbon-to-nitrogen ratio are preferentially used by fungi. Higher carbon substrate quality, i.e., more readily available and easily decomposable organic matter, has been found to lead to an increase in microbial biomass size whereas functional diversity decreased when studying the decomposition of different plant leaf components (Webster et al. 2000; Paterson et al. 2007) and increases in organic matter amendment had little influence on the diversity of rhizosphere communities (Ros et al. 2006). In contrast, Kubartová et al. (2007) showed that differences in carbon substrate supply resulting from changes in plant community composition (i.e., a change in substrate quality as well as quantity) impacted on microbial community composition while microbial abundance remained uninfluenced. More importantly, positive links have been established between microbial biomass and carbon mineralisation rates (Garcia-Pausas et al. 2008) (also cf. Fig. 4).

Due to experimental and methodological limitations, the effects of species richness and species diversity are often blurred and difficult to distinguish from each other. Generally richness effects on organic matter processes are less unequivocal than those of microbial abundance or species identity and composition of decomposers (Tiunov and Scheu 2005), and most current evidence suggests that decomposer community composition is of greater importance on ecosystem processes than species richness or diversity. As species richness is closely correlated with substrate quality of organic compounds, it is feasible that a community higher in species richness is also more diverse with regard to its decomposition capabilities (Wardle 2005; Bausenwein et al. 2008; Cookson et al. 2005; Orwin et al. 2006). Most research findings, however, suggest that decomposition activity is unaffected by increases, or decreases, in species richness (Nannipieri et al. 2003; Ekschmitt and Griffiths 1998; Wardle 2005; Ekschmitt et al. 2001) and it has been observed that microbial communities of different size or composition have similar or even identical carbon utilisation patterns (Buyer and Drinkwater 1997; Waldrop et al. 2000; Kemmitt et al. 2008; Chander et al. 2002). For example, mixtures of two to eight fungal species performed similar to monocultures when assessing respiration following wheat straw addition (Hedlund and Öhrn 2000), leaf litter mass loss (Dang et al. 2005; Duarte et al. 2006) or oxygen consumption (Tiunov and Scheu 2005) (Table 1). This lack of effect was often independent of variation in environmental conditions (e.g., Dang et al. 2005). Studying overall microbial diversity and its effects on soil processes before and after fumigation, Griffiths et al. (2000) found that a reduction in microbial diversity in some cases increased decomposition rates of plant residues. Similarly, Degens (1998) observed higher straw decomposition rates in soils with reduced microbial functional diversity.

Contrary to these conclusions, some positive links between species richness/ diversity and measures of decomposition have also been observed. Higher diversity of saprotrophic fungi has been shown to result in an increase in decomposition rates when assessing mass leaf mass loss (Bärlocher and Corkum 2003) or respiration (Robinson et al. 1993; Setälä and McLean 2004). Similarly, Tiunov and Scheu (2005) found that the rate of organic matter decomposition was positively associated with fungal species richness in sterile forest soils and even more so in a single-resource substrate (powdered cellulose). While competitive interactions can limit as well as increase decomposition activity (Jiang et al. 2008; Hättenschwiler et al. 2005), impacts of species richness and diversity on decomposition rates are generally believed to be more pronounced in species poor communities as species rich environments are characterised by overlaps of functional niches and higher functional redundancy (Ekschmitt et al. 2001; Tiunov and Scheu 2005; Setälä and McLean 2004).

Although biodiversity per se seems to be unaffected by decomposition, there are clear differences in decomposition patterns when the community shifts from fungal to bacterial dominated and vice versa depending on the age and complexity of the carbon compound (Wardle 2005; Six et al. 2006; Paterson et al. 2008a). Bacteria are generally more dominant in fertile soils that receive more labile organic matter with a low carbon-to-nitrogen ratio, under and fast growing plants with high leaf litter quality and in tilled soils that are characterised by the incorporation of residues into the soil profile (Wardle 2005; Scow 1997). In general, early stages of decomposition (14–28 days after substrate addition) are governed by bacterial activity (Bastian et al. 2009). Fungi, on the other hand, prefer more recalcitrant residues with a higher carbon-to-nitrogen ratio and dominate the later stages of decomposition (from 56 to 168 days) (Bastian et al. 2009). They are dominant in infertile soils that show slower plant growth and often low leaf litter quality known

Table 1 Relationship between fungal or bacterial decomposer diversity and organic matter decomposition based on experimental studies using single (cellulose) or complex (several compounds) carbon sources. Janzen et al. (1995) reported experiments with a neutral and a negative relationship, respectively. Table based on Jiang et al. (2008)

Decomposer type	Number of species tested	Substrate type added	Process measured	Process-diversity relationship	Source
Fungi	1–2 species	Cellulose	Cellulose mass loss	Positive	Deacon 1985
Fungi	1–5 species	Cellulose	O_2 consumption	Strong positive	Tiunov and Scheu 2005
Bacteria	1–8 species	Cellulose	Cellulose mass loss	Positive	Wohl et al. 2004
Bacteria	1–72 species	Leaf disk	CO_2 production	Positive	Bell et al. 2005
Fungi	1–5 species	Oak leaf	Leaf mass loss	Positive	Bärlocher and Corkum 2003
Fungi	1–2 species	Wheat straw	CO_2 production	Positive	Robinson et al. 1993
Fungi	1–43 species	Soil	CO_2 production	Positive	Setälä and McLean 2004
Fungi	1–5 species	Soil	O_2 consumption	Weak positive	Tiunov and Scheu 2005
Fungi	1–2 species	Alder leaf	Leaf mass loss	Positive	Treton et al. 2004
Fungi	1–8 species	Alder leaf	Leaf mass loss	Neutral	Dang et al. 2005
Fungi	1–4 species	Leaf	Leaf mass loss	Neutral	Duarte et al. 2006
Fungi	1–3 species	Soil + wheat straw	CO_2 production	Neutral	Hedlund and Öhrn 2000
Bacteria	1–4 species	Wheat seed	Wheat seed mass loss	Neutral	Jiang 2007
Fungi	1–2 species	Barley straw	CO_2 production	Neutral/negative	Janzen et al. 1995
Fungi	1–16 species	Wood	Wood mass loss	Negative	Toljander et al. 2006

to better utilise carbon sources with patchy distribution patterns, as is the case in no-till systems, where undecomposed residues are left on the soil surface. Consequently, more fungi are found in surface soils, while bacteria dominate in subsoils. Fungi thus mediate the translocation of nutrients between the surface and the underlying soil profile (Hendrix et al. 1986) and have a higher carbon storage capacity compared to bacteria.

The relationship between microbial diversity and decomposition rates is ambiguous at best. Many experimental studies have demonstrated that whenever diversity effects on the speed and extent of decomposition exist they are highly context specific and other, external, factors have proven to be equally, if not more important in governing soil processes or driving changes in microbial community composition (Loreau et al. 2001; Wardle 2005). The influencing factors range from environmental conditions, such as precipitation, nutrient availability or ecosystem properties (Loreau et al. 2001; Bärlocher and Corkum 2003; Schwartz et al. 2007), to seasonal effects (Kubartová et al. 2007) and experimental treatment (Jiang et al. 2008). According to Fierer et al. (2005), the physiological response observed in their study determining litter quality and temperature effects on decomposition was more strongly related to changes in temperature than in community composition. This effect increased with decreasing litter quality. Jiang et al. (2008) identified distinct differences in relationships between decomposer diversity and decomposition activity depending on the type and diversity of carbon sources used. They found primarily positive links in experiments using a single carbon source and more diverse links with relationships ranging from positive to neutral to negative in experiments using multiple carbon sources (Table 1). Moreover, specific traits of certain species rather than species number have been identified to exert greater influence on ecosystem functions and processes (Duarte et al. 2006; Jiang et al. 2008).

It is easy to see that interactions of above- with below-ground communities and the relationships between soil biodiversity and ecosystem processes are complex and diverse. A wide range of positive and negative feedback loops exists (Fig. 4). This complicates the interpretation of results and makes it difficult to test and understand cause-effect relationships, which makes it a "chicken and egg" situation. Whether higher biodiversity maintains soil functionality and sustainability or whether a healthier, more functional soil environment supports a more diverse microbial community cannot unambiguously be determined. Nonetheless, the very existence of feedbacks and co-dependencies strongly suggests that decomposer biodiversity is of similar significance to ecosystem processes as plant diversity. Overall, species diversity effects are most likely to occur at local scales and are greatest at low to intermediate diversity levels. That in many experimental situations microbial diversity changes are not reflected in changes in ecosystem process rates might be explained by the phenomenon of functional redundancy, whereby different species may takeover functions of lost microbial groups and a loss in diversity is thus compensated for. Alternatively, it is feasible that a microbial community expresses different functionality following perturbation, while its activity results in the same, pre-disturbance process rate, i.e., soil function and microbial activity do not closely correlate.

The exact nature and extent of the relationship between microbial diversity and carbon turnover remains opaque. It is obvious that microbial diversity and/or community composition have some effect on soil functionality, be it by changing ecosystem processes directly, or via interactions with above-ground communities, which impact on the availability of carbon and other essential nutrients and affect microbial metabolic activity. Improved knowledge on microbial community composition and function and their links with nutrient turnover will help achieve understanding of decomposition and other nutrient cycles at the process-level, and possibly pave the way to a more accurate and more sustainable management of agroecosystems.

Interactions Between Microorganisms and Soil Fauna

The role of microorganisms in formation and decomposition of soil organic matter cannot be discussed without reference to the interactions of microorganisms with the soil micro- and macrofauna. A diverse range of biota is present in soil, including micro-fauna (e.g., nematodes and protozoa), meso-fauna (e.g., Collembola, mites and acari) and macro-fauna (e.g., earthworms, molluscs and termites). The impact of invertebrates on soil organic matter turnover has been studied since Darwin's pioneering work (Darwin 1881). However, despite the importance of soil fauna in organic matter breakdown being well recognised, the complex interactions between soil fauna and microorganisms and the indirect effect on microfauna and microbial communities are less well understood. While soil invertebrates mediate about 15% of the carbon and 30% of the nitrogen turnover in a range of ecosystems (Anderson 1995), their indirect effect through activation of microflora is likely to be much greater. Microorganisms are the main agents of biochemical decomposition and turnover of organic matter but soil fauna enhance the fragmentation of coarse particulate matter into finer fractions and influence the distribution of organic matter in soil. Through comminution of organic matter, which increases the surface area available for attack by microbial decomposers, soil fauna directly stimulate microbial populations and activity.

The interactions between soil microflora and fauna have been most well studied in earthworms. However, other groups of soil-inhabiting invertebrates undoubtedly play a significant role in formation and turnover of organic matter, in particular mesoarthropods such as mites and Collembola. The extent to which these various groups of soil fauna control decomposition may depend on the ecosystem. For example, microarthropods that feed on fungi, especially Collembola, can play a key role in no-tillage agroecosystems, but are less important in conventionally tilled systems (Hendrix et al. 1986). Few studies have been carried out but it seems likely that, like the earthworms, microarthropods can play key roles in influencing the activities of microorganisms in organic matter processing in soil (Edwards 2000).

Although not numerically dominant, earthworms are major contributors to invertebrate biomass in soil because of their size. Earthworms are divided into three groups based on their feeding preferences and habitat. Epigeic earthworm species may feed directly on microorganisms or litter material, and dwell in the organic layer of soil. They have been shown to strongly affect decomposition processes (Sampedro and Dominguez 2008) and have been demonstrated to modify the fungal composition of forest soils (McLean and Parkinson 2000). Endogeic earthworms feed their way horizontally through the upper mineral layer of soil while aneic earthworms generally inhabit one single vertical burrow and can transport fresh organic detritus from the soil surface into burrows while mixing it with mineral soil. It is clear that all detritivorous earthworms can accelerate organic matter decomposition and nutrient release in soil. They exert a controlling activity through their strong interactions with microorganisms in the decomposition process; they strongly modify the physical, biochemical and biological properties of the substrates in which they live, including the structure and function of soil microbial communities.

The interactions between earthworms and microorganisms occur on several spatial scales but most research has been directed at microscale interactions which occur in the drilosphere (Brown and Doube 2004). The "drilosphere", as defined by Brown et al. (2000), includes the internal microenvironment of the earthworm gut; the earthworm surface in contact with the soil; surface and belowground casts; middens; and burrows and chambers made by the earthworm, all of which are microhabitats for a range of bacteria and fungi. Up to 60% of the carbon losses from earthworms during their life span can be in the form of mucous secretions, and this soluble organic carbon is an important microbial stimulant in the drilosphere (Brown and Doube 2004).

Different species and ecological categories of earthworms differ in their ability to digest various organic residues and assimilate nutrients from ingested organic matter (Lattaud et al. 1998, 1999). Many earthworm species consume a mixture of soil and organic matter, often choosing to feed in patches of soil that are relatively rich in organic matter, or microsites in soil that are enriched with bacteria and fungi (Wolter and Scheu 1999). Selective feeding on these sites may provide earthworms with additional soluble carbon sources such as those derived from microbial metabolism. Microorganisms themselves may be a source of food for earthworms but the amounts consumed and the ability of earthworms to digest and assimilate microbial biomass varies with earthworm species (Brown and Doube 2004). Based on limited data, it seems likely that at least some earthworm species have an indigenous, autochthonous gut flora. Lavelle and Gilot (1994) showed that several temperate earthworm species had a mutualistic digestive system in which the mixture of soluble organic carbon in the form of low molecular weight mucus with ingested organic matter, together with the moist conditions and neutral pH in the foregut, promoted the development of a microbial community that could digest cellulose and other substances that earthworms cannot typically digest. Essentially, the earthworm gut can act like a bioreactor where microbial activity and biomass are increased due to favourable conditions with readily available carbon of mucus and water.

All earthworms have high rates of consumption of soil and/or litter but their assimilation efficiencies vary widely depending on ecological grouping – 1% of ingested carbon for endogeic species *Aporrectodea rosea* through to 30–75% assimilation efficiencies in litter feeding species such as *Lumbricus rubellus* or *L. terrestris* (Brown and Doube 2004). Hence, earthworm casts may contain large amounts of organic matter that has not been assimilated but has been modified both physically and chemically during passage through the earthworm gut. Microorganisms that can survive passage through the earthworm gut (mainly fungal and protozoan spores and some resistant bacteria) provide inocula for microbial colonisation of newly formed earthworm casts (Brown 1995). These newly deposited casts are usually rich in ammonium-nitrogen and partially digested organic matter, providing a good substrate for growth of microorganisms. There is evidence to suggest that microorganisms may become activated or "primed" in the gut because of the abundance of soluble carbon and other nutrient resources (Lavelle and Gilot 1994) and this effect may also continue for a short time in the casts. It is well established that there are larger populations of fungi, bacteria and actinomycetes (Shaw and Pawluk 1986), and higher enzymatic activity in earthworm casts than in bulk soil. Higher proportions of cellulolytic, hemicellulolytic, nitrifying and denitrifying bacteria (Edwards et al. 2004) and larger, more diverse fungal populations have been recovered from casts than surrounding soil (Tiwari and Mishra 1993). Similarly, the walls of earthworm burrows typically support more diverse and larger microbial populations than surrounding uningested soil (Sampedro and Whalon 2007).

Given that increased microbial biomass, activity and diversity is often measured in earthworm casts, higher rates of organic matter decomposition might also be expected. However, balanced against the short term priming effects is the protection of the organic matter from microbial attack in compact, water stable castings of endogeic species, which may lower the organic matter decomposition rate (Marinissen and Hillenaar 1997). In addition, earthworms assimilate the more labile soil organic matter fractions preferentially, so that the more recalcitrant organic matter, with slower mineralisation kinetics, remains in the casts. However, it appears that in the drilosphere at least, the earthworm's priming of microorganisms leads to increased mineralization of nutrients. Some studies have demonstrated a short-term increase in carbon turnover in earthworm casts. For example, Coq et al. (2007) showed that casts extracted from mesocosms containing earthworms were slightly enriched in carbon and showed significantly higher mineralisation than soil which had not been ingested by earthworms, indicating that the carbon contained in the casts was not protected against mineralisation. Aira et al. (2008) characterized changes in the microfauna, microflora and biochemical properties of an organic substrate over a short (72 h) exposure to four densities of the epigeic earthworm *Eisinia fetida*. Carbon and nitrogen mineralisation increased with increasing earthworm density, as did microbial metabolic activity. The presence of earthworms enhanced the fungal populations and reduced the numbers of bacterivorous nematodes. The authors suggested that rapid changes in microflora and microfaunal communities and subsequent alterations in the decomposition rate of organic matter can be attributed to earthworm gut-associated processes.

The role of earthworms in soil carbon dynamics has studied in more detail in recent studies. Aneic earthworms, which inhabit a single vertical burrow for their lifespan, transport fresh organic detritus from the soil surface deep into their burrows while mixing it with mineral soil. Don et al. (2008) used measurement of enzyme activity, stable isotopes, carbon form (nuclear magnetic resonance spectroscopy) and carbon-14 ageing of burrow linings to test the hypothesis that persistent burrow structures provide space for additional carbon in soils and that the carbon turnover in the burrows should be decreased, due to physical stabilisation of organic carbon. They found that carbon stocks increased rapidly in burrow linings, but accumulated carbon could be mineralised rapidly, with turnover times of 3–5 years. Carbon stocks depended on earthworm activity within the burrow to maintain continuous carbon input. They concluded that earthworm activity does not substantially increase subsoil carbon stocks but the burrows facilitated movement of carbon into deep soil horizons.

Links between soil biota (in particular earthworms), soil aggregate dynamics, and soil organic matter decomposition and stabilisation are well-established (Six et al. 2004), although the precise mechanisms responsible are not fully understood. Earthworms are thought to enhance aggregation by stimulating formation of organo-mineral complexes through the fragmentation of organic residues and mineral material within the gut. Soil organic matter stored within aggregates often represents the vast majority of carbon within soils (Fonte et al. 2009). Aggregates are thought to protect soil organic matter physically by rendering it inaccessible to further decomposition. Differences in the binding mechanisms between aggregates of different sizes results in varying levels of aggregate, and hence soil organic matter, stability. Macroaggregates (>250 μm) are formed around fresh organic matter, using binding agents such as polysaccharides. More stable microaggregates (50–250 μm), consisting of mineral particles bound together by persistent binding agents such as complexes of clay and organic matter, are found within the macroaggregates. Microaggregates are thought to decrease the rate of soil organic matter turnover and protect carbon more effectively than macroaggregates (Fonte et al. 2009). By tracing carbon-13 labelled plant residue during formation of aggregates in soil with and without earthworms, Bossuyt et al. (2004) showed that earthworms helped to form large macroaggregates, which contained four times more stable microaggregates than those recovered from samples without earthworms. Larger amounts of organic matter were found within stable microaggregates in casts than in bulk soil after just 12 days, indicating these microaggregates are formed rapidly around freshly incorporated residues in casts.

While soil microorganisms are thought to be central to earthworm-facilitated aggregate formation, less is known about mechanisms by which earthworms drive microbial community structure and soil aggregation. As discussed in the previous section, soil organic matter decomposition rates can be strongly affected by the microbial community structure. It is now understood that bacteria are not randomly distributed throughout soil; there are variations in biomass and differential colonisation among different sizes of aggregates (Nunan et al. 2003; Schutter and Dick 2002). Recent studies have begun to examine the interactions between aggregate dynamics, differentiated microbial communities and activity and soil organic matter turnover (Noguez et al. 2008).

Differences in the distribution of bacterial species in various aggregate sizes have been revealed using molecular methods. For example, Mummey et al. (2006) used 16S rRNA gene-based terminal restriction fragment polymorphism (T-RFLP) analyses to examine bacterial communities associated with different aggregate size fractions in earthworm-worked soil relative to soil receiving only plant litter. Earthworms altered the bacterial community composition in all soil fractions analysed, with the greater changes seen in the macroaggregate than in the microaggregate communities. It appears that after earthworm ingestion, microbial substrates may rapidly become limited in microaggregates. For example, Bossuyt et al. (2005) found that activities of *Aporrectodea caliginosa* resulted in newly added plant residue carbon being protected in microaggregates, but there was almost no protection of recently added carbon in macroaggregates, where it was available for microbial decomposition and proliferation of the bacterial community.

The extent to which microbial activity and diversity vary between different soil aggregates and fractions appears to differ with soil type and management practice. For example, Kim et al. (2008) found that bacterial community structure and species composition for inner and outer-aggregate fractions of microaggregates of a desert agricultural soil were nearly identical; they suggested that much of the organic matter present was unprotected by association with clay surfaces within microaggregates. Thus, sandy desert agricultural soils with low organic matter content and high aggregate turnover are less likely to manifest distinct microbial communities in various soil fractions than is seen in soil types containing more stable aggregates.

Improved understanding of the complex linkages between soil organic matter turnover, microbial activity in soil aggregates, and soil biota is slowly developing. Progress in this area will require greater use of developing molecular techniques to characterise not just microbial diversity but also function and activity of microbial communities, including expression of genes involved in organic matter turnover. Coupled with these studies, further chemical analysis of soil organic matter fractions in micro- and macroaggregates across a range of soil types and in response in biota other than earthworms is warranted.

Impacts of Nutrient Inputs on Microbial Mineralisation

Over the last half century, human activity has dramatically altered the global nutrient cycle, and many ecosystems are undergoing increases in the input of anthropogenically derived nitrogen and phosphorus (Galloway et al. 2004; Vitousek 2004). One major consequence of increasing nutrient input into terrestrial ecosystems is associated changes in ecosystem functioning because ecosystem processes, such as decomposition of soil organic matter, likely depend on the soil nutrient concentrations (Vitousek 2004). Soils in terrestrial ecosystems are major carbon sinks. This pool of organic carbon is of particular interest because even small changes in flux rates into or out of pools of soil organic matter could have a strong influence on atmospheric carbon dioxide concentrations and associated climate change. The impacts of nutrient additions on the decomposition of soil organic matter continue to receive attention, but reports on changes in decomposition due to nutrient additions often lead to divergent conclusions, making it difficult to formulate a generalised model about the lasting effects of continued nutrient input on the decomposition of soil organic matter. Thus, the intention of this section is to synthesize present literature and to extract a possible common pattern describing the impact of nutrient additions on microbial mineralization of soil organic matter.

Microorganism-Nutrient Relationships

Microorganisms control the decomposition of soil organic matter, and it is conceivable that the biomass of fungi, bacteria and/or the microbial community composition would change following continued nutrient additions (Vitousek and Howarth 1991) (Fig. 4). For example, nitrogen additions may decrease soil pH leading to the leaching of micro-nutrients and organic carbon (Vitousek et al. 1997). Accordingly, microorganisms may become micronutrient- or carbon limited (Vitousek and Howarth 1991; Fontaine et al. 2007). Other mechanisms, which could elicit indirect effects of nutrient additions on soil microorganisms, include the decrease in soil carbon to nutrient ratios and an increase in melanoidins. The increase in melanoidins has been reported to accentuate carbon limitation of soil microorganisms (Fog 1988). Treseder (2008) synthesized responses of microbial biomass to nitrogen additions from 82 published field studies and found that microbial biomass decreased by an average of 15% under nitrogen fertilization. The extent of the decline depended on the loads and duration of nitrogen fertilisation as well as on the ecosystem itself. In addition to changes in biomass, a shift in soil microbial community composition from fungal to bacterial dominance has been observed under nitrogen addition resulting from decreased carbon-to-nitrogen ratio (Moore et al. 2003). Bacterial community composition changed in the nitrogen fertilized soil, with increases in the relative abundance of sequences related to the bacteroidetes and gemmatimonadetes, and decreases in the relative abundance of the verrucomicrobia (Nemergut et al. 2008). Decreases in abundance and diversity of soil archaea have also been reported in response to nitrogen amendments (He et al. 2007; Gattinger et al. 2007). Detailed reviews about microorganism-nitrogen relationships can also be found elsewhere (Treseder 2008).

Relative to nitrogen, the influence of phosphorus on microbial populations has been less extensively studied. In most studies, phosphorus fertilizer was applied in combination with nitrogen fertilizer or other nutrients (Cruz et al. 2009; He et al. 2008), which may obscure the phosphorus effect. In spite of this, studies imply that phosphorus additions could increase soil microbial biomass (Chu et al. 2007; Zhong and Cai 2007; Bünemann et al. 2004). Soil bacteria and actinomyces might have more evident responses to phosphorus fertilizer than fungal communities (Zhong and Cai 2007).

Mechanisms for Nutrient Effect on the Decomposition of Soil Organic Matter

Stoichiometric decomposition theory: Conceptual and analytical models developed from a long history of research suggest that decomposition is driven by the stoichiometry of substrates and microbial demands for resources, with maximal rates of decomposition observed when the supplies of carbon, nitrogen and phosphorus match the microbial demand (Hessen et al. 2004; Enríquez et al. 1993; Melillo et al. 1982). The theory predicts that microbial decomposition of soil organic carbon is constrained by the need of microorganisms to maintain their carbon to nutrient balance, and the soil will store more carbon in response to extra supply (Hessen et al. 2004; Dukes and Field 2000). Conversely, increasing nutrient availability would decrease soil carbon storage (Cleveland and Townsend 2006; Mack et al. 2004). The nutrient stimulation of decomposer activity often occurs in ecosystems where litter or substrate quality is low and carbon-to-nitrogen ratios are high (Berg and Matzner 1997). Indeed, increased carbon mineralization due to nutrient addition has been observed in a number of litterbag decomposition experiments (Conn and Day 1996; Downs et al. 1996; Prescott 1995), in the light fraction of soil organic matter (Neff et al. 2002) as well as in the whole organic matter (Cleveland and Townsend 2006; Mack et al. 2004; Bragazza et al. 2006; Paterson et al. 2008b). Although there is a general ignorance regarding phosphorus stoichiometry, in a highly weathered Hawaiian soil, where phosphorus availability limits net primary production, phosphorus additions have been reported to increase the rates of litter-mass loss (Hobbie and Vitousek 2000). It should be noted that in several situations, the decomposition of plant litter may be decoupled from nutrient additions. For example, in some tropical rain forests the decomposition or mass loss of plant litter is primarily attributed to leaching, not microbial carbon mineralization (Hobbie and Vitousek 2000). In this circumstance, the mass loss of plant litter may be more a product of litter solubility and rainfall. In addition, added nutrients may interact with aspects of litter carbon chemistry (Hobbie and Vitousek 2000). Decomposers may be unable to use additional nutrients due to low availability of labile carbon substrates (Fontaine et al. 2007). In the later stages of litter decomposition, where mass loss is dominated by lignin-degradation, other added nutrients may have a neutral or even retarding effect on decomposition (Berg and Matzner 1997).

Microbial nitrogen mining hypothesis: In sharp contrast to the basic stoichiometric decomposition theory, the microbial nitrogen mining hypothesis predicts that when mineral nitrogen becomes scarce in the soil, the population size of decomposers and the net decomposition rates of soil organic matter will increase (Fontaine and Barot 2005; Moorhead and Sinsabaugh 2006). Microbial nitrogen mining refers to the process whereby some microorganisms use energy-rich substrates (e.g., labile carbon) to actively degrade recalcitrant carbon compounds with their extracellular enzymes in order to acquire nitrogen (Moorhead and Sinsabaugh 2006). Based on the theory, soil carbon storage will increase with greater nitrogen availability as mining of recalcitrant carbon compounds for nitrogen is suppressed. Since the decomposition of recalcitrant carbon compounds yields little to no energy (Couteaux et al. 1995), the supply of fresh or labile carbon is important for various microorganisms to access recalcitrant nitrogen (Fontaine et al. 2004, 2007). This hypothesis might explain observed declines in the decomposition of soil humus (Berg and Matzner 1997; Michel and Matzner 2002) and increases in formation of refractory soil carbon (Neff et al. 2002; Resh et al. 2002; Dijkstra et al. 2004; Hagedorn et al. 2003) under conditions of high nitrogen deposition, fertilization, or fixation in terrestrial ecosystems. The increased in old and stable humus may significantly improve soil carbon storage in the long run (Hagedorn et al. 2003; Franklin et al. 2003) and decrease soil respiration in terrestrial ecosystems (Olsson et al. 2005; Billings and Ziegler 2008). However, there is no evidence for phosphorus mining as previous studies suggest that phosphorus additions would either increase short- and long-term mineralization of soil organic matter (Hobbie and Vitousek 2000; Craine et al. 2007) or have no effect (Cleveland and Townsend 2006; McGroddy et al. 2004) depending on the initial phosphorus availability in the substrate and soil.

Effects of Nutrient Addition on Decomposition of Soil Organic Matter Components

Lignin: Lignin has long been suspected to contribute substantially to the stable carbon pool in soils due to its chemical recalcitrance (Bahri et al. 2006). Studies have shown that long-term nitrogen additions would reduce decomposition rates of lignin components in soil organic matter (Fog 1988; Berg and Matzner 1997). The decreased rate of lignin decomposition by nitrogen addition could be attributed to a number of factors. Firstly, lignin degradation is carried out by a subset of the soil microbial community, which primarily consists of fungi. Nitrogen additions would cause a shift in fungal community composition. For example, decreases in the relative abundance of a well studied soil lignolytic fungus of the phylum basidiomycota have been observed in response to nitrogen amendments in both alpine tundra (Nemergut et al. 2008) and forest ecosystems (Edwards et al. 2004; Allison et al. 2007a). Basidiomycetes are key drivers of decomposition and soil carbon mineralization and responsible for the release of several enzymes that mediate lignin degradation, including laccase, manganese peroxidase and lignin peroxidase. Secondly, nitrogen fertilization can inhibit the formation of enzymes that break down lignin. Phenol oxidase, an important ligninolytic enzyme, has been shown to be greatly reduced by increased nitrogen availability during the decomposition of plant litter (Carreiro et al. 2000) and soil organic matter (Gallo et al. 2004; Saiya-Cork et al. 2002). However, it is still unclear how the nitrogen additions affect white rot fungi (*Phanerochaete chrysosporium*), the only microorganism known to produce phenol oxidase. Reduced phenol oxidase may contribute to either the suppressed production of this enzyme and/or to reduced competitiveness and hence decreased abundance of white rot fungi relative to other fungi in nitrogen-enriched plant litter and soil organic matter. Thirdly, nitrogen can react with lignin compounds in plant litter or soil organic matter to form more recalcitrant complexes. Reactions between nitrogen and phenolic compounds have been shown to occur (Haider et al. 1965; Zech and Kögel-Knabner 1994). For example, ammonium-nitrogen (Nömmik and Vahtras 1982) and amino acids (Kelley and Stevenson 1996) have been reported to be bound to aromatic ring structures. However, more recent studies have not detected such phenolic compounds in soil organic matter using nitrogen-15 nuclear magnetic resonance spectroscopy (Knicker et al. 1996; Knicker and Lüdemann 1995; Sjöberg et al. 2004).

Carbohydrates: The largest fraction of organic carbon entering the soil is that contributed by plant residues. Cellulose, the most abundant constituent of plant residues, is composed of glucose units with β(1→4) linkages. It has been reported that cellulose decay would be enhanced by nitrogen addition due to the increased cellobiases (Carreiro et al. 2000; Sjöberg et al. 2004). *Serpula lacrimans*, a cellulose decay fungus, will produce more mycelium and decay cellulose faster under nitrogen fertilization (Watkinson et al. 1981). Neff et al. (2002) determined the concentrations of two plant polysaccharide markers in soil samples (5-methyl-2-furanone and 2-hydroxy-3-methyl-2-cyclopentenone) and found that these two chemicals are substantially lower in the light fraction of fertilized plots than in that of unfertilized plots. The authors concluded that nitrogen addition led to significantly faster decomposition of polysaccharide markers in the light fraction of soil (Neff et al. 2002). Another experiment in a grassland suggested that decomposition of cellulose is much more rapid in fertilized plots, compared with unfertilized plots (Heal et al. 1981). Addition of nutrients produced no significant change in the rate of cellulose decomposition in the surface layer of soil, but at 4–8 cm, phosphorus fertilizer caused a threefold increase in decomposition rate compared with a two-fold increase caused by nitrogen addition (Heal et al. 1981).

Nitrogenous components: The majority of identifiable soil nitrogenous components occur as amino compounds, which include two main categories, the intact proteins released for various extracellular functions and detrital proteins – plant and microbial-derived residues in various stages of transformation. The protein content of soil organic matter may vary between 15% and 45% (Stevenson 1994). Recently, proteins are recognized increasingly as playing important roles in stabilization and destabilization of soil organic matter (Schindler et al. 2007). In soil ecosystems, most proteins are easily degraded although a significant amount of protein is stabilized for some time against microbial attack through interaction with other soil organic matter and/or physical protection (Rillig et al. 2007). Considerable effort has been invested in defining how increased nutrient input can influence decomposition of lignin and cellulose in litter and soil, but knowledge on decomposition of proteins is very limited. Antibus et al. (2006) investigated the decomposition processes of glomalin, a unique glycoprotein produced by arbuscular mycorrhizal fungi in three forest soils, and found a significant difference related to forest type but failed to detect significant effects of nitrogen amendment.

Lipids: Soil lipids, a complex series of 500 different fatty acids, are mostly derived from plants and microorganisms. The lipid content of soil organic matter ranges from 2% to 20% (Allison et al. 2007b). Assays of specific lipids such as ergosterol and the phospholipids are proving useful for quantitative evaluation of soil fungal biomass and for qualitative evaluation of microbial diversity (Bünemann et al. 2004; Allison et al. 2007b). In addition, characterization of the decomposition process of soil lipids often provides valuable biogeochemical information about the impact of vegetation, microorganisms, and abiotic factors on soil carbon sequestration (Almendros et al. 2001). Cutin-derived compounds, which originate from the waxy coating of leaves and roots are expected to remain stable on decadal to centennial timescales (Feng et al. 2008; Simpson et al. 2008; Otto et al. 2005).

This part of soil lipids is especially interesting because they are important components of recalcitrant alkyl carbon structures in soil organic matter and can potentially enhance carbon sequestration in the soil in the long-term (Lorenz et al. 2007; González-Pérez et al. 2008). However, like soil protein, limited studies on decomposition of soil lipids have been reported, and there is no literature on how nutrient additions affect the decomposition of soil lipids.

In summary, it has been shown that nitrogen addition can alter microbial decomposition of soil organic matter, but the direction and degree of response depend on addition rates, chemical composition of soil organic matter and supply of fresh carbon. While the responses of lipid and protein components of soil organic matter to nutrient amendment remains opaque, it is evident that in most ecosystems, long-term nitrogen addition would lead to a decline in lignin decomposition and an increase in cellulose decay. These trends, coupled with further studies on decomposition of soil lipids and proteins, may provide insight into the biogeochemical processes that create global patterns in soil organic matter storage.

Microbial Origin of Soil Organic Matter

There exists a considerable body of literature concerned with the nature of soil organic matter, its origin, composition and fate. Clearly, soil microorganisms and their interactions with soil fauna and responses to common management activities play an important role in the formation of soil organic matter. In addition, soil microorganisms can be significantly involved directly in terms of soil borne plant pathogens contributing to plant death and the subsequent inputs of organic matter of various qualities from above and belowground plant organs (see chapters elsewhere in this book). Soil microorganisms may also play a significant part in plant growth promotion either directly or indirectly through the production of plant growth promoting substances, nitrogen fixation, biological control of plant pathogens, improving soil structure, and increasing nutrient uptake (Alami et al. 2000; Barea 1997; Barea et al. 1998, 2002; Dakora and Tsavkelova et al. 2006; Vessey 2003; Postgate 1998) (see chapters elsewhere in this book).

This section will further examine the role of soil microorganisms as soil organic matter itself and examine the importance of the turnover of microbial biomass and the fate of the constituents of that same microbial biomass (Badalucco et al. 1992) together with the production of microbial metabolites that assist in the function and defence of soil microorganisms. This more passive activity represents the outcome of the normal function of soil microorganisms in contributing to carbon and nutrient cycling.

Our understanding of the carbon cycle has been dominated by several conceptual processes that have helped overshadow the role of role of microorganisms in the formation of soil organic matter. Firstly, it is commonly assumed that most organic matter enters the soil as fresh plant material (senescent or recently detached) or as dissolved organic carbon from root exudates and canopy leaching. This plant material or dissolved organic carbon then undergoes decomposition and stabilisation which can be defined as a decrease in the potential for soil organic matter loss by respiration, erosion or leaching (Sollins et al. 1996). Degradation and stabilisation of this material occurs by both biotic and abiotic processes. These processes result in the original plant material and dissolved organic carbon being decomposed to simpler organic molecules and ultimately carbon dioxide. Simultaneously, these breakdown products can be synthesised into a range of new larger molecules both within the original plant cell and outside it. Clearly at the heart of the release of carbon dioxide from the original material is the conversion of energy into microbial biomass (Sollins et al. 1996). What then happens to this microbial biomass is of some consequence for the role that soil microorganisms play in developing and increasing soil organic matter. That role is likely to be larger than previously considered and in fact soil microorganisms may be playing a controlling role in this globally important function of soil carbon storage (e.g., Simpson et al. 2007a; Kindler et al. 2009). There is an increasing body of evidence supporting the notion that soil microorganisms are important in forming and increasing soil organic matter.

Another reason that the role of soil microorganisms in soil organic matter formation has not been widely considered may be due a research focus on the fractionation of soil organic matter for the purposes of understanding its origin. There are many possible fractionation systems for soil organic matter, and some have been used to describe the nature of soil organic matter (von Lützow et al. 2007). However, relationships between operationally defined fractions and functional pools that determine the turnover of soil organic matter are not well understood.

Plant material on or beneath the soil surface is invaded and used by microorganisms and fauna, subsequently transformed by the physical and chemical environment and other organisms and translocated in the soil profile by leaching, animal movement and fungal hyphae. The fate and transformation of plant derived carbon has been the primary focus of research into soil organic matter dynamics (Kögel-Knabner 2002). This may have led to a greater focus on soil organic matter as a source for microbial growth (Kramer and Gleixner 2008) rather than on what happens to microbial biomass upon its death.

Clearly carbon inputs to soils are used by soil microorganisms, which acquire energy and nutrients from them to support growth and maintenance of metabolism. Estimates of microbial biomass size range from only 1% to 5% of total soil organic carbon that may have distracted from the importance of soil microorganisms in organic matter dynamics. As a result of predation by soil fauna, changes in soil conditions or natural death, microbially derived carbon will re-enter the soil organic matter pool. From this simple model, it can be concluded that the major sources of organic matter in soils are plants and microorganisms. Kindler et al. (2006, 2009) provide examples of how constituents of microbial cells can enter the soil organic matter pool, while the conclusions of Kelleher and Simpson (2006) support such a simple model of the origin of soil organic matter. They identified that nearly all of the nuclear magnetic resonance spectroscopy signals observed in traditional humic substances could be assigned to intact and degrading biopolymers from plants and microorganisms. This new insight should provide the starting point for any fractionation system of what will be a mixture of plant and microbially derived products reflecting the great diversity of these organisms and degrees of microbial utilization.

Along with fractionation, considerable attention has been given to researching the stabilization of organic matter in soils and the importance of this process to global carbon budgets. Recently, Ekschmitt et al. (2005) extended our understanding of the various processes involved in the stabilization of soil organic matter by suggesting that it is not only substrate quality or soil conditions that constrain decomposition rates but that the biology of the decay organisms also played an important part in determining decay rates. Clearly, soil microorganisms are not only an important source of carbon but they also have an important role in determining turnover rates, as directed by their physiological requirements and activities.

When bringing attention to microbial biomass in soils, we should not overlook that while carrying out their widely accepted role of organic matter decomposition, some microorganisms produce compounds that can fulfil a large range of roles from defence to energy storage (e.g., fatty acids, amino acids). The latter has received considerable attention due to the interest in the potential production of bioplastics (Luengo et al. 2003), but little attention has been paid to the ecological point of view of how this activity contributes to the formation of soil organic matter. Among the energy storage compounds, polyhydroxyalkanoic acids are synthesised by a wide variety of microorganisms, which use a variety of carbon sources and have been isolated from a wide range of soils. Microbial communities are diverse, so any activity to store energy, resist other microorganisms or carry out their function will result in a large mix of compounds produced which is clearly the case. Focusing on these compounds or biomarkers should provide new knowledge on the contribution of soil microorganisms to the development of soil organic matter (Badalucco et al. 1992; Kindler et al. 2006, 2009; Liang et al. 2007). The fact that a range of microbial compounds is found in soils despite their unstable nature suggests that considerable amounts are produced and subsequently protected within the soil matrix.

Considerable effort has been expended on the development and application of techniques to determine the ratio of fungi to bacteria in soil (Joergensen and Wichern 2008) so that the importance of the different functional groups under different land use and management practices can be determined. The use of certain compounds such as amino sugars or phospholipid fatty acids for this purpose has highlighted the degree to which some of these materials accumulate within soil organic matter. These observations suggest that there may be potential to manipulate and possibly increase the contribution of certain soil microbial assemblages to soil carbon sequestration (Bailey et al. 2002; Jastrow et al. 2007), which if successfully applied may add weight to the importance of soil microorganisms in controlling the formation of soil organic matter.

We should assume that the processes affecting the decomposition or stability of plant-derived soil organic matter will have similar affects on microbial derived carbon. Experiments that manipulate the conditions under which these processes operate will ultimately show if these two sources of soil organic matter are fundamentally different in their nature in terms of composition and turnover.

The simple question of what happens to the microbial biomass when it dies has been largely overlooked when considering the formation of soil organic matter. This has been described as the "eye of the needle" through which all nutrients and organic matter needs to pass. So although small in size, the turnover rate of the microbial biomass is seen as key to soil functioning because of its low carbon-to-nutrient ratio. Release of carbon and nitrogen from dead microbial cells further increases microbial activity and the outcome of the competitive interaction between plants and microorganisms determine rates of plant productivity. As indicated above, microbial cells consist of a wide range of compounds, some of which are not easily decomposed.

The incorporation of nitrogen-15 labelled compounds (Knicker et al. 1996; Knicker and Lüdemann 1995; Clinton et al. 1995) into the soil organic matter, or carbon-13 labelled glucose (Baldock et al. 1989) into structures resembling proteins and nucleic acids suggests a dominant biological function provided by the microbial community. Clinton et al. (1995) separated applied nitrogen-15 that was subsequently immobilised into humified organic matter and poorly transformed plant fragments and showed that most of the nitrogen-15 was incorporated into a pool of humified organic matter. The lack of a nitrogen-15 signal in the spectra assigned to poorly transformed plant fragments along with evidence that most of the nitrogen-15 shown to be in protein nitrogen was attributable to microbial activity, suggests that microbial material can contribute to a pool considered by some to be mainly derived from plant material. The carbon-13 spectra indicated that labelled glucose was incorporated into new compounds within microbial biomass. Longer term studies of nitrogen-15 recovery by plants (e.g., Mead et al. 2008; Mead and Preston 1994) have shown that the pool of applied nitrogen-15 immobilised in soils can be very stable suggesting that the contribution of microbial communities to the formation of soil organic may be significant and persistent.

The growth and incorporation of carbon, nitrogen and other nutrients into microbial biomass has been widely shown, so what happens to this biomass? A common assumption on the nature of soil organic matter is that it is consists of decaying plant matter. The soil biomass is commonly excluded from such definitions. However, it has been suggested that the proportion of decaying plant matter in soil organic matter that is soil microbial biomass is greater than expected. For example, using the glucosamine content of decaying wooden blocks colonised by specific wood decay fungi showed that mass loss of wood was in fact greater than indicated by the loss of mass alone because of the growth of wood decay fungi within the wood material which was included in the mass of the block (Jones and Worrall, 1995; Joergensen and Wichern 2008). Although Jones and Worrall (1995) did not demonstrate transfer of microbially derived molecules to the soil organic matter pool, the transfer of nutrients out of decaying wood blocks does suggest that ultimately microbially derived organic matter may be transferred into soils by hyphal growth (Wells et al. 1998). The significance of this observation is clearly overlooked in most organic matter decay studies although more recent studies support the transfer of microbial compounds to the soil organic matter pool (Kindler et al. 2009).

Few studies have determined quantitatively the contribution of microbially derived organic matter to total soil organic matter. As previously stated the microbial biomass is small and generally accepted as less than 5% of soil carbon. Engelking et al. (2007) recently showed that the incorporation of carbon into microbial residues can exceed the amount in the microbial biomass. Based on recent studies, between 30% and 80% of soil carbon may be derived from both live (biomass) and dead (necromass) soil microorganisms (Simpson et al. 2007a; Kindler et al. 2009; Appuhn et al. 2006). Clearly, the proportion will depend on a range of factors not yet fully studied, such as vegetation specific effects on microbial residues (Six et al. 2006; Liang et al. 2007). Mikutta et al. (2009) studied the ratio of plant-derived pentoses to microbial amino sugars in mineral organic associations in soils of different ages under the same dominant vegetation. Their results suggest that microbial growth and activity and the turnover of cells to form necromass may play an important role in the formation and control of mineral organic associations in some soils. Clearly, there is a view now with respect to the stabilisation of soil organic matter that microbially derived organic matter is important in this process. This new knowledge will challenge the importance of other sources of organic matter, such as roots, with regards to their contribution to the soil organic pool (Nierop 1998).

Recent studies confirm that we need to focus further on the factors controlling the soil microbial community (soils, vegetation and management) so that we can head towards a universally accepted model on the nature of soil organic matter formation. This in turn will allow us to manage this important pool of soil organic matter (Jastrow et al. 2007; Carletti et al. 2009; Chabbi and Rumpel 2009). For example, land use change can result in a shift from microbial-derived soil organic matter to plant derived organic matter with fungi being more sensitive to land-use change than bacteria (Jolivet et al. 2006). New methods and approaches will continue to improve our understanding of microbially derived soil organic matter (Nelson and Baldock 2005; Zhang et al. 2007; Simpson et al. 2007b; Glaser et al. 2006).

The implications of the significant role of soil microbes in soil organic matter formation are clearly aligned with emerging views on the usefulness of our current models of soil organic matter turnover. It may be timely to review the existing models and to consider more carefully other proposed models for understanding soil organic matter dynamics. In their review, Sutton and Sposito (2005) concluded that there was evidence to support a new view emerging on the nature of soil humic substances and their formation. If this view is eventually widely adopted, a strong case for the involvement of soil microorganisms in the formation of soil organic matter is verified as this group of organisms possess the physiological diversity to create the full range of smaller molecules required in the new model of soil organic matter formation.

We should examine more closely what we know about the inputs and transformations of organic matter and the processes responsible for this in the soil environment. Clearly, large amounts of organic material are added to the soil surface or incorporated into the soil profile by soil fauna or mechanical disturbance and this is well represented in a variety of models of carbon cycling that still form the conceptual basis explaining origin and fate of photosynthetically derived carbon. These models also include the important inputs from root turnover and root exudates. However, the number and complexity of models has lead to a lack of focus on the role of soil microbes in developing and increasing soil organic matter and for the time being the diversity of these models has become a research focus. It is clearly time for a major revamp of the black box approach to studying soil processes and recognise soil microbes both live and dead and their interactions with soil animals more routinely in models of soil organic matter dynamics. Increases in soil organic matter may be driven by the activities of a small microbial biomass over very long periods of time. If this was the case then a large pool of soil organic matter containing carbon of a range of ages and in a large diversity of molecules as well as fresh plant material derived from roots and aboveground plant parts would result. The challenge is to represent this complexity in the next generation of models of soil organic matter. It is thus necessary to study the factors controlling turnover of added and soil carbon in more detail with a particular focus on the role of soil microbial biomass and specific microbial assemblages and species.

References

Aira M, Sampedro L, Monroy F, Dominguez J (2008) Detritivorous earthworms directly modify the structure, thus altering the functioning of a microdecomposer food web. Soil Biol Biochem 40:2511–2516

Alami Y, Achouak W, Marol C, Heulin T (2000) Rhizosphere soil aggregation and plant growth promotion of sunflowers by an exopolysaccharide-producing *Rhizobium* sp. strain isolated from sunflower roots. Appl Environ Microbiol 66:3393–3398

Allison SD, Martiny JBH (2008) Resistance, resilience, and redundancy in microbial communities. Proc Natl Acad Sci 105:11512–11519

Allison SD, Hanson CA, Treseder KK (2007a) Nitrogen fertilization reduces diversity and alters community structure of active fungi in boreal ecosystems. Soil Biol Biochem 39:1878–1887

Allison VJ, Condron LM, Peltzer DA, Richardson SJ, Turner BL (2007b) Changes in enzyme activities and soil microbial community composition along carbon and nutrient gradients at the Franz Josef chronosequence, New Zealand. Soil Biol Biochem 39:1770–1781

Almendros G, Tinoco P, González-Vila FJ, Lüdemann HD, Sanz J, Velasco F (2001) ^{13}C-NMR of forest soil lipids. Soil Sci 166:186–196

Anderson J (1995) Soil organisms as engineers: microsite modulation of macroscale processes. In: Jones C, Lawton J (eds) Linking species and ecosystems. Chapman Hall, London, pp 94–106

Antibus RK, Lauber C, Sinsabaugh RL, Zak DR (2006) Responses of Bradford-reactive soil protein to experimental nitrogen addition in three forest communities in northern lower Michigan. Plant Soil 288:173–187

Appuhn A, Scheller E, Joergensen RG (2006) Relationships between microbial indices in roots and silt loam soils forming a gradient in soil organic matter. Soil Biol Biochem 38:2557–2564

Badalucco L, Gelsomino A, Dell'orco S, Grego S, Nannipieri P (1992) Biochemical characterization of soil organic compounds extracted by 0.5 M K_2SO_4 before and after chloroform fumigation. Soil Biol Biochem 24:569–578

Bahri H, Dignac MF, Rumpel C, Rasse DP, Chenu C, Mariotti A (2006) Lignin turnover kinetics in an agricultural soil is monomer specific. Soil Biol Biochem 38:1977–1988

Bailey VL, Smith JL, Bolton H Jr (2002) Fungal-to-bacterial ratios in soils investigated for enhanced C sequestration. Soil Biol Biochem 34:997–1007

Bais H, Weir T, Perry L, Gilroy S, Vivanco J (2006) The role of root exudates in rhizosphere interactions with plants and other organisms. Annu Rev Plant Biol 57:233–266

Baldock J (2007) Composition and cycling of organic carbon in soil. In: Marschner P, Rengel Z (eds) Nutrient cycling in terrestrial ecosystems. Springer, Berlin, Germany, pp 1–36
Baldock JA, Oades JM, Vassallo AM, Wilson MA (1989) Incorporation of uniformly labelled ^{13}C-glucose carbon into the organic fraction of a soil. Carbon balance and CP/MAS ^{13}C NMR measurements. Aust J Soil Res 27:725–746
Balser TC, Firestone MK (2005) Linking microbial community composition and soil processes in a California annual grassland and mixed-conifer forest. Biogeochemistry 73:395–415
Barea J (1997) Mycorrhiza/bacteria interactions on plant growth promotion. In: Ogoshi A, Kobayashi K, Homma Y, Kodama F, Kondo N, Akino S (eds) Plant growth-promoting rhizobacteria – present status and future prospects. Proceedings of the 4th international workshop on plant growth-promoting rhizobacteria, 5–10 October 1997, Sapporo, Japan. OECD, Paris
Barea JM, Andrade G, Bianciotto V, Dowling D, Lohrke S, Bonfante P et al (1998) Impact on arbuscular mycorrhiza formation of *Pseudomonas* strains used as inoculants for biocontrol of soil-borne fungal plant pathogens. Appl Environ Microbiol 64:2304–2307
Barea JM, Azcón R, Azcón-Aguilar C (2002) Mycorrhizosphere interactions to improve plant fitness and soil quality. Anton Leeuw Int J G 81:343–351
Bärlocher F, Corkum M (2003) Nutrient enrichment overwhelms diversity effects in leaf decomposition by stream fungi. Oikos 101:247–252
Bastian F, Bouziri L, Nicolardot B, Ranjard L (2009) Impact of wheat straw decomposition on successional patterns of soil microbial community structure. Soil Biol Biochem 41:262–275
Bastida F, Kandeler E, Moreno JL, Ros M, García C, Hernández T (2008) Application of fresh and composted organic wastes modifies structure, size and activity of soil microbial community under semiarid climate. Appl Soil Ecol 40:318–329
Bausenwein U, Gattinger A, Langer U, Embacher A, Hartmann HP, Sommer M et al (2008) Exploring soil microbial communities and soil organic matter: variability and interactions in arable soils under minimum tillage practice. Appl Soil Ecol 40:67–77
Bell T, Newman JA, Silverman BW, Turner SL, Lilley AK (2005) The contribution of species richness and composition to bacterial services. Nature 436:1157–1160
Berg B, Matzner E (1997) Effect of N deposition on decomposition of plant litter and soil organic matter in forest systems. Environ Rev 5:1–25
Billings SA, Ziegler SE (2005) Linking microbial activity and soil organic matter transformations in forest soils under elevated CO_2. Glob Change Biol 11:203–212
Billings SA, Ziegler SE (2008) Altered patterns of soil carbon substrate usage and heterotrophic respiration in a pine forest with elevated CO_2 and N fertilization. Glob Change Biol 14:1025–1036
Bossuyt H, Six J, Hendrix P (2004) Rapid incorporation of fresh residue-derived carbon into newly formed microaggregates within earthworm casts. Eur J Soil Sci 55:393–399
Bossuyt H, Six J, Hendrix P (2005) Protection of soil carbon by microaggregates within earthworm casts. Soil Biol Biochem 37:251–258
Bragazza L, Freeman C, Jones T, Rydin H, Limpens J, Fenner N et al (2006) Atmospheric nitrogen deposition promotes carbon loss from peat bogs. Proc Natl Acad Sci USA 103:19386–19389
Brant JB, Sulzman EW, Myrold DD (2006) Microbial community utilization of added carbon substrates in response to long-term carbon input manipulation. Soil Biol Biochem 38:2219–2232
Broughton LC, Gross KL (2000) Patterns of diversity in plant and soil microbial communities along a productivity gradient in a Michigan old-field. Oecologia 125:420–427
Brown G (1995) How do earthworms affect microfloral and faunal community diversity? Plant Soil 170:209–231
Brown G, Doube B (2004) Functional interactions between earthworms, microorganisms, organic matter and plants. Earthworm ecology, 2nd edn. London, CRC Press, pp 213–239
Brown G, Barois P, Lavelle P (2000) Regulation of sol organic matter dynamics and microbial activity in the drilosphere and the role of interactions with other edaphic functional domains. Eur J Soil Biol 36:177–198

Bünemann E, Condron L (2007) Phosphorus and sulphur cycling in terrestrial ecosystems. In: Marschner P, Rengel Z (eds) Nutrient cycling in terrestrial ecosystems. Springer, Berlin, Germany, pp 65–92
Bünemann EK, Bossio DA, Smithson PC, Frossard E, Oberson A (2004) Microbial community composition and substrate use in a highly weathered soil as affected by crop rotation and P fertilization. Soil Biol Biochem 36:889–901
Buyer JS, Drinkwater LE (1997) Comparison of substrate utilization assay and fatty acid analysis of soil microbial communities. J Microbiol Meth 30:3–11
Carletti P, Vendramin E, Pizzeghello D, Concheri G, Zanella A, Nardi S et al (2009) Soil humic compounds and microbial communities in six spruce forests as function of parent material, slope aspect and stand age. Plant Soil 315:47–65
Carney KM, Matson PA (2005) Plant communities, soil microorganisms, and soil carbon cycling: does altering the world belowground matter to ecosystem functioning? Ecosystems 8:928–940
Carreiro MM, Sinsabaugh RL, Repert DA, Parkhurst DF (2000) Microbial enzyme shifts explain litter decay responses to simulated nitrogen deposition. Ecology 81:2359–2365
Catovsky S, Bradford MA, Hector A (2002) Biodiversity and ecosystem productivity: implications for carbon storage. Oikos 97:443–448
Chabbi A, Rumpel C (2009) Organic matter dynamics in agro-ecosystems – The knowledge gaps: guest editors' introduction. Eur J Soil Sci 60:153–157
Chander K, Ulrich E, Jörgensen RG, Klein T (2002) Decomposition of ^{14}C labelled wheat straw in repeatedly fumigated and non-fumigated soils with different levels of heavy metal contamination. Biol Fert Soils 35:86–91
Chu H, Lin X, Fujii T, Morimoto S, Yagi K, Hu J et al (2007) Soil microbial biomass, dehydrogenase activity, bacterial community structure in response to long-term fertilizer management. Soil Biol Biochem 39:2971–2976
Cleveland CC, Townsend AR (2006) Nutrient additions to a tropical rain forest drive substantial soil carbon dioxide losses to the atmosphere. Proc Natl Acad Sci USA 103:10316–10321
Clinton PW, Newman RH, Allen RB (1995) Immobilization of ^{15}N in forest litter studies by ^{15}N CPMAS NMR spectroscopy. Eur J Soil Sci 46:551–556
Coleman D, Wall D (2007) Fauna: the engine for microbial activity and transport. In: Paul E (ed) Soil microbiology, ecology, and biochemistry. Academic Press, Amsterdam, The Netherlands, pp 163–194
Conn CE, Day FP (1996) Response of root and cotton strip decay to nitrogen amendment along a barrier island dune chronosequence. Can J Bot 74:276–284
Cookson WR, Abaye DA, Marschner P, Murphy DV, Stockdale EA, Goulding KWT (2005) The contribution of soil organic matter fractions to carbon and nitrogen mineralization and microbial community size and structure. Soil Biol Biochem 37:1726–1737
Coq S, Barthes B, Oliver R, Rabary B, Blanchart E (2007) Earthworm activity affects soil aggregation and organic matter dynamics according to the quality and localization of crop residues – An experimental study (Madagascar). Soil Biol Biochem 39:2119–2128
Couteaux MM, Bottner P, Berg B (1995) Litter decomposition climate and litter quality. Trend Ecol Evol 10:63–66
Craine JM, Morrow C, Fierer N (2007) Microbial nitrogen limitation increases decomposition. Ecology 88:2105–2113
Cruz AF, Hamel C, Hanson K, Selles F, Zentner RP (2009) Thirty-seven years of soil nitrogen and phosphorus fertility management shapes the structure and function of the soil microbial community in a Brown Chernozem. Plant Soil 315:173–184
Dakora FD, Phillips DA (2002) Root exudates as mediators of mineral acquisition in low-nutrient environments. Plant Soil 245:35–47
Dang CK, Chauvet E, Gessner MO (2005) Magnitude and variability of process rates in fungal diversity-litter decomposition relationships. Ecol Lett 8:1129–1137
Darwin C (1881) The formation of vegetable mould through the action of worms, with observations on their habit. In: Ridely M (ed) The essential Darwin. Allen and Unwin, London, pp 237–256

Deacon J (1985) Decomposition of filter paper cellulose by thermophilic fungi acting singly, in combination, and in sequence. T Brit Mycol Soc 85:663–669

Degens BP (1998) Decreases in microbial functional diversity do not result in corresponding changes in decomposition under different moisture conditions. Soil Biol Biochem 30: 1989–2000

Dijkstra FA, Hobbie SE, Knops JMH, Reich PB (2004) Nitrogen deposition and plant species interact to influence soil carbon stabilization. Ecol Lett 7:1192–1198

Don A, Steinberg B, Schöning I, Pritsch K, Joschko M, Gleixner G (2008) Organic carbon sequestration in earthworm burrows. Soil Biol Biochem 40:1803–1812

Downs MR, Nadelhoffer KJ, Melillo JM, Aber JD (1996) Immobilization of a ^{15}N-labeled nitrate addition by decomposing forest litter. Oecologia 105:141–150

Duarte S, Pascoal C, Cássio F, Bärlocher F (2006) Aquatic hyphomycete diversity and identity affect leaf litter decomposition in microcosms. Oecologia 147:658–666

Dukes JS, Field CB (2000) Diverse mechanisms for CO_2 effects on grassland litter decomposition. Glob Change Biol 6:145–154

Edwards C (2000) Soil invertebrate controls and microbial interactions in nutrient and organic matter dynamics in natural and agroecosystems. In: Coleman D, Hendrix P (eds) Invertebrates as webmasters in ecosystems. CAB International, Wallingford, UK, pp 141–159

Edwards IP, Cripliver JL, Gillespie AR, Johnsen KH, Scholler M, Turco RF (2004) Nitrogen availability alters macrofungal basidiomycete community structure in optimally fertilized loblolly pine forests. New Phytol 162:755–770

Ekschmitt K, Griffiths BS (1998) Soil biodiversity and its implications for ecosystem functioning in a heterogeneous and variable environment. Appl Soil Ecol 10:201–215

Ekschmitt K, Klein A, Pieper B, Wolters V (2001) Biodiversity and functioning of ecological communities – why is diversity important in some cases and unimportant in others? J Plant Nutr Soil Sci 164:239–246

Ekschmitt K, Liu M, Vetter S, Fox O, Wolters V (2005) Strategies used by soil biota to overcome soil organic matter stability – Why is dead organic matter left over in the soil? Geoderma 128:167–176

Engelking B, Flessa H, Joergensen RG (2007) Microbial use of maize cellulose and sugarcane sucrose monitored by changes in the $^{13}C/^{12}C$ ratio. Soil Biol Biochem 39:1888–1896

Enríquez S, Duarte CM, Sand-Jensen K (1993) Patterns in decomposition rates among photosynthetic organisms: the importance of detritus C:N:P content. Oecologia 94:457–471

Feng X, Simpson AJ, Wilson KP, Dudley Williams D, Simpson MJ (2008) Increased cuticular carbon sequestration and lignin oxidation in response to soil warming. Nature Geosci 1:836–839

Fierer N, Craine JM, McLauchlan K, Schimel JP (2005) Litter quality and the temperature sensitivity of decomposition. Ecology 86:320–326

Fog K (1988) The effect of added nitrogen on the rate of decomposition of organic matter. Biol Rev Camb Philos Soc 63:433–462

Fontaine S, Barot S (2005) Size and functional diversity of microbe populations control plant persistence and long-term soil carbon accumulation. Ecol Lett 8:1075–1087

Fontaine S, Bardoux G, Abbadie L, Mariotti A (2004) Carbon input to soil may decrease soil carbon content. Ecol Lett 7:314–320

Fontaine S, Barot S, Barré P, Bdioui N, Mary B, Rumpel C (2007) Stability of organic carbon in deep soil layers controlled by fresh carbon supply. Nature 450:277–280

Fonte S, Winsome T, Six J (2009) Earthworm populations in relation to soil organic matter dynamics and management in California tomato cropping systems. Appl Soil Ecol 41:206–214

Franklin O, Högberg P, Ekblad A, Ågren GI (2003) Pine forest floor carbon accumulation in response to N and PK additions: bomb ^{14}C modelling and respiration studies. Ecosystems 6:644–658

Frey S, Elliott E, Paustian K (1999) Bacterial and fungal abundance and biomass in conventional and no-tillage agroecosystems along two climatic gradients. Soil Biol Biochem 31:573–585

Gallo M, Amonette R, Lauber C, Sinsabaugh RL, Zak DR (2004) Microbial community structure and oxidative enzyme activity in nitrogen-amended north temperate forest soils. Microb Ecol 48:218–229

Galloway JN, Dentener FJ, Capone DG, Boyer EW, Howarth RW, Seitzinger SP et al (2004) Nitrogen cycles: past, present, and future. Biogeochemistry 70:153–226

Garcia-Pausas J, Casals P, Camarero L, Huguet C, Thompson R, Sebastià M-T et al (2008) Factors regulating carbon mineralization in the surface and subsurface soils of Pyrenean mountain grasslands. Soil Biol Biochem 40:2803–2810

Gattinger A, Höfle MG, Schloter M, Embacher A, Böhme F, Munch JC et al (2007) Traditional cattle manure application determines abundance, diversity and activity of methanogenic Archaea in arable European soil. Environ Microbiol 9:612–624

Glaser B, Millar N, Blum H (2006) Sequestration and turnover of bacterial- and fungal-derived carbon in a temperate grassland soil under long-term elevated atmospheric pCO_2. Glob Change Biol 12:1521–1531

González-Pérez JA, González-Vila FJ, González-Vázquez R, Arias ME, Rodríguez J, Knicker H (2008) Use of multiple biogeochemical parameters to monitor the recovery of soils after forest fires. Org Geochem 39:940–944

Grayston SJ, Vaughan D, Jones D (1996) Rhizosphere carbon flow in trees, in comparison with annual plants: the importance of root exudation and its impact on microbial activity and nutrient availability. Appl Soil Ecol 5:29

Griffiths BS, Ritz K, Bardgett RD, Cook R, Christensen S, Ekelund F et al (2000) Ecosystem response of pasture soil communities to fumigation-induced microbial diversity reductions; an examination of the biodiversity-ecosystem function relationship. Oikos 90:279–294

Griffiths BS, Ritz K, Wheatley R, Kuan HL, Boag B, Christensen S et al (2001) An examination of the biodiversity-ecosystem function relationship in arable soil microbial communities. Soil Biol Biochem 33:1713–1722

Gunapala N, Scow KM (1998) Dynamics of soil microbial biomass and activity in conventional and organic farming systems. Soil Biol Biochem 30:805–816

Hagedorn F, Spinnler D, Siegwolf R (2003) Increased N deposition retards mineralization of old soil organic matter. Soil Biol Biochem 35:1683–1692

Haider K, Frederick LR, Flaig W (1965) Reactions between amino acid compounds and phenols during oxidation. Plant Soil 22:49–64

Hamilton EW III, Frank DA (2001) Can plants stimulate soil microbes and their own nutrient supply? Evidence from a grazing tolerant grass. Ecology 82:2397–2402

Hättenschwiler S, Tiunov AV, Scheu S (2005) Biodiversity and litter decomposition in terrestrial ecosystems. Annu Rev Ecol Evol Syst 36:191–218

He JZ, Shen JP, Zhang LM, Zhu YG, Zheng YM, Xu MG et al (2007) Quantitative analyses of the abundance and composition of ammonia-oxidizing bacteria and ammonia-oxidizing archaea of a Chinese upland red soil under long-term fertilization practices. Environ Microbiol 9:2364–2374

He JZ, Zheng Y, Chen CR, He YQ, Zhang LM (2008) Microbial composition and diversity of an upland red soil under long-term fertilization treatments as revealed by culture-dependent and culture-independent approaches. J Soil Sediments 8:349–358

Heal OW, Flanagan PW, French DD, Maclean SF (1981) Decomposition and accumulation of organic matter in tundra. In: Bliss LC, Heal OW, Moore JJ (eds) Tundra ecosystems: a comparative analysis. Cambridge University Press, Cambridge, UK, pp 587–634

Hedlund K, Öhrn MS (2000) Tritrophic interactions in a soil community enhance decomposition rates. Oikos 88:585–591

Hendrix PF, Parmelee RW, Crossley DA Jr, Coleman DC, Odum EP, Groffman PM (1986) Detritus food webs in conventional and no-tillage agroecosystems. Bioscience 36:374–380

Hessen DO, Ågren GI, Anderson TR, Elser JJ, De Ruiter PC (2004) Carbon sequestration in ecosystems: the role of stoichiometry. Ecology 85:1179–1192

Hobbie SE, Vitousek PM (2000) Nutrient limitation of decomposition in Hawaiian forests. Ecology 81:1867–1877
Hopkins DW, Gregorich EG (2005) Carbon as a substrate for soil organisms. In: Bardgett RD, Usher MB, Hopkins DW (eds) Biological diversity and function in soils. Cambridge University Press, Cambridge, UK, pp 57–79
Howarth W (2007) Carbon cycling and formation of organic matter. In: Paul EA (ed) Soil microbiology, ecology, and biochemistry, 3rd edn. Academic Press, Amsterdam, The Netherlands, pp 303–340
Hutsch B, Augustin J, Merbach W (2002) Plant rhizodeposition – an important source for carbon turnover in soils. J Plant Nutr Soil Sci 165:397–407
Janzen HH (2004) Carbon cycling in earth systems–a soil science perspective. Agr Ecosys Environ 104:399–417
Janzen RA, Dormaar JF, McGill WB (1995) A community-level concept of controls on decomposition processes: decomposition of barley straw by *Phanerochaete chrysosporium* or *Phlebia radiata* in pure or mixed culture. Soil Biol Biochem 27:173–179
Jastrow JD, Amonette JE, Bailey VL (2007) Mechanisms controlling soil carbon turnover and their potential application for enhancing carbon sequestration. Climatic Change 80:5–23
Jiang L (2007) Negative selection effects suppress relationships between bacterial diversity and ecosystem functioning. Ecology 88:1075–1085
Jiang L, Pu Z, Nemergut DR (2008) On the importance of the negative selection effect for the relationship between biodiversity and ecosystem functioning. Oikos 117:488–493
Joergensen RG, Wichern F (2008) Quantitative assessment of the fungal contribution to microbial tissue in soil. Soil Biol Biochem 40:2977–2991
Johnson D, Booth R, Whiteley A, Bailey M, Read D, Grime J et al (2003) Plant community composition affects the biomass, activity and diversity of microorganisms in limestone grassland soil. Eur J Soil Sci 54:671–677
Johnston AE, Poulton PR, Coleman K, Donald LS (2009) Soil organic matter: its importance in sustainable agriculture and carbon dioxide fluxes. Adv Agron 101:1–57
Jolivet C, Angers DA, Chantigny MH, Andreux F, Arrouays D (2006) Carbohydrate dynamics in particle-size fractions of sandy spodosols following forest conversion to maize cropping. Soil Biol Biochem 38:2834–2842
Jones HL, Worrall JJ (1995) Fungal biomass in decayed wood. Mycologia 87:459–466
Kelleher BP, Simpson AJ (2006) Humic substances in soils: are they really chemically distinct? Environ Sci Technol 40:4605–4611
Kelley KR, Stevenson FJ (1996) Organic forms of N in soil. In: Piccolo A (ed) Humic substances in terrestrial ecosystems. Elsevier, Amsterdam, The Netherlands, pp 407–427
Kemmitt SJ, Lanyon CV, Waite IS, Wen Q, Addiscott TM, Bird NRA et al (2008) Mineralization of native soil organic matter is not regulated by the size, activity or composition of the soil microbial biomass–a new perspective. Soil Biol Biochem 40:61–73
Kennedy A (2005) Rhizosphere. In: Sylvia D, Fuhrmann J, Hartel P, Zuberer D (eds) Principles and applications of soil microbiology, 2nd edn. Pearson Prentice Hall, Upper Saddle River, NJ, pp 242–262
Kim J-S, Dungan R, Crowley D (2008) Microarray analysis of bacterial diversity and distribution in aggregates from a desert agricultural soil. Biol Fert Soils 44:1003–1011
Kindler R, Miltner A, Richnow HH, Kästner M (2006) Fate of gram-negative bacterial biomass in soil – Mineralization and contribution to SOM. Soil Biol Biochem 38:2860–2870
Kindler R, Miltner A, Thullner M, Richnow HH, Kästner M (2009) Fate of bacterial biomass derived fatty acids in soil and their contribution to soil organic matter. Org Geochem 40:29–37
Knicker H, Lüdemann HD (1995) N-15 and C-13 CPMAS and solution NMR studies of N-15 enriched plant material during 600 days of microbial degradation. Org Geochem 23:329–341
Knicker H, Almendros G, González-Vila FJ, Martin F, Lüdemann HD (1996) ^{13}C- and ^{15}N-NMR spectroscopic examination of the transformation of organic nitrogen in plant biomass during thermal treatment. Soil Biol Biochem 28:1053–1060

Kögel-Knabner I (2002) The macromolecular organic composition of plant and microbial residues as inputs to soil organic matter. Soil Biol Biochem 34:139–162
Kramer C, Gleixner G (2008) Soil organic matter in soil depth profiles: distinct carbon preferences of microbial groups during carbon transformation. Soil Biol Biochem 40:425–433
Kubartová A, Moukoumi J, Béguiristain T, Ranger J, Berthelin J (2007) Microbial diversity during cellulose decomposition in different forest stands: I. Microbial communities and environmental conditions. Microb Ecol 54:393–405
Lattaud C, Locati S, Mora P, Rouland C, Lavelle P (1998) The diversity of digestive systems in tropical geophagous earthworms. Appl Soil Ecol 9:189–195
Lattaud C, Mora P, MH G, Locati S, Rouland C (1999) Enzymatic digestive capabilities in geophagous earthworms – origin and activities of cellulolytic enzymes. Pedobiologia 43:842–850
Lavelle P, Gilot C (1994) Priming effect of macroorganisms on soil microflora: a key process of soil function? In: Ritz K, Dighton J, Giller K (eds) Beyond the biomass. Wiley, Chichester, UK, pp 173–180
Lehmann J (2007) Bio-energy in the black. Front Ecol Environ 5:381–387
Liang C, Fujinuma R, Wei L, Balser TC (2007) Tree species-specific effects on soil microbial residues in an upper Michigan old-growth forest system. Forestry 80:65–72
Loreau M, Naeem S, Inchausti P, Bengtsson J, Grime JP, Hector A et al (2001) Biodiversity and ecosystem functioning: current knowledge and future challenges. Science 294:804–808
Lorenz K, Lal R, Preston CM, Nierop KGJ (2007) Strengthening the soil organic carbon pool by increasing contributions from recalcitrant aliphatic bio(macro)molecules. Geoderma 142: 1–10
Luengo JM, García B, Sandoval A, Naharro G, Olivera ER (2003) Bioplastics from microorganisms. Curr Opin Microbiol 6:251–260
Mack MC, Schuur EAG, Bret-Harte MS, Shaver GR, Chapin FS III (2004) Ecosystem carbon storage in arctic tundra reduced by long-term nutrient fertilization. Nature 431:440–443
Marinissen J, Hillenaar S (1997) Earthworm-induced distribution of organic matter in macroaggregates from differently managed arable fields. Soil Biol Biochem 29:391–395
Marschner P, Yang CH, Lieberei R, Crowley DE (2001) Soil and plant specific effects on bacterial community composition in the rhizosphere. Soil Biol Biochem 33:1437–1445
McGrady-Steed J, Harris PM, Morin PJ (1997) Biodiversity regulates ecosystem predictability. Nature 390:162–165
McGroddy ME, Silver WL, De Oliveira Jr RC (2004) The effect of phosphorus availability on decomposition dynamics in a seasonal lowland Amazonian forest. Ecosystems 7:172–179
McGuire AD, Joyce LA, Kicklighter DW, Melillo JM, Esser G, Vorosmarty CJ (1993) Productivity response of climax temperate forests to elevated temperature and carbon dioxide: a North American comparison between two global models. Climatic Change 24:287–310
McLean MA, Parkinson D (2000) Field evidence of the effects of the epigeic earthworm *Dendrobaena octaedra* on the microfungal community in pine forest floor. Soil Biol Biochem 32:1671–1681
McNeill A, Unkovich M (2007) The nitrogen cycle in terrestrial ecosystems. In: Marschner P, Rengel Z (eds) Nutrient cycling in terrestrial ecosystems. Springer, Berlin, Germany, pp 37–64
Mead DJ, Preston CM (1994) Distribution and retranslocation of ^{15}N in lodgepole pine over eight growing seasons. Tree Physiol 14:389–402
Mead DJ, Chang SX, Preston CM (2008) Recovery of ^{15}N-urea 10 years after application to a Douglas-fir pole stand in coastal British Columbia. For Ecol Manage 256:694–701
Melillo JM, Aber JD, Muratore JF (1982) Nitrogen and lignin control of hardwood leaf litter decomposition dynamics. Ecology 63:621–626
Michel K, Matzner E (2002) Nitrogen content of forest floor Oa layers affects carbon pathways and nitrogen mineralization. Soil Biol Biochem 34:1807–1813
Mikutta R, Schaumann GE, Gildemeister D, Bonneville S, Kramer MG, Chorover J et al (2009) Biogeochemistry of mineral-organic associations across a long-term mineralogical soil gradient (0.3–4100 kyr), Hawaiian Islands. Geochim Cosmochim Acta 73:2034–2060

Moore JC, McCann K, Setälä H, De Ruiter PC (2003) Top-down is bottom-up: does predation in the rhizosphere regulate aboveground dynamics? Ecology 84:846–857
Moorhead DL, Sinsabaugh RL (2006) A theoretical model of litter decay and microbial interaction. Ecol Monogr 76:151–174
Mummey D, Matthias C, Six J (2006) Endogeic earthworms differentially influence bacterial communities associated with different soil aggregate size fractions. Soil Biol Biochem 38:1608–1614
Nannipieri P, Ascher J, Ceccherini MT, Landi L, Pietramellara G, Renella G (2003) Microbial diversity and functions. Eur J Soil Sci 54:655–670
Neff JC, Townsend AR, Gleixner G, Lehman SJ, Turnbull J, Bowman WD (2002) Variable effects of nitrogen additions on the stability and turnover of soil carbon. Nature 419:915–917
Nelson PN, Baldock JA (2005) Estimating the molecular composition of a diverse range of natural organic materials from solid-state ^{13}C NMR and elemental analyses. Biogeochemistry 72:1–34
Nemergut DR, Townsend AR, Sattin SR, Freeman KR, Fierer N, Neff JC et al (2008) The effects of chronic nitrogen fertilization on alpine tundra soil microbial communities: implications for carbon and nitrogen cycling. Environ Microbiol 10:3093–3105
Neumann G (2007) Root exudates and nutrient cycling. In: Marschner P, Rengel Z (eds) Nutrient cycling in terrestrial ecosystems. Springer, Berlin, Germany, pp 123–157
Nierop KGJ (1998) Origin of aliphatic compounds in a forest soil. Org Geochem 29:1009–1016
Noguez A, Escalante A, Forney LJ, Nava-Medoza M, Rosas I, Souza V et al (2008) Soil aggregates in a tropical deciduous forest: effects on C and N dynamics, and microbial communities as determined by t-RFLPs. Biogeochemistry 89:209–220
Nömmik H, Vahtras K (1982) Retention and fixation of ammonium and ammonia in soils. In: Stevenson FJ (ed) Nitrogen in agricultural soils. Agronomy Society of America, Madison, WI, pp 123–161
Nunan N, Wu K, Young I, Crawford I, Ritz K (2003) Spatial distribution of bacterial communities and their relationships with the micro-architecture of soil. FEMS Microbiol Ecol 44:203–215
Olsson P, Linder S, Giesler R, Högberg P (2005) Fertilization of boreal forest reduces both autotrophic and heterotrophic soil respiration. Glob Change Biol 11:1745–1753
Orwin KH, Wardle DA, Greenfield LG (2006) Ecological consequences of carbon substrate identity and diversity in a laboratory study. Ecology 87:580–593
Otto A, Shunthirasingham C, Simpson MJ (2005) A comparison of plant and microbial biomarkers in grassland soils from the Prairie Ecozone of Canada. Org Geochem 36:425–448
Parton W, Schimel D, Cole C, Ojima D (1987) Analysis of factors controlling soil organic-matter levels in great-plains grasslands. Soil Sci Soc Am J 51:1173–1179
Paterson E, Gebbing T, Abel C, Sim A, Telfer G (2007) Rhizodeposition shapes rhizosphere microbial community structure in organic soil. New Phytol 173:600–610
Paterson E, Osler G, Dawson LA, Gebbing T, Sim A, Ord B (2008a) Labile and recalcitrant plant fractions are utilised by distinct microbial communities in soil: independent of the presence of roots and mycorrhizal fungi. Soil Biol Biochem 40:1103–1113
Paterson E, Thornton B, Midwood AJ, Osborne SM, Sim A, Millard P (2008b) Atmospheric CO_2 enrichment and nutrient additions to planted soil increase mineralisation of soil organic matter, but do not alter microbial utilisation of plant- and soil C-sources. Soil Biol Biochem 40:2434–2440
Plassart P, Akpa Vinceslas M, Gangneux C, Mercier A, Barray S, Laval K (2008) Molecular and functional responses of soil microbial communities under grassland restoration. Agr Ecosys Environ 127:286–293
Postgate J (1998) Nitrogen fixation, 3rd edn. Cambridge University Press, Cambridge, UK
Powell J, Klironomos J (2007) The ecology of plant-microbial mutualisms. In: Paul EA (ed) Soil microbiology, ecology, and biochemistry, 3rd edn. Academic Press, Amsterdam, The Netherlands, pp 257–282
Prescott CE (1995) Does nitrogen availability control rates of litter decomposition in forests? Plant Soil 168–169:83–88

Reed HE, Martiny JBH (2007) Testing the functional significance of microbial composition in natural communities. FEMS Microbiol Ecol 62:161–170

Resh SC, Binkley D, Parrotta JA (2002) Greater soil carbon sequestration under nitrogen-fixing trees compared with Eucalyptus species. Ecosystems 5:217–231

Rillig MC (2004) Arbuscular mycorrhizae and terrestrial ecosystem processes. Ecol Lett 7:740–754

Rillig MC, Caldwell BA, Wösten HAB, Sollins P (2007) Role of proteins in soil carbon and nitrogen storage: controls on persistence. Biogeochemistry 85:25–44

Robinson C, Dighton J, Frankland J, Coward P (1993) Nutrient and carbon dioxide release by interacting species of straw-decomposing fungi. Plant Soil 151:139–142

Ros M, Pascual JA, Garcia C, Hernandez MT, Insam H (2006) Hydrolase activities, microbial biomass and bacterial community in a soil after long-term amendment with different composts. Soil Biol Biochem 38:3443–3452

Saiya-Cork KR, Sinsabaugh RL, Zak DR (2002) The effects of long term nitrogen deposition on extracellular enzyme activity in an *Acer saccharum* forest soil. Soil Biol Biochem 34: 1309–1315

Sampedro L, Dominguez J (2008) Stable isotope natural abundance (d13C and d15N) of the earthworm *Eisenia fetida* and other soil fauna living in two different vermicomposting environments. Appl Soil Ecol 38:91–99

Sampedro L, Whalon J (2007) Changes in fatty acid profiles through the digestive tract of the earthworm *Lumbricus terrestris* L. Appl Soil Ecol 35:226–236

Schimel JP, Gulledge J (1998) Microbial community structure and global trace gases. Glob Change Biol 4:745–758

Schindler FV, Mercer EJ, Rice JA (2007) Chemical characteristics of glomalin-related soil protein (GRSP) extracted from soils of varying organic matter content. Soil Biol Biochem 39:320–329

Schulze E-D, Mooney HA (1993) Biodiversity and ecosystem function. Springer, Berlin, Germany

Schutter M, Dick R (2002) Microbial community profiles and activities among aggregates of winter fallow and cover-cropped soil. Soil Sci Soc Am J 66:142–153

Schwartz E, Adair KL, Schuur EA (2007) Bacterial community structure correlates with decomposition parameters along a Hawaiian precipitation gradient. Soil Biol Biochem 39:2164–2167

Scow KM (1997) Interrelationships between soil microbial communities and carbon flow in agroecosystems. In: Jackson LE (ed) Ecology in agriculture. Academic Press, New York, pp 367–416

Setälä H, McLean M (2004) Decomposition rate of organic substrates in relation to the species diversity of soil saprophytic fungi. Oecologia 139:98–107

Shaw C, Pawluk S (1986) Faecal microbiology of *Octolasion tyrtaeum*, *Aporrectodea turgida* and *Lumbricus terrestris* and its relation to the carbon budgets of three artificial soils. Pedobiologia 29:377–389

Simpson AJ, Simpson MJ, Smith E, Kelleher BP (2007a) Microbially derived inputs to soil organic matter: are current estimates too low? Environ Sci Technol 41:8070–8076

Simpson AJ, Song G, Smith E, Lam B, Novotny EH, Hayes MHB (2007b) Unraveling the structural components of soil humin by use of solution-state Nuclear Magnetic Resonance spectroscopy. Environ Sci Technol 41:876–883

Simpson MJ, Otto A, Feng X (2008) Comparison of solid-state carbon-13 nuclear magnetic resonance and organic matter biomarkers for assessing soil organic matter degradation. Soil Sci Soc Am J 72:268–276

Singh G, Mukerji K (2006) Root exudates as determinant of rhizospheric microbial diversity. In: Mukerji K, Manoharachary C, Singh J (eds) Microbial activity in the rhizosphere. Springer, Berlin, Germany, pp 39–53

Six J, Elliott E, Paustian K (2000) Soil macroaggregate turnover and microaggregate formation: a mechanism for C sequestration under no-tillage agriculture. Soil Biol Biochem 32:2099–2103

Six J, Conant R, Paul EA, Paustian K (2002) Stabilization mechanisms for soil organic matter: implications for C-saturation of soils. Plant Soil 241:155–176
Six J, Bossuyt H, Degryze S, Denef K (2004) A history of research on the link between (micro) aggregates soil biota, and soil organic matter dynamics. Soil Till Res 79:7–31
Six J, Frey SD, Thiet RK, Batten KM (2006) Bacterial and fungal contributions to carbon sequestration in agroecosystems. Soil Sci Soc Am J 70:555–569
Sjöberg G, Knicker H, Nilsson SI, Berggren D (2004) Impact of long-term N fertilization on the structural composition of spruce litter and mor humus. Soil Biol Biochem 36:609–618
Smith JL, Paul EA (1990) The significance of soil microbial biomass estimations. In: Bollag J-M, Stotzky G (eds) Soil biochemistry. Marcel Dekker, New York, pp 357–393
Sollins P, Homann P, Caldwell BA (1996) Stabilization and destabilization of soil organic matter: mechanisms and controls. Geoderma 74:65–105
Stevenson F (1985) Geochemistry of soil humic substances. In: Aiken G, McKnight D, Wershaw R, MacCarthy P (eds) Humic substances in soil, sediment, and water. Wiley, New York, pp 13–52
Stevenson F (1994) Humus chemistry: genesis, composition, reactions. Wiley, New York
Sutton R, Sposito G (2005) Molecular structure in soil humic substances: the new view. Environ Sci Technol 39:9009–9015
Swift MJ, Heal OW, Anderson JM (1979) Decomposition in terrestrial ecosystems. Studies in ecology. Blackwell Scientific, Oxford, UK
Sylvia D (2005) Mycorrhizal symbioses. In: Sylvia D, Fuhrmann J, Hartel P, Zuberer D (eds) Principles and applications of soil microbiology, 2nd edn. Pearson Prentice Hall, Upper Saddle River, NJ, pp 263–282
Timonen S, Marschner P (2006) Mycorrhizosphere concept. In: Mukerji K, Manoharachary C, Singh J (eds) Microbial activity in the rhizosphere. Springer, Berlin, Germany, pp 155–172
Tiunov AV, Scheu S (2005) Facilitative interactions rather than resource partitioning drive diversity-functioning relationships in laboratory fungal communities. Ecol Lett 8:618–625
Tiwari S, Mishra R (1993) Fungal abundance and diversity in earthworm casts and in uningested soil. Biol Fert Soils 16:131–134
Toljander YK, Lindahl BD, Holmer L, Högberg NOS (2006) Environmental fluctuations facilitate species co-existence and increase decomposition in communities of wood decay fungi. Oecologia 148:625–631
Treseder KK (2008) Nitrogen additions and microbial biomass: a meta-analysis of ecosystem studies. Ecol Lett 11:1111–1120
Treton C, Chauvet E, Charcosset JY (2004) Competitive interaction between two aquatic hyphomycete species and increase in leaf litter breakdown. Microb Ecol 48:439–446
Tsavkelova EA, Klimova SY, Cherdyntseva TA, Netrusov AI (2006) Microbial producers of plant growth stimulators and their practical use: a review. Appl Biochem Microbiol 42:117–126
Vessey JK (2003) Plant growth promoting rhizobacteria as biofertilizers. Plant Soil 255:571–586
Vitousek PM (2004) Nutrient cycling and limitation: Hawai'i as a model system. Princeton University Press, Princeton, NJ
Vitousek PM, Howarth RW (1991) Nitrogen limitation on land and in the sea: how can it occur? Biogeochemistry 13:87–115
Vitousek PM, Aber JD, Howarth RW, Likens GE, Matson PA, Schindler DW et al (1997) Human alteration of the global nitrogen cycle: sources and consequences. Ecol Appl 7:737–750
von Lützow M, Kögel-Knabner I, Ekschmitt K, Flessa H, Guggenberger G, Matzner E et al (2007) SOM fractionation methods: relevance to functional pools and to stabilization mechanisms. Soil Biol Biochem 39:2183–2207
Waldrop MP, Balser TC, Firestone MK (2000) Linking microbial community composition to function in a tropical soil. Soil Biol Biochem 32:1837–1846
Wardle DA (2002) Communities and ecosystems – linking the aboveground and belowground components. Princeton University Press, Princeton, NJ
Wardle DA (2005) How plant communities influence decomposer communities. In: Bardgett RD, Usher MB, Hopkins DW (eds) Biological diversity and function in soils. Cambridge University Press, Cambridge, UK, pp 119–138

Wardle DA, Nilsson M-C, Zackrisson O (2008) Fire-derived charcoal causes loss of forest humus. Science 320:629

Watkinson SC, Davison EM, Bramah J (1981) The effect of nitrogen availability on growth and cellulolysis by *Serpula lacrimans*. New Phytol 89:295–305

Webster EA, Chudek JA, Hopkins DW (2000) Carbon transformations during decomposition of different components of plant leaves in soil. Soil Biol Biochem 32:301–314

Wells JM, Boddy L, Donnelly DP (1998) Wood decay and phosphorus translocation by the cord-forming basidiomycete *Phanerochaete velutina*: the significance of local nutrient supply. New Phytol 138:607–617

Wertz S, Degrange V, Prosser JI, Poly F, Commeaux C, Freitag T et al (2006) Maintenance of soil functioning following erosion of microbial diversity. Environ Microbiol 8:2162–2169

White P, Rice C (2009) Tillage effects on microbial and carbon dynamics during plant residue decomposition. Soil Sci Soc Am J 73:138–145

Wohl DL, Arora S, Gladstone JR (2004) Functional redundancy supports biodiversity and ecosystem function in a closed and constant environment. Ecology 85:1534–1540

Wolf D, Wagner G (2005) Carbon transformations and soil organic matter formation. In: Sylvia D, Fuhrmann J, Hartel P, Zuberer D (eds) Principles and applications of soil microbiology. Pearson Prentice Hall, Upper Saddle River, NJ, pp 285–332

Wolter C, Scheu S (1999) Changes in bacterial numbers and hyphal lengths during the gut passage through *Lumbricus terrestris* (Lumbricidae, Oligochaeta). Pedobiologia 43:891–900

Zech W, Kögel-Knabner I (1994) Patterns and regulation of organic matter transformation in soils: litter decomposition and humification. In: Schulze E-D (ed) Flux control in biological systems: from enzymes to populations and ecosystems. Academic Press, San Diego, CA, pp 35–73

Zhang X, He H, Amelung W (2007) A GC/MS method for the assessment of ^{15}N and ^{13}C incorporation into soil amino acid enantiomers. Soil Biol Biochem 39:2785–2796

Zhong WH, Cai ZC (2007) Long-term effects of inorganic fertilizers on microbial biomass and community functional diversity in a paddy soil derived from quaternary red clay. Appl Soil Ecol 36:84–91

Chapter 5
Intimate Associations of Beneficial Soil Microbes with Host Plants

Oksana Y. Shtark, Alexey Y. Borisov, Vladimir A. Zhukov, Nikolay A. Provorov, and Igor A. Tikhonovich

Introduction

Symbiosis, or "coexistence of diverse organisms", is the term which was proposed firstly by Anton de Bary (1879). Roughly, there are three types of symbiosis: commensalism (an interaction which is beneficial for one symbiotic partner but neutral for the other), antagonism (or parasitism, where one symbiont develops to the detriment of another – see elsewhere in this Book) and mutualism (where both symbionts benefit from mutual coexistence). Mutualism is a special ecological and evolutionary strategy for two (or more) dissimilar organisms to share (allocate) all of their physiological

O.Y. Shtark (✉)
All Russia Research Institute for Agricultural Microbiology (ARRIAM),
Laboratory of Genetics of Plant-Microbe Interactions, Department of Biotechnology,
Podbelsky chausee 3, St. Petersburg, Pushkin 8, 196608, Russia
e-mail: oshtark@yandex.ru

A.Y. Borisov
All Russia Research Institute for Agricultural Microbiology (ARRIAM),
Laboratory of Genetics of Plant-Microbe Interactions, Dept of Biotechnology,
Podbelsky chausee 3, St. Petersburg, Pushkin 8, 196608, Russia
e-mail: ayborisov@yandex.ru

V.A. Zhukov
All Russia Research Institute for Agricultural Microbiology (ARRIAM),
Laboratory of Genetics of Plant-Microbe Interactions, Department of Biotechnology.
Podbelsky chausee 3, St. Petersburg, Pushkin 8, 196608, Russia
e-mail: vladimir.zhukoff@gmail.com

N.A. Provorov
All Russia Research Institute for Agricultural Microbiology (ARRIAM),
Podbelsky chausee 3, St. Petersburg, Pushkin 8, 196608, Russia
e-mail: provorov@newmail.ru

I.A. Tikhonovich
All Russia Research Institute for Agricultural Microbiology (ARRIAM),
Podbelsky chausee 3, St. Petersburg, Pushkin 8, 196608, Russia
e-mail: contact@arriam.spb.ru

G.R. Dixon and E.L. Tilston (eds.), *Soil Microbiology and Sustainable Crop Production*,
DOI 10.1007/978-90-481-9479-7_5,

functions between them rather than them being separated. This is distinguished principally from individual adaptations which are dependent on morpho-physiological and behavioural reactions of one free-living organism. Here there is an evolutionary opportunity to develop certain functions and possibly lose others leading to genetic and metabolic integration according to a dialectical law "negation of negation" which may be applicable to evolution. As a result some intimate genetic and metabolic cooperation between the symbiotic partners of complicated super-organismic systems develop resulting in new adaptations (functions) with their environments. In mutualistic (mutually beneficial) symbioses between plants and microbes, novel types of cells and tissues (physiologically and/or structurally) and even organs have often evolved (Douglas 1994; Tikhonovich and Provorov 2007).

In the process of evolution, symbiosis plays a key role in developing highly organized lifestyles and possibly new types of ecological interactions (Douglas 1994; Margulis 1998; Margulis and Sagan 2002). The eukaryotic cell in particular, is considered to be an evolutionary product of beneficial symbioses with different prokaryotes (Taylor 1974). In accordance with this theory, cell organelles such as mitochondria and photosynthesising plastids originated from free-living bacteria which turned to a symbiotic life style and lost their capabilities for autonomous existence (Margulis 1970; Margulis 1988; Dyall and Johnson 2000; Margulis and Sagan 2002; Embley et al. 2003). Lichens for example are the products of beneficial symbiotic interactions between algae and/or cyanobacteria with fungi. Moreover, it has been shown that in the geological time in which plants achieved a terrestrial existence symbiotic associations with arbuscular mycorrhizal fungi also developed (Pirozynski and Dalpé 1989; Remy et al. 1994; Redecker et al. 2000).

According to contemporary nomenclature, symbioses may be categorized in accordance with a range of indicators (Lewis 1985; Provorov et al. 2008). According to the level of dependence of the partners on each other, symbioses can be generally considered as either obligatory or facultative in terms of their ecology and genetics. The concept of "ecological obligation" means that organisms are only able to occupy new ecological niches in the symbiotic state (Smith and Read 1997), while "genetic obligation" indicates a state where autonomic existence of the symbionts is impossible because they have lost genes encoding for products with essential functions (Provorov 2005). If one of the partners penetrates into the other, this is endosymbiosis, which can be intra-cavernous, intratissular or intracellular; whereas ectosymbiosis is coexistence without deep penetration of one partner into the other. Trophic interactions between partners also show a range of dependencies from: biotrophy (where the nutrient source is a living partner's cells) through to necrotrophy (where the nutrient source is a partner's dead cells).

Symbioses of microbes with plants (as well as animals) are quite widespread in nature and enable the higher organisms to occupy particular ecological niches (Douglas 1994). In the case of facultative or ecologically obligatory interactions, the microsymbiont through intimate association with the host achieves ecological advantages compared with free-living forms. In particular the microsymbiont benefits from priority utilization of host metabolites or protection from predators. There are however, concomitant disadvantages since genetically obligatory microsymbionts are not able to survive outside the symbiotic system wherein their reproduction is strictly controlled by the host (Provorov et al. 2008). With regard to the host plant, mutually beneficial plant–microbe symbioses are generally facultative or ecologically obligatory and usually subdivided according to nutritional (microbes providing the host with nutrients, in particular with nitrogen or phosphorus) or defensive (microbes supply the host with the resistance to pathogens) benefit (Vance 2001; Lum and Hirsch 2003; Provorov 2009). Usually, in natural environments both types of interactions are available simultaneously for the same plant due to presence of different types of microsymbionts.

The taxonomy, morphology and function of mutually beneficial plant–microbe symbioses is diverse; they include intracellular, intratissular (endophytic) ones or associative ectosymbiosis (epiphytic) forms. Nevertheless, despite superficial differences between the intracellular and intratissular forms of endophytic symbiosis and associative ectosymbiosis or epiphytic forms, all beneficial plant–microbe symbioses are based on similar genetic, cellular and molecular mechanisms. Moreover, they have a series of traits in common with plant–pathogenic systems and represent an evolutionary plant–microbe continuum (Jones et al. 2007; Provorov 2009).

This chapter focuses on the ecologically and agriculturally most important beneficial plant–microbe endosymbiotic systems, such as arbuscular mycorrhiza (AM) and root-nodule (RN) symbiosis, or root nodules. These are characterized by the formation of special complex symbiotic compartments, as well as by interactions with plant growth-promoting rhizosphere bacteria (PGPR) and/or beneficial endophytic types, where from the outset of interaction special symbiotic structures are not formed (Fig. 1). These groups of beneficial microbes improve host mineral nutrition, acquisition of water, promote plant development and offer protection from pathogens and pests.

The taxonomy of the partners, the processes of the formation of symbioses and their functions will be reviewed. Development of symbiotic root nodules will be described using pea (*Pisum sativum* L.) as an example of an evolutionary advanced model of nodulation. Special attention will be given to developmental genetics, evolution of the symbioses and their inter-relationships, and then, using this perspective the prospects for improving effectiveness of beneficial plant–microbe systems in sustainable agriculture will be discussed.

Mycorrhizas

The Main Types of Mycorrhizas

Mutually beneficial symbioses of plant roots with fungi are referred to as mycorrhizas (from the Greek 'mycos', meaning fungus and 'rhiza', meaning root), this term was coined by Frank in 1885 (Frank 1885). Several forms of mycorrhiza are recognized: endomycorrhiza, including arbuscular, orchid and ericoid mycorrhiza, when part of the fungal hyphae penetrates into the plant cell, and ectomycorrhiza, when the fungal mycelium remains outside plant cells. As arbuscular mycorrhiza is

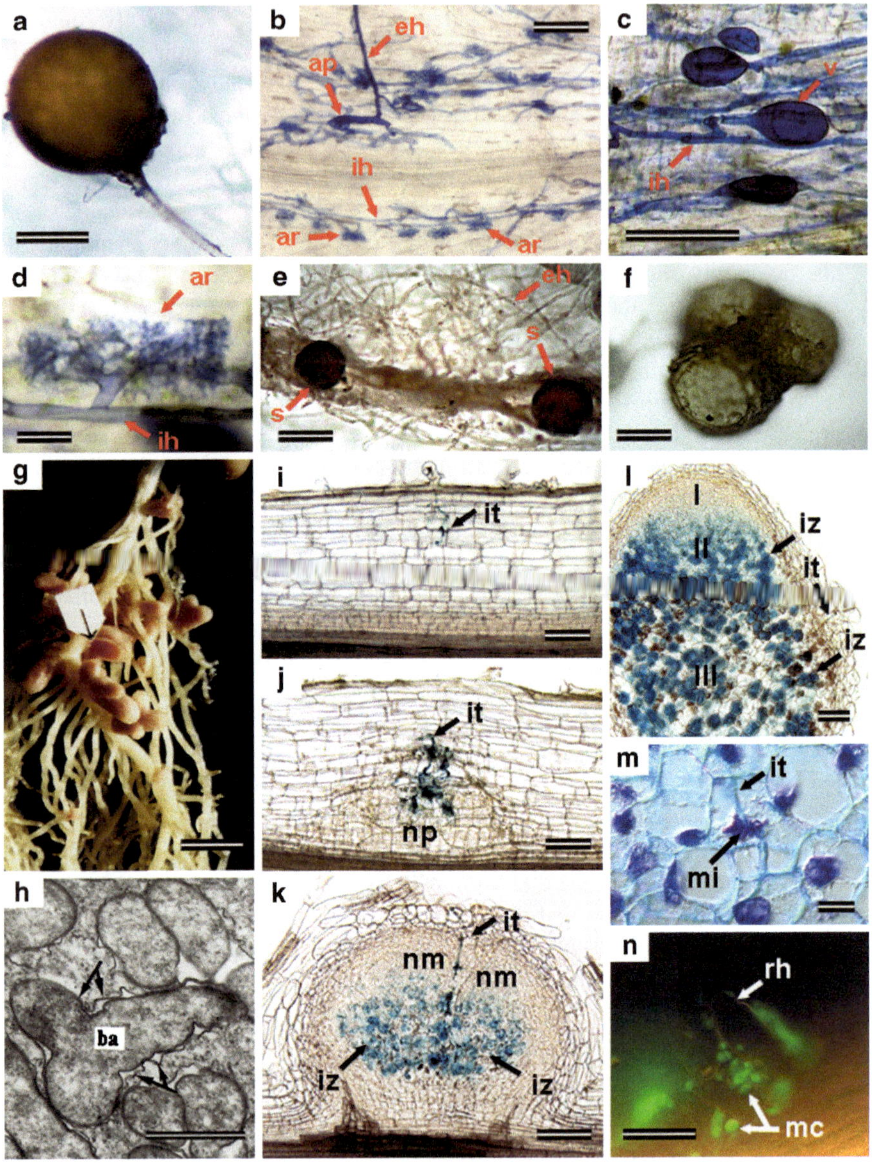

Fig. 1 Mutually beneficial plant-microbe symbioses. **a–f** – Arbuscular mycorrhiza (AM), **g–m** – Root-nodule (RN) symbiosis, **n** – Symbiotic association of root with plant growth-promoting rhizobacteria (PGPR). **a, e, f** – Spores of *Glomus* sp. **a** – Extraradical spore, **e** – Intraradical spore, **f** – Sporocarp. **b–d** – Light microscopy of macerated root fragments stained with black ink according to Vierheilig et al., 1998. **b, d** – Roots of *P. sativum* colonized by *Glomus intraradices*. **c** – Root of chives (*Allium schoenoprasum* L.) colonized by *G. fasciculatum*. eh – Extraradical hypha, ap – Appressorium, ih – Intraradical hypha, ar – Arbuscule, v – Vesicle, s – Spore. **g** – Root nodules in pea (*Pisum sativum*). A nodule is indicated by an arrowhead. **i–m** – Indeterminate symbiotic root nodule development in *P. sativum* inoculated with a derivative of *Rhizobium leguminosarum* bv. *viciae* strain containing a Tn*5*-*gusA* insertion constitutively expressing *gusA*

probably the most widespread terrestrial symbiosis (Fitter 2005) and the most important form of mycorrhizas in agriculture it will be described detail. Other mycorrhizas will be treated in less depth.

Arbuscular Mycorrhiza (AM)

Arbuscular Mycorrhiza (AM) is an ancient mutually beneficial plant–fungal symbiotic system formed by at least 80% of terrestrial plants, including the majority of angiosperms and gymnosperms. This symbiotic association has existed for 400–500 millions years, and was considered to play a decisive role in plants achieving a terrestrial existence (Pirozynski and Dalpé 1989; Remy et al. 1994; Redecker et al. 2000). The evolutionary age, life-style and genetic structure of AM fungi is unusual and because they have existed in a morphologically unaltered state they could therefore possibly qualify as living fossils (Parniske 2008). The taxonomy of fungi able to form AM has been revised several times (reviewed in Koide and Mosse 2004). Latterly, after analysis of 18S rRNA genes sequences the AM fungi have been placed in a new phylum Glomeromycota (Schüßler et al. 2001) which replaced the zygomycete order Glomales. The new phylum contains the genera *Glomus*, *Acaulospora*, *Scutellospora*, *Gigaspora*, *Paraglomus* and *Archaeospora* together with *Geosiphon* (which is non-mycorrhizal but is macrosymbiont in an association with photosynthetic nitrogen-fixing cyanobacteria *Nostoc*). The Glomeromycota are unique as the only monophyletic mycorrhizal fungus lineage that has co-evolved with land plants throughout their history (Brundrett 2002). This is why AM fungi are considered as obligate biotrophs that colonize plant roots obtaining photosynthates, such as carbohydrates (hexoses), and niches for both their growth and reproduction (Smith and Read 1997; Bago and Becard 2002). Although it should be noted that according to recent studies, some AM fungi can be cultivated *in vitro*, in the absence of plant, in a dixenic system with the bacterium *Paenibacillus validus* (Hildebrandt et al. 2006).

The hyphae of AM fungi are usually aseptate and coenocytic, with hundreds of nuclei located in the cytoplasm (Smith and Read 1997). Spores of AM fungi are

Fig. 1 (continued) (from Voroshilova et al., 2009, modified). **i–l** – Longitudinal vibratome sections of roots and nodules, with GUS staining of bacteria in infection threads (ITs) and colonized nodule cells. **m** – Cross-section of nodule with Feulgen/alcian blue staining. **i** – Transcellular IT growth in the outer root cortex and resumption of cell proliferation in the inner cortex cells 5 days after inoculation (DAI). **j** – Nodule primordium at 9 DAI with proliferating cells harbouring ITs. **k** – An emerging juvenile nodule at 12 DAI. **l** – A nodule at 28 DAI showing normal histological organization. (I), nodule meristem; (II), colonization zone; (III), nitrogen fixation zone. The 'IT entrance' to the nodule usually appeared at an epicentric position. **m** – Mitosis in the nodule meristem cell that forms transcellular ITs. Neighbouring cells have the preinfection structures. it – infection thread (IT), iz – infection zone, np – nodule primordium; nm – nodule meristem, mi – dividing plant cells (mitoses). **h** – Transmission electronic micrograph of ultrathin section of *P. sativum* root nodule demonstrating the symbiosome, or bacteroid (ba). **n** – Fluorescent microscopy of wheat (*Triticum monococcum* L.) root colonized by *Pseudomonas chlororaphis* strain carrying a constitutive reporter fusion with Green Fluorescent Protein (GFP) gene. rh – root hair, mc – micro-colony of bacteria. Bars: **g** – 5 mm; **e, k, l** – 100 μm; **a-c, f, i, j, n** – 50 μm; **d, m** – 10 μm; **h** – 2 μm

relatively large in size (up to 400 μm) and are the main reserve of genetic material and carbon sources for subsequent generation of mycelium (Smith and Read 1997). Individual spores contain numerous nuclei and this raises the question of how the different polymorphic DNA-sequence variants that are present within a single cell are distributed between genomes or nuclei. Not surprisingly, this question is subject to continuing debate (Smith and Read 1997; Pawlowska and Taylor 2004; Hijri and Sanders 2005; Rosendahl 2008). Although there is no confirmed report of a sexual stage in the life-cycle of AM fungi, it is possible that genetic material is exchanged and recombined through mechanisms such as anastomosis between hyphae (Giovannetti et al. 2004; De la Providencia et al. 2005). Anastomosis allows the exchange of nuclei (De la Providencia et al. 2005), but has so far only been observed between hyphae of closely related fungal isolates. It would be interesting to determine the level of relatedness that is required for these fusions to occur. Molecular evidence for recombination in AM fungi has been controversial (Vandenkoornhuyse et al. 2001; Gandolfi et al. 2003; Rosendahl 2008). As an important step towards the genetic manipulation of these fungi, transient transformation by particle bombardment has been achieved (Helber and Requena 2008).

Highly differentiated intracellular structures called arbuscules are formed on colonization of roots by AM fungi (Smith and Read 1997; Harrison 2005). In addition to growth within the root cortex, they develop an extensive extraradical mycelium in the surrounding soil. Different AM fungi have different hyphal growth patterns, anastomoses and branching frequencies. Such differences probably reflect varying strategies and the occupation of several niches within the soil. For example, many *Glomus* species form highly branched and anastomosing hyphal networks. These networks are more resistant to soil disturbance than the mycelia of species of *Scutellospora* or *Gigaspora*, which form longer hyphae and can, probably, explore soil at greater distance from the root (De la Providencia et al. 2005; Voets et al. 2006). Most of the fungal biomass in members of the Gigasporaceae family is found in the hyphae that are located outside the plant root, whereas in members of the Glomeraceae family, most of the hyphal biomass is inside the root (Maherali and Klironomos 2007).

The intraradical and extraradical growth phases are a single continuum via which the fungus is able to translocate mineral nutrients, primarily phosphates and water from the soil to the interior of root systems; thus, by living in association with AM fungi the host plant becomes capable of existing in nutrient-poor and drought-affected soils (Ames et al. 1983; Jakobsen 1995; Ruiz-Lozano and Azcon 1996; Harrison 1997; Smith et al. 2001; Liu et al. 2002; Quilambo 2003). This is why AM can be considered as nutritional plant–microbe symbioses. In addition, well developed AM-symbiosis improves host resistance against biotic for example phytopathogenic attack (Cordier et al. 1998; Dassi et al. 1998; Hause et al. 2002; Fritz et al. 2006; Liu et al. 2007), and abiotic stresses, in particular heavy metal contamination of soils (Turnau et al. 2007). An important function of extraradical mycelium in ecosystems is the formation of a common 'hyphal web', linking different plants (Molina et al. 1992). Also extraradical mycelium has an active role in soil structure formation (Celik et al. 2004; Rillig 2004). The AM fungi can be involved in bioremediation including phytoremediation, a technique based on the use of plants for the remediation of contaminated soil (Requena et al. 2001; Celik et al. 2004; Turnau et al. 2007).

Ectomycorrhiza (ECM)

Ectomycorrhiza (ECM) is predominantly found on trees in temperate forests, such as: *Betula, Fagus, Picea, Pinus, Populus, Quercus* and *Salix*. It has also been found in a few plant families of the southern hemisphere and the monsoon forests of south-east Asia (Jackson and Mason 1984; Smith and Read 1997). A wide range of fungi predominantly belonging to the phyla Basidiomycota and Ascomycota form ECM (Smith and Read 1997). These fungi are polyphyletic (they have four or more origins) (LoBuglio et al. 1996; Hibbert et al. 2000). Most ECM fungi are epigeous (having above-ground fruiting bodies), but up to a quarter of them are hypogeous (with underground fruiting bodies) (Molina et al. 1992). The ECM fungi do not show a high degree of host specificity in the colonization of the secondary or tertiary roots of woody species. The hyphae of ECM fungi cover the root like a sheath and grow inwards between epidermal and cortical cells forming a Hartig net. Penetration of the root cortex is achieved through mechanical force and secretion of pectinases (Werner 1992). The hyphae never penetrate into cell lumens or into the stele. A complex network of fungal hyphae is the site of nutrient exchange (metabolic integration) between the fungus and host trees; predominantly nitrogen is exchanged for photosynthates, but due to their polyphyletic origins it has been suggested that ECM fungi have considerable functional diversity (Brundrett 2002).

Orchid Mycorrhiza

Orchids have mycorrhizal associations with soil fungi that are considered to be essential for seed germination and assist with the growth of adult plants (Rasmussen 1995; Currah et al. 1997). Most orchids have fairly specific fungal associates that vary between host species and habitat (Currah et al. 1997; Sen et al. 1999). Most of these fungi are assigned to the anamorphic form genus *Rhizoctonia* (Currah et al. 1997), some species of which are close relatives of pathogenic forms (Pope and Carter 2001). The benefits provided by orchids to mycorrhizal fungi, if any, are not clear, as these fungi seem to grow as well without their hosts as they do with them. Saprophytic (myco-heterotrophic) orchids without chlorophyll have fully exploitative mycorrhizal associations that supply both the energy and nutrient requirements of the host (Leake 1994). Many of these plants associate with fungi that are not related to the mycorrhizal fungi of chlorophyllous orchids, including ECM associates of trees, wood-rotting and parasitic fungi. These associations have a high degree of host-fungal specificity and species of *Corallorhiza, Gastrodia* and *Galeola* may only associate with a single fungal genus (Leake 1994; Currah et al. 1997).

Ericoid Mycorrhiza

Mycorrhizal fungi that associate with members of the Ericaceae and Epacridaceae include several groups of ascomycetes which generally do not form mycorrhizas with other vascular plants (Smith and Read 1997). Studies of DNA sequences of fungi from these plants in different Continents have revealed that two or more distantly related groups of fungi are involved in ericoid mycorrhizas (McLean et al. 1999; Monreal et al. 1999; Sharples et al. 2000). *Hymenoscyphus*-like fungi associate with the Ericales and bryophytes throughout the world, but other taxa are more restricted to specific geographic regions (Chambers et al. 1999; Read et al. 2000). It is not certain whether ericoid mycorrhizal fungi exist primarily as soil saprophytes, or as mycorrhizal associates of plants. Ericoid mycorrhizal associations are considered to be capable of neutralizing highly acidic soils and acquiring organic nutrients (Smith and Read 1997). Substantial nutritional benefits accrued by the host have been shown in some experiments, but not in others (Bell and Pate 1996; Jansa and Vosatka 2000) which may indicate facultative associations.

Individual members of Ericales form monotropoid or arbutoid mycorrhizas (ECM-like associations). They generally have much greater host-fungus specificity than other ECM associations (Smith and Read 1997).

Arbuscular Mycorrhiza (AM): Development and Function

Two principal morphological types of AM are known, the *Paris* and *Arum* types, named after the two plant species (*Paris quadrifolia* and *Arum maculatum*) in which these types were first described more than 100 years ago (Gallaud 1905). In the *Arum* type, the intraradical fungal hyphae grow linearly in intercellular spaces forming fine intracellular arbuscules, whereas, in the *Paris* type, cortical cells are colonized by thick, coiled and fine arbusculate, coiled hyphae. When the hypha of the coiled type are growing through longitudinally continuous air spaces they assume a typical linear morphology, but in the absence of abundant aerenchyma coils are formed (Brundrett 2002). Both types are widely distributed throughout the plant kingdom (Smith and Read 1997; Smith and Smith 1997). In spite of some AM fungi forming one specific morphological pattern of AM association, other fungi have been demonstrated to be capable of forming both types of colonization depending on the host involved (Smith and Read 1997; Brundrett 2002). While the *Paris* type is most common in contemporary bryophytes, ferns and gymnosperms, the *Arum* type is found predominantly in cultivated herbaceous plants, and has been more frequently studied than the *Paris* type (Smith and Read 1997; Smith and Smith 1997). Therefore, in this chapter attention is mainly focused on the *Arum* type of interaction.

Cycle of AM Development

Development of an AM association involves two major stages: pre-symbiotic and symbiotic (Bago and Becard 2002). The first stage starts when the dormant fungal spores germinate and unbranched germ tubes appear (Fig. 2). This is a reversible process: spore germination can be induced by several abiotic stimuli (Powell 1976; Koske 1981), but where root exudates are absent development of the germ tube stops and it

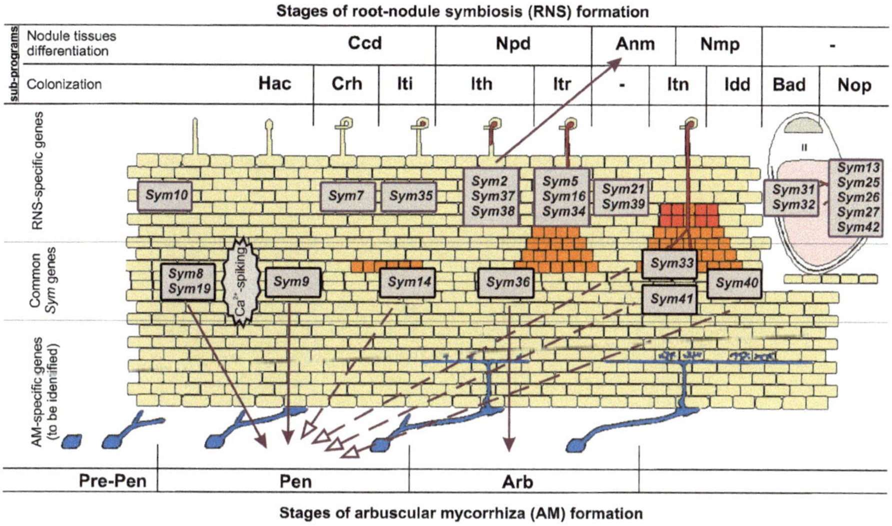

Fig. 2 Control of development of root nodule (RN) symbiosis and arbuscular mycorrhiza (AM) in pea (*Pisum sativum*) by the sequential functioning of regulatory symbiotic genes.The programme of RN symbiosis formation (according to (Borisov et al. 2007)) involves: (1) sub-programme of nodule tissue differentiation: Ccd – cortical cell division, Npd – nodule primordium development; Anm – apical nodule meristem development, Nmp – nodule meristem persistence; (2) sub-programme of colonization (infection and bacteria differentiation): Hac – root hair curling, Crh – colonization of curled root hair, Iti – infection thread growth initiation, Ith – infection thread growth inside root hair cells, Itr – infection thread growth in root tissue, Itn – infection thread formation in nodule primordium, Idd – infection droplet differentiation, Bad – bacteroid differentiation, Nop – nodule persistence. The programme of AM formation (according to (Marsh and Schultze 2001)) involves: Pre-Pen – pre-penetration; Pen - root penetration; Arb - arbuscule development. Stages of RN symbiosis formation are located directly above the responsible genes. Sign '–' means that genes which are located beneath do not involve in corresponding sub-programme of root nodule development. The continuous arrows point to the stages of RN symbiosis or AM formation which are blocked by the gene mutations; broken arrows point to the stages of AM formation which are partially repressed

is autolysed or invaginated back into the spore (Giovannetti et al. 1993; Giovannetti et al. 1996; Bago et al. 1998). Spores may germinate several times (Koske 1981).

The role of the root exudates in AM fungal mycelial growth from germinated spores has been well reported. Flavonoids (Fig. 3a) and polyamines have been found to provide induction, but are not major stimuli for mycelial growth (Nair et al. 1991; Tsai and Phillips 1991). During the few last years a considerable change in our understanding of signalling process at the pre-symbiotic stage of AM development has happened. The compound inducing primarily ramification of hyphae, the so-called, branching factor, has been isolated recently from the root exudates of *Lotus japonicus* (Akiyama et al. 2005) and identified as 5-deoxy-strigol, a strigolactone with sesquiterpene structure (Fig. 3b). Strigolactones were shown to stimulate AM fungi by activating their respiration (Besserer et al. 2006). Plant root strigolactones are also chemical signals which induce seed germination of parasitic weeds (Akiyama and Hayashi 2006). This fungus also produces diffusible

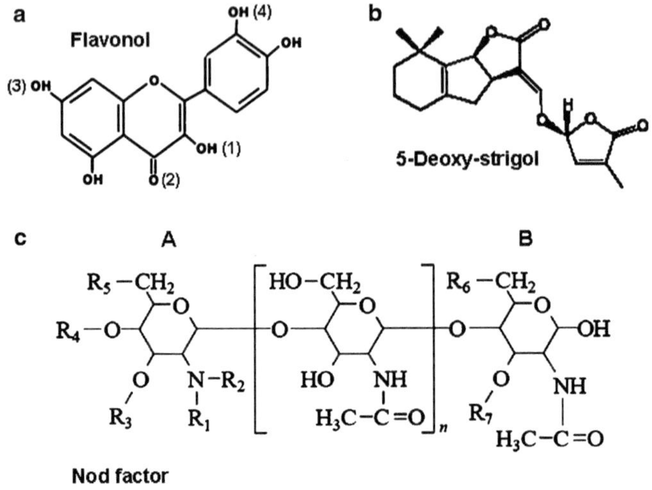

Fig. 3 Signal molecules essential for the earliest stages of arbuscular mycorrhiza (AM) and root nodule (RN) symbiosis formation. **a** – The plant molecule, flavonoid (flavonol). The OH-group may be removed or replaced with different substituents to produce various flavonoid compounds, for example:- removed at (1) (flavone); replaced at (1) with 3rd ring (isoflavone); removed at (1) and (4) and replaced at (3) with glucose (apigenin-7-monoglucoside). The O may by replaced at (2) with H (anthocyanin). **b** – The plant molecule, sesquiterpene lactone, strigolactone (5-Deoxy-strigol). **c** – Rhizobial lipo-chito-oligosaccharide (LCO) molecule, Nod factor. The possible replacements include: R2, methyl; R3, R4, carbomoyl; R5, carbomoyl or acetyl; R6, sulfate, acetyl, fucosyl, methylfucosyl, acetylmethylfucosyl, or sulfomethylfucosyl (this position is crucial for a specific recognition of "Afghan" pea by *Rhizobium legumonosarum* bv. *viciae*; see Sections 3 and 4); R7, arabinosyl. A – non-reducing terminus of Nod factor; B – reducing terminus of Nod factor

signal molecules called Myc factor (Mycorrhization factor) (Kosuta et al. 2003). The chemical nature of the Myc factor has not yet been identified although the molecule is thought to have non-proteinaceous nature because of its thermostability, amphiphilic properties, and small mass (< 3 kD).

The branched primary hyphae reach the root surface and form attachment structures called appressoria (Fig. 1b, 2). Their formation is evidence of a successful recognition of a host-plant (Nagahashi and Douds 1997). After contact by the fungus with the root surface the symbiotic stage of AM development starts: from the apressorium the colonising hyphae grow through the epidermis and into the root. It was demonstrated recently that the host plant plays an active role in fungal growth into both the epidermal layer (Genre et al. 2005; Genre et al. 2008) and cortex (Genre et al. 2009) forming a new intracellular structure, termed pre-penetration cytoskeleton apparatus. It represents a 'tunnel' resembling the cortical pre-infection threads formed during nodulation (see later in Section 4.3).

Passing from the epidermis, the intraradical hypha grows through the cortex where it branches and forms intercellular mycelia. In the inner cortical layer(s) close to endodermis as was shown for *Lotus* (Demchenko et al. 2004), the fungus invades into cells and there differentiates into highly dichotomically ramified tree-liked structures, the arbuscules (Fig. 1b, 1d, 2). The intercellular hyphae continuously form new arbuscules which are ephemeral structures: several days after emergence they are degraded in the plant cell (Gianinazzi-Pearson 1996; Bonfante et al. 2009).

The arbuscules are subcellular compartments which ensure close metabolic integration between the partners. This is manifested as an intensive exchange of metabolites, primarily in transport to the host of phosphates absorbed by the fungus from soil. This function is fulfilled despite a separation of fungal and plant cytoplasms. The arbuscule remains separated from the plant cytoplasm by an extension of the plasma membrane, called the periarbuscular membrane, similar to peribacteroid membrane formed during nodulation (Provorov et al. 2002) (see later in Section 3.2). This envelopment of the arbuscule also results in the formation of a new apoplastic space between the periarbuscular membrane and the arbuscule, called the periarbuscular space which contains a structurally differentiated matrix containing polysaccharides and enzymes of both plant and fungal origin (Smith and Read 1997; Harrison et al. 2002).

A pronounced differentiation is characteristic for the plant cells containing arbuscules. The cytoskeleton of the invaded cells indicates extensive, dynamic rearrangements of the cortical microtubules, which presumably enables the trafficking of membrane and cell wall precursors into the extending membrane and new apoplastic compartment. The vacuole is reduced or degraded, the amount of α-tubulin increases, the nucleus is deformed (sometimes it is surrounded by the branches of arbuscule) and chromatin is decondensed due to a high transcriptional activity. The number of ribosomes and Golgi vesicles is greatly increased correlating with the synthesis of periarbuscular membrane (Takemoto and Hardham 2004; Genre and Bonfante 2005).

Intraradical mycelium formed inside the cortex creates a network of channels in which carbon is transported from the plant to the fungus (Smith and Read 1997). When the carbon supply is sufficient, lipid-rich vesicles (Fig. 1c) are formed intercellularly within the cortex, as fungal storage organs. Hexoses, in particular glucose, were found as the major form in which carbon is taken up and metabolized by AM fungi (Pfeffer et al. 1999; Douds et al. 2000). Glucose can then be directly incorporated into trehalose and glycogen, the first substantial compound in the fungal carbon pool (Shachar-Hill et al. 1995). Later on, when the carbon supply is sufficient storage lipids are synthesized, predominantly triacylglycerols in AM fungi (Bago et al. 2002), which are also the principal form of carbon translocated from hyphae present in the root cortex to more distal regions (Smith and Read 1997; Smith and Smith 1997). As a result of biotrophic carbon assimilation, growth of AM fungi is activated and an abundant net of extraradical mycelium is developed due to active ramification and anastomosis of hyphae coming out from the root into the rhizosphere (Smith and Read 1997; Bago et al. 1998; De la Providencia et al. 2005). In the symbiotic stage, carbohydrates can only be taken up within intraradical structures, no uptake of hexoses by the extraradical mycelium has been detected (Pfeffer et al. 1999; Douds et al. 2000).

A growing colony of AM fungus includes intraradical and extraradical mycelia and is in an assimilative condition, when most of nutrients taken up from the soil are used for growth. Mature AM with dense extraradical mycelium, actively taking up minerals, is characterized by a transfer into dissimilative stage, when a portion of these nutrients is transported from the fungus to the host playing an important role in plant development (Smith and Read 1997; Bago et al. 1998).

Spore formation is a turning-point in AM life cycle. They are formed on the extraradical mycelium (Smith and Read 1997; Bago et al. 1998) (Fig. 1a), but in several AM fungi spores are formed on intraradical mycelium as well (Bago et al. 1998) (Fig. 1e). Spores contain reserve nutrients predominantly lipids and smaller amounts of glycogen (Sancholle et al. 2001). Some fungi can form sporocarps, which may be as simple as a few spores surrounded with mycelium (Fig. 1f). Extra- and intraradical spores, sporocarps, root fragments containing mycelium with vesicles as well as living extraradical hypae can all operate as propagules, starting new root colonization (Smith and Read 1997; Daplé et al. 2005; Souza et al. 2005). In addition to propagating the AM fungus, spores probably contain cells of permanent bacterial endosymbionts, therefore single-spore cultures do not guarantee truly monoxenic culture of AM fungi (see below).

Improvement of Plant Phosphorus Uptake by AM Fungi

Phosphorus (P) is one of the mineral nutrients essential for plant growth (constituting up to 0.2% of the dry weight of the plant cell) and development. It has diverse regulatory, structural, and energy transfer roles and consequently is required in significant quantities (Bieleski 1973; Schachtman et al. 1998). Biochemical data support the suggestion that AM fungi improve phosphorus uptake from soils (mainly as the water soluble $H_2PO_4^-$ ion, referred to as P_i) since the process involves several fungal transport systems some of which have an extremely high affinity for P_i (Ezawa et al. 2002). After transport into the hyphae, a major part of P_i is transformed into linear polyphosphate chains (polyP) through reactions catalyzed by a polyphosphate kinase. Granules rich in polyP, accompanied by P-containing esters, are packed into cylindrical vacuoles for translocation within hyphae along tubulin fibrils (Ashford 2002). After reaching the arbuscules, P-compounds are degraded by phosphatases allowing the release of P_i across the partners' interfaces. The crucial role of arbuscules in P_i export was proved by the dependence of their development on the soil phosphorus status: arbuscules are formed actively under P_i limiting conditions to moderate its supply, while an absence or excess of P_i inhibits their formation.

Molecular mechanisms controlling subsequent phosphate transport from extraradical hyphae to intraradical ones and efflux from the arbuscule are largely unknown, but it is well documented that plants possess many classes of phosphate transport proteins, including those which are expressed only in AM symbiosis (Karandashov and Bucher 2005; Javot et al. 2007). On the fungal side, the presence of a *G. mosseae* P_i transporter was shown in cells containing arbuscules (Balestrini et al. 2007). The import of P_i by plant phosphate transporter is suggested to be a signal which permits continuous development of the arbuscule and consequently sustains the fungal persistence within the root (Bonfante et al. 2009).

Increase of Plant Drought Tolerance

Drought stress is a major agricultural constraint in the semi-arid tropics. In most cases symbiosis with AM fungi has been shown to increase host plant growth rates during drought stress and plant resistance to drought. Several mechanisms explaining this phenomenon have been proposed: an influence of AM on plant hormone profiles, increasing intensity of gaseous exchange and photosynthesis in leaves, direct water transport via fungal hyphae from soil into the host plant, enhanced water uptake through improved hydraulic conductivity and increasing leaf conductance and photosynthetic activity, nitrate assimilation by fungal hyphae, enhanced activity of plant enzymes involved in defence against oxidative stress, plant osmosis regulation, and changes in cell-wall elasticity (reviewed in: Quilambo 2003; Augé 2001; Augé et al. 2004).

Using bean mutants not forming AM associations, it was determined that about half of the considerable promotion of stomatal conductance by AM fungi was attributable to soil colonization by fungal hyphae and about half to plant colonization by AM fungus (Augé et al. 2004). A path analysis modelling approach revealed that soil hyphal colonization had larger direct and total effects on the of tolerance of beans to dehydration compared with plant root AM colonization or several other soil or plant variables (Augé et al. 2003).

Soil structure refers to pore space as well as to aggregates, while the moisture characteristic of a soil depends on the size and distribution of its pores. Fungi may be the most effective soil organisms in stabilizing soil structure, and AM fungi often comprise the largest portion of the soil microbial biomass (reviewed in: Augé et al. 2003). The AM fungal hyphae grow into the soil matrix and create conditions conducive to the formation of microaggregates and then their packing into macroaggregates (Celik et al. 2004; Rillig 2004) due to production copious amounts of the glycoprotein glomalin (Rillig 2004). Through AM fungi-mediated effects on soil structure, it seems logical to suggest that AM colonization of a soil might affect its moisture retention properties and, in turn, the behaviour of plants growing in the soil, particularly when it is relatively dry (Augé et al. 2004).

A comparative analysis of the results from a wide series of studies embracing interactions between more than 60 genera of plants and five genera of AM fungi (reviewed in: Augé 2001) showed that different effects of AM on host-plant/water relations either under normal environmental conditions or during drought stress are either related or unrelated to improved mineral nutrition of the host by fungi. Usually the type of response depended on the prevailing environmental conditions and the particular plant–microbe association studied. It was generally concluded that mycorrhizal effects on plant/water relations are not as dramatic and consistent as those relating to phosphorus acquisition and host growth.

Increase of Plant Resistance to Pathogens

Plants colonized by AM fungi have been demonstrated to manifest increasing resistance to some pathogens. In particular, AM fungi have been observed to reduce the susceptibility of tomato to pathogenic fungi such as *Phytophthora parasitica* (Cordier et al. 1998; Dassi et al. 1998) and *Alternaria solani* (Fritz et al. 2006) as well as leading to an inhibition of the bacterial leaf pathogen *Xanthomonas campestris* in *Medicago truncatula* (Liu et al. 2007). Whether such increased resistance to pathogens is a consequence of improved plant fitness or is due to a specific defence response induced by AM fungi is unknown. At the same time AM formation has been shown to be accompanied by a complex pattern of local and systemic changes in plant cell defence reactions (Ruiz-Lozano and Azcon 1996) and gene expression, including the induction of a functional defence response (Liu et al. 2007). Most probably, systemic plant defence reactions caused by AM fungi occur as a type of SAR (systemic acquired resistance) or ISR (induced systemic resistance) (see for review: Pieterse et al. 2001; Vallad and Goodman 2004) as several authors refer to an induction of the biosynthesis of signalling molecules in arbuscular mycorrhizal plants typical of the signal transduction pathways such as pathogenesis related proteins (PR-proteins) (Dassi et al. 1998) and jasmonate (Hause et al. 2002).

A range of processes occurring in plants as a result of pathogen invasion (plant defence responses) also take place during interactions with AM fungi. These include signal perception, signal transduction and defence-related gene activation. Hypersensitive responses have been observed to take place with compatible and non-compatible combinations of plants with AM fungi (Salzer et al. 1999). Reactions similar to an 'oxidative burst' are typical for the AM fungus penetration into epidermal cells (Salzer et al. 1999). Some elements of signal transduction pathways activated during pathogenesis are also activated at the early stages of AM development. During appressorial formation in different plants catalase and peroxidase activities are increased and are associated with salicylic acid accumulation (Spanu and Bonfante-Fasolo 1988; Blilou et al. 2000a; Lambais 2000). The process of salicylic acid accumulation also correlates with expression of genes encoding for the lipid transport protein (LTP) and phenylalanine-ammonium-lyase (PAL) (Blilou et al. 2000b), the later being of relevance to lignification of the plant cell wall. Localized expression of defence-related genes or an accumulation of gene products in arbuscule-containing cells has also been observed (Harrison 1999; Bonanomi et al. 2001).

In AM, as in other compatible biotrophic interactions, the defence response appears to be weak and occurs transitorily during the early phases of colonization (Harrison 1999). By the time mycorrhizal fungi are well established in the roots, responses are often similar to or lower than those in non-mycorrhizal control plants, suggesting that the suppression of plant defence responses may contribute to successful, compatible AM fungal colonization (Harrison 1999; Garcia-Garrido and Ocampo 2002).

Important evidence supporting the defence-related reactions has been obtained from AM-defective mutants induced in legumes, such as *Pisum*, *Medicago*, and *Lotus* spp. While in the wild-type plants the AM development is characterized by weak or suppressed defence responses (possibly, due to special fungal signals), in mutants they appear stronger and involve changes at several different levels, including increased production of the signalling compound salicylic acid (Blilou et al. 1999), expression of defence-related genes (Harrison 1999), and accumulation of defence-associated proteins (Dassi et al. 1998) or phytoalexins (Bonanomi et al. 2001; Blilou et al. 1999; Gollotte et al. 1993). Cytological studies have identified the cell wall appositions on the inside of root cell walls adjacent to the appressoria of pea mutants (Gollotte et al. 1993) and death of invaded epidermal cells of *Lotus* mutants (Bonfante et al. 2000).

Bacterial Endosymbionts of Mycorrhizal Fungi

A few fungi, including some *Glomeromycota* species (AM fungi and *Geosiphon pyriforme*) (Scannerini and Bonfante 1991; Bianciotto et al. 1996; Bianciotto et al. 2000; Schüßler and Kluge 2001; De Boer et al. 2005) and in the ECM basidiomycete *Laccaria bicolor* (Bertaux et al. 2003) are reported to have endocellular bacteria. In the cytoplasm of several AM fungal species (*Glomus versiforme*, *Acaulospora laevis*, *Gigaspora margarita*) are Gram-negative rod-shaped bacteria, which have not been grown on cell-free media (Scannerini and Bonfante 1991; Bianciotto et al. 1996; Mosse 1959; MacDonald and Chandler 1981; Bianciotto et al. 2004; Jargeat et al. 2004). In the most of fungal species evaluated intracellular bacteria were detected through all the stages of the fungal life cycle, i.e. spores, germtubes, and extra- and intraradical hyphae, except arbuscules (Bianciotto et al. 1996). It has been shown that a *G. margarita* spore contains an average about 20,000 bacteria (Bianciotto et al. 2004; Jargeat et al. 2004). These bacteria were initially assigned to the genus *Burkholderia* on the basis of their 16S ribosomal RNA gene sequence, but were more recently reassigned to a new taxon named *Candidatus Glomeribacter gigasporarum* (Bianciotto et al. 2003); the provisional *Candidatus* designation is given to uncultured bacteria (Murray and Schleifer 1994; Murray and Stackebrandt 1995).

The physiological role of the endosymbiotic bacteria in AM fungi is unknown (Jargeat et al. 2004), but two bacterial genes have been found in the genomic library developed from *G. margarita* spores including a putative phosphate transporter gene (*pst*) and a *vacB*-like gene involved in host cell colonization by enteroinvasive, pathogenic bacteria such as *Shigella flexneri* and *Escherichia coli* es components of the bacterial endosymbiont genome (van Buuren et al. 1999). It is assumed that these genes originate not from *G. margarita*, but endosymbiotic bacteria, and might be of particular interest for future determination of the potential role of the bacterium in mycorrhizal symbiosis (Jargeat et al. 2004; Ruiz-Lozano and Bonfante 1999; Ruiz-Lozano and Bonfante 2000).

The bacterium *Candidatus Glomeribacter gigasporarum* is supposed to be vertically transmitted within AM fungi (Artursson et al. 2006). Cells of the bacteria were demonstrated to be transmitted through five fungal vegetative generations of asexual *G. margarita* spores (Bianciotto et al. 2004). The asexual reproduction typical of AM fungi and the coenocytic nature of their mycelium may facilitate the migration of the endosymbiotic bacteria from spores into hyphae, and thereby allowing the vertical transmission to take place (Artursson et al. 2006). Additional support to the hypotheses regarding the vertical transmission of *Candidatus Glomeribacter gigasporarum* bacteria within AM fungi, and their obligate endocellular nature comes from the fact that *Candidatus Glomeribacter gigasporarum* has a surprisingly small genome size for a bacterium of only around 1.4 Mb in total and consisting of a chromosome of approximately 750-kb and an additional replicon of approximately 650 kb (Jargeat et al. 2004). Small genomes are however, often a feature of obligate endocellular bacterial species, for example beneficial endosymbionts of insects: *Buchnera* (aphid's symbiont) (Shigenobu et al. 2000; Hosokawa et al. 2006) and other gut endocellular bacteria of different incects (Hosokawa et al. 2006), *Sitofhilus zeamais* primarily endosymbiont (SZPE) (Dale et al. 2002), as well as animal endocellular parasitic bacteria: *Mycoplasma*, *Rickettsia* (reviewed in: Provorov et al. 2008). Considering processes like reductive evolution where only those genes absolutely essential for survival in an intracellular environment are retained (Dale et al. 2002), *Candidatus Glomeribacter gigasporarum* represents a typical candidate for a permanent endosymbiont with its small genome size (Artursson et al. 2006). One of the major future challenges within this research area is to understand the functional significance of AM fungal endosymbiotic bacteria. One important step in this direction will be the ability to remove the bacteria from the fungal cytoplasm, enabling comparisons of fungal effects on plants, in the presence and absence of the bacterial symbiont.

Legume-Rhizobia Root-Nodule (RN) Symbiosis

Leguminous plants can grow in the complete absence of combined nitrogen in soil/substrate due to the fixation of atmospheric nitrogen by symbiotic nodule bacteria as described elsewhere in this Book. Leguminous plants in collaboration with rhizobia make a large contribution to the global nitrogen balance in natural and agricultural ecosystems (Vance 2001). Nitrogen fixation occurs within special plant organs, typically root nodules (Fig. 1g) but in some associations stem nodules are also formed. The ontogeny of nodulation represents a well-organized process based on the tightly coordinated expression of specialized plant and bacterial genes. In the nodules bacteria differentiate into a specialized symbiotic organelle-like form, termed bacteroids (Fig. 1h). The main enzyme of nitrogen fixation is a nitrogenase that has a complex structure (Hirsch 1992; Sprent 2001). The legume nodules provide an ecological niche for bacteria, as well as structure for metabolic/signal exchange between the partners and for the control of symbionts by the hosts (Brewin 2004).

Specificity of RN Symbiosis

Leguminous plants have a world-wide distribution and are classified in one of the biggest angiosperm families, the Fabaceae, which contains 17,000–19,000 species divided between three sub-families (Ceasalpinioideae, Mimosoideae, Papilionoideae) and more than 700 genera (Allen and Allen 1981). With a single exception (*Parasponia*: Ulmaceae), the ability for symbioses with rhizobia is restricted to Fabaceae, although in eight related dicotyledonous families (Rosid I clade) an ability to form nodules with the nitrogen-fixing actinobacteria *Frankia* is known (Wall 2000).

By contrast to legumes, their nitrogen-fixing microsymbionts do not constitute a taxonomically coherent group of organisms. The majority of rhizobia belong to the α-proteobacteria assigned into the family Rhizobiaceae solely on the basis of their ability to nodulate legumes. A range of diverse genera (e.g., *Azorhizobium*, *Bradyrhizobium*, *Mesorhizobium*, *Rhizobium*, *Sinorhizobium*) are distinguished among these bacteria. Moreover, recently some β-proteobacteria (close to *Burkholderia*, *Cupriavidus*, *Pseudomonas* and *Ralstonia*) and even some γ-proteobacteria have been discovered that can form nitrogen-fixing nodules with legumes (Balachandar et al. 2007). All these bacteria (collectively called "rhizobia") vary enormously in their overall genome organization, location of "symbiotic" (*Sym*) genes and their molecular organization and regulation (Spaink et al. 1998; MacLean et al. 2007).

According to the early surveys of symbiotic specificity (Fred et al. 1932), legumes were suggested to comprise a range of taxonomically restricted Cross-Inoculation Groups (CIGs) within which cross-inoculation can freely occur, while species from different groups do not cross-inoculate. The best studied examples of this classification are represented by four CIGs: "*Trifolium* – *Rhizobium leguminosarum* bv. *trifolii*", "*Pisum, Vicia, Lathyrus, Lens* – *R. leguminosarum* bv. *viciae*", "*Galega* – *R. galegae*", "*Medicago, Melilotus, Trigonella* – *Sinorhizobium meliloti*".

It was later demonstrated however (Provorov 1994; Broughton and Perret 1999) that such strictly defined specificity is limited to the herbage papilionoid legumes living in temperate zones and representing the "galegoid complex". For many tropical and subtropical legumes a strict separation into CIGs is not typical, and a variable degree of symbiotic promiscuity is observed. It is important to note that *Lotus* sp. demonstrates promiscuity with respect to different rhizobia. Also, the species of tribe Phaseoleae (*Phaseolus vulgaris, Vigna unguiculata, Glycine max*) may be inoculated with strains representing different rhizobia genera isolated from the taxonomically distant legume species. Moreover, some legumes may have the taxonomically distant symbionts (soybean – *Sinorhizobium fredii* and *Bradyrhizobum japonicum*) which share the genes required to nodulate a common host. Many tropical rhizobia can inoculate very broad spectra of hosts: e.g., strain NGR234 (*Sinorhizobium fredii*) inoculates plants from 112 legume genera and also the non-legume *Parasponia*.

The specificity of legume-rhizobia interactions is expressed only during the pre-infection interactions when rhizobia recognize the roots of appropriate host plants and colonize their surfaces. The interaction starts when the root-derived signals, in particular, flavonoids (Fig. 3a), activate the rhizobial nodulation (*nod*/*nol*/*noe*) genes (stage Ngi – nodulation gene induction) (Ovtsyna and Staehelin 2005). These genes control the synthesis of lipo-chito-oligosaccharide (LCO) Nod factors (nodulation factors) (Fig. 3c) which induce the early stages of nodule development (Spaink 1995; Schultze and Kondorosi 1998a; D'Haeze and Holsters 2002).

With the exception of the genes *nodA*, *B*, *C* and *D*, the majority of nodulation genes are host specific: they are responsible for affinities of different bacterial species and strains to the particular legume hosts. When rhizobia are entering into the vicinities of host roots, the *nod/nol/noe* genes are activated by the NodD protein (a sensor for root-derived signals) interacting with the universal *nod*-box sequences located in the promoters of other nodulation genes (Long 1989). The products of genes *nodA*, *nodB* and *nodC* catalyze the biosynthesis of Nod factor core structures which are common for all rhizobia species (Spaink et al. 1991). In turn, the products of other *nod* genes perform a decoration of this core with different substitutes, which could be unique to a specific strain.

Nod factors represent a unique group of bacterial signalling molecules not known outside legume–rhizobia associations. They are among the most potent developmental regulators: their effect is expressed at concentrations as low as of 10^{-8}–10^{-12} M. Nod factors cause varying changes in plants leading to the formation of a root-nodule symbiosis; therefore they can be considered as hormone-like molecules. The core structure of these molecules, common for all rhizobia species, consists of 3–6 residues of N-acetylglucosamine and a fatty acid (acyl) chain (Fig. 3b).

The symbiotic specificity is dependent mainly on the chemical modifications in Nod factor structures (Ovtsyna and Staehelin 2005). Narrow specificity typical for rhizobia inoculating the galegoid legumes is usually associated with the presence of unusual highly unsaturated acyl groups (2–4 double C=C bounds per 16–20 carbon atoms in the chain) while broad host specificity is correlated with the presence of saturated mono-unsaturated acyl groups (none or only one double C=C bound). Greater fine tuning of specificity towards the particular host species and even towards some genotypes is correlated with chemical substitutions in the oligo-chitin part of Nod factor. For example, the sulphation of the R6 position at the reducing terminus encoded by *nodP*, *nodQ*, *nodH* genes is required for interaction between *S. meliloti* and alfalfa (lucerne) (*Medicago*) while the acetylation encoded by *nodX* is necessary for *R. leguminosarum* bv. *viciae* to inoculate the "Afghan" pea (*Pisum sativum*) lines (Broughton and Perret 1999). Interestingly, for interaction with "Afghan" pea, *nodX* may be substituted functionally by *nodZ* gene, controlling a dissimilar modification in R6 (fucosylation instead of acetylation) (Ovtsyna et al. 1999; Ovtsyna et al. 2000), suggesting a flexibility in the mechanisms for signal reception (see also Section 4.3). These recognition and early signalling processes are controlled in a "gene-for-gene" manner indicating the similarities between early symbiosis development and host–pathogen interactions (Tikhonovich et al. 2000).

It is likely that the control of specificity of interactions between legume plants and nodule bacteria is much more complex than current evidence suggests. Recognition of the structural differences in the Nod factor molecule may not turn out to be the one and only mechanism controlling the specificity of legume–rhizobial interactions. Further control of host specificity may result from interactions between the bacterial surface molecules (some polysaccharides and proteins) (Becker and Pühler 1998; Lugtenberg 1998) and the lectins located on root hair surfaces (Jones et al. 2008). Transfer of the pea lectin gene *Psl* into the clover roots *via* the *Agrobacterium rhizogenes* mediated transformation resulted in an ability for clover to be nodulated by pea rhizobia (Diaz et al. 1995). Similar results were obtained after transfer of the soybean lectin gene into trefoil plants (van Rhijn et al. 1998). Recently, a rhizobia–legume signalling system, alternative to the "flavonoid/Nod factor" circuit has been reported. Other low-molecular weight compounds having non-flavonoid structures play important roles in RN symbiosis development determining polarization of a bacterial cell and controlling adhesive properties of rhizobia (J.A. Downie, 2006 personal communication).

Giraud et al. (Giraud et al. 2007) have recently reported results of a complete genome sequencing of two symbiotic, photosynthetic, *Bradyrhizobium* strains, BTAi1 and ORS278. It was shown that canonical *nodABC* genes and typical lipochitooligosaccharidic Nod factors are not required for symbiosis in some legumes (especially, aquatic legumes *Aeschynomene sensitiva* (from Africa) and *A. indica* (from North America)). Mutational analyses indicated that these unique rhizobia use an alternative pathway to initiate symbioses, here a purine derivative may play a key role in triggering nodule formation. These results suggest that signal interactions and specificity of symbiosis formation are more complicated and variable than it was supposed; but in any event, the activity of signalling cascades results in transcriptional changes in the root, nodule morphogenesis and nitrogen fixation.

Development and Functioning of RN

Unfortunately, only very limited numbers of *Rhizobium*–legume interactions have been studied in detail (Hadri et al. 1998) but it is possible to identify at least two major types of root nodule, namely those with indeterminate and those with determinate nodule meristems (NMs) (reviewed in: Sprent and James 2007). In this section, the development of symbiotic structures in the best-studied indeterminate nodules formed by evolutionarily advanced legumes from the "galegoid complex" (e.g. *Pisum, Medicago, Trifolium*) is described. These plants use one of the most complicated and evolutionarily advanced nodulation programmes which may provide a conduit for a greater variety of simpler nodulation programmes used by several representatives of the Fabaceae (Sprent 2001).

Early cytological responses to Nod factors in root hair cells have been described in several legumes (Cardenas et al. 1998; Esseling et al. 2003). The first event, detectable only a few seconds after Nod factor application, is a Ca^{2+} ion influx at the root hair tip, causing a several-fold increase in cytosolic calcium concentration (Felle et al. 1999). This is associated with a transient depolarization of the plasma membrane (Lhuissier et al. 2001; Shaw and Long 2003a). Within 20–30 min following application of Nod factors, there is suppression of peroxide release from the roots (Shaw and Long 2003b), and a reorganisation of the actin and microtubule networks in the root hair cells (Timmers et al. 1999; Sieberer and Emons 2000; Sieberer et al. 2002). There is also a second calcium event, the induction of periodic oscillations of cytoplasmic calcium concentration (Ca^{2+} spiking), which is observed in the perinuclear area (Ben Amor et al. 2003; Oldroyd and Long 2003; Shaw and Long 2003a). Following application of Nod factors, the earliest observed morphological event is a cessation of tip-growth of root hairs which occurs within one hour. The arrest of apical growth is associated with the swelling of root hair tips (Lhuissier et al. 2001).

Nodule morphogenesis in clover (Nutman 1969), pea (Bond 1948; Newcomb 1976; Brewin 1998; Tsyganov et al. 2002; Voroshilova et al. 2009) and alfalfa (lucerne) (Vasse et al. 1990) starts on rhizobial colonization of roots with strong deformation and curling of root hairs (stage Hac – Hair Curling) (Fig. 1i, 2), this reaction is mediated by Nod factor action. Then the curl is colonized by bacteria at high density (stage Crh – Colonization of the pocket in the curled root hair) (Tsyganov et al. 2002) (Fig. 2). Within the curls, the root hair cell wall is "weakened" and the plasmalemma invaginates initiating growth of a special tunnel structure – Infection Thread (IT, stage Iti – Infection Thread Growth Initiation) – which starts growing through the root hair cell first (stage Ith – Infection Thread Growth in root Hair cell) (Fig. 2). These processes mainly involve the plant molecular machinery changing cell wall composition and biogenesis of intracellular membranes (endoplasmic reticulum and Golgi vesicles) ensuring construction of a pathway for bacteria into the host (Gualtieri and Bisseling 2000; Brewin 2004).

Bacteria move, thereafter into the root cortex through IT growing into the root perpendicular to the root surface in a "cell-to-cell" or "transcellular" manner (Brewin 1998; Brewin 2004) and a certain amount of IT branching can be observed (Fig. 1-i). The transcellular growth is associated with the development in cortical cells of "Pre-Infection Threads" (PITs) which represent the cytoplasmic zones with specially reorganized cytoskeletal components which facilitate a rapid passage of ITs (van Brussel et al. 1992). In IT the walls are built up using plant cell wall polymers, while the internal space contains a martix synthesised by both partners. Several proteins are secreted by the plant into the IT matrix controlling the rhizobial activity, among them extensin-like and proline-rich proteins are most abundant (Brewin 2004). Although IT develops either between or within the plant cells, the rhizobia inside IT have a topologically intercellular location (Brewin 1998).

Simultaneously with the root hair curling and IT growth initiation, and on Nod factor action, cell division in the inner cortex of the root is initiated due to re-activation of the mitotic cycle in G0/G1-arrested plant root cells that leads to formation of Nodule Primordium (NP) (Fig. 1j) this later develops into the nodule meristem (NM) (Fig. 1k) (Vasse et al. 1990; Hadri et al. 1998; Foucher and Kondorosi 2000; Demchenko and Demchenko 2001; Young et al. 2003; Voroshilova et al. 2009). Consequently, the mitotic reactivation (Fig. 1m), dedifferentiation and proliferation of the cortical cells induced by Nod factors (stage Ccd – Cortical Cell Division) (Fig. 2) leads to formation of a mature persistent nodule meristem through a series of successive developmental stages. Then the advancing IT enters the NP where plant cells with endocytosed bacteria can be found.

Further growth of the NP tissues and ITs results in a formation of juvenile nodules as they emerge from the root (Fig. 1k). At this stage, the nodules do not yet have the histological differentiation characteristic for mature indeterminate nodules described by Vasse et al. (1990). Nevertheless, the transient histological patterns in juvenile nodules are in preparation for the future developmental pathways of mature nodule structures. First, the IT is visible which in combination with the surrounding cells provides growth and forms the "entrance" for bacteria to the root. Second, peripheral layers of NP cells which are not penetrated by ITs can be distinguished. Furthermore, dividing cells on the distal side of the NP form a hemispherical structure located above the plant cells already colonized by bacteria and surrounding the original "entrance" pathway for the IT into the root. At the distal part of this hemispherical structure actively dividing plant cells are observed while in the proximal cell layers of this hemispherical structure simultaneous plant cell divisions and growth of branching ITs occur. At the same time the orientation of IT growth is completely reversed and turned backwards toward the nodule apex (Libbenga and Harkes 1973; Timmers et al. 1999). It is proposed that the presence of proliferating cells harbouring infection threads is a prerequisite for normal formation of the Apical Nodule Meristem (Stage Anm) (Voroshilova et al. 2009) (Fig. 2). Therefore, only after completing the colonization of the plant cells with ITs, the plant cells leave the mitotic cycle under the influence of at least one plant cell cycle regulator (Cebolla et al. 1999) and undergo several rounds of endoreduplication in preparation for hosting intracellular bacteria and the control of their differentiation passing to the symbiotic (bacteroid) state (Mergaert et al. 2006). In parallel to establishment of the final nodule anatomy, the histological zones (I, II, III, IV) of the inner tissue of indeterminate nodules can be observed as described by Vasse et al. (1990) (Fig. 1l).

The key stage of endosymbiosis development is represented by the bacterial "release" from infection droplets into the plant cytoplasm *via* an endocytosis-like process. Infection droplets are "unwalled" regions of the ITs (they usually arise at the sides or at the tips of growing of ITs) which invaginate into the plant cells and in which the bacteria come into "direct" contact with the plant cell plasma membrane. At these contacts, the membrane vesicles which contain single bacterial cells are segregated into the plant cell cytoplasm. Therefore, from the start of the "release", bacteria inside plant cytoplasm are surrounded by specialised (PeriBacteroid) Membranes (PBMs) which originate from the plasmalemma of infection droplets and are completed by products from endoplasmic reticulum and Golgi vesicles. The bacterial cell(s) surrounded by PBM represent a major intracellular symbiotic compartment – the symbiosome (Roth and Stacey 1989). Within symbiosomes the bacteria differentiate into bacteroids (stage Bad – Bacteroid Differentiation) (Fig. 1h, 2). They are considerably larger than free-living bacteria and have an altered shape (in pea nodules the bacteroids are of X- or Y-shape) (Fig. 1h).

Bacteroid differentiation is correlated with a repression of many bacterial genes, with the exception of oxygen-regulated symbiotic genes, the products of which are directly involved in the processes of symbiotic nitrogen-fixation (e.g. *fixNc*) (Tsyganov et al. 2003). Irrespective of their related to symbiosis, the expression of bacterial genes gradually declines as the programme of bacteroid differentiation advances (Jimenez and Casadesus 1989; Quispel 1998; Perret et al. 1999; Margolin 2000; Tsyganov et al. 2003).

Decline occurs in parallel with the gradual morphological changes observed in the bacterial cells during such developmental process. Even actively expressed genes at the beginning and/or intermediate stages of symbiotic interactions, such as the flavonoid-induced nodulation genes and *dctA* appear to be subject to this decline in expression during bacteroid development (Tsyganov et al. 2003).

These observations (Tsyganov et al. 2003) agree with previous reports (Vasse et al. 1990) suggesting that development of symbiosomes after release ("endocytosis") of bacteria into plant nodule cells is not a simple result of endocytosis, but rather a gradual process involving several additional stages, each controlled by different plant genes. Within this perspective, the process of Bacteroid Differentiation (a separate stage in symbiosis development, Bad) could resemble an adjustable process of adaptation to the symbiosome environment, believed to be of a stressful nature (Brewin 1998; Santos et al. 2000; Nogales et al. 2002). The final product of this adaptive process could be the mature, actively nitrogen-fixing bacteroid, which is transformed into an organelle-like state and has lost the capacity to resume a free-living growth (Jimenez and Casadesus 1989; Quispel 1998; Tsyganov et al. 2003).

Synthesis of nitrogenase (the enzyme catalysing reduction of N_2 into NH_4^+) and other proteins involved in nitrogen fixation is induced in bacteroids. Therefore, symbiosomes are organelle-like units of plant cell responsible for nitrogen fixation (Brewin 1998; Tsyganov et al. 2003). A pronounced differentiation is typical for rhizobia-infected plant cells, such as an increase in internal membrane structures participating in the PBM formation and biosynthetic processes. Polyploidization and chromatin decondensation are typical for these cells correlating with an elevated transcription activity (Pawlowski and Bisseling 1996). Biochemically plant cell differentiation is expressed as a de novo synthesis of many proteins including leghaemoglobin and nodule-specific isozymes of carbon and nitrogen metabolism (Vance and Heichel 1991). Leghaemoglobin binds oxygen actively ensuring its transport towards symbiosomes (which are characterised by the intensive respiration necessary to support energy consuming nitrogen fixation) and microaerobic conditions inside the nodules (required for the nitrogenase activity). The carbon and nitrogen metabolic enzymes responsible for the energy supply to nitrogenase and for the assimilation of fixed nitrogen are nodule specific (Fedorova et al. 1999).

The processes described above result in formation of a complex nodule of indeterminate type (Fig. 1g). Its major structures are: (1) central rhizobia-infected tissue where nitrogen fixation occurs (histological zone III); (2) peripheral vascular bundles which transport photosynthates into and nitrogen-compounds out of the nodule; (3) apical meristem which ensures nodule growth (histological zone I) (Fig. 1l). The latter is responsible for a continuous renovation of "infected" (colonised) tissue and its differentiation into zones corresponding to the different stages of endosymbiosis development (Brewin 1991, 1998, 2004; Mergaert et al. 2006).

The final stage of symbiosis development is represented by the senescence of nodule tissue in the basal histological zone IV of the nodule where the symbiosomes are transformed into lytic compartments, bacteroids and other symbiotic sub-cellular compartments are digested by the host cells. Moreover, in alfalfa (lucerne) the release of bacteria into the decaying nodular tissue was observed followed by intensive propagation of bacteria without any signs of bacteroid differentiation (Timmers et al. 2000). This process, localized in a special histological zone V (occasionally, even in the actively nitrogen-fixing nodules), may be responsible for an increase in the number of nodule bacteria cells in the environment after the decay of the nodule and for the movement of bacteria towards an improved nitrogen-fixing activity in symbiosis with the host plant.

A fine balancing of partners' interactions is also highly dependent on exopolysaccharides (EPS), lipopolysaccharides and cyclic glucans located on bacterial surfaces (Breedveld and Miller 1994; Leigh and Walker 1994; Brewin 1998; Cheng and Walker 1998; Niehaus et al. 1998; Broughton and Perret 1999). The surface rhizobia components are suggested either to suppress the plant defence systems or to make bacteria tolerant of them but in either case they play an important role in all stages of the colonisation process: during infection thread formation and growth, endocytosis of bacteria and symbiosome differentiation.

In addition to the local mechanisms of endosymbiont control, legumes possess a systemic autoregulation of nodulation by signals circulating between roots and shoots. Mutants impaired in autoregulation were selected as supernodulating (phenotype Nod^{++}). Usually these mutants retain nodulation in the presence of high nitrate doses inhibiting symbiosis in wild-growing plants (phenotype Nts – Nitrate tolerant symbiosis). Therefore, autoregulation represents a strategy through which the host plant can balance the symbiotrophic nitrogen nutrition with the energetically cheaper combined nitrogen nutrition. Involvement of the shoots in control of Nod^{++} phenotypes was demonstrated by means of shoot/root reciprocal grafting (Caetano-Anolles and Gresshoff 1991; Sagan and Duc 1996). The autoregulatory response may be mediated by an abortion of excessive IT in the cortical cells *via* a mechanism similar to hypersensitive reaction between plants and some pathogens (Vasse et al. 1993).

The ontogeny of nodulation exists along with other developmental and regulatory plant processes (van Buuren et al. 1999; Beveridge et al. 2007) in which phytohormones, play important roles. Physiological and developmental effects of phytohormones on nodulation have long been recognized and are extensively described (Guinel and LaRue 1992; Penmetsa and Cook 1997; Guinel and Geil 2002), however, the mechanisms of phytohormone regulation of nodule initiation remain relatively obscure. The gaseous plant hormone ethylene and its precursor 1-aminocyclopropane-1-carboxylic acid (ACC) are some of the earliest phytohormonal signals associated with the 'negative' regulation of nodule number. For example, treatment with ethylene or ACC inhibited nodulation in a wide range of legumes. In contrast, exposure to inhibitors of ethylene activity, for example aminoethoxyvinylglycine (AVG), and silver ions increased nodulation in many of the legumes tested. Such inhibitors also partially restored nodulation in selected low-nodulating pea mutants (Guinel and LaRue 1992). Ethylene is supposed to regulate the number of developing nodules by controlling the position of nodule initiation and the ability of plant cells to initiate nodule foci (Nukui et al. 2000; Lohar et al. 2009). It has been reported that transgenic ethylene and ACC in sensitive *L. japonicus* lines increased nodule number per plant by comparison with wild type (Lohar et al. 2009). The *Lotus* transgenic lines described were not, however, super- or hypernodulated over a large portion of the root, and did not exhibit nitrate tolerance in nodulation, demonstrating that ethylene regulates nodulation at other sites and is not directly functional in autoregulation of nodulation and nitrate inhibition of nodulation.

Immediately after induction of nodulation the processes aimed at regulating the performance of the endosymbiont start inside the host. The importance of these processes is evident from a much greater potential for propagation of bacteria with respect to plant cells. The perisymbiotic interfaces (IT walls, symbiosome membranes) ensure separation of the partners and play an important role in regulation of the internal bacterial population. In addition to these "physical" barriers, a range of processes occur inside nodules which are similar to the defence reactions induced by pathogens. They include synthesis of flavonoids, phenolics, chitinases, callose, peroxidases, extensins and some pathogen-regulated (PR) proteins (reviewed in: Tikhonovich and Provorov 2007). In nodules however, these reactions are not as intensive as those occurring during pathogenesis because the level of defence compounds in plants undergoing nodulation is much lower than in plants responding to pathogen attack. It is likely therefore, that these reactions are probably regulated in a different way, by regulatory symbiotic genes that control the number of symbiotic microorganisms and the development of symbiotic compartments rather than determining their activity.

Expectedly, early stages of both determinate and indeterminate nodules are encoded by orthologous genes in different legumes. Moreover, recent achievements in molecular cloning and sequencing the legume genes responsible for nodule development have shown that some of these genes are involved not only in nodulation, but also in mycorrhization. The developmental genetics of early stages of RN and AM symbioses is described in the next section.

Developmental Genetics of RN and AM Symbioses

In RN and AM symbioses the genetic systems of the partners are highly integrated because the complex developmental processes which lead to the formation of intercellular and sub-cellular symbiotic compartments are controlled by both organisms. The development and function of the symbioses is reliant to greatest extent on the plant. Developmental genetics in RN is now well described because both plants and nodule bacteria can be subjected to genetic analysis during nitrogen-fixing nodule formation and functioning. There has been less investigation of AM systems. Mainly this is due to the difficulties encountered in culturing AM-fungi, caused by their obligate symbiotic lifestyle and impossibility of using selective media. Additionally, genetic analysis of AM fungi is more complex because of their heterokaryotic nature and lack of sexual process (Smith and Read 1997; Pawlowska and Taylor 2004).

The plant genes involved in development of RN and AM symbioses may be divided into two groups, depending on how they are identified. The genes of one group, *Sym* genes (from symbiosis, the term was suggested by Lie (1971), have been identified with the use of formal genetic analysis (commencing with the selection of plant mutants defective in nodule development). The other group of genes comprises nodulins (from *nodule*) (van Kammen 1984), mycorrhizins (from *mycorrhiza*) (Gianinazzi-Pearson 1996; Niehaus et al. 1998), and symbiosins (from *symbiosis*) (Kistner et al. 2005; Küster et al. 2007). These genes were identified by molecular genetic methods, through identification of proteins and mRNAs synthesized *de novo* in root nodules (nodulins) or roots colonized by AM fungi (mycorrhizins). The genes whose expression is induced during the development of both RN and AM are called symbiosins (Küster et al. 2007). Genes of these groups are suspected to be playing different roles in the processes which may be referred to as the "management of microsymbionts" within plant roots.

Most of the known nodulins/mycorrhizins/symbiosins are cloned and sequenced; however, functions for many of them have been identified only preliminarily using the sequence data of the encoding genes and location of gene products in the symbiotic compartments.

The products of several nodulin genes represent the structural elements of newly constructed temporary compartments (ITs, symbiosomes) developed during symbiosis (Albrecht et al. 1998). Also, some of nodulins may play several roles in modulating the hormonal status of developing nodules (Kumagai et al. 2006; Wan et al. 2007). Recently *in silico* and microarray-based transcriptome profiling has allowed the identification of nodulins and of other classes of these genes, mycorrhizins and symbiosins. All of these are activated in response to an AM fungal signal, or by either rhizobial or AM fungal stimuli, respectively (Kistner et al. 2005). Several hundred genes were found to be activated in different stages of either symbiosis, with almost 100 genes being co-induced during nodulation and in AM formation. These co-induced genes can be associated with different cellular functions required for symbiotic efficiency, such as facilitating transport processes across the perisymbiotic membranes that surround the endosymbiotic bacteroids in root nodules and the arbuscules in AM roots (Küster et al. 2007).

Most of the nodulins, mycorrhizins and symbiosins seem to play subordinate roles in symbioses, having an influence on the functionality and stability. In turn, the main, regulatory role controlling symbiotic programmes is attributed to the *Sym* genes. These genes are usually not expressed outside of the symbiotic structures and there are many examples of high levels of functional and sequence homologies between them in different legumes. The first genes of this group were identified in spontaneous mutants obtained from natural populations of legumes (Nutman 1946) and afterwards in experimentally induced mutants deficient in nodulation and nitrogen fixation (Nod$^-$ phenotype) (Jacobsen 1984). Subsequently, it was demonstrated that mutations in some of these genes also affect the ability of plant to form AM (Gianinazzi-Pearson 1996). The existence of common genes necessary for both AM and RN symbioses development suggests that both symbioses were more closely linked than was previously suspected. Cloning and sequencing of the common symbiotic genes helped to show that AM and root-nodule symbioses share a common signalling pathway, which probably was established during evolution of the plant–AM symbiosis and was recruited afterwards into RN development (Parniske 2008).

Developmental Genetics of RN Symbiosis

Mutant plants which cannot form nodules (Nod$^-$, no nodules) or form nodules devoid of nitrogenase activity (Fix$^-$, no nitrogen fixation) may be induced either by physical or by chemical mutagens or be present in natural populations or in cultivars. Usually the mutants are selected on the basis of impaired growth in N-free substrates. More than 80 genes have been identified which are responsible for nodule development (Hadri et al. 1998; Borisov et al. 2007) by using these mutants in different legumes (*Pisum sativum* L., *Glycine max* (L.) Merr., *Clover* spp., *Medicago sativa* L., *Phaseouls vulgaris* L., *Cicer arietinum* L., *Vicia faba* L.). More than 40 genes were identified in pea (*Pisum sativum* L.) which is still one of the most convenient models for the genetic analysis of nodule development. As a rule, plant alleles determining Nod$^-$ phenotype are recessive, while Fix$^-$ alleles may be either recessive (in pea, clover, alfalfa) or dominant (in soybean). Some "symbiotic" mutations alter the root or shoot morphology, while the majority of these mutations do not affect development of "non-symbiotic" organs.

Analysis of genetic variability of the symbiotic properties of pea plants was pioneered by Govorov (1928) and Razumovskaya (1937). They found that some genotypes of pea originating from Afghanistan (which is part of the Central-Asian centre of origin for these plants) fail to form nodules with root nodule bacteria taken from European soils. Although some plants from the region with different genotypes, did form the nodules when inoculated with bacteria from Europe. Genetic analysis of local Afghan host strains by Lie (1971), identified the first symbiotic (*Sym*) pea genes.

The later genetic analysis of experimentally induced mutants in pea allowed the dissection of the major stages of symbiosis development into simple (elementary) steps, each step being controlled by a few genes or even single ones (Borisov et al. 1997, 2007; Tsyganov et al. 1998, 2002). To identify the order of expression of genes controlling the sequence of developmental stages an analysis of single and double mutant lines has been made. The developmental epistasis detected in some cases permitted the genes to be arranged in their order of activity.

The results of phenotypic characterization of single and double mutants suggested that the integrated programme of nodule development is composed not from a chain of successive stages (each begining only when the preceeding stages had been completed) rather by two sub-programmes which: (a) are implemented to some extent independently from each other; and (b) may be dissected into succeeding steps (Fig. 2).

The stages in the sub-programme of plant tissue colonization and differentiation of bacteria into symbiotic forms (the so-called bacteroids) are as follows: Hac – root hair curling; Crh – colonization of the pocket formed by curled root hair cells; Iti – infection thread growth initiation; Ith – growth and development of the infection thread in the root hair cell; Itr – growth and development of the infection thread in the root tissue; Itn – growth and development of the infection thread in the young root nodule tissue; Idd – infection droplet differentiation; Bad – bacteroid differentiation, i.e., transformation of bacteria into organelle-like symbiotic forms; and Nop – structural and functional nodule persistence. The sequence of stages in nodule tissue differentiation is defined as follows: Ccd – cortical cell divisions, division of the cells of internal root cortex; Npd – nodule primordium development; Anm – apical nodule meristem formation; and Nmp – nodule meristem persistence.

These sub-programmes are controlled by the same sets of plant *Sym* genes (Fig. 2), but, their functional orders in the colonisation (infection) and symbiosome differentiation and nodule tissue development subprograms may be not identical (Tsyganov et al. 2002; Borisov et al. 2007). Selected stages of both sub-programmes are described in previously in detail, here the order of pea (*P. sativum*) regulatory gene functioning identified after experimental mutagenesis followed by classical genetic analysis and phenotypic characterisation is given (Fig. 2).

Plant Genes Implicated in AM Development

The most productive approach allowing the identification of plant mycorrhization regulatory *Sym* genes is to search among those of legumes implicated in nodulation. Duc et al. (1989) were first to demonstrate that some legume mutants that were impaired in nodule development also blocked AM formation. As a result of analysis of diverse collections of legume mutants with impaired nodulation several AM developmental stages controlled by common symbiotic genes were identified. In particular, for pea eight common *Sym* genes (Fig. 2) and two developmental stages were identified: Pen – root penetration, and Arb – arbuscule development (according to Marsh and Schultze 2001). Concurrently two additional stages were found for *Lotus japonicus*: Coi (Marsh and Schultze 2001) – cortex invasion; Ici (Senoo et al. 2000a) – inner cortex invasion.

To date the vast majority of mutants isolated in different legumes are defective in the early stage of AM formation and block fungal penetration (Pen$^-$) into the root epidermis (previously, the phenotype has been referred as Myc^{-1} (Gianinazzi-Pearson 1996)). These include the first AM mutants isolated from populations of *P. sativum* and *Vicia faba* (Duc et al. 1989), as well as mutants identified in *Medicago truncatula* (Sagan et al. 1995), *Medicago sativa* (Bradbury et al. 1991), and more recently in *L. japonicus* (Senoo et al. 2000b). All these mutants are Nod$^-$ and carry mutations in early symbiotic genes, for example *PsSym8*,[1] *PsSym9* and *PsSym9* for pea (Fig. 2).

Commonly associated with this type of mutantion is the formation of complex and somewhat abnormal appressoria as compared with the wild-type (WT). This can be seen clearly in the *mcbep* (mycorrhizal colonization blocked at epidermis) mutants of *L. japonicus* (Senoo et al. 2000b), the alfalfa (*M. sativa*) line MN-NN1008 (Bradbury et al. 1991), and the P2 (*Pssym9*) mutant of pea (Duc et al. 1989).

---

[1] *Ps, Lj, Mt* are abbreviations to the Latin plant names *Pisum sativum*, *Lotus japonicus*, *Medicago truncatula*, respectively.

Irrespective to fungal genotype, each of these mutations prevents penetration of the root epidermis and results in the formation of complex hyphal branching, multiple and unusually swollen appressoria. This effect has been suggested to be a consequence of the continuing but unsuccessful attempts by the fungus to penetrate through the root surface (Bradbury et al. 1991; Senoo et al. 2000a). This is possibly why it was found that the number of appressoria in the roots of Pen$^-$ mutants is often increased as compared with the wild-type (Bradbury et al. 1991; Senoo et al. 2000b) although sometimes there may be decreases (Duc et al. 1989) or no difference (Bradbury et al. 1991; Bonfante et al. 2000; Senoo et al. 2000b). Also the number of appressoria varies between Pen$^-$ mutants depending on the species of the AM fungi (Bradbury et al. 1991).

It has been shown that plant defence responses and signalling events are altered in Pen$^-$ mutants, for instance: cell wall thickening and increases in the deposition of callose and phenolics by epidermal and hypodermal cells at the points of contact with appressoria (Gollotte et al. 1993); increased levels of endogenous salicylic acid and transcripts of a number of defence related genes in the mutant roots compared with the wild-type plants (Blilou et al. 1999; Ruiz-Lozano and Bonfante 1999). By using the *M. truncatula* mutant TRV25 (Sagan et al. 1995) possessing an epidermal block, a common *MtDMI3* gene (orthologue for *PsSym9* pea gene) (see later in Section 4.3.1) has been found which is essential for the assembly of the pre-penetration apparatus in epidermal cells (Genre et al. 2005).

By contrast to the mutants in which an infection process aborts after appressorial formation, a large collection of mutants were described in *L. japonicus* which are blocked prior to cortex invasion (Coi$^-$) although they form normal appressoria and are capable of penetrating the rhizodermis (Wegel et al. 1998; Bonfante et al. 2000; Parniske 2000). The mutants having phenotypes (Ici$^-$) that are affected to varying degrees in the rate of intercellular progression by hyphae. These include phenotypes in which inner cortical invasion does not occur as in the case of the *mcbee* (mycorrhizal colonization blocked between epidermis and exodermis] and the *mcbex* (mycorrhizal colonization blocked in exodermis] mutants of *L. japonicus* (Senoo et al. 2000a). It should be noted, that in the case of the *mcbex* mutants, although they predominantly express an Ici$^-$ phenotype, the overproduction of deformed appressoria and occasional formation of abnormal arbuscules is found. In pea and alfalfa, the stages Coi or Ici was not identified, and in all mutants blocked for cortex colonization then epidermis colonization is similarly affected (Duc et al. 1989; Bradbury et al. 1991; Sagan et al. 1995). Very likely, these stages are absent in such plants because they lack an exodermal layer which distinguishes them from *L. japonicus*. An exodermis is known to be an additional barrier for fungal penetration (Brundrett 2002).

The cloned legume genes controlling Pen and Coi stages are shown to be responsible for the signal transduction cascade inside the root which is common for both RN and AM (see later in Section 4.3).

Some mutants permit the colonization of the inner cortex but do not form arbuscules, as exemplified by R69 of *P. vulgaris* (Shirtliffe and Vessey 1996) and the Nod$^-$/Fix$^-$, MN-IN3811 mutant of alfalfa (Bradbury et al. 1991), or form truncated (stumpy) arbuscules (RisNod24, RisNod26 (*Pssym36*) isolated in *P. sativum* (Gianinazzi-Pearson 1996)), whereas appressoria formation, root penetration and hyphal proliferation within the cortex proceed normally. The phenotype was designated as Arb$^-$ (Marsh and Schultze 2001), while previously it was denoted as Myc^{-2} (Gianinazzi-Pearson 1996). This is the latest stage of AM development which can be revealed at present using the legume mutants defected in nodulation.

The two *P. sativum* mutants SGEFix$^-$-1 (*Pssym40*), SGEFix$^-$-2 (*Pssym33*), forming root nodules which do not fix atmospheric nitrogen, were characterized for the dynamics of AM development. Mutation in the *Pssym33* gene caused decreased mycorrhizal colonization and delayed arbuscule development (Rmd$^-$ phenotype), whereas mutation in *Pssym40* accelerated AM colonization and arbuscule turnover (Rmd^{++} phenotype) (Jacobi et al. 2003). Since the both mutants produce leaky phenotypes dependent on temperature, it seems likely that these genes are involved in some kind of hormonal regulation of the host plant and in symbiosis development.

Interestingly, several hypernodulating mutants, defective in systemic autoregulation of nodulation, are also characterized by increased mycorrhizal colonization and arbuscule development. The examples of such mutants are P88 (*Pssym29*) induced in *P. sativum*, those of *M. truncatula* impaired in the *MtSUNN* gene – Myc^{A++} phenotype (Morandi et al. 2000; Schnabel et al. 2005a), as well as those of *L. japonicus* mutated in the *LjHAR1* – Arb^{++} phenotype (Solaiman et al. 2000; Nishimura et al. 2002a), similar to those mentioned above. Hence both local signal transduction and systemic autoregulation may be common in nodular and AM symbioses.

The AM mutants obtained by using the approach described above allows the identification of genes common in both AM and RN symbioses, but not AM-specific genes. The direct screening of mutants with impaired AM development was until recently, only possible for non-legumes including tomato (Barker et al. 1998; Gao et al. 2001; David-Schwartz et al. 2001, 2003) and maize (*Zea mays* L.) (Paszkowski et al. 2006). One tomato *rmc* mutant demonstrated Pen$^-$, Coi$^-$ and Rmc (reduced mycorrhizal colonization) phenotypes which were fungus-specific (Gao et al. 2004). The phenotypes with fungal penetration into epidermal or hypodermal cells were shown to be associated with an enhanced and more prolonged defence gene expression (Gao et al. 2004). So, it has been supposed that defence-like responses may be the basis of the mechanisms underlying the specificity of interactions between AM fungi and the *rmc* mutant. The phenotypes were similar to those of Pen$^-$, Coi$^-$ and Rmd$^-$, respectively, observed in legume mutants. The same is true for the maize mutant phenotypes *taci1* (decreased mycorrhization level) and *Pram1* (hypermycorrhization) (Paszkowski et al. 2006). A rice (*Oryza sativa*) mutant affected in *OsIPD3* gene manifested *rmc* phenotype (Chen et al. 2008). The *OsIPD3* gene is determined to be an orthologue of *M. truncatula MtIPD3* gene common in both AM and RN symbioses (Chen et al. 2008).

Although phenotypically similar to other Pen$^-$ mutants selected from pre-existing Nod$^-$ mutants, the *rmc* mutant was identified by direct screening for abnormal AM. It may therefore harbour a mutated gene specific to the mycorrhizal symbiosis. At the same time, nearly all cloned legume genes required for nodulation and AM development have their putative homologues in the non-leguminous plants (Zhu et al. 2005; Chen et al. 2008) including *Arabidopsis thaliana*, which forms neither RN nor AM sybmioses (Zhu et al. 2005). These data suggest strongly that the pre-existing genes have been recruited from the diverse cellular processes to function as the symbiotic (AM) genes.

The highly interesting group of the mutants induced in non-legume plants represented by *pmi* and *pmi2* of tomato (David-Schwartz et al. 2001, 2003) and *nope1* of maize (Paszkowski et al. 2006) mutant lines is those impaired in the pre-symbiotic stage Pre-Pen (pre-penetration) (according to (Marsh and Schultze 2001)). These mutants are principally distinguished from those described in legumes as their mutations are unique to the AM symbiosis.

Only recently were the first AM-specific legume mutants (*ram1*, *ram2*) characterized in *M. truncatula* by using direct screening of a mutated plant population (Marsh et al. 2006, 2008). These mutants have been shown to represent at least two complementation groups, which are not allelic to any of the known symbiotic mutants. These mutants had normal nodulation phenotypes. By contrast to the non-nodulating and non-mycorrhizal *M. truncatula dmi* mutants, which form extensive and complex appressoria on the root epidermis (Sagan et al. 1995), these new mutants exhibited no substantial contact with the root surface. It is supposed therefore, that these mutants may be altered in their reception of fungal signals required in order to establish normal symbiosis (Marsh et al. 2008). Also the direct screening of a mutagen-treated pea (*P. sativum*) population has been initiated and a series of putative mutants presumably defective in AM development were isolated (Shtark et al. 2007). A "nurse plant" inoculation system (Rosewarne et al. 1997) has several advantages by comparison with conventional methods of inoculation with AM-fungi, and is supposed to be a potential tool for the isolation of mycorrhizal mutants from different legumes allowing the identification of new AM-specific plant genes. The genetic analysis of the mutants which is now in progress will identify new pea genes controlling the development of AM symbiosis.

Therefore, based on the analysis of all the diverse mutant phenotypes, the following stages of AM development can be distinguished : (1) Pre-Pen, (2) Pen, (3) Coi, and (4) Arb. The present classification is preliminary and will be modified along with accumulation of novel data on genetic and molecular control of symbiosis. In the next section, the comparative analysis of present knowledge of the molecular mechanisms of AM and RN symbioses in the leguminous (crop and model) plants is presented.

The Use of Model Legumes for Studying Molecular Genetics of Symbioses

The large sizes of genomes in crop legumes (e.g., in soybean and pea) as well as their low capabilities for transformation, greatly complicates the cloning of symbiotic genes, analysis of primary structures and their subsequent genetic manipulation. In the early 1990s other legume species, such as: *Lotus japonicus* (Regel.) Larsen (Handberg and Stougaard 1992) and *Medicago truncatula* Gaertn. (Barker et al. 1990; Cook 1999) were used as a consequence as model plants for such studies. These species are characterized by small genomes (~ 470–500 Mb, according to (Young et al. 2003)) and can be genetically transformed with relative ease (Cook et al. 1997; Cook 1999; Stougaard 2001; Udvardi 2001). In addition, their short lifecycles and high volumes of seed production make them attractive and convenient models for studying the molecular basis of RN and AM symbioses.

Genetic analysis in model and crop legumes was started using experimental mutagenesis. Large-scale programmes of insertion, chemical and X-rays mutagenesis, performed by several research groups, resulted in the identification of numerous symbiotic mutations in *L. japonicus* and *M. truncatula* (Penmetsa and Cook 1997; Schauser et al. 1998). The genes involved in early stages of RN symbiosis (so-called early *Sym* genes) were of primary interest in these studies, particularly because dissecting the mechanisms by which the Nod factor signal is perceived and transduced in host plants were considered to be the main targets for symbiosis research (Albrecht et al. 1999). The results of this research showed that after the perception of Nod factor RN symbiosis uses the same signalling pathway as does AM (so-called "common symbiosis pathway" (CSP)), though with slight differences. The general scheme of CSP work is described below.

General Scheme of Functioning of the Common Symbiosis Pathway

The data obtained during the last 10 years allows the reconstruction of the symbiotic signalling pathway which starts in RN and AM symbioses with recognition of the Nod and Myc factor, respectively, and goes on as signal transduction inside the root (Fig. 4). In legumes Nod and Myc factors are most likely perceived by specific receptor complexes, though the structure of such complexes remains unknown for the possession of the Myc factor (Parniske 2008). In turn, the receptor for the Nod factor is thought to be a heterodimer composed of at least two LysM containing receptor kinases (LjNFR5/MtNFP/PsSYM10 and LjNFR1/MtHCL/PsSYM37) (see below). This receptor complex (at least in pea and possibly also in *M. truncatula*) is thought to be needed at the start of interactions and for penetration by the bacteria into the cortex through root hairs (Zhukov et al. 2008). At this later stage, the presence of some alternative homologous proteins able to substitute for these kinases in the receptor complex is suggested (Smit et al. 2007; Zhukov et al. 2008). Probably, the diversity of receptor kinases could compensate for the variability of *Rhizobium* strains and increase the specificity of their interactions with rhizobia.

Reception of the Myc factor is less specific, but it is absent in the non-mycotrophic plants (Navazio et al. 2007). The Myc factor was demonstrated to induce expression of symbiosis-specific nodulin (symbiosin) *MtENOD11* in the roots of *Medicago truncatula* (Kosuta et al. 2003). At the same time plant defence responses are not induced by this factor (Navazio et al. 2007). It is supposed that

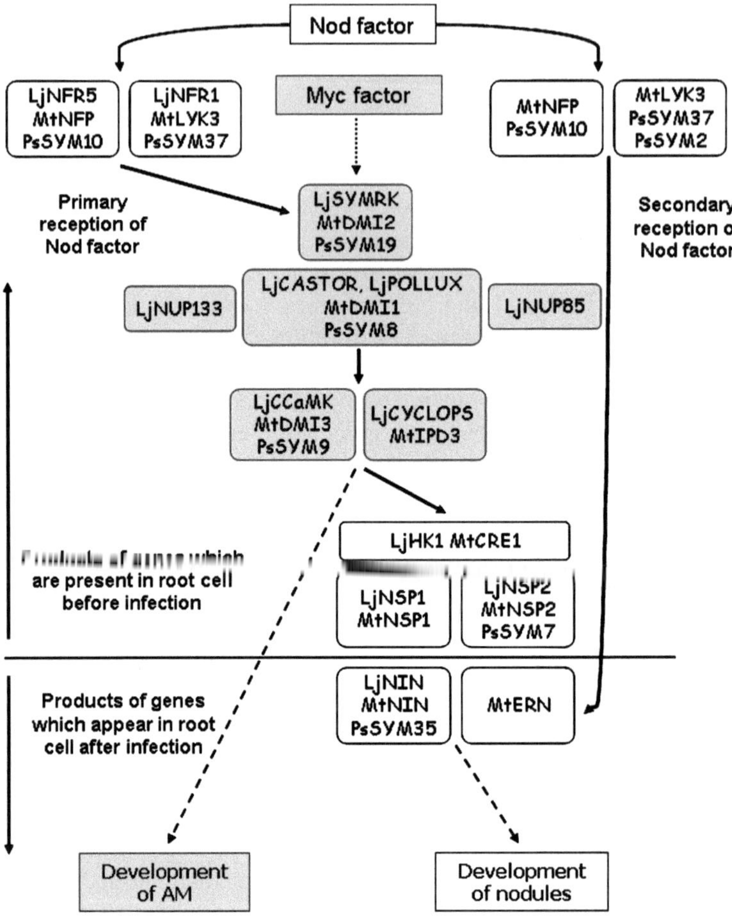

Fig. 4 The schematic representation of symbiotic signal transmission by the products of legume *Sym* genes.Products of genes involved in nodule formation only are given in white boxes. Products of common genes which are involved in both arbuscular mycorrhiza (AM) and root nodule (RN) formation are in grey boxes. Products of genes which expression is induced *de novo* by signal from microsymbionts are below the horizontal line separating the picture, whereas the products of genes which are already present in root cells are above this line. Products of common genes which are in the same box are homologues in different plant species (Lj – *Lotus japonicus*, Mt – *Medicago truncatula*, Ps – *Pisum sativum*). The continuous arrows represent a direct transmission of signal; dotted arrows mean its indirect transmission (e.g. in the case of Myc factor reception the signal goes indirectly to SYMRK)

the Myc factor could be produced constitutively by the fungus, without any physical contact with the plant partner (Navazio et al. 2007).

After the first step in reception of Nod or Myc factor, the signal is transmitted to the common signalling pathway. The first element in this pathway is an LRR-receptor kinase, or SYMRK (symbiotic receptor kinase) (LjSYMRK/MtDMI2/PsSYM19), which is required for both RN and AM development (Endre et al. 2002;

Stracke et al. 2002). Earlier it was suspected that LRR-receptor kinase could be a Myc factor receptor, but now it is thought more likely that this kinase indirectly perceives the signal from the fungus. Interestingly, the activity of this kinase is also required for a complete development of late stages of the symbioses, at least for rhizobial infection, since silencing of *MtDMI2* by inducible anti-sense mRNA at a later stage, during nodule formation, results in Fix$^-$ phenotype (Limpens et al. 2005).

The symbiosis receptor kinase SYMRK acts upstream of the Nod and Myc factor-induced Ca^{2+} spiking in the perinuclear region of root hairs within a few minutes after the application of the Nod factor (Wais et al. 2000). Perinuclear calcium spiking involves the release of calcium from a storage compartment (probably the nuclear envelope) through as-yet-unidentified calcium channels. Currently, it is known that the potassium-permeable channels (LjCASTOR and LjPOLLUX/MtDMI1/PsSYM8) (Ane et al. 2004; Imaizumi-Anraku et al. 2005; Edwards et al. 2007) may compensate for the resulting charge imbalance and could regulate calcium channels in plants (Peiter et al. 2007; Riely et al. 2007). Also, nucleoporins LjNUP85 and LjNUP133 (to date described only in *Lotus*) are required for calcium spiking, although their mode of involvement is unknown. Probably, LjNUP85 and LjNUP133 may be a part of specific nuclear pore subcomplex that plays a crucial role in the signalling process requiring interaction at the cell plasma membrane and at nuclear and plastid organelle-membranes to induce Ca^{2+} spiking (Kanamori et al. 2006; Saito et al. 2007).

Calcium spiking is characteristic for both RN and AM formation, but the frequency and amplitude of oscillations differ in these processes (Oldroyd and Downie 2004). These Ca^{2+} spikes are supposed to activate a calcium- and calmodulin-dependent protein kinase (LjCCAMK/MtDMI3/PsSYM9) that is also necessary for Nod factor signalling and AM development (Catoira et al. 2001). This kinase contains an autoinhibition domain, which if removed leads to a spontaneous activation of downstream transcription events and the induction of nodule formation in absence of rhizobia (Gleason et al. 2006). Thus, CCaMK appears to be a general regulator in both symbioses activating different cascades of signals for RN and AM in response to different Ca^{2+} spiking, because the next steps of nodulation signalling are independent from those of AM: the mutations in downstream *Sym* genes do not change the mycorrhizal phenotype of legume plant. Interestingly, the mutations in any *Sym* genes do not influence the defence reactions, suggesting that signalling pathways of mutualistic symbioses and pathogenesis are substantially different.

The calcium–calmodulin-dependent protein kinase of *Lotus japonicus* (LjCCaMK) is known to form a complex with LjCYCLOPS, a phosphorylation substrate, within the nucleus (Parniske 2008). *Ljcyclops* mutants severely impair the infection processes of bacterial or fungal symbionts (Albrecht et al. 1998). During RN symbiosis, *Ljcyclops* mutants exhibit specific defects in their infection-thread initiation, but not in nodule organogenesis (Yano et al. 2008), indicating that LjCYCLOPS acts in an infection-specific branch of the symbiotic signalling network (Parniske 2008). *LjCYCLOPS* encodes for a protein with no overall sequence similarity to proteins of known function, but contains a functional nuclear localization signal and a carboxy-terminal coiled-coil domain. Further work is needed to shed light upon the role(s) of *LjCYCLOPS* (as well as its homologue *MtIPD3* in *Medicago* (Messinese et al. 2007) and possible a homologue in pea) during development of symbioses. No other *Sym* genes cloned to date have been found to be involved in the mycorrhization process in legumes.

It is supposed that CCaMK probably phosphorylates specific transcription factors already present in cell: LjNSP1/MtNSP1 and LjNSP2/MtNSP2/PsSYM7, which influences changes of expression in several genes related to symbiosis development (Kalo et al. 2005; Smit et al. 2005). The activity of these proteins leads to transcriptional changes in root tissues, for instance, increasing the level of early nodulins *ENOD40*, *ENOD11*, *ENOD12*, *ENOD5*, which are known to be potential regulators of infection thread growth and nodule primordial formation (Albrecht et al. 1999; Heckmann et al. 2006; Murakami et al. 2006). Also, the changes in cytokinin status of host plants are detected, followed by up-regulation of genes encoding for nitrogen-fixing symbiosis-specific cytokinin receptors *LjLHK1* and *MtCRE1* (Gonzalez-Rizzo et al. 2006; Murray et al. 2007; Tirichine et al. 2007). Moreover, transcription regulators LjNIN/MtNIN/PsSYM35 and MtERN, are induced specifically downstream of the early Nod factor signalling pathway in order to coordinate and regulate the correct temporal and spatial formation of root nodules (Schauser et al. 1999; Borisov et al. 2003; Marsh et al. 2007; Middleton et al. 2007).

The presently identified genes are responsible for a signal cascade which aims to induce the nodulin, mycorrhizin and symbiosin genes responsible for building the symbiotic structures and initiating their biochemical functions. It is supposed that the signalling pathway did not appear *de novo* in legumes when they become able to form nodules, but developed from a pre-existing system of AM formation into which the novel, nodule specific genes were recruited. New genes had been involved in RN development, especially those encoding receptors recognizing hormones (e.g. cytokinins) and hormone-like molecules (Nod factors).

Plant Receptors Involved in Nodule Formation

The spectrum of secreted signal molecules – Nod factors – appears to be a unique feature of strains of nodule forming bacteria (Geurts et al. 1997). In turn, the specificity of interactions is determined by plant receptors recognizing the structure of the Nod factor molecule (Limpens et al. 2003; Radutoiu et al. 2003, 2007; Oldroyd and Downie 2008). Intensive work in the past 10 years has led to the identification and cloning of legume genes encoding for receptor kinases perceiving a Nod factor during signal interactions with nodule bacteria.

In *Lotus japonicus* two genes *LjNFR1* and *LjNFR5* (Nod factor receptor) were identified mutations of which elicited no response to *Mesorhizobium loti* inoculation or to Nod factor application, while remaining able to develop AM (Schauser et al. 1998). Positional cloning of these genes performed by fine genetic mapping followed by analysis of genomic libraries showed the respective nucleotide and putative protein structures of these genes (Radutoiu et al. 2003; Madsen et al. 2003). Both *LjNFR1* and *LjNFR5* genes appeared to encode receptor kinases with

3 LysM domains presented in the extracellular C-terminal part of the putative protein. LysM domains are found among enzyme proteins of some prokaryotes and eukaryotes, while receptor kinases with the LysM motif have been found only in plants. At a moment, LysM motifs have been shown to bind oligosaccharide molecules, therefore LysM-containing receptor kinases are supposed to represent the molecules capable of recognizing a bacterial Nod factor, although predicting the structure for a putative Nod factor-binding site needs further research (Radutoiu et al. 2003, 2007).

The transmembrane domain in both LjNFR1 and LjNFR5 putative proteins suggests that receptor kinases are located in the plasma membrane of the root cell, and when LysM domains perceive the Nod factor, Serine/Threonine kinase domain gives a start to a signal transduction cascade. It was found that kinase domain in LjNFR5 lacks an "activation loop" the absence of which makes kinase incapable of auto-phosphorylation. It is supposed that LjNFR5 needs the other protein, LjNFR1, for proper functioning, and therefore, they might work in tandem.

In *Medicago* and pea, the system recognizing Nod factor appears to be more complex. For *LjNFR5*, orthologues in *Pisum* and *Medicago* were found among known *Sym* genes (*PsSym10* and *MtNFP*, respectively) (Amor et al. 2003; Madsen et al. 2003; Arrighi et al. 2006). Mutations in both *Pssym10* and *Mtnfp* leads to an absence of any reactions to Nod factors but does not affect mycorrhization, indicating that receptors for Nod factors are placed upstream of the common signal transduction pathway. Obviously, if LjNFR5/PsSYM10/MtNFP needs the second component of the Nod factor recognition complex in order to work properly then it might be the protein product of an orthologue of *LjNFR1*. The phenotype of pea and *Medicago* mutants in homologous genes however, turned out to differ from that of *Lotus nfr1* mutants.

Historically, the first example of legume symbiotic genes involved in the recognition of rhizobia and perception of a Nod factor was pea gene *PsSym2* identified more than 30 years ago as a gene causing resistance in pea lines from Afghanistan (*Pssym2*^{A} allele) to infection by European strains of *Rhizobium leguminosarum* bv. *viciae* (Lie 1984). This resistance could be overcome by *R. l.* bv. *viciae* strains found in soils from the Middle East and North-West Russia that produce a Nod factor with an acetyl moiety substituted on a reducing terminus of a lipo-chito-oligosaccharide chain (Götz et al. 1985). The strains of *R. l.* bv. *viciae* from Europe producing a Nod factor without this substitution failed to nodulate "Afghan" peas (Fig. 3c). This situation illustrates the "gene-for-gene" interactions, when one gene of microorganism (*nodX*, encoding acetyltransferase, performing acetylation of Nod factor (Davis et al. 1988) interacts with one gene of plant (*PsSym2*, probably encoding a Nod factor receptor).

For years, *PsSym2* has been a model for researchers working with unique phenotypes of "Afghan" peas, but the DNA sequence of this gene remains unknown. The attempts to analyze the region of the *M. truncatula* genome orthologous to the region of localization for *PsSym2* in pea by means of genetic mapping of the phenotypic manifestations of *PsSym2* lead to the identification a series of genes encoding for receptor-like kinases containing LysM-domains. These studies were switched from pea to *M. truncatula* and as a result one of these kinases, *MtLYK3*, has been shown to be a Nod factor receptor, highly homologous to *LjNFR1* of *L. japonicus* (Limpens et al. 2003).

Medicago hcl mutants carrying mutations in *MtLYK3*, as well as *LYK3*-silenced transformants, have phenotypes characterized either by defects in root hair curling and an absence of the bacterial penetration into the root cortex or by successful root hair curling and the infection thread growth initiation (Wais et al. 2000; Catoira et al. 2001; Limpens et al. 2003; Smit et al. 2007). A similar phenotype was also described for "Afghan" pea carrying *Pssym2*^{A} allele and pea *Pssym37* mutants (Kozik et al. 1995; Zhukov et al. 2008). Both *Medicago* and pea plants were mycorrhized normally suggesting that *MtLYK3/HCL*, *PsSym2* and *PsSym37* are important only for RN symbiosis.

The map position of *PsSym37* appeared to be similar to that of *PsSym2* and syntenic (i.e. it is found on the same chromosome) to the location of *MtLYK3/HCL* in *M. truncatula* and *LjNFR1* of *L. japonicus*. Sequencing a pea homologue of *LjNFR1* in *Pssym37* mutants allowed the finding of nucleotide substitutions, suggesting an homology of *PsSym37* with *MtLYK3/HCL* and *LjNFR1*. Intriguingly one pea *Pssym37* mutant with an amino acid change in its receptor portion may be complemented with the Nod^{+} phenotype by the double acetylated Nod factor structure, like "Afghan" pea lines (Zhukov et al. 2008). The putative protein PsSYM37 in "Afghan" lines does not contain any amino acid changes compared with one in "European" lines with normal nodulation, suggesting that *PsSym37* could only be a functional analog of *PsSym2* gene, but not of *PsSym2* itself.

The region of the *Medicago* genome containing *MtLYK3/HCL* includes about 10 homologous genes encoding for receptor kinases with similar structures (Limpens et al. 2003). In such a region, new variants of receptor molecules could appear, for instance, as a result of non-reciprocal recombination events (Michelmore and Meyers 1998). Accordingly, the gene *MtLYK4* is highly homologous to the neighbouring genes *MtLYK3* and *MtLYK5*, whereas its receptor part is homologous to that of *MtLYK3* and the kinase part is homologous to that of *MtLYK5*. Among these genes, probably only *MtLYK3* participates in interactions with rhizobia, as its expression is specific to roots and nodules (Limpens et al. 2003). Presumably, the role of LysM-containing kinases encoded by other genes of this region provides recognition of chitin elicitors similar to Nod factor in their structure. The orthologous region of the pea genome must also contain a series of homologous kinase genes, one of which could be the elusive *PsSym2* gene. Sequencing one such gene, namely *PsK1*, showing the differences between "Afghan" and "European" pea lines in receptor part of the putative PsK1 protein allows supposition that *PsK1* or another gene of this region could actually be *PsSym2* encoding the changeable component of receptor complex recognizing the Nod factor structure (Zhukov et al. 2008).

Several years ago, a "two-receptor model" was proposed based on studies of bacterial mutants (Ardourel et al. 1994). This model suggests a signalling receptor inducing early responses with low specificity towards Nod factor structure, and an entry receptor that controls infection with more stringent specificity and, in the case of successful recognition the Nod factor, allows rhizobia to enter the root hair and root tissues. Indeed, blocks in infection occurring at this stage in pea *Pssym37* mutants and "Afghan" pea lines interacting with "European" rhizobia, as well as *Medicago hcl* mutants, tend to corroborate this model.

It could be concluded that the system of Nod factor recognition works in pea and *M. truncatula* in a similar fashion, and PsSYM10/MtNFP plays the role of a low-specific signal receptor, whereas PsSYM37/MtLYK3 performs more stringent recognition of Nod factor as an entry receptor. In fact, other LysM-containing receptor kinases potentially capable of binding lipochito-oligosaccharide molecules (and therefore for determining the structure of a Nod factor) could function as alternative components of Nod factor receptor complex, substituting *PsSym37/MtLYK3* (Smit et al. 2007; Zhukov et al. 2008). The adaptive role in the functioning of such LysM-containing kinases as exchangeable subunits of receptor complex could be explained by a necessity for interactions with several strains of rhizobia, which appear to be highly variable based on the structure of their secreted Nod factors.

In *Lotus*, which is evolutionarily remote from pea and *Medicago*, mutations in both *LjNFR1* and *LjNFR5* genes lead to an absence of any responses to a Nod factor present in the rhizosphere. This fact shows that mechanisms of Nod factor reception in *Lotus* differ from those of pea and *Medicago*, despite genes encoding for receptor kinases which recognize Nod factors being highly homologous in *Lotus*, *Medicago* and pea.

Cytokinin Signalling and Reception in Legumes

Nodule primordium formation and development is regulated in legume plants by changes in local concentrations of plant hormones – auxins, cytokinins and ethylene (Schultze and Kondorosi 1998b). Cytokinin appears to play a key regulatory role in this process, for the following reasons. Several legumes (e.g. *Medicago sativa* L. and *Trifolium pratense* L.) are reported to form spontaneous nodules in the absence of rhizobia which are structurally similar to nitrogen-fixing nodules (Truchet et al. 1989; Blauenfeldt et al. 1994). Additionally, experimental mutagenesis in *Lotus japonicus* revealed mutants which spontaneously form nodules (Tirichine et al. 2006). It was shown that spontaneously nodulating mutant lines carry mutations in a known gene for the calcium/calmodulin-dependent kinase *LjCCaMK* (Tirichine et al. 2006) and in a novel symbiotic gene for the cytokinin receptor *LjLHK1* (Tirichine et al. 2007).

LjCCaMK is known to detect the "calcium spikes" and activate the programmes of symbiosis development; such mutations which are associated with gains in functions leading to a loss of auto-inhibition of the activity of this kinase and spontaneous nodule formation (Gleason et al. 2006; Tirichine et al. 2006). Similarly, *LjLHK1* (*Lotus* histidine kinase) perceiving the changes in cytokinin concentrations could normally activate cell divisions in the root cortex leading to nodule primordium formation, while the mutant with an altered histidine kinase receptor has cytokinin-independent activity (Tirichine et al. 2007). It was concluded that cytokinin signalling is required for cell divisions that initiate nodule development and defines an autoregulated process where cytokinin induction of nodule stem cells is controlled by shoots.

In *Medicago*, homologous gene *MtCRE1* (cytokinin receptor 1) has also been shown to be involved in nodulation signalling. RNA interference of this gene leads to significant increases in lateral root development and a reduction in nodulation, associated with perturbations in both infection and nodule primordia formation (Gonzalez-Rizzo et al. 2006). Similarly, "loss-of-function" mutants in *Ljlhk1* of *Lotus*, exhibit abundant infection-thread formation but fail to initiate timely cortical cell divisions in response to rhizobial signalling (Murray et al. 2007). Conclusively, cytokinin signalling plays an important role in plant meristem formation and is directly involved in initiating root nodule organogenesis.

Intriguingly, cytokinin signalling plays a key role in Nod factor-independent nodulation. Recently, it was shown that nodulation of some legumes by rhizobia could occur in the absence of the *nodABC* genes and lipochito-oligosaccharidic Nod factors, indicating that other signalling strategies could trigger nodule organogenesis in some legumes (Giraud et al. 2007). Indeed, two symbiotic and photosynthetic *Bradyrhizobium* sp. strains specifically induce nodules on both the root and stem of the aquatic legume *Aeschynomene* sp., although the nodulation ability of these strains is restricted to a few species, including *A. sensitiva* and *A. indica* (Giraud and Fleischman 2004). These *Bradyrhizobium* strains produce a purine derivative, which is a cytokinin-like molecule (or a cytokinin precursor) and may act as a signalling molecule (Giraud et al. 2007). It is supposed that this molecule activates the cytokinin receptor and starts the Nod factor-independent nodulation. During "conventional" infection, Nod factors induce root hair deformation, leading to bacterial entrapment. In turn, root hair curling and infection thread formation are not initiated in Nod factor-independent nodulation and bacteria enter the plant tissue through cracks in the epidermis, often at the sites of lateral root emergence (Downie 2007).

Autoregulation of RN and AM Symbioses Development

Beside mutual recognition and signalling, another important feature of the development of symbioses is autoregulation by the host plant. The plant is known to govern the growth and differentiation of the microsymbionts in roots. For RN symbiosis, it is considered that host legumes control root nodule numbers by sensing external and internal cues. A major external cue is soil nitrate content, there is a feedback regulatory system by which earlier formed nodules suppress further nodulation through shoot–root communication using an important internal cue. The latter is known as autoregulation of nodulation (AUT), and is believed to consist of two long-distance signals: a root-derived signal that is generated in infected roots and transmitted to the shoot; and a shoot-derived signal that systemically inhibits nodulation.

It has been shown that in *Lotus japonicus*, the leucine-rich repeat receptor-like kinase, *HYPERNODULATION ABERRANT ROOT FORMATION 1* (*LjHAR1*), mediates AUT and nitrate inhibition of nodulation, and is thought to recognize the root-derived signal. Some cloned orthologous genes in *M. truncatula* and pea

(*MtSUNN* and *PsSym29*) are supposed to serve a similar autoregulation process as mutations in all these genes lead to supernodulating phenotype (Nod^{++}) (Krusell et al. 2002; Nishimura et al. 2002a; Schnabel et al. 2005). The root signal is suspected to be a small peptide related to the *CLAVATA3/ESR-related* (*CLE*) family of peptides, and it could be speculated that *LjHAR1* might recognize a *Lotus CLE* peptide induced by rhizobial infection as the root-derived AUT signal (Okamoto et al. 2009). Two *L. japonicus* genes *CLERoot Signal 1* (*LjCLE-RS1*) and *LjCLE-RS2* are strong candidates to encode for this root-derived signal. Another gene, *LjASTRAY*, encoding for transcription activator, is also supposed to play some role in autoregulation of the nodulation process (Nishimura et al. 2002b).

Very little is known about plant regulation of the mycorrhization process. It is demonstrated that in the case of *LjHAR1*, mutation also affects mycorrhizal formation (Myc^{++}) suggesting the shared role of *LjHAR1/MtSUNN/PsSYM29* in controlling the rate of root colonization by microsymbionts. Therefore, not only the local signal transduction but the systemic autoregulation may be common for the RN and AM symbioses. The systemic regulation of symbioses formation and maintenance also need further investigations.

The next step in unravelling the developmental genetics of symbioses is the study of gene networks at an intergenomic level i.e., coordinated expression by plant and microbe genes. For AM, the use of new molecular approaches, in particular transcriptomics, have identified a series of AM fungal genes showing altered expression levels during the symbiosis formation (Grunwald et al. 2004; Gianinazzi-Pearson and Brechenmacher 2004; Seddas et al., 2009 personal communication). The stages of fungal–plant interaction at which induction or repression commence are not well understood, and so the use of plant mutants impaired at different stages of AM development might be a useful approach to uncover patterns in the plant and fungal genetic relationship (Kuznetsova et al. 2010). Work aimed at identifying plant–rhizobial gene interactions using plant and microbe mutants is in progress.

Associations of Roots with Plant Growth-Promoting Rhizobacteria (PGPR)

Plant Growth Promoting Rhizobacteria (PGPR) are the most widespread plant co-habitants. They are taxonomically diverse including assorted groups of Gram-negative bacteria (e.g., *Azospirillum, Burkholderia, Enterobacter, Pseudomonas*), Gram-positive bacteria (*Bacillus* (and new genus *Paenibacillus*), *Streptomyces*) and even some archaea (Lugtenberg et al. 2001). The PGPR are inhabitants of soil in the vicinity of plant roots, many are able to attach to root surfaces and to AM and other fungal hyphae.

The PGPR provide several benefits (affecting the host plant either directly or indirectly) including: nitrogen fixation, stimulation of root development (due to phytohormone production), solubilization of soil phosphates, defence of plants from soil borne pathogens and improving host tolerance to abiotic stresses.

Nutritional Associations

The nutritional root-PGPR interactions were discovered and initially characterized by J. Dobereiner who identified a broad distribution of *Azospirillum* – cereal associations and described their basic physiology and ecology descriptions, as well as demonstrating a substantial agronomic potential (reviewed in: Döbereiner 1988). In her early papers, plant growth promoting activity was attributed mainly to nitrogen fixation which was estimated to provide 40–60 kg ha^{-1} and in some circumstances up to 100 kg ha^{-1} per season in cereal–*Azospirillum* associations. It was demonstrated later, however that a partial role in these plant–PGPR associations was due to phytohormone (auxin) synthesis (Costacurta and Vanderleyden 1995) which improves the root growth and assimilatory capabilities and hence aids nitrogen uptake by plants.

Despite the absence of visible anatomic differentiation in root-*Azospirillum* associations, their development involves a range of molecular interactions some of which may have commonalities with endosymbiotic associations. For example, in the genomes of *A. brasilense* and *A. lipoferum* DNA sequences were found which are homologous with genes *nodABC* and *nodQPCEFH* of *Sinorhizobium meliloti* (Fogher et al. 1985; Vieille and Elmerich 1990, 1992). By contrast to rhizobia, the expression of *nod* genes in *Azospirillum* does not require the presence of root exudates or their flavonoid components. This is consistent with the absence of promoters of *Azospirillum nod* genes in the *nod*-box sequences which are the targets of the NodD regulator activating these genes in rhizobia in the presence of root-exuded flavonoids (see Section 3). The presence of some elements from the system for Nod factor synthesis suggests however, that *Azospirillum* can elicit some signalling processes in the roots which are common to the components of nodulation/AM formation cascades.

Furthermore, genes homologous with chromosomal virulence genes of agrobacteria (*chvA* and *chvB*) were found in *Azospirillum* (Waelkens et al. 1987). Close homologues of these genes (*ndvA* and *ndvB*) were identified in rhizobia and found to participate in the synthesis of cyclic β-glucans involved in the formation of surface structures which are important for nodule development (Breedveld and Miller 1994). Genes *fixABC* were found in azospirilla which have homologues in rhizobia (Fogher et al. 1985) suggesting similarities in the regulation of associative and endo-symbiotic nitrogen fixation.

An important prerequisite of high nitrogen-fixing activity in azospirilla may be the secretion of auxins which can be synthesized via different biochemical pathways (Costacurta and Vanderleyden 1995). The direct involvement of bacterial auxins in the stimulation of root growth was demonstrated using IAA-deficient mutants of *A. brasilense*. The production of IAA by azospirilla may be based on root-exuded tryptophane and be responsible for promoting root growth and an increase in root exudation which supports rhizospheric nitrogen-fixation. The efficiency of the conversion of tryptophane into IAA by azospirilla however, does not exceed 1% while the amount of synthesized phytohormone is only about 10^{-11} M per seedling per 24 h (Kravchenko et al. 1994). The role of these synthetic processes in the development of associative symbiosis needs to be clarified further.

It has been reported that PGPR which produce ACC deaminase (which is involved in the catabolism of ACC (1-aminocyclopropane-1-carboxylate) – a precursor of plant hormone, ethylene), can lower ethylene concentration in a developing or stressed plant, protecting it against the deleterious effects of ethylene induced stress and facilitating the formation of longer roots (Penrose and Glick 2003). Also it has been demonstrated using transformants of *Azospirillum brasilense* Cd in which ACC deaminase gene (acdS) from *Enterobacter cloacae* UW4 was expressed under the control of different promoters that transformants having lower ACC deaminase activity showed significantly increased IAA synthesis as well. In addition, the result was increased ability to promote the growth of tomato seedlings (Holguin and Glick 2003).

The amount and composition of plant exudates which support the development of *Azospirillum*-root associations are dependent greatly on host genotype and environmental growth conditions. It has been demonstrated that components of wheat root exudates which support growth and nitrogen-fixing activity of *Azospirillum* include organic acids (initially citrate, succinate and malate) but not necessarily sugars (glucose, sucrose, xylanose) (Kravchenko and Leonova 1993). Interestingly, in wild wheat genotypes organic acids are dominant components of exudates, while in agronomically advanced cultivars sugars dominate demonstrating that during the domestication and breeding of this crop there has been a decrease in the potential for the development of beneficial PGPR communities in the rhizosphere.

Another mechanism ensuring the effective interaction between the host plants and their rhizospheric associates is the signalling process which attracts azospirilla into the rhizosphere and stimulates their attachment to the roots (plant function). This is dependent on the activities of bacterial flagella and on the surface molecules, initially, exo- and lipopolysaccharides. The effective attachment of azospirilla to the root surfaces is dependent on plant lectins which may be responsible also for the expression of properties which are beneficial to the bacteria. For example, purified wheat lectins were demonstrated to enhance the nitrogen-fixing activity of azospirilla and the accumulation of ammonium in gnotobiotic systems (Antonyuk et al. 1993). It is interesting to note that some azospirilla produce their own lectins which may mediate the adsorption of bacteria to root surfaces and also improve root susceptibility to inoculation (Nikitina et al. 1996) suggesting an operation of effective positive feedbacks between the partners in associative symbiosis.

Defensive Associations

A very efficient defence from the phytopathogens may be provided by microbes creating protective barriers solely on root surfaces. The best studied examples are: *Pseudomonas* (*P. fluorescens, P. chlororaphis, P. putida*), some *Serratia* (*S. marcescens*) and *Bacillus* (*B. cereus, B subtilis*) species. Many of these bacteria are capable of preventing attacks by pathogenic fungi (such as: *Fusarium, Rhizoctonia,* and *Verticillium*). Diverse mechanisms may be involved in host protection offered by PGPR (Lugtenberg et al. 2001).

The best studied mechanisms are the competitive exclusion of pathogens often related to their direct suppression by the bacterial antibiotic substances. For example, many *Pseudomonas* strains produce phenazines, e.g. phenazine-1-carboxamide (PCN), which are active against *Fusarium oxysporum* (Chin-A-Woeng et al. 2000) and *F. culmorum* (Shtark et al. 2003). The bacterial mutations leading to the PCN$^-$ phenotype usually results in the loss of biocontrol activity. Genes for PCN synthesis are transcriptionally activated in the host rhizosphere as a result of root exudates (host plant function). The bacterial synthesis of antibiotics may be accompanied by production of volatile dyes, for example of cyanides, their involvement in antifungal activities was demonstrated during the cultivation of PGPR-*Fusarium* combinations (Thomashow and Weller 1996). Cyclic lipopeptides, for example: tensine, syringomycin, syringopeptide, produced by *Pseudomonas* species also exhibit potent surfactant and broad-spectrum antibiotic properties (Raaijmakers et al. 2006).

An important mechanism for the suppression of pathogens by biocontrol microbes may result from competition for nutrients. Among the processes involved in this competition are bacterial siderophores which may possess much greater affinities for ferric ions than those for fungal siderophores (Miyazaki et al. 1995). The direct evidence for the importance of bacterial synthesis of siderophores in the suppression of pathogens was obtained by using genetically modified PGPR strains with increased or arrested siderophore production (Buysens et al. 1994; Raaijmakers et al. 1995). The value of siderophores in biocontrol effects under natural conditions is currently predominantly associated with their ability to induce forms of systemic resistance in plant (Preston 2004). See below and other chapters in this Book.

Competitive exclusion of pathogens by PGPR is best achieved when the bacteria exhibit high root-colonizing activity. Application of the technique of genetic labelling with Green Fluorescent Protein (GFP) suggested that these bacteria do not regularly result in the colonization of root interiors, and only sometimes may be observed inside the outer root tissues (Bloemberg et al. 1997, 2000). Root exudates may account for up to 30% of the net plant photosynthetic output and the optimal ecological niches for PGPR are zones of greatest rhizo-deposition. Most PGPR cells are concentrated on the root surface where the micro-colonies or bio-films form (O'Toole and Kolter 1998). Since the interactions of plants with root-associated bacteria are not specialised, the bacteria will colonize the roots of a broad spectrum of hosts. Specificity of the defensive association may be expressed however, at the point when antimicrobial compounds are being synthesized and this does not always correlate with bacteria taxonomy; many strains of *Bacillus* and *Pseudomonas* which have plant-protective properties have close relatives amongst phytopathogenic types (Stephens and Murray 2001; Catara 2007).

The importance of root colonization in the expression of beneficial defensive properties in symbionts is evident from research demonstrating that genes encoding for bacteria surface components that are required for the root adhesion (lipopolysaccharides, flagella) are also very important in this type of defensive process (Whipps 2001). Inactivation of these genes usually results in the loss of plant-protecting properties in PGPR strains while increasing the gene activities (*via* gene amplifications or fusions to constitutive "strong" promoters) improves host protection markedly (Lugtenberg et al. 2001).

Microscopic observations demonstrated that suppression of *Fusarium* may be correlated with *P. chloloraphis* attachment to the root surfaces and to pathogen hyphae (Bolwerk et al. 2004). As a result of this attachment, some PGPR strains commence their biocontrol functions by behaving as hyper-parasites of pathogenic fungi. This suppression may be correlated with the production of bacterial enzymes which destroy the pathogen cell walls. Some *Serratia* and *Bacillus* strains produce extracellular chitinases which inhibit the pathogen during developmental stages such as conidial germination and the growth of hyphae (Popova and Khatskevich 2004).

Several mechanisms which increase bacterial variability in the genes controlling surface components may be important in the enhancement of root colonizing activity in defensive PGPR. For example, one of the Tn5 insertions leading to enhancement of root-colonizing activity in *Pseudomonas* was located in the gene *mutY* encoding for repair of A:G base pairs in DNA after mutagen treatments (Lugtenberg et al. 2004). It was found that *mutY*:Tn5 mutation does not provide an immediate increase in root colonization. This is due to the secondary mutations which are generated in the background of *mutY*:Tn5 insertion since mutants with improved root adsorption may be selected positively by plants during the interaction of bacterial inoculum with root surface.

Another example of the involvement of enzymes from DNA metabolism in the control of plant-protective activities of PGPR was found in gene *sss* coding for site-specific recombinase in *P. fluorescens*. Inactivation of this gene resulted in an impairment of bacterial traits related to biocontrol of *Fusarium* (root colonization, synthesis of phenazines, cyanide and chitinases). The Sss recombinase appeared to be responsible for the "phase variation" in *Pseudomonas* resulting from the rearrangements of genes involved in biocontrol (Lugtenberg et al. 2001). Introduction of the multiple *sss* copies into *Pseudomonas* strains possessing a low biocontol activity may improve them significantly (Dekkers et al. 2000).

Sometimes the biocontrol activities in PGPR do not correlate with intensive colonization of host roots and plant protection results from only a small number of bacteria cells. This occurs when PGPR inoculation results in the induction of systemic resistance mechanisms that make the root non-accessible by pathogens. Initially this effect of PGPR was called ISR (Induced Systemic Resistance) and was attributed exclusively to nonpathogenic systems (Van Loon et al. 1998); SAR reactions, by contrast, were considered to be typical of the interactions with plant pathogens. Nevertheless, it was later found that the reactions of both types occur in either pathogen or nonpathogen systems and are distinguished by the nature of their endogenous elicitors (reviewed in: Preston 2004; Vallad and Goodman 2004). The conventional SAR reaction is characterized by an accumulation of salicylic acid signalling molecules and pathogenesis-related proteins (PR-proteins), whereas ISR reaction is based on signal transduction pathways regulated by jasmonates and ethylene. The systemic defence responses of both types may be elicited exogenously by PGPR cells attached to the roots or penetrating their outer tissues. Some molecules produced by PGPR (cell wall and cyclic lipopolysaccharides, flagella components, exoenzymes, phytohormones, type III secretion system (TTSS) effectors, siderophores, salicylic acid, and toxins) may be perceived by the plant and elicit a defensive response (Preston 2004). Pronounced plant species-specificity has been observed in the manifestation of ISR reactions, caused by PGPR (Liu et al. 1995; Van Wees et al. 1997).

The activities of plant-protecting PGPR are under direct control of the host which provides nutrients and energy for the bacteria existing in the rhizosphere. The greatest expression of plant protective activities by PGPR is typically seen in the zone of root elongation, where these bacteria are also the most numerous (Kamilova et al. 2006a). The importance of nutritional interactions by the symbiotic partners in providing biocontrol of pathogens is evident from the prevalence of genes, whose products are responsible for producing different catabolic enzymes, among the *rhi* genes of *P. fluorescens* which are activated specifically in the host rhizosphere (Rainey 1999).

Active growth, expression of the biocontrol-related genes, and production of antifungal factors such as antibiotics and siderophores are induced in plant-protecting PGPR by root-secreted organic acids (malate, citrate, succinate, fumarate) while the majority of mono- and disaccharides are less important for the host-beneficial activities (Lugtenberg and Dekkers 1999; Lugtenberg et al. 2001; [illegible] et al. 2002, 2003) in a comparable manner to *Azospirillum*-root associations. The difference between organic acid and sugar formation was confirmed *via* mutational analysis: the root colonization activity of *P. fluorescens* mutants with impaired utilization efficiency of organic acids (e.g., defective malate dehydrogenase) is sharply decreased, while the mutants for sugar utilization (e.g., devoid of glucose-6-phosphate dehydrogenase) are unaffected (Lugtenberg and Dekkers 1999; Lugtenberg et al. 2001).

Highly effective plant defence may be due to an ability of the host to regulate PGPR biocontrol functions by modulating the composition of root exudates. When plants invaded by a pathogen are inoculated with PGPR, the amount of organic acids in exudates may increase and stimulate the growth of bacteria and antibiotic production (Kamilova et al. 2006b). Additionally, some plants (including the legumes, pea and alfalfa [lucerne]) regulate their PGPR functions by exudating specialised signals from the roots which mimic the bacterial "quorum sensing" regulators required for root colonization and antifungal activities (Teplitski et al. 2000). These observations suggest that improvement of biocontrol functions in root-PGPR associations may be achieved *via* manipulations of the bacterial genotypes and host genotypes.

Bacterially-mediated synthesis of phytohormones, mainly auxins (IAA) may produce improvements in plant nutrition and in plant protection. For example, combinations of enhanced IAA synthesis and biocontrol activities were detected in *Pseudomonas* strains grown in rhizosphere of radish plants that exuded 30–100 times more tryptophane (the precursor of auxins) compared with wheat or tomato.

An important role in the expression of bacterial defensive functions may be provided by ACC-deaminase (Glick 2004). Investigations in the ability of PGPR to use ACC as a sole source of nitrogen demonstrated that only some strains possess ACC-deaminase. Transfer of the gene coding for ACC-deaminase from *Erwinia cloacae* to *P. fluorescens* resulted in an increase in ACC utilization efficiency sufficient to enhance the recombinants' ability to suppress phytopathogenic fungi. Therefore, in defensive associations ethylene has a role as a negative regulator similar to its effects in plant associations with azospirilla (Penrose and Glick 2003) and legume nodulation (Guinel and Geil 2002), although mechanisms of the regulation may be different.

Summarizing what is known about nitrogen-fixing and defensive root-PGPR associations it may be concluded that a range of common molecular mechanisms are available: (i) root-exuded organic acids, but not sugars, supporting different types of symbiotic bacteria; (ii) ethylene acting as a negative regulator for these symbioses; (iii) surface components, namely, EPS, LPS, lectins and cyclic β-glucans involved in the associations. A range of genes common to rhizobia are induced in root associated azospirilla. Moreover, inoculation of the model legume *Medicago truncatula* by *Pseudomonas fluorescens* results in induction of some plant genes common to nodulation/AM formation signalling cascades, e.g., gene *MtDMI3* (Sanchez et al. 2005) coding for calcium calmoduline-dependent kinase (CCaMK) (Catoira et al. 2000) (Section.4.3.1). There is a possibility of finding some sequence homologous in *Pseudomonas* similar to rhizobial *nod* genes (as were found in *Azospirillum* genomes).

Despite relatively low specificity of plant associations with PGPR, plant genotype has been shown to influence their effectiveness (i.e. genetic integration exists between the partners), and a series of genome loci (QTL) was identified controlling its quantitative variation (Wu et al. 1995; Smith and Goodman 1999; Preston 2004).

Mutually Beneficial Associations of Plants with Endophytic Bacteria

Healthy naturally propagated plants grown in the field or in pot cultures are colonized by populations of endophytic bacteria, averaging 10^3–10^6 cfu g^{-1} fresh plant weight and sometimes reaching 10^{10} cfu g^{-1} (reviewed in: Hallmann and Berg 2006). The spectrum of endophytic bacteria isolated from the roots of various plants covers a wide range of species from about 70 genera; representatives of the genera *Pseudomonas*, *Bacillus* or *Streptomyces* are most frequently encountered as endophytes (reviewed in: Hallmann and Berg 2006). Newly developed molecular methods enable complete analyses of the diversity of culturable and non-culturable bacteria (Van Overbeek et al. 2006). Although most of the known genera include some phytopathogenic endophytes, many of the individual species are known to be beneficial to higher plants (Sturz et al. 2000; Rosenblueth and Martínez-Romero 2006).

Both individually and collectively, the associations of endophytic bacteria represent a continuum from mutualism, commensalisms and latent pathogenesis through to active pathogenesis (Schulz et al. 2006). According to a working hypothesis proposed by the current authors, asymptomatic colonization (without causing visible plant disease symptoms) is a balance of antagonisms between host and endophytes. Endophytes and pathogens both possess many similar virulence factors. Indeed the loss of a virulence factor in model experiments resulted in transforming a pathogen into a harmless endophyte (reviewed in: Hallmann and Berg 2006). It is vexed question, however, as to which form found in Nature was transformed into the other: pathogen into endophyte or vice versa?

Some endophytes are seed-borne, but others have mechanisms for colonizing plants that have yet to be elucidated (Schulz et al. 2006). Although there are occasional poorly substantiated reports of intracellular colonization of bacteria providing a consistent and effective increase in the productivity of crops, it is still considered that the intercellular apoplastic space is the most suitable niche for endophytes (Sturz et al. 2000). It is suggested that many bacterial 'endophytes' may not colonize living tissues per se, but occupy protective niches in dead surface tissues or closely adhering soil of rhizosheaths. Opportunities for passive 'crack entry' into healthy, undisturbed roots under field conditions have been presumed but after studies with laboratory-grown plants these may not be demonstrated to be feasible. Consistent entry of endophytes into living root tissues in the field is supposed to require a bacterial capability to hydrolyse the hydrophobic incrustations of the walls of epidermal, hypodermal, endodermal, and other cortical cells. The lumen apoplast of xylem is considered as an unsuitable niche for endophytes, especially in grasses, because of the reduced fitness that such colonization imposes on plants when subjected to stresses in the field (McCully 2001). Although this niche, is of course, frequently colonized by plant pathogens, notably wilt pathogens such as fungal *Verticillium* and *Fusarium* spp., as well as bacterial *Ralstonia spp.* (Denny 2006) and *Erwinia spp.*

Plant associations with endophytic bacteria can increase plant growth and promote general development or improve plant resistance to pathogens and other environmental stresses enhancing the host's ability to acquire nutrients, or by production of plant growth-regulating, allelopathic or antibiotic compounds (Sturz et al. 2000; Berg and Hallmann 2006). Sometimes improved plant resistance can be linked to induced systemic resistance caused by bacterial elicitors coming from the endophyte (Kloepper and Ryu 2006). Certain cropping sequences have been shown to favour the accumulation of particular plant growth-promoting bacterial endophyte populations. These may lead to the development of beneficial host-endophyte allelopathies, with implications for the formation and maintenance of fertile, disease-suppressive soils. Manipulating bacterial populations in soils and within crops will be crucial if endophytes are to be utilized in crop production systems, and specialised techniques will be required to do so. It is necessary to study the natural associations between bacterial endophytes and their hosts for the purposes of employing such systems most productively in sustainable agriculture (Sturz et al. 2000). Some human pathogens, such as *Salmonella* spp., have been found as endophytes, and these bacteria are not removed by disinfection procedures that eliminate superficially occurring bacteria. Delivery of endophytes to the environment or agricultural fields should be carefully evaluated to avoid introducing plant, animal and human pathogens (Rosenblueth and Martínez-Romero 2006).

Synergism of the Beneficial Soil Microbes in the Rhizosphere

The rhizosphere is a dynamic environment where the interactions between plants and microbes are profoundly influenced by carbon fluxes (Toal et al. 2000). In addition to the interactions between plants and microbes there are also interactions within the rhizosphere microbial community (Barea et al. 2005). The AM fungi and bacteria can interact synergistically with each other to stimulate plant growth through a range of mechanisms that include improved nutrient acquisition and the inhibition of fungal plant pathogens.

With regard to AM symbiosis, these interactions result in modification of the environment in the vicinity of AM mycelia to such an extent that the term 'mycorrhizosphere' has been adopted (Linderman 1988). Both ECM (Garbaye 1994) and AM (reviewed in: Artursson et al. 2006; Barea et al. 2005) fungi can interact with different bacterial species which frequently attach to fungal mycelium. For those bacteria known to stimulate mycelial growth of mycorrhizal fungi and/or enhance root mycorrhization the term 'mycorrhiza-helper-bacteria' has been proposed (Garbaye 1994). This effect on mycorrhiza establishment is associated with the secretion of bacterial hormones and increasing the rates of root exudation.

Rhizospheric bacteria considered to be examples of PGPR are also known to affect the pre-symbiotic stage of AM development, namely germination rate and mycelium growth (Barea et al. 2005). Furthermore, according to new information some AM fungi are capable of forming fertile spores even in the absence of plant roots when they are cultivated dixenically together with bacteria *P. validus* isolated as AM fungal spore contaminants (Hildebrandt et al. 2006). In turn, AM formation is accompanied with alterations to plant physiology which can affect the rhizospheric microbial populations both quantitatively and qualitatively (Barea et al. 2005; Artursson et al. 2006). It was shown that co-inoculation with AM fungi improves the establishment of both introduced and indigenous PGPR (Barea et al. 2005). Simultaneous inoculation with AM fungi and phosphate-solubilising PGPR has been shown to be a most effective plant treatment by comparison with mono-inoculation variants (Barea et al. 2005). Multi-microbial interactions, including those between locally isolated AM fungi, phosphate-solubilising bacteria and *Azospirillum*, have also be reported, which indicate clearly that microbes act synergistically when inoculated simultaneously (Muthukumar et al. 2001).

Several studies have demonstrated that bacterial antagonists of fungal pathogens do not exert any anti-microbial effect against AM fungi (reviewed in: Barea et al. 2005). The results from *in vitro* or in situ laboratory experiments demonstrated no negative effects from *Pseudomonas* strains including those over-producing the antifungal compound 2,4-diacetylphloroglucinol on *G. mossea* spores germination. Moreover, the *Pseudomonas* strains improved plant growth and nutrient (N and P) acquisition by mycorrization (Barea et al. 1998) under field conditions. Nevertheless, in several cases of double inoculation with AM fungi and PGPR being used as biocontrol agents great variability in plant growth-promoting effects were observed between experiments indicating that more studies are needed for elucidation and effective use of whole plant–growth substrate-AM fungus-RGPR systems (Barea et al. 2005).

The establishment of AM has been shown to improve nodulation and nitrogen-fixation (reviewed in: Barea et al. 2005). In addition to the existence of common genetic system controlling the development of RN symbiosis and AM (genetic interference) (Section 4), there is trophic interference in symbioses. The AM can enhance the activity of rhizobia through a generalized stimulation of host nutrition. Nodulation processes have a significant requirement for phosphorus (Barea and Azcon-Aguilar 1983), so it would seem logical to need phosphorus-acquiring symbioses such as AM (Sprent and James 2007). But some localized effects may also occur at the root or nodule level (Barea et al. 2005). Also AM inoculation improves RN symbiosis development and functions at low levels of water potential and compensates for the negative effects of salinity on them (Quilambo 2003; Barea et al. 2005).

The common mechanisms available for RN symbioses and PGPR associations may result in their synergistic effects on plant growth. There is a general experience of enhanced nodulation in legumes by simultaneous inoculation with *Azospirillum* and PGPR (Okon et al. 1995; Sessitsch et al. 2002). Some mechanisms of synergism between rhizobia and PGPR were studied in pea where inoculation with the root growth promoting actinobacteria *Streptomyces lydicus* resulted in increased nodule mass, size and biochemical activities of the bacteroids as well as an accumulation of ferric ions in the nodules (Tokala et al. 2002).

Thus, there have been many studies concerning interactions between AM fungi and bacteria, but the underlying mechanisms of the functional properties behind these associations still require much further experimental confirmation (Artursson et al. 2006). These interactions are crucially important for sustainable, low-input agricultural cropping systems which rely on biological processes rather than agrochemicals in order to maintain soil fertility and plant health. The potential of microbial synergisms suggest that the prospects for establishing multi-partite symbioses which improve different agronomic traits in leguminous crops (Barea et al. 2005; Artursson et al. 2006) are high. The results of experiments with multi-microbial systems, including AM fungi, rhizobia and PGPR, support the importance of encouraging the physiological and genetic adaptation of microbes to the environment (Requena et al. 1997). Thus, locally sourced isolates are recommended for use in biotechnological applications.

Adaptive Evolution of the Mutually Beneficial Plant–Microbe Symbioses

It is generally accepted that the capacities of terrestrial plants for regulating the functions of microorganisms colonizing root systems developed during the earliest stages of evolution. Comparative data obtained by various methods may be used to probe the origins of plant–microbe symbiotic interactions and lead to the conclusion that AM is "the mother of plant root endosymbioses" (Parniske 2008) and, probably, some other symbioses as well. Therefore, most probably, mutualism was an early form of plant–microbe symbiotic interaction, and not antagonism as is sometimes postulated (Dyakov et al. 2007).

Fossilised plants have been found (Kidston and Lang 1921; Remy et al. 1994) which contain structures similar to those of present-day AM and are related to Silurian and Devonian flora. This was the time (400–500 millions years ago) when terrestrial plants appeared (Schüßler et al. 2001; Redecker 2002). The importance of AM associations for these plants is evident since they lacked conventional root systems, but were equipped with rhizomes and rhizoids which could anchor them to substrates but not take up nutrients effectively (see for review: Brundrett 2002). Possibly true roots co-evolved with AM-fungal partners, and anatomy of mycotrophic plants tends to confirm this (Brundrett 2002). Most of contemporaneous plants still have a dependence on AM; where this is lacking then such individuals are thought to result from later evolutionary events (Brundrett 2002).

Discussion continues as to which morphological form of AM development (linear, so-called *Arum* type, or coiled, *Paris* type) is the original ancestor. Indeed, it is difficult to detect an evolutionary trend in AM types, since they are found throughout the plant kingdom (Smith and Smith 1997; Dickson et al. 2007). Based on the analysis of fossils the *Arum* type could be the oldest mycorrhizal form (Kidston and Lang 1921; Remy et al. 1994). It has been suggested that coiling AM is more advanced than linear types, since the former appears to allow the host greater control of the fungus. Additionally, linear colonization is more rapid and efficient, but this may result in greater energy costs for the plant (see for review: Brundrett 2002). On the other hand, coiling AM is the most common form in contemporary bryophytes, ferns and gymnosperms and thus might be considered as an ancestral condition (Smith and Smith 1997).

Studies of molecular phylogeny based on the conserved genes (16S or 18S rDNA nucleotide sequences) using polymorphism analyses provides new views on the origins and evolution of bacterial and fungal symbiotic organisms and their interactions with hosts. Apparently AM fungi are a monophyletic group Glomeromycota (Schüßler et al. 2001), indicating that they have originated from a common ancestor. Interestingly this phylum includes along with obligatory biotrophic AM fungi, the genus *Geosiphon* represented by the single species *G. pyriformis* which is both free-living and able to form symbioses with the photosynthetic (and also nitrogen-fixing) cyanobacteria *Nostoc*. Possibly this symbiosis might in evolutionary time have preceded AM (Schüßler 2002; Parniske 2008), but not those marine ancestors of terrestrial plants and unicellular or possibly parasitic fungi as was suggested earlier (Parniske 2000; Schüßler et al. 2001). Thus, it is possible that Glomeromycota evolved before the land plants. The low specificity of AM fungi in relation to putative host plants demonstrates that their gene systems, which control AM development, are rather conservative and might have been only slightly reorganized for different symbioses in various host groups. Curiously, a recent study demonstrates that some AM fungi are capable of completing their life cycles in the absence of plant roots, but rely instead on the simultaneous growth of bacteria *Paenibacillus validus* which is supposed to be the natural partner of these fungi (Hildebrandt et al. 2006).

In contrast to AM fungi, many other microsymbionts including nodule bacteria (Sprent 2001) and ectomycorrhizal fungi (LoBuglio et al. 1996; Hibbert et al. 2000; Moncalvo et al. 2000) are polyphyletic in their origin. Consequently, these plant–microbe symbioses evolved in parallel and most probably after AM. It is known that legumes evolved much later than AM (Lavin et al. 2005); hence, their symbiosis with nitrogen-fixing nodule bacteria are considerably younger (60–70 million years) (Sprent 2001; Sprent and James 2007).

Parallel evolution and convergence plays an important role in the evolution of plant–microbe interactions, as plants possess universal gene systems for the control of basic functions required in symbiosis, including the maintenance of interior homeostasis, novel tissue and cellular structures (symbiotic compartments) and the utilization of nutrients provided by the partner organisms (Gualtieri and Bisseling 2000). Based on the universal organization and regulatory mechanisms of the plant genetic material (Cronk et al. 2002), the gene systems could have been rearranged in similar manners during co-evolution with different microbes. Various microorganisms when adapting to an existence within host plants, obviously, followed convergent evolutionary pathways developing similar mechanisms for dialogue with their defensive (regulatory) systems (Provorov 2009).

According to the phylogeny of flowering plants developed on the basis of plastid gene *rbcL* structure (Doyle 1998), which differs from earlier systems based on plant morphology, all plants that are able to form nitrogen-fixing nodules with rhizobia and actinobacteria in the genus *Frankia*, belong to a monophyletic Rosid I clade. A common ancestor of Rosid I, obviously, could not form nodules with rhizobia but had some pre-adaptations to form nitrogen-fixing symbioses which developed in a parallel manner in several plant families which evolved later. Those pre-adaptations can be subdivided into two groups: the first includes concerns the capacity for being colonized by AM fungi and of controlling the mycobiont once inside host tissues; the second includes capacities which determine the formation of special cellular, tissue and organ structures characteristic of nitrogen-fixing nodules (symbiotic organs and compartments).

Developmental genetics using structural analyses, cell biology, genetics and functional comparative genomics gives many clues for reconstructing the origins and evolutionary pathways of plant–microbe symbioses since they reveal broad involvement of common plant genes and stages in the developmental programmes of root symbiotic nodules and ancestral AM symbiosis. These commonalities suggest that AM was a source of pre-adaptations for other forms of interactions (Sprent 2001; Kistner and Parniske 2002; Provorov et al. 2002; Sprent and James 2007; Parniske 2008) responsible for host regulatory reactions. These in turn may be controlled by plant genes recruited from AM programmes into those for nodule developmental and possibly, into those of plant resistance to pathogens.

Regarding the second type of these pre-adaptations contributing to nodulation, it is possible that the ancestor of Rosid I was able to form outgrowths (nodule-like structures) which carried features of depository and systems of metabolic exchange with aerial plant organs. These were used for the integration of mechanisms of nitrogen-fixation with consequential implications for the derivation of energy and nitrogen-carbon exchange. Evidence for this may be deduced from the occurrence of nodule formation induced by some rhizobia non-secreting host-specific Nod factors (through cracks in the root epidermis or on the stem from rudimental air-roots) (Downie 2007; Giraud et al. 2007). Furthermore, spontaneous formation of pseudo-nodules has been found in some current native legumes (Bonnier 1961) and some alfalfa (lucerne) and clover lines (Truchet et al. 1989). Several genes were found in those leguminous lines controlling the formation of pseudo-nodules either under sterile conditions or in the presence of avirulent, Nod factor defective rhizobial mutants (Caetano-Anollés et al. 1992).

Summarizing the knowledge available so far permits the development of a hypothesis for consequential placement during evolution of different types of microorganisms which can occupy intercellular and sub-cellular symbiotic compartments within the root cortex, as a result of which dicotyledonous plants have acquired an ability to form nitrogen-fixing symbioses with nodule bacteria (Provorov 2009). Hence, nitrogen-fixing prokaryotes exploited an ancestral programme in the plant for microsymbiont "hosting" which appeared during co-evolution of most ancient plants with AM fungi. The ability to form both symbioses was incorporated in parallel into different families of Rosid I under natural selection pressures acting in favour of the ability for symbiotrophic nitrogen nutrition. The first step could be a replacement of AM fungi by *Frankia* spp. which have a similarly mycelial-like growth habit and in turn was replaced later by another nitrogen-fixing prokaryote, rhizobia.

These analyses also enable us to suggest a new insight into the origins of legume nitrogen fixing symbionts (rhizobia). Conventionally, their origins were considered mainly to have come from plant pathogens or from beneficial plants symbionts (PGPR, endophytes) in which the systems for Nod factor synthesis arose before later distribution into a broad spectrum of soil or plant-associated microbes *via* horizontal gene transfer (Provorov 1998; Terefework et al. 2000; Suominen et al. 2001). For example, slow-growing rhizobia might have arisen from *Azospirillum*-like strains since there is a visible taxonomic relatedness between *Azospirillum* and *Bradyrhizobium* genera, and in the genomes of some *Azospirillum* strains there are close homologues to some rhizobial *nod* and *fix* genes (Fogher et al. 1985; Vieille and Elmerich 1990, 1992; Rivas et al. 2002). This suggestion is consistent with numerous findings demonstrating that rhizobia can effectively operate as the beneficial PGPRs or as endophytic symbionts in "non-host" (legume and non-legume) plants (Sessitsch et al. 2002). Although the mechanisms of beneficial effects on plants may be atypical for rhizobia (involving not nitrogen fixation but rather the synthesis of hormones, defence of plants from pathogens and solubilization of soil phosphates), these endo- and epiphytic associations of rhizobia with plants may be considered as a vestige from the early stages of evolution in morphologically different nodular symbioses.

Here we would like to introduce an alternative hypothesis concerning the origins of rhizobia and postulate that these bacteria came first from micro-symbionts associated with AM fungi. Some arguments in favour of this hypothesis are summarized below:

1. The root systems of current land plants evolved in the presence of AM fungi (Brundrett 2002) suggesting that the genetic systems of plants for controlling root development and mineral nutrition can overlap with the genetic system controlling AM development at the large scale.
2. AM (as well as the other mycorrhizal fungi) are associated with different types of endo- and ecto-symbiotic bacteria (including nitrogen-fixers). Moreover these fungi may operate as effective vectors for introducing bacteria into the plants (Artursson et al. 2006). Therefore, the multipartite symbiotic communities (plants + fungi + bacteria) may be considered as being ancestral with respect to the two-component systems (e.g., to root symbiotic nodules). It is important to note that colonization of roots by mycorrhizal fungi is not associated normally with induction of bacterial diseases, and therefore, a certain degree of selectivity occurs with respect to the bacterial satellites of fungi which invade roots. This selectivity may be an important factor for the maintenance of multipartite symbiotic communities otherwise they would have been eliminated by natural selection.
3. There are common regulatory plant genes controlling the development of AM and nitrogen-fixing nodules (reviewed in: Borisov et al. 2008 and this Chapter]; numerous common plant genes are expressed during both symbioses (Küster et al. 2007); furthermore some common plant genes have been found which are expressed in the process of establishing AM and mutually beneficial associations with rhizospheric bacteria (Sanchez et al. 2005, Vivienne Gianinazzi-Pearson, 2008 personal communication).
4. Stimulation of AM formation by Nod factors produced by rhizobia has been observed (Xie et al. 1997). Therefore, a primary function of Nod factors in the evolution of rhizobia could be the stimulation of penetration by fungal vector and only later were Nod factors acquired with a function for the independent induction of bacterial symbiotic developmental programmes.
5. Taxonomic relatedness was found between some fungal symbionts and some rhizobia, for example, *Burkholderia*-like strains are found among endo-symbionts of AM fungi (Minerdi et al. 2002) and of ectomycorrhizal fungi (Izumi et al. 2006) as well as amongst legume nodular symbionts (Balachandar et al. 2007).

Of course, not all of the numerous and diverse rhizobia species might have arisen from the bacterial satellites of the mycorrhizal fungi. Nonetheless, the bacterial co-habitants of mycorrhizal fungi may be seen as possible precursors of the first rhizobia in which the genes for Nod factor synthesis developed primarily since Nod factors are chitin-like compounds which can mimic fungal cell wall intermediates and therefore they can be very useful tools for the colonization of fungal hosts by associated bacteria. Moreover, intimate interactions between bacteria and fungi might allow the acquisition by ancient rhizobia of some genes involved in the synthesis of chitin-like compounds, (e.g., accepting of *nodA* by horizontal gene transfer from fungi was suggested as a mechanism by Hirsch et al. 2001). An extended comparative study of the rhizobia species and of bacterial satellites of mycorrhizal fungi is required to identify their inputs into the evolution of nodular symbiosis in legumes.

Since fossil land plants were, possibly, obligate biotrophs, maintenance of symbiotic fungi in roots was a more ancient function compared with the independent assimilation of nutrients from soil (Provorov 2009). Two opposite strategies in the evolution of plant symbiotrophy can be suggested: a change to autotrophy (for example, loss of mycotrophy in Brassicaceae (previously Cruciferae) or symbiotrophy in commercial legume cultivars grown by the elevated use of mineral fertilizers) or phototrophic denial and formation of myco-heterotrophic associations, as for instance, in orchids as well as in some dicotyledonous plants, ferns, bryophytes and liverworts (Bruns and Bidartondo 2002).

From the perspective of an entire plant–microbe system, mutually beneficial symbioses are represented as the products of interspecies (reciprocal) altruism which resemble to some extent antagonistic (pathogenic) interactions at the molecular and cytological levels (Provorov and Vorobyov 2009a, b). Plant hosts, possessing putative gene networks for regulating beneficial microbes and for defence against pathogens, with direct parallels between mutualism and antagonism are evident logical suggestions. But viewed from the microbial perspective this is unlikely, since beneficial and pathogenic microbes often demonstrate independent origins (Provorov and Vorobyov 2010). The evolutionary divergence of beneficial and deleterious microsymbionts is suggested to result from specific selective pressures (different forms of group selection) maintaining genes for beneficial traits in microbial populations (Provorov and Vorobyov 2009a, b). Using the legume–rhizobia symbiosis as a model, it was suggested that this selection might have been induced by two factors. The first one is the positive feedbacks operating between partners at the level of symbiotic metabolism (carbon and nitrogen compounds) and of partners' co-evolution (synergistic selective pressures supporting mutualism in plants and in microbes). A second one is clonal structures of microbial populations in plants which result in selection favouring symbiotically effective strains.

New Approaches of Application of Mutually Beneficial Plant–Microbe Systems in Sustainable Agriculture

Firstly, an existence of plant genes (Duc et al. 1989; Gianinazzi-Pearson 1996; Küster et al. 2007) and their molecular products (Frühling et al. 1997) common for both AM and RN led to a conclusion that legumes possess common genetic systems controlling the development of a tripartite symbiosis (legume plant + AM fungi + rhizobia). This fact along with the demonstration of synergistic activity in beneficial soil microbes (reviewed in: Barea et al. 2005) and a suggestion that plant genetic systems controling the development of RN and, probably, of some other beneficial plant–microbe associations evolved on the basis of that present in AM (Parniske 2008; Provorov 2009) has great importance for the application of tripartite or even multi-partite symbiotic systems in low-input sustainable environmentally-friendly agrotechnologies.

Development of New Types of Microbial Inocula

The use in sustainable agriculture of inocula based on beneficial soil microbes as described above allows the improvement crop productivity with decreased doses of mineral fertilizers and pesticides (reviewed in: Xavier et al. 2004; Rai 2006). Currently the majority of commercial inocula contain pure cultures of single microorganisms and only occasionally multiple combinations. There are several objections to the use of mono-inoculation. Firstly, endemic microbial communities are stable and although plastic the introduced microbe may be allowed to occupy a very small niche in the whole community as a whole. Secondly, genetic material in microbes is very plastic, and consequently strains introduced into natural ecosystems can rapidly lose their beneficial traits. Thirdly, the existence of microbial cooperation in the rhizosphere (Barea et al. 2005) as well as in natural synergistic associations of different microbes including those between AM fungi and their endocellular or superficial symbionts (Artursson et al. 2006; Barea et al. 2005) question the possibility and expediency of applying mono-inoculants and even use of the term ‘mono-inoculation’ itself. Finally, plants possess relatively stable genomes and this fact contributes significantly to the effectiveness of symbiosis (Tikhonovich and Provorov 2007). Therefore, for industrial plant production in sustainable systems we should use plants having highly effective interactions with all forms of beneficial soil microbes, which can encourage the development of multiple niches hosting microbes and regulating their activity. For this it is necessary to develop new multi-component microbial inocula which increase the content and biodiversity of beneficial soil microbes in agricultural land.

There is experimental evidence of the effectiveness of simultaneous inoculation of legumes with AM fungi and nodule bacteria leading to increased productivity and quality of the yield, e.g. groundnut (Ibrahim et al. 1995), pea (Jacobi et al. 1999; Borisov et al. 2002, 2004; Shtark et al. 2006), albaida (*Anthyllis cytisoides*) (Requena et al. 2001), and soybean (Labutova et al. 2004). The effect achieved equalled or exceeded that achieved with mineral fertilizers (Borisov et al. 2002, 2004; Shtark et al. 2006). The effect also exceeded that of mono-inoculation with AM fungi or with rhizobia either in model experiments or under field conditions (Jacobi et al. 1999; Borisov et al. 2002; Labutova et al. 2004). In long-term experiments in a desertified Mediterranean ecosystem, it was found that simultaneous inoculation with AM fungi and rhizobia enhanced the establishment of key plant species and increased soil fertility and quality; increased soil nitrogen content, organic matter content, and soil aggregate hydrostability and enhanced nitrogen transfer from nitrogen-fixing to non-fixing species associated with the natural succession of the plants (Requena et al. 2001).

There is an example of application of triple inoculum (AM fungi, rhizobia and PGPR) to the legume *A. cytisoides* which was successful only when the microorganisms used were isolated from local environment (Requena et al. 1997). In collaboration with an innovation company “Bisolbi-Inter” (Russian Federation) the All-Russia Research Institute for Agricultural Microbiology (ARRIAM), Saint-Petersburg,

Russian Federation, has developed technology for the production and application of a new multifunctional biopreparation "BisolbiMix" (Chebotar et al. 2008) containing a complex of the most effective isolates of endosymbiotic microbes (AM fungi and rhizobia) and associative bacteria (PGPR) from the collection held at ARRIAM. A non-sterile substrate-carrier which is derived from washing-filtration by-products of a sugar-beet factory contains its own microbial community including all the above groups of beneficial microbes. The preparation can be formulated into a seed dressing (not effective for all the crop plants tested) or granules. The efficacy of "BisolbiMix" was demonstrated in field trials with legumes, e.g. pea (Shtark et al. 2006) or non-legumes such as wheat, pumpkin and potato (unpublished results). The use of microbial formulations containing rhizobia for non-legumes seems to be sensible because it is known that nodule bacteria which do not form nodules on a non-host legume as well as non-legume roots can operate as PGPR (Prévost and Antoun 2005; Hossain and Mårtensson 2008). Thus, the selection of rhizobia with both PGPR activity and efficient symbiotic nitrogen fixation should be advantageous in crop rotations or intercropping systems using legumes and non-legumes.

It is possible, therefore, to develop effective multi-microbial inoculants, but it is necessary to use local communities of beneficial microbes because this exploits the natural biological and genetical adaptations of the partners to their environment (Requena et al. 1997; Gentili and Jumpponen 2006).

Legumes in Sustainable Agriculture

During development of plant–microbe systems for low-input sustainable ecologically friendly plant cultivation it is necessary to be guided by conclusions of European Community (EC) experts about global productivity of legumes (http://www.grainlegumes.com/aep/; http://ec.europa.eu/research/biosociety/food_quality/projects/002_en.html) for sustainable agriculture. The use of legumes in agriculture is leading to: improved soil fertility and increased diversity of crops and soil microbial communities; reductions in the use of non-renewable natural resources; decreased negative effects from intensive agrotechnologies on the natural environment due to decreased requirement for mineral fertilizers and pesticides and decreased production of animal protein and associated wastes; local production of pollution-free food and forage; and a more stable income for the agricultural producers. This is why it is necessary to breed legumes which have highly effective interactions with beneficial soil microbes.

Legume Breeding to Improve Their Symbiotic Effectiveness

For more than 25 years the authors' laboratory has specialized in the genetics of plant–microbe interactions using pea (*P. sativum* L.) as a model plant. Our experience for improving the effectiveness of beneficial plant–microbe systems with pea is consequently given as an example. At the same time, the authors' team knows only one record of other activity of a similar nature viz. genetic variability in onion (*Allium* spp.) has been shown with respect to its responsiveness to AM fungal inoculation which indicates that onion breeding for improving efficacy of association with AM fungi is possible (Galvan et al. 2007). The necessity for this sort of plant breeding is also considered in isolation and mainly with respect to the effectiveness of RN symbiosis (Herridge and Rose 2000; Rengel 2002; Graham et al. 2004; Howieson et al. 2008).

Analysis of Genetic Variability of Pea with Respect to its Effectiveness of Interactions with Beneficial Soil Microbes

A high level of genetic variability was demonstrated in analyses of the symbiotic effectiveness under double inoculation with AM fungi and nodule bacteria of 99 land-races and outclassed heritage cultivars of *P. sativum* from the collection N.I. Vavilov's All Russia Research Institute of Plant Industry, Saint-Petersburg, Russian Federation, with different geographical origins (Jacobi et al. 1999; Borisov et al. 2004). In a few genotypes considerable increases in plant dry weight (about 300%), seed productivity (more than 650%), phosphorus and nitrogen content (more than 900 and more than 300%, respectively) were observed. The most promising highly symbiotically effective genotypes and those with low symbiotic potential were included in the Pea Genetic Collection (ARRIAM) to be used for experiments studying the functioning of tripartite/multipartite symbiosis. Types identified as highly symbiotically effective genotypes were involved in breeding programmes to create commercial pea cultivars with great potential for interactions with beneficial soil microbes were used in collaboration with All-Russia Institute of Leguminous and Groat Crops (ARILGC), Orel, Russian Federation.

The most promising highly symbiotically effective pea genotypes previously selected and different commercial pea cultivars created without consideration of symbiotic effectiveness were involved in 3-year field trials (Orel district) (Shtark et al. 2006). Seed productivity and plant dry weight were chosen as the main criteria for the evaluation of symbiosis effectiveness in legume crops. The double (actually multiple, see above comments on the nature of AM fungi) inoculation was shown to increase seed productivity and plant dry weight in most of the pea genotypes studied and sometimes this could exceed the effect of mineral fertilizers. The effectiveness of legume breeding to improve symbiotic potential of legume cultivars was proven therefore, under field conditions and the genotypes to be used in such breeding programmes were identified. The genotype K-7284 (non-commercial) was selected as a standard of symbiotic effectiveness. Additionally, it was demonstrated that highly effective genotypes can be also found among commercial pea cultivars created without consideration for effectiveness of interactions with beneficial soil microbes. Taking into account that most commercial legume cultivars have accidentally lost their abilities for symbiotrophic nutrition without selective pressure during breeding of intensive crops, the latter constitutes a very important finding for plant breeders and gives them the possibility for concurrent generation of cultivars with required pea plant architecture, other agriculturally important traits and high effectiveness of interactions with all types of beneficial soil microbes in a single breeding programme.

Breeding to Improve Pea Symbiotic Effectiveness

In order to cultivate plants with improved symbiotic potential a special breeding nursery was created in the experimental trials ground of ARILGC on land where for the preceeding 5 years before nursery establishment mineral fertilizers had not been applied. To reduce the incidence and severity of root pathogens a six-field crop rotation was used where cultivation of winter wheat was followed by peas. The multi-component preparation "BisolbiMix" was used for the inoculation of test plants.

Using the breeding nursery as well as a breeding protocol developed from long-term collaboration of ARRIAM with ARILGC the first (in the whole history of legume breeding) pea cultivar "Triumph" having increased potential of interactions with beneficial soil microbes was intentionally bred (Borisov et al. 2008). It arose as a result of crossing a commercial cultivar 'Classic' (donor of agriculturally important traits) and the genotype K-8274 (donor of symbiotic effectiveness trait) and subsequent individual selection of genotypes with high productivity and capacity for supporting various beneficial microbes.

The cultivar "Triumph" is of middle stem height, semi-leafless and has stable productivity under different climate conditions, it is comparatively resistant to root rotting pathogens and pests. Its productivity is not lower than those of the productivity standards for Orel district using conventional production technologies and 10% greater in comparison with standard cultivars when inoculated with "BisolbiMix". As a result of 2-year state trials (2007–2008) the productivity of "Triumph" was shown to be comparable with that of standard regional cultivars enabling recommendation for commercial cultivation in the Central region of Russian Federation (unpublished results). Thus, the innovative concept of the authors' research team for plant breeding (applicable not only for legumes, but also for non-legumes) is bearing its first fruits.

Conclusions

Intimate associations of beneficial soil microbes with the host plants described above in detail are applicable in sustainable crop production if taken either separately or in combination. Many authors are now recognizing the need for using multi-microbial plant inoculants and the advantages of using indigenous plants (or cultivars of local breeding) and microbes.

The authors' team proposes its own concept which offers fundamentally new approaches to plant production. Firstly, it is necessary to consider plant genetic systems controlling interactions with different beneficial soil microbes in unison. Secondly, plants used as a component of this complex plant–microbe system controlling its effectiveness should be bred to improve the effectiveness of interactions with all types of beneficial soil microbes. Increases of plant biomass production due to plant–microbe symbiosis should be used as the main parameter for an evaluation of plant effectiveness in interactions with beneficial soil microbes. The plant production should be done with inoculation composed of multi-component microbial inocula consisting of AM fungi, rhizobia, PGPR and/or beneficial endophytic bacteria. Finally, taking into consideration the importance of legumes to global agriculture, greater emphasis should be placed on plant–microbial systems in the development of low-input agro-biotechnologies enabling wider cultivation of leguminous crops.

Acknowledgements We are grateful to Timofey A. Nemankin and Alexandr I. Zhernakov for their help in preparing the figures. We thank Dr. Margarita A. Vishnyakova (N.I. Vavilov All-Russia Research Institute of Plant Industry, RF), Dr. Tatiana S. Naumkina (All-Russia Institute of Legumes and Groat Crops, RF) and Dr. Vladimir K. Chebotar ("Bisolbi-Inter" Ltd., RF) for the long-term collaboration in the fields of legume breeding and microbial inocula development. This work was supported by the grants of RFBR (09-04-00907, 09-04-13895, 09-04-91054, 09-04-91293, 10-04-00961, 10-04-01146), NWO 047.018.001, Grant to support leading Russian science school 3440.2010.4, Governmental contracts for research with Russian Federal Agency of Science and Innovations (02.512.11.2280, 02.740.11.0276).

References

Akiyama K, Hayashi H (2006) Strigolactones: chemical signals for fungal symbionts and parasitic weeds in plant roots. Ann Bot (Lond) 97:925–931

Akiyama K, Matsuzaki K, Hayashi H (2005) Plant sesquiterpenes induce hyphal branching in arbuscular mycorrhizal fungi. Nature 435:824–827

Albrecht C, Geurts R, Lapeyrie F, Bisseling T (1998) Endomycorrhizae and rhizobial Nod factors both require *SYM8* to induce the expression of the early nodulin genes *PsENOD5* and *PsENOD12A*. Plant J 15:605–614

Albrecht C, Geurts R, Bisseling T (1999) Legume nodulation and mycorrhizae formation; two extremes in host specificity meet. EMBO J 18:281–288

Allen ON, Allen EK (1981) The leguminosae. A source book of characteristics, uses and nodulation. The University of Wisconsin Press, Madison

Ames RN, Reid CPP, Porter L, Cambardella C (1983) Hyphal uptake and transport of nitrogen from two ^{15}N-labelled sources by *Glomus mosseae*, a vesicular-arbuscular mycorrhizal fungus. New Phytol 95:381–396

Amor BB, Shaw SL, Olroyd GE, Maillet F, Penmetsa RV, Cook D, Long SR, Denarie J, Gough C (2003) The NFP locus of *Medicago truncatula* controls an early step of Nod factor signal transduction upstream of a rapid calcium flux and root hair deformation. Plant J 34: 495–506

Ane JM, Kiss GB, Riely BK, Penmetsa RV, Oldroyd GED, Avax C, Levy J, Debelle F, Baek JM, Kalo P, Rosenberg C, Roe BA, Long SR, Denarie J, Cook DR (2004) *Medicago truncatula DMI1* required for bacterial and fungal symbioses in legumes. Science 303:1364–1367

Antonyuk LP, Fomina OR, Galkin MA, Ignatov VV (1993) The effect of wheat germ agglutinin on dinitrogen fixation, glutamine synthetase activity and ammonia excretion in *Azospirillum brasilense* Sp245. FEMS Microbiol Lett 11:285–290

Ardourel M, Demont N, Debelle F, Maillet F, de Billy F, Prome JC, Denarie J (1994) *Rhizobium meliloti* lipooligosaccharide nodulation factors. Plant Cell 6:1357–1374

Arrighi JF, Barre A, Amor BB, Bersoult A, Soriano LC, Mirabella R, de Carvalho-Niebel F, Journet EP, Gherardi M, Huguet T, Geurts R, Denarie J, Rouge P, Gough C (2006) The *Medicago truncatula* lysin motif-receptor-like kinase gene family includes NFP and new nodule-expressed genes. Plant Physiol 142:265–279

Artursson V, Finlay RD, Jansson JK (2006) Interactions between arbuscular mycorrhizal fungi and bacteria and their potential for stimulating plant growth. Environ Microbiol 8:1–10

Ashford A (2002) Tubular vacuoles in arbuscular mycorrhizas. New Phytologist 154:545–547

Augé RM (2001) Water relations, drought and vesicular-arbuscular mycorrhizal symbiosis. Mycorrhiza 11:3–42

Augé RM, Moore JL, Cho K, Stutz JC, Sylvia DM, Al-Agely A, Saxton AM (2003) Relating dehydration resistance of mycorrhizal *Phaseolus vulgaris* to soil and root colonization by hyphae. J Plant Physiol 160:1147–1156

Augé RM, Moore JL, Sylvia DM, Cho K (2004) Mycorrhizal promotion of host stomatal conductance in relation to irradiance and temperature. Mycorrhiza 14:85–92

Bago B, Becard G (2002) Bases of obligatory biotrophy of arbuscular mycorrhizal fungi. In: Gianninazzi S, Schupp H, Barea JM, Hasalwander K (eds), Mycorrhizal technology in agriculture. Birkhauser, Basel, pp 33–48

Bago B, Azcon-Aguilar C, Piche Y (1998) Architecture and developmental dynamics of the external mycelium of the arbuscular mycorrhizal fungus *Glomus intraradices* growth under monoxenic conditions. Mycologia 90:52–62

Bago B, Zipfel W, Williams RM, Jun J, Arreola R, Lammers PJ, Pfeffer PE, Shachar-Hill Y (2002) Translocation and utilization of fungal storage lipid in the arbuscular mycorrhizal symbiosis. Plant Physiol 128:108–124

Balachandar D, Raja P, Kumar K, Sundaram SP (2007) Non-rhizobial nodulation in legumes. Biotechnol Molec Biol Rev 2:49–57

Balestrini R, Gomez-Ariza J, Lanfranco L, Bonfante P (2007) Laser microdissection reveals that transcripts for five plant and one fungal phosphate transporter genes are contemporaneously present in arbusculated cells. Molec Plant-Microbe Interact 20:1055–1062

Barea JM, Azcon-Aguilar C (1983) Mycorrhizas and their significance in nodulating nitrogen-fixing plants. Adv Agron 36:1–54

Barea JM, Andrade G, Bianciotto VV, Dowling D, Lohrke S, Bonfante P, O'Gara F, Azcon-Aguilar C (1998) Impact on arbuscular mycorrhiza formation of pseudomonas strains used as inoculants for biocontrol of soil-borne fungal plant pathogens. Appl Environ Microbiol 64:2304–2307

Barea JM, Pozo MJ, Azcon R, Azcon-Aguilar C (2005) Microbial cooperation in the rhizosphere. J Exp Botany 56:1761–1778

Barker D, Bianchi S, Blondon F, Dattee Y, Duc G, Essad S, Flament P, Gallusci P, Genier G, Guy P, Muel X, Tourneur J, Denarie J, Huguet T (1990) *Medicago truncatula*, a model plant for studying the molecular genetics of the *Rhizobium*-legume symbiosis. Plant Mol Biol Rep 8:40–49

Barker SJ, Stummer B, Gao L, Dispain I, O'Connor PJ, Smith SE (1998) A mutant in *Lycopersicon esculentum* Mill. with highly reduced VA mycorrhizal colonization: isolation and preliminary characterization. Plant J 15:791–797

Bary de A (1879) A Die Erscheinung der Symbiose. Von Karl J, Trubner, Strassburg

Becker A, Pühler A (1998) Production of exopolysaccharides. In: Spaink HP, Kondorosi A, Hooykaas PJJ (eds) The rhizobiaceae. Molecular biology of model plant-associated bacteria. Kluwer, Dordrecht/Boston/London, pp 87–118

Bell TL, Pate JS (1996) Nitrogen and phosphorus nutrition in mycorrhizal Epacridaceae of Southwest Australia. Annals Bot 77:389–397

Ben Amor B, Shaw SL, Oldroyd GED, Maillet F, Penmetsa RV, Cook D, Long SR, Denarie J, Gough C (2003) The *NFP* locus of *Medicago truncatula* controls an early step of Nod factor signal transduction upstream of a rapid calcium flux and root hair deformation. Plant J 34:495–506

Berg G, Hallmann J (2006) Control of plant pathogenic fungi with bacterial endophytes. In: Schulz B, Boyle S, Sieber T (eds) Microbial root endophytes. Springer, Berlin, pp 53–70

Bertaux J, Schmid M, Prevost-Boure NC, Churin JL, Hartmann A, Garbaye J, Frey-Klett P (2003) *In situ* identification of intracellular bacteria related to *Paenibacillus* spp. in the mycelium of the ectomycorrhizal fungus *Laccaria bicolor* S238N. Appl Environ Microbiol 69:4243–4248

Besserer A, Puech-Page V, Kiefer P, Gomez-Roldan V, Jauneau A, Roy S, Portais JC, Roux C, Becard G, Sejalon-Delmas N (2006) Strigolactones stimulate arbuscular mycorrhizal fungi by activating mitochondria. PLoS Biol 4:1239–1247

Beveridge CA, Mathesius U, Rose R, Gresshoff PM (2007) Common threads of meristem development and homeostasis: branching, nodules and lateral roots. Curr Opin Plant Biol 10: 44–51

Bianciotto V, Bandi C, Minerdi D, Sironi M, Tichy HV, Bonfante P (1996) An obligately endosymbiotic mycorrhizal fungus itself harbors obligately intracellular bacteria. Appl Environ Microbiol 62:3005–3010

Bianciotto V, Lumini E, Lanfranco L, Minerdi D, Bonfante P, Perotto S (2000) Detection and identification of bacterial endosymbionts in arbuscular mycorrhizal fungi belonging to the family *Gigasporaceae*. Appl Environ Microbiol 66:4503–4509

Bianciotto V, Lumini E, Bonfante P, Vandamme P (2003) '*Candidatus Glomeribacter gigasporarum*' gen. nov., sp. nov., an endosymbiont of arbuscular mycorrhizal fungi. Int J Syst Evol Microbiol 53:121–124

Bianciotto V, Genre A, Jargeat P, Lumini E, Becard G, Bonfante P (2004) Vertical transmission of [illegible] in the arbuscular mycorrhizal fungus *Gigaspora margarita* through generation of vegetative spores. Appl Environ Microbiol 70:3600–[illegible]

Bieleski RL (1973) Phosphate pools, phosphate transport and phosphate availability. Annu Rev Plant Physiol 24:225–252

Blauenfeldt J, Joshi PA, Gresshoff PM, Caetano-Anollés G (1994) Nodulation of white clover (*Trifolium repens*) in the absence of *Rhizobium*. Protoplasma 179:106–110

Blilou I, Ocampo JA, Garcia-Garrido JM (1999) Resistance of pea roots to endomycorrhizal fungus or *Rhizobium* correlates with enhanced levels of endogenous salicylic acid. J Exp Bot 50: 1663–1668

Blilou I, Bueno P, Ocampo JA, Garcia-Garrido JM (2000a) Induction of catalase and ascorbate peroxidase activities in tobacco roots inoculated with the arbuscular mycorrhizal fungus *Glomus mossea*. Mycol Res 104:722–725

Blilou I, Ocampo JA, Garcia-Garrido JM (2000b) Induction of Ltp (Lipid transfer protein) and Pal (phenylalanine ammonia-lyase) gene expression in rice roots colonized by the arbuscular mycorrhizal fungus *Glomus mosseae*. J Exp Bot 51:1969–1977

Bloemberg GV, O'Toole GA, Lugtenberg BJJ, Kolter R (1997) Green fluorescent protein as a marker for *Pseudomonas* spp. Appl Environ Microbiol 63:4543–4551

Bloemberg GV, Wijfjes AHM, Lamers GEM, Stuurman N, Lugtenberg BJJ (2000) Simultaneous imaging of *Pseudomonas fluorescens* WCS365 populations expressing three different autofluorescent proteins in the rhizosphere: new perspectives for studying microbial communities. Mol Plant-Microbe Interact 13:1170–1176

Bolwerk A, Lagopodi AL, Wijfjes AHM, Lamers GEM, Lugtenberg BJJ, Bloemberg GV (2004) Interactions between *Pseudomonas* biocontrol strains and *Fusarium oxysporum* f. sp. *radicis-lycopersici* in the tomato rhizosphere. In: Tikhonovich IA, Lugtenberg BJJ Provorov NA (eds) Biology of plant-microbe interactions. IS-MPMI, St.-Petersburg 4: 323–326

Bonanomi A, Wiemken A, Boller T, Salzer P (2001) Local induction of a mycorrhiza-specific class III chitinase gene in cortical root cells of *Medicago truncatula* containing developing or mature arbuscules. Plant Biol 3:94–199

Bond L (1948) Origin and developmental morphology of root nodules of *Pisum sativum*. Bot Gazette 109:411–434

Bonfante P, Genre A, Faccio A, Martini I, Schauser L, Stougaard J, Webb J, Parniske M (2000) The *Lotus japonicus LjSym4* gene is required for the successful symbiotic infection of root epidermal cells. Mol Plant-Microbe Interact 13:1109–1120

Bonfante P, Balestrini R, Genre A, Lanfranco L (2009) Establishment and functioning of arbuscular mycorrhizas. In: Deising H (ed) The Mycota V. Plant relationship, 2nd edn. Springer, Berlin/Heidelberg, pp 259–274

Bonnier C (1961) Spontaneous tumors on the roots of germ-free Leguminosae. Their possible significance in the *Rhizobium*-Leguminosae symbiosis. Ann Inst Past 100:358–367

Borisov AY, Rozov SM, Tsyganov VE, Morzhina EV, Lebsky VK, Tikhonovich IA (1997) Sequential functioning of *Sym-13* and *Sym-31*, two genes affecting symbiosome development in root nodules of pea (*Pisum sativum* L.). Mol Gen Genet 254:592–598

Borisov AY, Tsyganov VE, Shtark OY, Jacobi LM, Naumkina TS, Serdyuk VP, Vishnyakova MA (2002) The catalogue of world-wide collection. Issue 728. Pea (Symbiotic effectiveness). In: Tikhonovich IA, Vishnyakova MA (eds) VIR, Saint Petersburg

Borisov AY, Madsen LH, Tsyganov VE, Umehara Y, Voroshilova VA, Batagov AO, Sandal N, Mortensen A, Schauser L, Ellis N, Tikhonovich IA, Stougaard J (2003) The *Sym35* gene required for root nodule development in pea is an ortholog of *Nin* from *Lotus japonicus*. Plant Physiol 131:1009–1017

Borisov AY, Shtark OY, Danilova TN, Tsyganov VE, Naumkina TS (2004) Effectiviness of combined inoculation of field peas with arbuscular mycorrhizal fungi and nodule bacteria. Russian Agricultural Sciences (Doklady Rossiiskoi Akademii Sel'skohozyaistvennykh Nauk) 4:5–7

Borisov AY, Danilova TN, Koroleva TA, Kuznetsova EV, Madsen L, Mofett M, Naumkina TS, Nemankin TA, Ovchinnikova ES, Pavlova ZB, Petrova NE, Pinaev AG, Radutoiu S, Rozov SM, Rychagova TS, Shtark OY, Solovov II, Stougaard J, Tikhonovich IA, Topunov AF, Tsyganov VE, Vasil'chikov AG, Voroshilova VA, Weeden NF, Zhernakov AI, Zhukov VA (2007) Regulatory Genes of Garden Pea (*Pisum sativum* L.) Controlling the development of nitrogen-fixing nodules and arbuscular mycorrhiza: a review of basic and applied aspects. Russ J Appl Biochem Microbiol 43:265–271

Borisov AY, Danilova TN, Shtark OY, Solovov II, Kazakov AE, Naumkina TS, Vasilchikov AG, Chebotar VK, Tikhonovich IA (2008) Tripartite symbiotic system of pea (*Pisum sativum* L.): applications in sustainable agriculture. In: FD Dakora, BM Chimphango, AJ Valentine, C Elmerich, WE Newton (eds) Biological nitrogen fixation: towards poverty alleviation through sustainable agriculture. Proceedings of 15th International Congress on Nitrogen Fixation & 12th International Conference of the African Association for Biological Nitrogen Fixation. Springer Science and Business Media BV, Berlin/Heidelberg pp 15–17

Bradbury SM, Peterson RL, Bowley SR (1991) Colonisation of three alfalfa (*Medicago sativa* L.) nodulation genotypes by indigenous vesicular-arbuscular mycorrhizal fungi from soil. Symbiosis 15:207–215

Breedveld MJ, Miller KJ (1994) Cyclic β-glucans of members of the family Rhizobiaceae. Microbiol Rev 58:145–161

Brewin NJ (1991) Development of the legume root nodule. Annu Rev Cell Biol 7:191–226

Brewin NJ (1998) Tissue and cell invasion by *Rhizobium*: the structure and development of infection threads and symbiosomes. In: Spaink HP, Kondorosi A, Hooykaas PJJ (eds) The Rhizobiaceae. Molecular biology of model plant-associated bacteria. Kluwer, Dordrecht, pp 417–429

Brewin NJ (2004) Plant cell wall remodeling in the *Rhizobium*-legume symbiosis. Crit Rev Plant Sci 23:1–24

Broughton WJ, Perret X (1999) Genealogy of legume-*Rhizobium* symbiosis. Curr Opin Plant Biol 2:305–311

Brundrett MC (2002) Coevolution of roots and mycorrhizas of land plants. New Phytol 154: 275–304

Bruns TD, Bidartondo MI (2002) Molecular windows into the below-ground interactions of ectomycorrhizal fungi. Mycologist 16(2):47–50

Buysens S, Poppe J, Hofte M (1994) Role of siderphores in plant growth stimulation and antagonism by *Pseudomonas aeruginosa* 7NSK2. In: Ryder MH, Stephens PM, Bowen GD (eds) Improving plant productivity with rhizosphere bacreria. Commonwealth Scientific and Industrial Research Organization, Adelaide, Australia, pp 187–189

Caetano-Anolles G, Gresshoff PM (1991) Plant genetic control of nodulation. Annu Rev Microbiol 45:345–382

Caetano-Anollés G, Wrobel-Boerner E, Bauer WD (1992) Growth and movement of spot inoculated *Rhizobium meliloti* on the root surface of alfalfa. Plant Physiol 98:1181–1189

Cardenas L, Vidali L, Dominguez J, Perez H, Sanchez F, Hepler PK, Quinto C (1998) Rearrangement of actin microfilaments in plant root hairs responding to *Rhizobium etli* nodulation signals. Plant Physiol 116:871–877

Catara V (2007) *Pseudomonas corrugata*: plant pathogen and/or biological resource? Mol Plant Pathol 8:233–244

Catoira R, Galera C, de Billy F, Penmetsa RV, Journet EP, Maillet F, Rosenberg C, Cook D, Gough C, Denarie J (2000) Four genes of *Medicago truncatula* controlling components of a Nod factor transduction pathway. Plant Cell 12:1647–1665

Catoira R, Timmers AC, Maillet F, Galera C, Penmetsa RV, Cook D, Denarie J, Gough C (2001) The HCL gene of *Medicago truncatula* controls *Rhizobium*-induced root hair curling. Development 128:1507–1518

Cebolla A, Vinardell JM, Kiss E, Olah B, Roudier F, Kondorosi A, Kondorosi E (1999) The mitotic inhibitor *ccs52* is required for endoreduplication and ploidy-dependent cell enlargement in plants. EMBO J 18:101–109

Celik I, Ortas I, Kilic S (2004) Effects of compost, mycorrhiza, manure and fertilizer on some physical properties of a Chromoxerert soil. Soil Tillage Res 78:59–67

Chambers SM, Williams PG, Seppelt RD, Cairney JWC (1999) Molecular identification of *Hymenoscyphus* sp. from the rhizoids of the leafy liverwort *Cephalozilla exiflora*. Australia and Antarctica. Mycol Res 103:286–288

Chebotar VK, Kazakov AE, Erofeev SV, Danilova TN, Naumkina TS, Shtark OY, Tikhonovich IA, Borisov AY (2008) Method of production of complex microbial fertilizer. Patent No 2318784

Chen C, Ane JM, Zhu H (2008) OsIPD3, an ortholog of the *Medicago truncatula* DMI3 interacting protein IPD3, is required for mycorrhizal symbiosis in rice. New Phytol 180:311–315

Cheng HP, Walker GC (1998) Succinoglycan is required for initiation and elongation of infection threads during nodulation of alfalfa by *Rhizobium meliloti*. J Bacteriol 180:5183–5191

Chin-A-Woeng TFC, Bloemberg GV, Mulders IHM, Dekkers LC, Lugtenberg BJJ (2000) Root colonization by phenazine-1-carboxamide producing bacterium *Pseudomonas chlororaphis* PCL1391 is essential for biocontrol of tomato foot and root rot. Mol Plant-Microbe Interact 13:340–1345

Cook DR (1999) *Medicago truncatula* – a model in the making. Curr Opin Plant Biol 2:301–304

Cook DR, Vandenbosch K, de Brujn FJ, Huguet T (1997) Model legumes get the nod. Plant Cell 9:275–281

Cordier C, Pozo MJ, Barea JM, Gianinazzi S, Gianinazzi-Pearson V (1998) Cell defense responses associated with localized and systemic resistance to *Phytophthora parasitica* induced in tomato by an arbuscular mycorrhizal fungus. Mol Plant-Microbe Interact 11:1017–1028

Costacurta A, Vanderleyden J (1995) Synthesis of phytohormones by plant-associated bacteria. Crit Rev Microbiol 21:1–18

Cronk QCB, Bateman RM, Hawkins JA (2002) Developmental genetics and plant evolution. CRC Press, London

Currah RS, Zelmer CD, Hambleton S, Richardson KA (1997) Fungi from orchid mycorrhizas. In: Arditti J, Pridgeon AM (eds) Orchid biology: reviews and perspectives VII. Kluwer, Dordrecht, The Netherlands, pp 117–170

Dale C, Plague GR, Wang B, Ochman H, Moran NA (2002) Type III secretion systems and the evolution of mutualistic endosymbiosis. Proc Natl Acad Sci USA 99:12397–12402

Daplé Y, Souza FA, Declerck S (2005) Life cycle of *Glomus* species in monoxenic culture. In: Declerck S, Strullu DG, Fortin JA (eds) *In vitro* culture of mycorrhizas. Springer, Berlin/Heidelberg/New York, pp 49–71

Dassi B, Dumas-Gaudot E, Gianinazzi S (1998) Do pathogenesis-related (PR) proteins play a role in bioprotection of mycorrhizal tomato roots towards *Phytophthora parasitica*. Physiol Mol Plant Pathol 52:167–183

David-Schwartz R, Badani H, Wininger S, Levy AA, Galili G, Kapulnik Y (2001) Isolation of a novel genetically controlled step in mycorrhizal colonozation: plant resistance to infection by fungal spores but not extra-radical hyphae. Plant J 27:561–569

David-Schwartz R, Rakefet, Gadkar V, Wininger S, Bendov R, Galili G, Levy AA, Kapulnik Y (2003) Isolation of a premycorrhizal infection (pmi2) mutant of tomato, resistant to arbuscular mycorrhizal fungal colonization. Molec Plant-Microbe Interact 16(5):382–8

Davis EO, Evans AWB, Johnston IJ (1988) Identification of *nodX*, a gene that allows *Rhizobium leguminosarum* bv. *viciae* strain TOM to nodulate Afghanistan peas. Mol Gen Genet 212:531–535

De Boer W, Folman LB, Summerbell RC, Boddy L (2005) Living in a fungal world: impact of fungi on soil bacterial niche development. FEMS Microbiol Rev 29:795–811

De la Providencia EG, de Souza FA, Fernandez F, Delmas NS, Declerck S (2005) Arbuscular mycorrhizal fungi reveal distinct patterns of anastomosis formation and hyphal healing mechanisms between different phylogenic groups. New Phytol 165:261–271

Dekkers LC, Mulders IHM, Phoelich CC, Chin-A-Woeng T, Wijfjes AH, Lugtenberg BJJ (2000) The *sss* colonization gene of the tomato-*Fusarium oxysporum* f. sp. *radicis-lycopersici* biocontrol strain *Pseudomonas fluorescens* WCS365 can improve root colonization of the wild type *Pseudomonas* bacteria. Mol Plant-Microbe Interact 13:1177–1183

Demchenko NP, Demchenko KN (2001) Resumption of DNA synthesis and cell division in wheat roots as related to lateral root initiation. Russ J Plant Physiol 48:755–763

Demchenko K, Winzer T, Stougaard J, Parniske M, Pawlowski K (2004) Distinct roles of *Lotus japonicus* SYMRK and SYM15 in root colonisation and arbuscule formation. New Phytol 163:381–392

Denny TP (2006) Plant pathogenic *Ralstonia* species. In: Gnanamanickam SS (ed) Plant-associated bacteria. Springer, Dordrecht, The Netherlands, pp 573–644

D'Haeze W, Holsters M (2002) Nod factor structures, responses, and perception during initiation of nodule development. Glycobiology 12:79–105

Diaz CL, Spaink HR, Wijffelman CA, Kijne JA (1995) Genomic requirements of *Rhizobium* for nodulation of white clover hairy roots transformed with the pea lectin gene. Mol Plant Microbe Interact 8:348–356

Dickson S, Smith FA, Smith SE (2007) Structural differences in arbuscular mycorrhizal symbioses: more than 100 years after Gallaud, where next? Mycorrhiza 17(5):375–393

Döbereiner J (1988) Isolation and identification of root associated diazotrophs. Plant Soil 110:207–212

Douds D, Pfeffer P, Shachar-Hill Y (2000) Application of *in vitro* methods to study carbon uptake and transport by AM fungi. Plant Soil 226:255–261

Douglas AE (1994) Symbiotic interactions. Oxford University Press, Oxford

Downie JA (2007) Infectious Heresy. Science 316(5829):1296–1297

Doyle JJ (1998) Phylogenetic perspectives on nodulation: evolving views of plants and symbiotic bacteria. Trends in Plant Sci 2:473–478

Duc G, Trouvelot A, Gianinazzi-Pearson V, Gianinazzi S (1989) First report of non-mycorrhizal plant mutants (Myc$^-$) obtained in pea (*Pisum sativum* L.) and fababean (*Vicia faba* L.). Plant Sci 60:215–222

Dyakov YT, Dzhavakhiya V, Korpela T (eds) (2007) Comprehensive and molecular phytopathology. Elsevier, Amsterdam

Dyall SD, Johnson PJ (2000) Origins of hydrogenosomes and mitochondria: evolution and organelle biogenesis. Curr Opin Microbiol 3:404–411

Edwards A, Heckmann AB, Yousafzai F, Duc G, Downie JA (2007) Structural implications of mutations in the pea SYM8 symbiosis gene, the DMI1 ortholog, encoding a predicted ion channel. Mol Plant Microbe Interact 20(10):1183–1191

Embley TM, van der Giezen M, Horner DS, Dyal PL, Foster P (2003) Mitochondria and hydrogenosomes are two forms of the same fundamental organelle. Phil Trans R Soc Lond B 358:191–202

Endre G, Kereszt A, Kevei Z, Mihacea S, Kalo P, Kiss GB (2002) A receptor kinase gene regulating symbiotic nodule development. Nature 417:962–966

Esseling JJ, Lhuissier FGP, Emons AMC (2003) Nod factor-induced root hair curling: continuous polar growth towards the point of nod factor application. Plant Physiol 132:1982–1988

Ezawa S, Smith SE, Smith FA (2002) P metabolism and transport in AM fungi. Plant Soil 244:221–230

Fedorova M, Tikhonovich IA, Vance CP (1999) Expression of C-assimilating enzymes in pea (*Pisum sativum* L.) root nodules. *In situ* localization in effective nodules. Plant Cell Environ 22:1249–1262

Felle HH, Kondorosi E, Kondorosi A, Schultze M (1999) Elevation of the cytosolic free [Ca^{2+}] is indispensable for the transduction of the Nod factor signal in alfalfa. Plant Physiol 121:273–279

Fitter AH (2005) Darkness visible: reflections on underground ecology. J Ecol 93:231–243

Fogher C, Dusha I, Barbot P, Elmerich C (1985) Heterologous hybridization of *Azospirillum* DNA to *Rhizobium nod* and *fix* genes. FEMS Microbiol Lett 30:245–249

Foucher F, Kondorosi E (2000) Cell cycle regulation in the course of nodule organogenesis in *Medicago*. Plant Mol Biol 43:773–786

Frank AB (1885) Über die auf Würzelsymbiose beruhende Ehrnдhrung gewisser B um durch unterirdische Pilze. Berichte der Deutschen Botanischen Gesellschaft 3:128–145

Fred EB, Baldwin IL, McCoy E (1932) Root nodule bacteria and leguminous plants. Univ Wisconsin Stud Sci, Madison

Fritz M, Jakobsen I, Lyngkjaer MF, Thordal-Christensen H, Pons-Kühnemann J (2006) Arbuscular mycorrhiza reduces susceptibility of tomato to *Alternaria solani*. Mycorrhiza 16:413–419

Frühling M, Roussel H, Gianinazzi-Pearson V, Pühler A, Perlick AM (1997) The *Vicia faba* leghemoglobin gene VfLb29 is induced in root nodules and in roots colonized by the arbuscular mycorrhizal fungus *Glomus fasciculatum*. Mol Plant-Microbe Inter 10(1):124–131

Gallaud I (1905) Études sur les mycorrhizes endotrophes. Rev Gén Bot 17:5–48; 66–83, 123–135; 223–239; 313–325; 425–433; 479–500

Galvan GA, Burger-Meijer K, Kuiper TW, Kik C, Scholten OE (2007) Breeding for improved responsiveness to arbuscular mycorrhizal fungi in onion. Proceedings of 3rd International Congress of the European Integrated Project Quality Low Input Food (QLIF) Congress, Hohenheim, Germany, March 20–23, 2007. (Online) http://orgprints.org/view/projects/int_conf_qlif2007.html

Gandolfi A, Sanders IR, Rossi V, Menozzi P (2003) Evidence of recombination in putative ancient asexuals. Mol Biol Evol 20:754–761

Gao LL, Delp G, Smith SE (2001) Colonization patterns in a mycorrhiza-defective mutant tomato vary with different arbuscular–mycorrhizal fungi. New Phytol 151:477–491

Gao LL, Knogge W, Delp G, Smith FA, Smith SE (2004) Expression patterns of defense-related genes in different types of arbuscular mycorrhizal development in wild-type and mycorrhiza-defective mutant tomato. Molec Plant-Microbe Interact 17:1103–1113

Garbaye J (1994) Helper bacteria: a new dimension to the mycorrhizal symbiosis. New Phytol 128:197–210

Garcia-Garrido JM, Ocampo JA (2002) Regulation of the plant defence response in arbuscular mycorrhizal symbiosis. J Exp Bot 53:377–1386

Genre A, Bonfante P (2005) Building a mycorrhiza cell: how to reach compatibility between plants and arbuscular mycorrhizal fungi. J Plant Interact 1:3–13

Genre A, Chabaud M, Timmers T, Bonfante P, Barker DG (2005) Arbuscular mycorrhizal fungi elicit a novel intracellular apparatus in *Medicago truncatula* root epidermal cells before infection. Plant Cell 17:3489–3499

Genre A, Chabaud M, Faccio A, Barker DG, Bonfante P (2008) Prepenetration apparatus assembly precedes and predicts the colonization patterns of arbuscular mycorrhizal fungi within the root cortex of both *Medicago truncatula* and *Daucus carota*. Plant Cell 20: 1407–1420

Genre A, Ortu G, Bertoldo C, Martino E, Bonfante P (2009) Biotic and abiotic stimulation of root epidermal cells reveals common and specific responses to arbuscular mycorrhizal fungi. Plant Physiol 149:1424–1434

Gentili F, Jumpponen A (2006) Potential and possible uses of bacterial and fungal biofertilizers. In: Rai MK (ed) Handbook of microbial biofertilizers. Haworth Press, Technology & Engineering, New York, pp 1–28
Geurts R, Heidstra R, Hadri AE, Downie A, Franssen H, van Kammen A, Bisseling T (1997) *Sym2* of *Pisum sativum* is involved in Nod factor perception mechanism that controls the infection process in the epidermis. Plant Physiol 115:351–359
Gianinazzi-Pearson V (1996) Plant cell responses to arbuscular mycorrhizal fungi: getting to the roots of the symbiosis. Plant Cell 8:1871–1883
Gianinazzi-Pearson V, Brechenmacher L (2004) Functional genomics of arbuscular mycorrhiza: decoding the symbiotic cell programme. Canad J Bot 82:1228–1234
Giovannetti M, Sbrana C, Sitemesi AS, Logi C (1993) Differential hyphal morphogenesis in arbuscular mycorrhizal fungi during pre-infection stages. New Phytol 125:587–593
Giovannetti M, Sbrana C, Sitemesi AS, Avio L (1996) Analysis of factors involved in fungal recognition response to host-derived signals by arbuscular mycorrhizal fungi. New Phytol 133:65–71
Giovannetti M, Sbrana C, Avio L, Strani P (2004) Patterns of below-ground plant interconnections established by means of arbuscular mycorrhizal networks. New Phytol 164:175–181
Giraud E, Fleischman D (2004) Nitrogen-fixing symbiosis between photosynthetic bacteria and legumes. Photosynth Res 82:115–130
Giraud E, Moulin L, Vallenet D, Barbe V, Cytryn E, Avarre JC, Jaubert M et al (2007) Legumes symbioses: absence of *Nod* genes in photosynthetic Bradyrhizobia. Science 316:1307–1312
Gleason C, Chaudhuri S, Yang T, Munoz A, Poovaiah BW, Oldroyd GE (2006) Nodulation independent of rhizobia induced by a calcium-activated kinase lacking autoinhibition. Nature 441:1149–1152
Glick BR (2004) The role of bacterial ACC deaminase in promoting the plant growth. In: Tikhonovich IA, Lugtenberg BJJ Provorov NA (eds) Biology of plant-microbe interactions. IS-MPMI, St.-Petersburg 4:557–560
Gollotte A, Gianinazzi-Pearson V, Giovannetti M, Sbrana C, Avio L, Gianinazzi S (1993) Cellular localization and cytochemical probing of resistance reactions to arbuscular mycorrhizal fungi in a 'locus *a*' Myc$^-$ mutant of *Pisum sativum* L. Planta 191:112–122
Gonzalez-Rizzo S, Crespi M, Frugier F (2006) The *Medicago truncatula* CRE1 cytokinin receptor regulates lateral root development and early symbiotic interaction with *Sinorhizobium meliloti.* Plant Cell 18(10):2680–2693
Götz R, Evans IJ, Downie JA, Johnton AWB (1985) Identification of the host-range DNA which allows *Rhizobium leguminosarum* strain TOM to nodulate cv Afghanistan peas. Mol Gen Genet 201:296–300
Govorov LI (1928) Afghan peas. Proceedings on applied botany, genetics and breeding (Gorokh Afganistana. Trudy po prikladnoi botanike, genetike i selektsii) 19:497–522 (In Russian)
Graham PH, Hungria M, Tlusty B (2004) Breeding for better nitrogen fixation in grain legumes: where do the rhizobia fit in? Crop Management (Online). doi:10.1094/CM-2004-0301-02-RV
Grunwald U, Nyamsuren O, Tarnasloukht M, Lapopin L, Becker A, Mann P, Gianinazzi-Pearson V, Krajinski F, Franken P (2004) Identification of mycorrhiza-regulated genes with arbuscule development-related expression profile. Plant Mol Biol 55:553–566
Gualtieri G, Bisseling T (2000) The evolution of nodulation. Plant Mol Biol 42:181–194
Guinel FC, Geil RD (2002) A model for the development of the rhizobial and arbuscular mycorrhizal symbioses in legumes and its use to understand the roles of ethylene in the establishment of these two symbioses. Can J Bot 80:695–720
Guinel FC, LaRue TA (1992) Ethylene inhibitors partly restore nodulation to pea mutant E107 (brz). Plant Physiol 99:515–518
Hadri AE, Spaink HP, Bisseling T, Brewin NJ (1998) Diversity of root nodulation and rhizobial infection processes. In: Spaink HP, Kondorosi A, Hooykaas PJJ (eds) The Rhizobiaceae. Molecular biology of model plant-associated bacteria. Kluwer, Dordrecht, pp 347–360
Hallmann J, Berg G (2006) Spectrum and population dynamics of bacterial root endophytes. In: Schulz B, Boyle S, Sieber T (eds) Microbial root endophytes. Springer, Berlin, pp 15–32

Handberg K, Stougaard J (1992) *Lotus japonicus*, an autogamous, diploid legume species for classical and molecular genetics. Plant J 2:487–496

Harrison MJ (1997) The arbuscular mycorrhizal symbiosis: an underground association. Trends Plant Sci 2:54–56

Harrison MJ (1999) Molecular and cellular aspects of the arbuscular mycorrhizal symbiosis. Ann Rev Plant Physiol Plant Mol Biol 50:361–389

Harrison MJ (2005) Signaling in the arbuscular mycorrhizal symbiosis. Annu Rev Microbiol 59:19–42

Harrison MJ, Dewbre GR, Liu J (2002) A phosphate transporter from *Medicago truncatula* involved in the acquisition of phosphate released by arbuscular mycorrhizal fungi. Plant Cell 14:2413–2429

Hause B, Maier W, Miersch O, Kramell R, Strack D (2002) Induction of jasmonate biosynthesis in arbuscular mycorrhizal barley roots. Plant Physiol 130:1213–1220

Heckmann AB, Lombardo F, Miwa H, Perry JA, Bunnewell S, Parniske M, Wang TL, Downie JA (2006) *Lotus japonicus* nodulation requires two GRAS domain regulators, one of which is functionally conserved in a non-legume. Plant Physiol 142:1739–1750

Helber N, Requena N (2008) Expression of the fluorescence markers DsRed and GFP fused to a nuclear localization signal in the arbuscular mycorrhizal fungus *Glomus intraradices*. New Phytol 177:537–548

Herridge D, Rose I (2000) Breeding for enhanced nitrogen fixation in crop legumes. Field Crops Res 65:229–248

Hibbert DS, Gilbert LB, Donoghue M (2000) Evolutionary instability of ectomycorrhizal symbiosis in basidiomycetes. Nature 407:506–508

Hijri M, Sanders IR (2005) Low gene copy number shows that arbuscular mycorrhizal fungi inherit genetically different nuclei. Nature 433:160–163

Hildebrandt U, Ouziad F, Marner FJ, Bothe H (2006) The bacterium *Paenibacillus validus* stimulates growth of the arbuscular mycorrhizal fungus *Glomus intraradices* up to the formation of fertile spores. FEMS Microbiol Lett 254:258–267

Hirsch AM (1992) Developmental biology of legume nodulation. New Phytol 122:211–237

Hirsch AM, Lum MR, Downie JA (2001) What makes the rhizobia-legume symbiosis so special? Plant Physiol 127:1484–1492

Holguin G, Glick BR (2003) Transformation of *Azospirillum brasilense* Cd with an ACC deaminase gene from *Enterobacter cloacae* UW4 fused to the Tet^r gene promoter improves its fitness and plant growth promoting ability. Microb Ecol 46(1):122–133

Hosokawa T, Kikuchi Y, Nikoh N, Shimada M, Fukatsu T (2006) Strict host-symbiont cospeciation and reductive genome evolution in insect gut bacteria. PLoS Biol 4(10):e337. doi:10.1371/journal.pbio.0040337

Hossain MS, Mårtensson A (2008) Potential use of *Rhizobium* spp. to improve fitness of non-nitrogen-fixing plants. Acta Agriculturae Scandinavica, Section B – Plant Soil Science 58(4):352–358

Howieson JG, Yates RJ, Foster KJ, Real D, Besier RB (2008) Prospects for the future uses of legumes. In: Dilworth MJ, James EK, Sprent JI, Newton WE (eds) Springer Science+Business Media BV, pp 363–394

Ibrahim KK, Arunachalam V, Rao PSK, Tilak KVBR (1995) Seasonal response of groundnut genotypes to arbuscular mycorrhiza – *Bradyrhizobium* inoculation. Microbiol Res 150:218–224

Imaizumi-Anraku H, Takeda N, Charpentier M, Perry J, Miwa H, Umehara Y, Kouchi H, Murakami Y, Mulder L, Vickers K, Pike J, Downie JA, Wang T, Sato S, Asamizu E, Tabata S, Yoshikawa M, Murooka Y, Wu GJ, Kawaguchi M, Kawasaki S, Parniske M, Hayashi M (2005) Plastid proteins crucial for symbiotic fungal and bacterial entry into plant roots. Nature 433:527–531

Izumi H, Anderson IC, Alexander IJ, Killham K, Moore ERB (2006) Diversity and expression of nitrogenase genes (*nifH*) from ectomycorrhizas of Corsican pine (*Pinus nigra*). Environ Microbiol 8:2224–2230

Jackson RM, Mason PA (1984) Mycorrhiza. Edward Arnold, London, pp 13

Jacobi LM, Kukalev AS, Ushakov KV, Tsyganov VE, Provorov NA, Borisov AY, Tikhonovich I (1999) Genetic variability of garden pea (*Pisum sativum* L.) for symbiotic capacities. Pisum Genet 31:44–45

Jacobi LM, Zubkova LA, Barmicheva EM, Tsyganov VE, Borisov AY, Tikhonovich IA (2003) Effect of mutations in the pea genes *Sym33* and *Sym40*. II. Dynamics of arbuscule development and turnover. Mycorrhiza 13:9–16

Jacobsen E (1984) Modification of symbiotic interaction of pea (*Pisum sativum* L.) and *Rhizobium leguminosarum* by induced mutations. Plant Soil 82(3):427–438

Jakobsen I (1995) Transport of phosphorus and carbon in VA mycorrhizas. In: Varma A, Hock B (eds) Mycorrhiza structure, function. Molecular biology and biotechnology. Springer, Berlin, pp 297–325

Jansa J, Vosatka M (2000) In vitro and post vitro inoculation of micropropagated Rhododendrons with ericoid mycorrhizal fungi. Appl Soil Ecol 15:125–136

Jargeat P, Cosseau C, Ola'h B, Jauneau A, Bonfante P, Batut J, Becard G (2004) Isolation, free-living capacities, and genome structure of '*Candidatus Glomeribacter gigasporarum*', the endocellular bacterium of the mycorrhizal fungus *Gigaspora margarita*. J Bacteriol 186:6876–6884

Javot H, Pumplin N, Harrison MJ (2007) Phosphate in the arbuscular mycorrhizal symbiosis: transport properties and regulatory roles. Plant Cell Environ 30:310–322

Jimenez J, Casadesus J (1989) An altruistic model of *Rhizobium*-legume association. J Heredity 80:335–337

Jones KM, Kobayashi H, Davies BW, Taga ME, Walker GC (2007) How rhizobial symbionts invade plants: the *Sinorhizobium-Medicago* model. Nat Rev Microbiol 5:619–633

Jones KM, Sharopova N, Lohar DP, Zhang JQ, VandenBosch KA, Walker GC (2008) Differential response of the plant *Medicago truncatula* to its symbiont *Sinorhizobium meliloti* or an exopolysaccharide-deficient mutant. Proc Natl Acad Sci USA 105(2):704–709

Kalo P, Gleason C, Edwards A, Marsh J, Mitra RM, Hirsch S, Jakab J, Sims S, Long SR, Rogers J, Kiss GB, Downie JA, Oldroyd GE (2005) Nodulation signaling in legumes requires *NSP2*, a member of the GRAS family of transcriptional regulators. Science 308:1786–1789

Kamilova F, Kravchenko LV, Shaposhnikov AI, Azarova T, Makarova N, Lugtenberg B (2006a) Organic acids, sugars and L-tryptophane in exudates of vegetables growing on stonewool and their effects on activities of rhizosphere bacteria. Mol Plant-Microbe Interact 9:250–256

Kamilova F, Kravchenko LV, Shaposhnikov AI, Makarova N, Lugtenberg B (2006b) Effect of tomato pathogen *Fusarium oxysporum* f. sp. *radicis-lycopersici* and biocontrol bacterium *Pseudomonas fluorescens* WCS635 on the composition of organic acids and sugars in tomato root exudates. Mol Plant-Microbe Interact 19:1121–1126

Kanamori N, Madsen LH, Radutoiu S, Frantescu M, Quistgaard EMH, Miwa H, Downie JA, James EK, Felle HH, Haaning LL, Jensen TH, Sato S, Nakamura Y, Tabata S, Sandal N, Stougaard J (2006) A nucleoporin is required for induction of Ca^{2+} spiking in legume nodule development and essential for rhizobial and fungal symbiosis. PNAS 103:359–364

Karandashov V, Bucher M (2005) Symbiotic phosphate transport in arbuscular mycorrhizas. Trends Plant Sci 10:22–29

Kidston R, Lang WH (1921) Old Red Sandstone plants showing structure, from the Rhynie chert bed, Aberdeenshire. Part 5. The Thallophyta occurring in the peat-bed; the succession of the plants throughout a vertical section of the bed, and the conditions of accumulation and preservation of the deposit. Trans Roy Soc Edinburgh 52:855–902

Kistner C, Parniske M (2002) Evolution of signal transduction in intercellular symbiosis. Trends in Plant Sci 7:511–518

Kistner C, Winzer T, Pitzschke A, Mulder L, Sato S, Kaneko T, Tabata S, Sandal N, Stougaard J, Webb KJ, Szczyglowski K, Parniske M (2005) Seven *Lotus japonicus* genes required for transcriptional reprogramming of the root during fungal and bacterial symbiosis. Plant Cell 17(8):2217–2229

Kloepper JW, Ryu CM (2006) Bacterial endophytes as elicitors of induced systemic resistance. In: Schulz B, Boyle S, Sieber T (eds) Microbial root endophytes. Springer, Berlin, pp 33–52

Koide RT, Mosse B (2004) A history of research on arbuscular mycorrhiza. Mycorrhiza 14:145–163

Koske RE (1981) *Gigaspora gigantea*: observation of spore germination of VA-mycorrhizal fungus. Mycologia 73:288–300

Kosuta S, Chabaud M, Lougnon G, Gough C, Dénarié J, Barker DG, Bécard G (2003) A diffusible factor from arbuscular mycorrhizal fungi induces symbiosis-specific *MtENOD11* expression in roots of *Medicago truncatula*. Plant Physiol 131:952–962

Kozik AV, Heidstra R, Horvath B, Kulikova OA, Tikhonovich IA, Ellis THN, van Kammen A, Bisseling T (1995) Pea lines carrying *sym1* or *sym2* can be nodulated by *Rhizobium* strains containing *nodX; sym-1* and *sym-2* are allelic. Plant Sci 108:41–49

Kravchenko LV, Leonova EI (1993) Use of tryptophane from root exometabolites for biosynthesis of indolil-3-acetic acid by the root-associated bacteria. Russian J Microbiol 62:453–459

Kravchenko LV, Leonova EI, Tikhonovich IA (1994) Effect of root exudates of non-legume plants on the response of auxin production by associated diazotrophs. Microbial Releases 2:267–271

Krusell L, Madsen LH, Sato S, Aubert G, Genua A, Szczyglowski K, Duc G, Kaneko T, Tabata S, de Bruijn F, Pajuelo E, Sandal N, Stougaard J (2002) Shoot control of root development and nodulation is mediated by a receptor-like kinase. Nature 420(6914):422–6

Kumagai H, Kinoshita E, Ridge RW, Kouchi H (2006) RNAi knock-down of *ENOD40*s leads to significant suppression of nodule formation in *Lotus japonicus*. Plant Cell Physiol 47(8):1102–1111

Küster H, Vieweg MF, Manthey K, Baier MC, Hohnjec N, Perlick AM (2007) Identification and expression regulation of symbiotically activated legume genes. Phytochemistry 68:8–18

Kuznetsova E, Seddas-Dozolme PMA, Arnould C, Tollot M, van Tuinen D, Borisov A, Gianinazzi S, Gianinazzi-Pearson V (2010) Symbiosis-related pea genes modulate fungal and plant gene expression during the arbuscule stage of mycorrhiza with *Glomus intraradices*. Mycorrhiza. doi:10.1007/s00572-009-0292-8

Labutova NM, Polyakov AI, Lyakh VA, Gordon VL (2004) Influence of inoculation with nodule bacteria and endomycorrhizal fungus *Glomus intraradices* on yield and seed protein and oil content of different soybean cultivars. Russian Agricultural Sciences (Doklady Rossiiskoi Akademii Sel'skohozyaistvennykh Nauk) 4(2):2–4

Lambais MR (2000) Regulation of plant defence-related genes in arbuscular mycorrhizae. In: Podila GK, Douds DD (eds) Current advances in mycorrhizae research. The American Phytopathological Society, Minnesota, pp 45–59

Lavin M, Herendeen PS, Wojciechowski MF (2005) Evolutionary rates analysis of Leguminosae implicates a rapid diversification of lineages during the tertiary. Syst Biol 54:574–594

Leake JL (1994) The biology of myco-heterotrophic ('saprophytic') plants. New Phytologist 127:171–216

Leigh JA, Walker GC (1994) Exopolysaccharides of *Rhizobium*: synthesis, regulation and symbiotic function. Trends Genet 10:63–67

Lewis DH (1985) Symbiosis and mutualism: crisp concept and soggy semantics. In: Boucher DH (ed) The biology of mutualism: ecology and evolution. Croom Helt, London/Sydney

Lhuissier FGP, De Ruijter NCA, Sieberer BJ, Esseling JJ, Emons AMC (2001) Time course of cell biological events evoked in legume root hairs by *Rhizobium* Nod factors: state of the art. Ann Bot 87:289–302

Libbenga KR, Harkes PAA (1973) Initial proliferation of cortical cell the formation of root nodules in *Pisum sativum* L. Planta 114:17–28

Lie TA (1971) Temperature-dependent root-nodule formation in pea cv. Iran Plant Soil 34:751–752

Lie TA (1984) Host genes of *Pisum sativum* L. conferring resistance to European *Rhizobium leguminosarum* strains. Plant Soil 82:462–465

Limpens E, Franken C, Smit P, Willemse J, Bisseling T, Geurts R (2003) LysM-domain receptor kinases regulating rhizobial Nod factor-induced infection. Science 302:630–633

Limpens E, Mirabella R, Fedorova E, Franken C, Franssen H, Bisseling T, Geurts R (2005) Formation of organelle-like N_2-fixing symbiosomes in legume root nodules is controlled by *DMI2*. Proc Natl Acad Sci USA 102:10375–10380

Linderman RG (1988) Mycorrhizal interactions with the rhizosphere microflora: the mycorrhizosphere effect. Phytopathology 78:366–371
Liu L, Kloepper JW, Tuzun S (1995) Induction of systemic resistance in cucumber by plant growth-promoting rhizobacteria: duration of protection and root colonization. Phytopathol 85:1064–1068
Liu A, Hamel C, Elmi A, Costa C, Ma B, Smith DL (2002) Concentrations of K, Ca and Mg in maize colonised by arbuscular mycorrhizal fungi under field conditions. Can J Soil Sci 82:271–278
Liu J, Maldonado-Mendoza I, Lopez-Meyer M, Cheung F, Town CD, Harrison MJ (2007) Arbuscular mycorrhizal symbiosis is accompanied by local and systemic alterations in gene expression and an increase in disease resistance in the shoots. Plant J 50:529–544
LoBuglio KF, Berbee ML, Taylor JW (1996) Phylogenetic origins of the asexual mycorrhizal symbiont *Cenococcum geophilum* Fr. and other mycorrhizal fungi among Ascomycetes. Mol Phylogen Evol 6:287–294
Lohar D, Stiller J, Kam J, Stacey G, Gresshoff PM (2009) Ethylene insensitivity conferred by a mutated Arabidopsis ethylene receptor gene alters nodulation in transgenic Lotus japonicus. Annals Bot 104:277–285
Long SR (1989) *Rhizobium*-legume nodulation: life together in the underground. Cell 56:203–214
Lugtenberg BJJ (1998) Outer membrane proteins. In: Spaink HP, Kondorosi A, Hooykaas PJJ (eds) The Rhizobiaceae. Molecular biology of model plant-associated bacteria. Kluwer, Dordrecht/Boston/London, pp 45–53
Lugtenberg BJJ, Dekkers LC (1999) What makes *Pseudomonas* bacteria rhizosphere competent? Environ Microbiol 1:9–13
Lugtenberg BJJ, Dekkers L, Bloemberg G (2001) Molecular determinants of rhizosphere colonization by *Pseudomonas*. Annu Rev Phytopathol 39:461–490
Lugtenberg BJJ, Bloemberg GV, Bolwerk A, van den Broek D, Cazorla-Lopez F, Chin-A-Woeng TFC, Eijkemans K, Kamilova FD, Kuiper I, Mulders IHM, van Rijn ET, de Weert S (2004) Microbial control of tomato foot and root rot. In: Tikhonovich IA, Lugtenberg BJJ Provorov NA (eds) Biology of plant-microbe interactions. IS-MPMI, St.-Petersburg 4:305–309
Lum MR, Hirsch AM (2003) Roots and their symbiotic microbes: strategies to obtain nitrogen snd phosphorus in a nutrient-limited environment. J Plant Growth Regul 21:368–382
MacDonald RM, Chandler MR (1981) Bacterium-like organelles in the vesicular-arbuscular mycorrhizal fungus *Glomus caledonius*. New Phytol 89:241–246
MacLean AM, Finan T, Sadowsky MJ (2007) Genomes of the symbiotic nitrogen-fixing bacteria of legumes. Plant Physiol 144:615–622
Madsen EB, Madsen LH, Radutoiu S, Olbryt M, Rakwalska M, Szczyglowski K, Sato S, Kaneko T, Tabata S, Sandal N, Stougaard J (2003) A receptor kinase gene of the LysM type is involved in legume perception of rhizobial signals. Nature 425:637–640
Maherali H, Klironomos JN (2007) Influence of phylogeny on fungal community assembly and ecosystem functioning. Science 316:1746–1748
Margolin W (2000) Differentiation of free-living rhizobia into endosymbiotic bacteroids. In: Brun YV, Shimkets LJ (eds) Prokaryotic development. Amer Soc Microbiol, Washington, DC, pp 441–466
Margulis L (1970) Origin of eukaryotic cell. Yale Univ Press, New Haven
Margulis L (1998) Symbiotic planet: a new look at evolution. Perseus Books Group, Amherst, MA
Margulis L, Sagan D (2002) Acquiring genomes: a theory of the origins of species. Perseus Books Group, Amherst, MA
Marsh JF, Schultze M (2001) Analysis of arbuscular mycorrhizas using symbiosis-defective plant mutants. New Phytol 150:525–532
Marsh JF, Shultze M, Oldroyd GEF (2006) Isolation and analysis of *Medicago truncatula* mutants defective in the arbuscular mycorrhizal symbiosis. Handbook & Abstracts of 3rd International Conference on Legume Genomics & Genetics, Brisbane, Australia, April 9–13, 2006, p 52
Marsh JF, Rakocevic A, Mitra RM, Brocard L, Sun J, Eschstruth A, Long SR, Schultze M, Ratet P, Oldroyd GE (2007) *Medicago truncatula NIN* is essential for rhizobial-independent nodule organogenesis induced by autoactive calcium/calmodulin-dependent protein kinase. Plant Physiol 144:324–335

Marsh JF, Gobbato E, Schultze M, Oldroyd GED (2008) Ram1 and Ram2: signaling specificity in mycorrhizal symbiosis. Abstract Book of 8th European Nitrogen Fixation Conference, Gent, Belgium, August 30–September 3, 2008, p 26

McCully ME (2001) Niches for bacterial endophytes in crop plants: a plant biologists's view. Aust J Plant Physiol 28:983–990

McLean CB, Cunnington JH, Lawrie AC (1999) Molecular diversity within and between ericoid endophytes from the Ericaceae and Epacridaceae. New Phytol 144:351–358

Mergaert P, Uchiumi T, Alunni B, Evanno G, Cheron A, Catrice O, Mausset AE, Barloy-Hubler F, Galibert F, Kondorosi A, Kondorosi E (2006) Eukaryotic control on bacterial cell cycle and differentiation in the *Rhizobium*–legume symbiosis. Proc Natl Acad Sci USA 103: 5230–5235

Messinese E, Mun JH, Yeun LH et al (2007) A novel nuclear protein interacts with the symbiotic DMI3 calcium- and calmodulin-dependent protein kinase of *Medicago truncatula*. Mol Plant Microbe Interact 20(8):912–921

Michelmore RW, Meyers BC (1998) Clusters of resistance genes in plants evolve by divergent selection and a birth-and-death process. Genome Res 8:1113–1130

Middleton PH, Jakab J, Penmetsa RV, Starker CG, Doll J, Kalo P, Prabhu R, Marsh JF, Mitra RM, Kereszt A, Dudas B, VandenBosch K, Long SR, Cook DR, Kiss GB, Oldroyd GE (2007) An ERF transcription factor in *Medicago truncatula* that is essential for Nod factor signal transduction. Plant Cell 19:1221–1234

Minerdi D, Bianciotto V, Bonfante P (2002) Endosymbiotic bacteria in mycorrhizal fungi: from their morphology to genomic sequences. Plant Soil 244:211–219

Miyazaki H, Kato H, Nakazawa T, Tsuda M (1995) A positive regulatory gene, pvdS, for expression of pyoverdin biosynthetic genes in *Pseudomonas aeruginosa* PAO. Mol Gen Genet 248(1):17–24

Molina R, Massicotte H, Trappe JM (1992) Specificity phenomena in mycorrhizal symbioses: community-ecological consequences and practical implications. In: Allen MJ (ed) Mycorrhizal functioning: an interactive plant-fungal process. Chapman & Hall, London, pp 357–423

Moncalvo JM, Lutzoni FM, Rehner SA, Johnson J, Vilgalys R (2000) Phylogenetic relationships of agaric fungi based on nuclear large subunit ribosomal DNA sequences. Syst Biol 49:278–305

Monreal M, Berch SM, Berbee M (1999) Molecular diversity of ericoid mycorrhizal fungi. Can J Bot 77:1580–1594

Morandi D, Sagan M, Prado-Vivant E, Duc G (2000) Influence of genes determining supernodulation on root colonisation by the mycorrhizal fungus *Glomus mosseae* in *Pisum sativum* and *Medicago truncatula* mutants. Mycorrhiza 10:37–42

Mosse B (1959) The regular germination of resting spores and some observations on the growth requirements of an *Endogone* sp. causing vesicular-arbuscular mycorrhiza. T Brit Mycol Soc 42:273–286

Murakami Y, Miwa H, Imaizumi-Anraku H, Kouchi H, Downie JA, Kawasaki KMS (2006) Positional cloning identifies *Lotus japonicus NSP2*, a putative transcription factor of the GRAS family, required for *NIN* and *ENOD40* gene expression in nodule initiation. DNA Res 13:255–265

Murray RG, Schleifer KH (1994) Taxonomic notes: a proposal for recording the properties of putative taxa of procaryotes. Int J Syst Bacteriol 44:174–176

Murray RG, Stackebrandt E (1995) Taxonomic note: implementation of the provisional status *Candidatus* for incompletely described procaryotes. Int J Syst Bacteriol 45:186–187

Murray JD, Karas BJ, Sato S, Tabata S, Amyot L, Szczyglowski K (2007) A cytokinin perception mutant colonized by *Rhizobium* in the absence of nodule organogenesis. Science 315(5808):101–104

Muthukumar T, Udaiyan K, Rajeshkannan V (2001) Response of neem (*Azadirachta indica* A. Juss) to indigenous arbuscular mycorrhizal fungi, phosphate-solubilizing and asymbiotic nitrogen-fixing bacteria under tropical nursery conditions. Biol Fertil Soil 34:417–426

Nagahashi G, Douds JDD (1997) Appressorium formation by AM fungi on isolated cell walls of carrot roots. New Phytol 136:299–304

Nair MG, Safir GR, Siqueira JO (1991) Isolation and identification of vesicular-arbuscular mycorrhiza-stimulatory compounds from clover (*Trifolium repens*) roots. Appl Environ Microbiol 57:434–439

Navazio L, Moscatiello R, Genre A, Novero M, Baldan B, Bonfante P, Mariani P (2007) A diffusible signal from arbuscular mycorrhizal fungi elicits a transient cytosolic calcium elevation in host plant cells. Plant Physiol 144:673–681

Newcomb WE (1976) A correlated light and electron microscopic study of symbiotic growth and differentiation in *Pisum sativum* root nodules. Canad J Bot 54:2163–2186

Niehaus K, Lagares A, Puhler A (1998) A *Sinorhizobium meliloti* lipopolysaccharide mutant induces effective nodules on the host plant *Medicago sativa* (alfalfa) but fails to establish a symbiosis with *Medicago truncatula*. Mol Plant-Microbe Interact 11:906–914

Nikitina VE, Alen'kina SA, Ponomareva EG, Savenkova NI (1996) Analysis of the role of lectins from the surface of azospirilla in the interaction with wheat roots. Russian J Microbiol 65:12–19

Nishimura R, Hayashi M, Wu GJ, Kouchi H, Imaizumi-Anraku H, Murakami Y, Kawasaki S, Akao S, Ohmori M, Nagasawa M, Harada K, Kawaguchi M (2002a) HAR1 mediates systemic regulation of symbiotic organ development. Nature 420:426–429

Nishimura R, Ohmori M, Fujita H, Kawaguchi M (2002b) A *Lotus* basic leucine zipper protein with a RING-finger motif negatively regulates the developmental program of nodulation. Proc Natl Acad Sci USA 99:15206–15210

Nogales J, Campos R, BenAbdekhalek H, Olivares J, Lluch C, Sanjuan J (2002) *Rhizobium tropici* genes involved in free-living salt tolerance are required for the establishment of efficient nitrogen-fixing symbiosis with *Phaseolus vulgaris*. Mol Plant-Microbe Interact 15:225–232

Nukui N, Ezura H, Yuhashi K-I, Yasuta T, Minamisawa K (2000) Effects of ethylene precursor and inhibitors for ethylene biosynthesis and perception on nodulation in *Lotus japonicus* and *Macroptilium atropurpureum*. Plant Cell Physiol 41:893–897

Nutman PS (1946) Genetic factors concerned in the symbiosis of clover and nodule bacteria. Nature, London 151:463–465

Nutman PS (1969) Symbiotic nitrogen fixation: legume nodule bacteria. Rothamsted Exper Stat Rep 2:141–167

Okamoto S, Ohnishi E, Sato S, Takahashi H, Nakazono M, Tabata S, Kawaguchi M (2009) Nod factor/nitrate-induced CLE genes that drive HAR1-mediated systemic regulation of nodulation. Plant Cell Physiol 50(1):67–77

Okon Y, Itzigsohn R, Burdman S, Hampel M (1995) Advances in agronomy and ecology of the *Azospirillum*/plant association. In: Tikhonovich IA, Provorov NA, Romanov VI, Newton WE (eds) Nitrogen fixation: fundamentals and applications. Kluwer, Dordrecht/Boston/London, pp 635–640

Oldroyd GE, Downie JA (2004) Calcium, kinases and nodulation signalling in legumes. Nat Rev Mol Cell Biol 5(7):566–576

Oldroyd GE, Downie JA (2008) Coordinating nodule morphogenesis with rhizobial infection in legumes. Annu Rev Plant Biol 59:519–546

Oldroyd GE, Long SR (2003) Identification and characterization of nodulation-signaling pathway 2, a gene of *Medicago truncatula* involved in Nod factor signaling. Plant Physiol 131:1027–1032

O'Toole GA, Kolter R (1998) Initiation of biofilm formation in *Pseudomonas fluorescens* WCS365 proceeds via multiple, convergent signalling pathways: a genetic analysis. Mol Microbiol 28(3):449–61

Ovtsyna AO, Staehelin C (2005) Bacterial signals required for the *Rhizobium*-legume symbiosis. Recent Res Develop Microbiol 7:631–648

Ovtsyna AO, Rademaker GJ, Esser E, Weinman J, Lugtenberg B, Tihonovich IA (1999) Comparison of characteristics of the *nodX* genes from various *Rhizobium leguminosarum* strains. Molec Plant Microbe Interact 12:252–258

Ovtsyna AO, Schultze M, Tikhonovich IA, Spaink HP, Kondorosi E, Kondorosi A, Staehelin C (2000) Nod factors of *Rhizobium leguminosarum* bv. *viceae* and their fucosylated derivatives stimulate a Nod factor cleaving activity in pea roots and are hydrolysed *in vitro* by plant chitinases at different rates. Molec Plant-Microbe Interact 13:799–807

Parniske M (2000) The *Lotus japonicus* LjSym4 gene is required for the successful symbiotic infection of root epidermal cells. Molec Plant-Microbe Iinteract 13:1109–1120

Parniske M (2008) Arbuscular mycorrhiza: the mother of plant root endosymbioses. Nature Rev Microbiol 6:763–775

Paszkowski U, Jakovleva L, Boller T (2006) Maize mutants affected at distinct stages of the arbuscular mycorrhizal symbiosis. Plant J 47:165–173
Pawlowska TE, Taylor JW (2004) Organization of genetic variation in individuals of arbuscular mycorrhizal fungi. Nature 427:733–737
Pawlowski K, Bisseling T (1996) Rhizobial and actinorhizal symbioses: what are the shared features? Plant Cell 8:1899–1913
Peiter E, Sun J, Heckmann AB, Venkateshwaran M, Riely BK, Otegui MS, Edwards A, Freshour G, Hahn MG, Cook DR, Sanders D, Oldroyd GE, Downie JA, Ané JM (2007) The *Medicago truncatula* DMI1 protein modulates cytosolic calcium signaling. Plant Physiol 145(1):192–203
Penmetsa RV, Cook DR (1997) A legume ethylene-insensitive mutant hyperinfected by its rhizobial symbiont. Science 275:527–530
Penrose DM, Glick BR (2003) Methods for isolating and characterizing ACC deaminase-containing plant growth-promoting rhizobacteria. Physiologia Plantarum 118:10–15
Perret X, Freiberg C, Rosenthal A, Broughton WJ, Fellay R (1999) High-resolution transcriptional analysis of the symbiotic plasmid of *Rhizobium* sp. NGR234. Mol Microbiol 32:415–425
Pfeffer PE, Douds DDJ, Bécard G, Shachar-Hill Y (1999) Carbon uptake and the metabolism and transport of lipids in an arbuscular mycorrhiza. Plant Physiol 120:587–598
Pieterse CMJ, Ton J, Van Loon LC (2001) Cross-talk between plant defence signaling pathways: boost or burden? Ag Biotech Net 3:1–7
Pirozynski KA, Dalpé Y (1989) The geological history of the Glomaceae with particular reference to mycorrhizal symbiosis. Symbiosis 7:1–36
Pope EJ, Carter DE (2001) Pylogenetic placement and host specificity of mycorrhizal isolates belonging to AG-6 and AG-12 in the *Rhizoctonia solani* species complex. Mycologia 93:712–719
Popova EV, Khatskevich LK (2004) In: Tikhonovich IA, Lugtenberg BJJ Provorov NA (eds) Biology of plant-microbe interactions. IS-MPMI, St.-Petersburg 4:315–318
Powell CV (1976) Development of mycorrhiza infection from *Endogone* spores and infected root fragments. Trans Br Mycol Soc 66:439–445
Preston GM (2004) Plant perceptions of plant growth-promoting *Pseudomonas*. Phil Trans R Soc Lond B 359:907–918
Prévost D, Antoun H (2005) Potential Use of *Rhizobium* as PGPR with Non-Legumes. Abstracts from the (2005), Inoculant Forum, March 17–18. Saskatoon, Saskatchewan, Canada
Provorov NA (1994) The interdependence between taxonomy of legumes and specificity of their interaction with rhizobia in relation to evolution of the symbiosis. Symbiosis 17:183–200
Provorov NA (1998) Coevolution of rhizobia with legumes: facts and hypotheses. Symbiosis 24:337–367
Provorov NA (2005) Molecular basis of symbiogenic evolution: from free-living bacteria towards organelles. Zh Obshch Biol (Russian J Gen Biol) 66(5):371–388, In Russian
Provorov NA (2009) Plant-microbe symbioses as an evolutionary continuum. Zh Obshch Biol (Russian J Gen Biol) 70(1):10 –34, In Russian
Provorov NA, Vorobyov NI (2009a) Host plant as an organizer of microbial evolution in the beneficial symbioses. Phytochem Rev 8:519–534
Provorov NA, Vorobyov NI (2009b) Interspecies altruism in plant-microbe symbioses: use of group selection models to resolve the evolutionary paradoxes. In: Azcón-Aguilar C, Barea JM, Gianinazzi S, Gianinazzi-Pearson V (eds) Mycorrhizas – Functional processes and ecological impact. Springer, Berlin/Heidelberg/Germany, pp 17–32
Provorov NA, Vorobyov NI 2010 Evolutionary genetics of plant-microbe symbioses. Ed. by Igor A. Tikhonovich. NOVA Science, New York. 290 pp
Provorov NA, Borisov AY, Tikhonovich IA (2002) Developmental genetics and evolution of symbiotic structures in nitrogen-fixing nodules and arbuscular mycorrhiza. J Theor Biol 214:215–232
Provorov NA, Vorobyov NI, Andronov EE (2008) Macro- and microevolution of bacteria in symbiotic systems. Russian J Genet 44(1):6–20
Quilambo OA (2003) The vesicular-arbuscular mycorrhizal symbiosis. African J Biotechnol 2:539–546

Quispel A (1998) Evolutionary aspects of symbiotic adaptations: *Rhizobium's* contribution to evolution of associations. In: Spaink HP, Kondorosi A, Hooykaas PJJ (eds) The Rhizobiaceae. Molecular biology of model plant-associated bacteria. Kluwer, Dordrecht, pp 487–507
Raaijmakers JM, Leeman M, van Oorschot MMP et al (1995) Dose-response relationships in biological control of fusarium wilt of radish by *Pseudomonas* spp. Phytopathology 85:1075–1081
Raaijmakers JM, de Bruijn I, de Kock MJD (2006) Cyclic lipopeptide productionby plant-associated *Pseudomonas* spp.: diversity, activity, biosynthesis, and regulation. MPMI 19(7):699–710
Radutoiu S, Madsen LH, Madsen EB, Felle HH, Umehara Y, Gronlund M, Sato S, Nakamura Y, Tabata S, Sandal N, Stougaard J (2003) Plant recognition of symbiotic bacteria requires two LysM receptor-like kinases. Nature 425:585–592
Radutoiu S, Madsen LH, Madsen EB, Jurkiewicz A, Fukai E, Quistgaard EM, Albrektsen AS, James EK, Thirup S, Stougaard J (2007) LysM domains mediate lipochitinoligosaccharide recognition and Nfr genes extend the symbiotic host range. EMBO J 26(17):3923–3935
Rai MK (ed) (2006) Handbook of microbial biofertilizers. Haworth Press, Technology & Engineering, New York
Rainey PB (1999) Adaptation of *Pseudomonas fluorescens* to the plant rhizosphere. Environ Microbiol 1:243–257
Rasmussen HN (1995) Terrestrial orchids from seed to mycotrophic plant. Cambridge University Press, Cambridge
Razumovskaya ZG (1937) Nodule formation in various pea cultivars. Russian J Microbiol 6:321–328
Read DJ, Duckett JG, Francis R, Ligrone R, Russell A (2000) Symbiotic fungal associations in lower land plants. Phil Trans Roy Soc Lond Series B 355:815–831
Redecker D (2002) Molecular identification and phylogeny of arbuscular mycorrhizal fungi. Plant Soil 244(1–2):67–73(7)
Redecker D, Kodner R, Graham LE (2000) Glomalean fungi from the Ordivician. Science 289:1920–1921
Remy W, Taylor TN, Hass H, Kerp H (1994) Four hundred-million-year-old vesicular arbuscular mycorrhizas. Proc Natl Acad Sci USA 91:11841–11843
Rengel Z (2002) Breeding for better symbiosis. Plant Soil 245:147–162
Requena N, Jimenez J, Toro M, Barea JM (1997) Interactions between plant-growth-promoting rhizobacteria (PGPR), arbuscular mycorrhizal fungi and *Rhizobium* spp. in the rhizosphere of *Anthyllis cytisoides*, a model legume for revegetation in Mediterranean semi-arid ecosystems. New Phytologist 136:667–677
Requena N, Perez-Solis E, Azcón-Aguilar C, Jeffries P, Barea JM (2001) Management of indigenous plant-microbe symbioses aids restoration of desertified ecosystems. Appl Environ Microbiol 67(2):495–498
Riely BK, Lougnon G, Ané JM, Cook DR (2007) The symbiotic ion channel homolog DMI1 is localized in the nuclear membrane of *Medicago truncatula* roots. Plant J 49(2):208–216
Rillig MC (2004) Arbuscular mycorrhizae, glomalin and soil aggregation. Can J Soil Sci 84:355–363
Rivas R, Velazquez E, Willems A, Vizcaino N, Subba-Rao NS, Mateos PF, Gillis M, Dazzo F, Martinez-Molina E (2002) A new species of *Devosia* that forms a unique nitrogen-fixing root-nodule symbiosis with the aquatic legume *Neptunia natans* (L.f.) Druce. Appl Environ Microbiol 68(11):5217–5222
Rosenblueth M, Martнnez-Romero E (2006) Bacterial endophytes and their interactions with hosts. MPMI 19:827–837
Rosendahl S (2008) Communities, populations and individuals of arbuscular mycorrhizal fungi. New Phytol 178:253–266
Rosewarne G, Barker SL, Smith SE (1997) Production of near synchronous colonisation in tomato for developmental and molecular analysis of mycorrhiza. Mycol Res 101:966–970
Roth LE, Stacey G (1989) Bacterium release into host cells of nitrogen-fixing soybean nodules: the symbiosome membrane comes from three sources. Eur J Cell Biol 49:13–23

Ruiz-Lozano JM, Azcon R (1996) Mycorrhizal colonization and drought stress as factors affecting nitrate reductase activity in lettuce plants. Agric Ecosys Environ 60:175–181

Ruiz-Lozano JM, Bonfante P (1999) Identification of a putative P-transporter operon in the genome of a *Burkholderia* strain living inside the arbuscular mycorrhizal fungus *Gigaspora margarita*. J Bacteriol 181:4106–4109

Ruiz-Lozano JM, Bonfante P (2000) A *Burkholderia* strain living inside the arbuscular mycorrhizal fungus *Gigaspora margarita* possesses the *vacB* gene, which is involved in host cell colonization by bacteria. Microb Ecol 39:137–144

Sagan M, Duc F (1996) *Sym*-28 and *Sym*-29, two new genes involved in regulation of nodulation in pea (*Pisum sativum* L.). Symbiosis 20:229–245

Sagan M, Morandi D, Tarenghi E, Duc G (1995) Selection of nodulation and mycorrhizal mutants in the model plant *Medicago truncatula* (Gaertn.) after x-ray mutagenesis. Plant Sci 111:63–71

Saito K, Yoshikawa M, Yano K, Miwa H, Uchida H, Asamizu E, Sato S, Tabata S, Imaizumi-Anraku H, Umehara Y, Kouchi H, Murooka Y, Szczyglowski K, Downie JA, Parniske M, Hayashi M, Kawaguchi M (2007) NUCLEOPORIN85 is required for calcium spiking, fungal and bacterial symbioses, and seed production in *Lotus japonicus*. Plant Cell 19(2):610–624

Salzer P, Corbiere H, Boller T (1999) Hydrogen peroxide accumulation in *Medicago truncatula* roots colonized by the arbuscular mycorrhiza-forming fungus *Glomus mosseae*. Planta 208:319–325

Sanchez L, Weidmann S, Arnould C, Bernard AR, Gianinazzi S, Gianinazzi-Pearson V (2005) *Pseudomonas fluorescens* and *Glomus mosseae* trigger *DMI3*-dependent activation of genes related to a signal transduction pathway in roots of *Medicago truncatula*. Plant Physiol 139:1065–1077

Sancholle M, Daplé Y, Grandmougin-Ferjani, (2001) A lipids of mycorrhiza. In: Hock B (ed) The Mycota IX: fugal association. Springer, Berlin/Heidelberg, pp 63–93

Santos R, Herouart D, Puppo A, Touati D (2000) Critical protective role of bacterial superoxide dismutase in *Rhizobium*-legume symbiosis. Molec Microbiol 38:750–759

Scannerini S, Bonfante P (1991) Bacteria and bacteria like objects in endomycorrhizal fungi (*Glomaceae*). In: Margulis L, Fester R (eds) Symbiosis as a source of evolutionary innovation: speciation and morphogenesis. MIT Press, Cambridge, MA, pp 273–287

Schachtman DP, Reid RJ, Ayling SM (1998) Phosphorus uptake by plants: from soil to cell. Plant Physiol 116:447–453

Schauser L, Handberg K, Sandal N, Stiller J, Thykjaer T, Pajuelo E, Nielsen A, Stougaard J (1998) Symbiotic mutants deficient in nodule establishment identified after T-DNA transformation of *Lotus japonicus*. Mol Gen Genet 259:414–423

Schauser L, Roussis A, Stiller J, Stougaard J (1999) A plant regulator controlling development of symbiotic root nodules. Nature 402:191–195

Schnabel E, Journet EP, de Carvalho-Niebel F, Duc G, Frugoli J (2005) The *Medicago truncatula* SUNN gene encodes a CLV1-like leucine-rich repeat receptor kinase that regulates nodule number and root length. Plant Mol Biol 58:809–822

Schultze M, Kondorosi A (1998) Regulation of symbiotic root nodule development. Annu Rev Genet 32:33–57

Schulz B, Boyle S, Sieber T (2006) What are endophytes? In: Schulz B, Boyle S, Sieber T (eds) Microbial root endophytes. Springer, Berlin, pp 1–14

Schüßler A (2002) Molecular phylogeny, taxonomy, and evolution of *Geosiphon pyriformis* and arbuscular mycorrhizal fungi. Plant Soil 244:75–83

Schüßler A, Kluge M (2001) *Geosiphon pyriforme*, and endocytosymbiosis between fungus and cyanobacteria, and its meaning as a model system for arbuscular mycorrhizal research. In: Hock B (ed) The Mycota IX. Springer, Berlin, Germany, pp 151–161

Schüßler A, Schwarzott D, Walker C (2001) A new fungal phylum, the Glomeromycota: phylogeny and evolution. Mycol Res 105:1413–1297

Sen R, Hietala AM, Zelmer C (1999) Common anastomosis and internal transcribed spacer RFLP groupings in binucleate *Rhizoctonia* isolates representing root endophytes of *Pinus sylvestris*. *Ceratorhiza* spp. from orchid mycorrhizas and a phytopathogenic anastomosis group. New Phytol 144:331–341

Senoo K, Solaiman ZM, Kawaguchi M, Imaizumi-Anaraku H, Akao S, Tanaka A, Obata H (2000a) Screening and characterization of mycorrhizal mutants in *Lotus japonicus*. In: Molecular Genetics of Model Legumes: Impact for Legume Biology and Breeding. Meeting of Euroconference Project HPCF-CT-1999-00015, Abstract 28. John Innes Center, Norwich, UK.

Senoo K, Solaiman Z, Kawaguchi M, Imaizumi-Anraku H, Akao S, Tanaka A, Obata H (2000b) Isolation of two different phenotypes of mycorrhizal mutants in the model legume plant *Lotus japonicus* after EMS treatment. Plant Cell Physiol 41:726–732

Sessitsch A, Howieson JG, Perret X, Antoun H, Martinez-Romero E (2002) Advances in *Rhizobium* research. Crit Rev Plant Sci 21:323–378

Shachar-Hill Y, Pfeffer PE, Douds D, Osman SF, Doner LW, Ratcliffe RG (1995) Partitioning of intermediary carbon metabolism in vesicular-arbuscular mycorrhizal leek. Plant Physiol 108:7–15

Sharples JM, Chambers SM, Meharg AA, Cairney JWG (2000) Genetic diversity of root-associated fungal endophytes from *Calluna vulgaris* at contrasting field sites. New Phytol 148:153–162

Shaw SL, Long SR (2003a) Nod factor elicits two separable calcium responses in *Medicago truncatula* root hair cells. Plant Physiol 131:976–984

Shaw SL, Long SR (2003b) Nod factor inhibition of reactive oxygen efflux in a host legume. Plant Physiol 132:2196–2204

Shigenobu S, Watanabe H, Hattori M, Sakaki Y, Ishikawa H (2000) Genome sequence of the endocellular bacterial symbiont of aphids *Buchnera* sp. APS Nature 407:81–86

Shirtliffe SJ, Vessey JK (1996) A nodulation (Nod^{-}/Fix^{-}) mutant of *Phaseolus vulgaris* L. has nodule-like structures lacking peripheral vascular bundles (Pvb^{-}) and is resistant to mycorrhizal infection (Myc^{-}). Plant Sci 118:209–220

Shtark OY, Azarova TC, Shaposhnikov AI, Makarova NM, Bloemberg GV, Lugtenberg BJJ, Tikhonovich IA, Kravchenko LV (2002) Effect of growth media and wheat root exudates components on antibiotic production by root-colonizing pseudomonads. In: Modern Fungicides and Antifungal Compounds. Proceedings of 13th International Reinhardsbrunn Symposium, May 14–18, 2001, Friedrichroda, Germany, pp 419–428

Shtark OY, Shaposhnikov AI, Kravchenko LV (2003) The production of antifungal metabolites by *Pseudomonas chlororaphis* grown on different nutrient sources. Russian J Microbiol 72(5):574–578

Shtark OY, Danilova TN, Naumkina TS, Vasilchikov AG, Chebotar VK, Kazakov AE, Zhernakov AI, Nemankin TA, Prilepskaya NA, Borisov AY, Tikhonovich IA (2006) Analysis of pea (*Pisum sativum* L.) source material for breeding of cultivars with high symbiotic potential and choice of criteria for its evaluation. Ecological genetics ("Ekologicheskaja genetika") 4(2):22–28 (In Russian)

Shtark OY, Ovchinnikova ES, Zhukov VA, Nemankin TA, Titov VS, Borisov AY, Gianinazzi-Pearson V, Seddas P, Ambrose M, Ellis N, Tikhonovich IA (2007) Isolation of pea (*Pisum sativum*) mutants impaired in arbuscular mycorrhiza development, using a direct screening. Pisum Genet 39:26–27

Sieberer B, Emons AMC (2000) Cytoarchitecture and pattern of cytoplasmic streaming in root hairs of *Medicago truncatula* during development and deformation by nodulation factors. Protoplasma 214:118–127

Sieberer BJ, Timmers ACJ, Lhuissier FGP, Emons AMC (2002) Endoplasmic microtubules configure the subapical cytoplasm and are required for fast growth of *Medicago truncatula* root hairs. Plant Physiol 130:977–988

Smit P, Raedts J, Portyanko V, Debelle F, Gough C, Bisseling T, Geurts R (2005) *NSP1* of the GRAS protein family is essential for rhizobial Nod factor-induced transcription. Science 308:1789–1791

Smit P, Limpens E, Geurts R, Fedorova E, Dolgikh E, Gough C, Bisseling T (2007) *Medicago LYK3*, an entry receptor in rhizobial nodulation factor signaling. Plant Physiol 145:183–191

Smith KP, Goodman RM (1999) Host variation for interactions with beneficial plant-associated microbes. Annu Rev Phytopathol 37:473–491

Smith SE, Read DJ (1997) Mycorrhizal symbiosis, 2nd edn. Academic Press, San Diego/London/New York/Boston/Sydney/Tokyo/Toronto

Smith FA, Smith SE (1997) Structural diversity in vesicular-arbuscular mycorrhizal symbioses. New Phytol 137:373–388
Smith SE, Dickson S, Smith FA (2001) Nutrient transfer in arbuscular mycorrhizas: how are fungal and plant processes integrated? Aust J Plant Physiol 28:683–694
Solaiman ZM, Senoo K, Kawaguchi M, Imaizumi-Anraku H, Akao S, Tanaka A, Obata H (2000) Characterization of mycorrhizas formed by *Glomus* sp. on roots of hypernodulating mutants of *Lotus japonicus*. J Plant Res 113:443–448
Souza FA, Daplé Y, Declerck S, de la Providencia IE, Séjalon-Delmas N (2005) Life history strategies in Gigasporaceae: insight from monoxenic culture. In: Declerck S, Strullu DG, Fortin JA (eds) *In vitro* culture of mycorrhizas. Springer/Berlin/Heidelberg/New York, pp 73–91
Spaink HP (1995) The molecular basis of infection and nodulation by rhizobia: the ins and outs of sympathogenesis. Ann Rev Phytopathol 33:345–368
Spaink HP, Sheeley DM, van Brussel AA, Glushka J, York WS, Tak T, Geiger O, Kennedy EP, Reinhold VN, Lugtenberg BJ (1991) A novel highly unsaturated fatty acid moiety of lipo-oligosaccharide signals determines host specificity of *Rhizobium*. Nature 354(6349):125–130
Spaink HP, Kondorosi A, Hooykaas PJJ (eds) (1998) The rhizobiaceae. Molecular biology of model plant-associated bacteria. Kluwer, Dordrecht/Boston/London
Spanu P, Bonfante-Fasolo P (1988) Cell wall-bound peroxidase activity in roots of mycorrhizal *Allium porrum*. New Phytol 109:119–124
Sprent JI (2001) Nodulation in legumes. Cromwell Press Ltd, Royal Botanical Gardens, Kew
Sprent JI, James EK (2007) Legume evolution: where do nodules and mycorrhizas fit in? Plant Physiol 144:575–581
Stephens C, Murray W (2001) Pathogen evolution: how good bacteria go bad. Curr Biol 11:53–56
Stougaard J (2001) Genetics and genomics of root symbiosis. Curr Opin Plant Biol 4:328–335
Stracke S, Kistner C, Yoshida S, Mulder L, Sato S, Kaneko T, Tabata S, Sandal N, Stougaard J, Szczyglowski K, Parniske M (2002) A plant receptor-like kinase required for both bacterial and fungal symbiosis. Nature 417:959–962
Sturz AV, Christie BR, Nowak J (2000) Bacterial endophytes: potential role in developing sustainable systems of crop production. Cr Rev Plant Sci 19(1):1–30
Suominen L, Roos C, Lortet G, Paulin L, Lindström K (2001) Identification nd structure of the *Rhizobium galegae* common nodulation genes: evidence for horizontal gene transfer. Mol Biol Evol 18:907–916
Takemoto D, Hardham R (2004) The cytoskeleton as a regulator and target of biotic interactions in plants. Plant Physiol 136:3864–3876
Taylor FJR (1974) Implications and extensions of the serial endosymbiosis theory of the origin of eukaryotes. Taxon 23:229–258
Teplitski M, Robinson JB, Bauer WD (2000) Plants secrete substances that mimic bacterial N-acyl homoserine lactone signal activities and affect population density-dependent behaviors in associated bacteria. Mol Plant-Microbe Interact 13:637–648
Terefework Z, Lortet G, Souminen L, Lindström K (2000) Molecular evolution of interactions between rhizobia and their legume hosts. In: Procaryotic Nitrogen Fixation: a Model System for Analysis of a Biological Process. Horizon Sci Press, Wymondham, pp 187–206
Thomashow LS, Weller DM (1996) Current concepts in the use of introduced bacteria for biological disease control: mechanisms and antifungal metabolites. In: Stacey G, Keen N (eds) Plant-microbe interactions. Chapman & Hall, New York, pp 187–235
Tikhonovich IA, Provorov NA (2007) Cooperation of plants and microorganisms: getting closer to the genetic construction of sustainable agro-systems. Biotechnol J 2(7):833–48
Tikhonovich IA, Kozhemyakov AP, Ovtsyna AO, Provorov NA (2000) Construction of highly efficient symbiotic systems. In: Provorov NA, Tikhonovich IA, Veronesi F (eds) New approaches and techniques in breeding sustainable fodder crops and amenity. Grasses Biont, St.-Petersburg, pp 131–135
Timmers ACJ, Auriac MC, Truchet G (1999) Refined analysis of early symbiotic steps of the *Rhizobium-Medicago* interaction in relationship with microtubular cytoskeleton rearrangements. Development 126:3617–3628

Timmers ACJ, Soupene E, Auriac MC, de Billy F, Vasse J, Boistard P, Truchet G (2000) Saprophytic intracellular rhizobia in alfalfa nodules. Mol Plant Microbe Interact 13:1204–1213

Tirichine L, Imaizumi-Anraku H, Yoshida S, Murakami Y, Madsen L, Miwa H, Nakagawa T, Sandal N, Albrektsen AS, Kawaguchi M, Downie A, Sato S, Tabata S, Kouchi H, Parniske M, Kawasaki S, Stougaard J (2006) Deregulation of a Ca^{2+}/calmodulin-dependent kinase leads to spontaneous nodule development. Nature 441:1153–1156

Tirichine L, Sandal N, Madsen LH, Radutoiu S, Albrektsen AS, Sato S, Asamizu E, Tabata S, Stougaard J (2007) A gain-of-function mutation in a cytokinin receptor triggers spontaneous root nodule organogenesis. Science 315(5808):104–107

Toal ME, Yeomans C, Killham K, Meharg AA (2000) A review of rhizosphere carbon flow modeling. Plant Soil 222:263–281

Tokala RK, Strap JL, Jung CM, Crawford DL, Salove MH, Deobald LA, Bailey JF, Morra MJ (2002) Novel plant-microbe rhizosphere interaction involving *Streptomyces lydicus* WYEC108 and the pea plant (*Pisum sativum*). Appl Environ Microbiol 68:2161–2171

Truchet G, Barker DG, Camut S, de Billy F, Vasse J, Huguet T (1989) Alfalfa nodulation in the absence of *Rhizobium*. Mol Gen Genet 219:65–68

Tsai SM, Phillips DA (1991) Flavonoids released naturally from alfalfa promote development of symbiotic *Glomus* spores *in vitro*. Appl Environ Microbiol 57:1485–1488

Tsyganov VE, Morzhina EV, Stefanov SY, Borisov AY, Lebsky VK, Tikhonovich IA (1998) The pea (*Pisum sativum* L.) genes *sym33* and *sym40* control infection thread formation and root nodule function. Mol Gen Genet 259:491–503

Tsyganov VE, Voroshilova VA, Priefer UB, Borisov AY, Tikhonovich IA (2002) Genetic dissection of the initiation of the infection process and nodule tissue development in the *Rhizobium*-pea (*Pisum sativum* L.) symbiosis. Ann Bot 89:357–366

Tsyganov VE, Voroshilova VA, Herrera-Cervera JA, Sanjuan-Pinilla JM, Borisov AY, Tikhonovich IA, Priefer UB, Olivares J, Sanjuan J (2003) Developmental down-regulation of rhizobial genes as a function of symbiosome differentiation in symbiotic root nodules of *Pisum sativum* L. New Phytol 159:521–530

Turnau K, Orlowska E, Ryszka P, Zubek S, Anielska T, Gawronski S, Jurkiewicz A (2007) Role of mycorrhizal fumgi in phytoremediation and toxity monitoring of heavy metal rhich industrial wastes in southern Poland. In: Twardowska I, Allen HE, Höggblom MM, Stefaniak S (eds) NATO Science Series. Soil and Water Pollution Monitoring, Protection and Remediation Springer, Netherlands, 69:533–551

Udvardi MK (2001) Legume models strut their stuff. Mol Plant-Microbe Inter 14:6–9

Vallad E, Goodman RM (2004) Systemic acquired resistance and induced systemic resistance in conventional agriculture. Crop Sci 44:1920–1934

van Brussel AAN, Bakhuizen R, van Spronsen PC, Spaink HP, Tak TK, Lugtenberg BJJ, Kijne JW (1992) Induction of pre-infection thread structures in the leguminous host plant by mitogenic lipooligosaccharides of *Rhizobium*. Science 257:70–72

van Buuren ML, Lanfranco L, Longato S, Minerdi D, Harrison MJ, Bonfante P (1999) Construction and characterization of genomic libraries of two endomycorrhizal fungi: *Glomus versiforme* and *Gigaspora margarita*. Mycol Res 103:955–960

van Kammen A (1984) Suggested nomenclature for plant genes involved in nodulation and symbiosis. Plant Mol Biol Rep 2:43–45

Van Loon LC, Bakker PA, Pieterse CMJ (1998) Systemic resistance induced by rhizosphere bacteria. Annu Rev Phytopathol 36:453–483

Van Overbeek LS, van Vuurde J, van Elsas JD (2006) Application of molecular fingerprinting techniques to explore the diversity of bacterial endophytic communities. In: Schulz B, Boyle S, Sieber T (eds) Microbial root endophytes. Springer, Berlin, pp 337–354

van Rhijn P, Goldberg RB, Hirsch AM (1998) *Lotus corniculatus* nodulation specificity is changed by the presence of a soybean lectin gene. Plant Cell 10:1233–1249

Van Wees SCM, Pieterse CMJ, Trijssenaar A, van Westende TAM, Hartog F, van Loon LC (1997) Differential induction of systemic resistance in *Arabidopsis* by biocontrol bacteria. Mol Plant-Microbe Interact 10:716–724

Vance CP (2001) Symbiotic nitrogen fixation and phosphorus acquisition. Plant nutrition in the world of declining renewable resources. Plant Physiol 127:390–397

Vance CP, Heichel GH (1991) Carbon in N_2 fixation: limitation or exquisite adaptation? Annu Rev Plant Physiol 42:373–392

Vandenkoornhuyse P, Leyval C, Bonnin I (2001) High genetic diversity in arbuscular mycorrhizal fungi: evidence for recombination events. Heredity 87:243–253

Vasse J, De Billy F, Camut S, Truchet G (1990) Correlation between ultra-structural differentiation of bacteroids and nitrogen fixation. J Bacteriol 172:4295–4306

Vasse J, De Billy F, Truchet G (1993) Abortion of infection during the *Rhizobium meliloti*-alfalfa symbiotic interaction is accompanied by a hypersensitive reaction. Plant J 3:555–566

Vieille C, Elmerich C (1990) Characterization of two *Azospirillum brasilense* Sp7 plasmid genes homologous to *Rhizobium meliloti nodPQ*. Mol Plant-Microbe Inter 6:389–400

Vieille C, Elmerich C (1992) Characterization of an *Azospirillum brasilense* Sp7 plasmid gene homologous to *Alcaligenes eutrophis phbB* and *Rhizobium meliloti nodG*. Mol Gen Genet 231:375–384

Voets L, de la Providencia EG, Declerck S (2006) Glomeraceae and Gigasporaceae differ in their ability to form hyphal networks. New Phytol 172:185–188

Voroshilova VA, Demchenko KN, Borisov AY, Priefer UB, Tikhonovich IA (2009) Functioning of *Pisum sativum* genes *Sym33, Sym40 and Sym41* with respect to coordinated infection thread and meristem development in symbiotic root nodules. New phytol 181(4):913–923

Waelkens F, Marris M, Verreth C, Vanderleyden J, Matassi G, Van Gool A (1987) *Azospirillum* DNA shows homologies with *Agrobacterium* chromosomal virulence genes. FEMS Microbiol Lett 43:241–246

Wais RJ, Galera C, Oldroyd G, Catoira R, Penmetsa RV, Cook D, Gough C, Dénarié J, Long SR (2000) Genetic analysis of calcium spiking responses in nodulation mutants of *Medicago truncatula*. Proc Natl Acad Sci USA 97:13407–13412

Wall LG (2000) The actinorhizal symbiosis. J Plant Growth Regul 1:167–182

Wan X, Hontelez J, Lillo A, Guarnerio C, van de Peut D, Fedorova E, Bisseling T, Franssen H (2007) Medicago truncatula *ENOD40-1* and *ENOD40-2* are both involved in nodule initiation and bacteroid development. J Exp Bot 58(8):2033–2041

Wegel E, Schauser L, Sandal N, Stougaard J, Parniske M (1998) Mycorrhiza mutants of *Lotus japonicus* define genetically independent steps during symbiotic infection. Mol Plant-Microbe Interact 11:933–936

Werner D (1992) Symbiosis of plants and microbes. Chapman & Hall, London

Whipps JM (2001) Microbial interactions and biocontrol in the rhizosphere, J Experim Botany 52:487–511

Wu P, Zang G, Ladha JK, McCouch SR, Huang N (1995) Molecular-marker-facilitated investigation on the ability to stimulate N_2 fixation in the rhizosphere by the irrigated rice plants. Theor Appl Genet 91:1177–1183

Xavier IJ, Holloway G, Leggett M (2004) Development of rhizobial inoculant formulations. Crop Management (Online). doi:10.1094/CM-2004-0301-06-RV

Xie ZP, Muller J, Wiemken A, Broughton W, Boller T (1997) Nod factors and *tri*-iodobenzoic acid stimulate mycorrhizal colonization and affect carbohydrate partitioning in mycorrhizal roots of *Lablab purpureus*. New Phytol 139:361–366

Yano K, Yoshida S, Muller J, Singh S, Banba M, Vicker K (2008) CYCLOPS, a mediator of symbiotic intracellular accommodation. Proc Natl Acad Sci USA 105:20540–20545

Young ND, Mudge J, Ellis THN (2003) Legume genomes: more than peas in a pod. Curr Opin Plant Biol 6:199–204

Zhu H, Choi HK, Cook DR, Shoemaker RC (2005) Bridging model and crop legumes through comparative genomics. Plant Physiol 137:1189–1196

Zhukov VA, Radutoiu S, Madsen LH, Rychagova T, Ovchinnikova E, Borisov A, Tikhonovich I, Stougaard J (2008) The pea *Sym37* receptor kinase gene controls infection thread initiation and nodule development. Mol Plant-Microbe Inter 21(12):1600–1608

Chapter 6
Soil-Borne Pathogens and Their Interactions with the Soil Environment

Geoffrey R. Dixon and Emma L. Tilston

Introduction

Wheat yields in thirteenth century Europe have been estimated at 385 kg ha^{-1} (Pretty 1990; Houghton 1996), more than half a millennium later, by 1939 they had been increased to little more than 2 t ha^{-1}. Subsequently, in the period 1952–1986 scientific and technologically based innovation applied to farming increased yields by an average value of 2.6% pa. It is predicted that wheat yields will rise to 10.48–13.69 t ha^{-1} by 2015, with a current theoretical biological ceiling of 19.2 t ha^{-1} (Britton 1990). These rising yields have been accompanied by a tenfold increase in the amount of nitrogen applied to wheat in 1943/1945–1994 (Houghton 1996). They represent one example of what has been achieved by a combination of genotype (plant breeding) and environmental modification increasing the nutrients available to the plant in step with its physiological demands and fending-off consequential invasions by pests and pathogens attendant on rapid growth and high yields. Burgeoning human populations, reaching an estimated 10 billion by 2050, demand that similar yield increases are continued and accelerated. The environmental and climatic consequences arising from a policy of raising yield solely by increasing inputs are becoming apparent as dangerously unsustainable. Dramatic changes in the manner by which crops are husbanded are required with the aim of achieving increased yield but with minimal damage to the world's ecosystems (Dixon and Margerison 2009). Concomitant with the need to continue and accelerate yield

G.R. Dixon (✉)
Centre for Horticulture and Landscape, School of Biological Sciences,
Whiteknights, University of Reading, Berkshire RG6 6AS, UK
e-mail: geoffrdixon@btinternet.com

E.L. Tilston
Scottish Universities Environmental Research Centre, Scottish Enterprise Technology Park,
East Kilbride G75 0QF, UK
e-mail: e.tilston@surec.gla.ac.uk

G.R. Dixon and E.L. Tilston (eds.), *Soil Microbiology and Sustainable Crop Production*,
DOI 10.1007/978-90-481-9479-7_6, © Springer Science+Business Media B.V. 2010

increases in order to fill empty stomachs and enhance lifestyles far greater control of pests and pathogens using intelligent, integrated and environmentally benign methods based on sound scientific knowledge of the intimate relationships of beneficial microbes, roots, soil and pest and pathogen biology is required for the health of our planet's environment and biodiversity. So far, our knowledge of soil microbiology is at best scant and fragmentary and at worst non-existent. Yet as shown in this book the agronomic opportunities for enhanced productivity offered by the soil and its inhabitants are huge.

Soils contain very substantial numbers of phytopathogenic microbes capable of devastating crops worldwide. These include representatives of the bacteria, fungi, mycoplasma, protista and viruses. Some pathogens have very limited and highly specific host ranges while others are generalists, to a lesser or greater degree, causing diseases across many host taxa. Similarly, their geographical ranges may be restricted or alternatively spread around the entire globe. The intensity of pathogenesis varies also from those which devastate crops by ultimately killing their hosts in the processes of colonisation and reproduction to those which are only mildly aggressive and possibly almost commensalists. The forms of disease syndrome generated in crop hosts vary from simple root invasion and rotting through to altered root growth, e.g. root galling and clubbing, to collar rotting and damping-off into vascular wilting and colonisation throughout all the aerial organs of the plant. Some pathogens invade early in the host life-cycle and may then enter a quiescent or dormant phase from which they re-emerge devastating maturing flowers and fruit or destroying products stored after harvesting from the crop. Indeed some pathogens are only apparent from the toxins which they elicit in stored products.

The extensive range of pathogenic lifestyles available to soil-borne microbes increases the problems facing plant pathologists striving to unravel their biology. These difficulties are compounded by the sheer physical obscurity resulting from dwelling either wholly or at least partially in soil. Attempting to ascribe taxonomic identity and to understand the life-cycle biology of organisms that conduct all or part of their life-cycles in soil and subsequently are enclosed in the bodies of their host plants is a complex and at times confusing task. Adding to these problems are barriers in understanding the physiology and metabolism of host roots which are the primary targets for invasion by soil-borne pathogens. Roots possess a modular structure as described by Hodge (Hodge 2009) permitting responses to their soil environment and adaptation to changes, some of which are a result of the presence of pathogenic microbes. In achieving adaptation the roots differentiate between adopting partnership modes with benign microbes which enhance the potential efficiency for resource capture (see other chapters in this volume) and adopting defensive modes as a reaction to the presence of pathogens. The root-cap region appears to be the main environmental sensing and response control centre. Recent research indicates apparent root-to-root interactions and the capacity for recognising 'self' and non-self' roots (Falik et al. 2003). Root exudates possibly form one means for positive communication with benign microbes and negative responses to pathogens (Badri and Vivanco 2009).

Developing husbandries which are both economically and environmentally sustainable demands a thorough understanding of the impact and interactions between host roots, benign microbes and pathogenic organisms. Pathogens are capable of wreaking either rapid crop destruction or causing the long-term degeneration and decline of hosts possibly without their presence being easily obvious. This chapter discusses the abilities of pathogenic microbes to cause diseases as moderated by the soil environment and influenced by husbandry practices. Soil-borne microbial pathogens are well evolved and efficiently fitted for their ecological niches, for example as described by Dixon (Dixon 2009a) for *Plasmodiophora brassicae*, the causal agent of Clubroot Disease in the Brassicaceae. Ecologically soil-borne microbes exploit their edaphic and *in planta* environments with high degrees of effectiveness. Their traits of pathogenic success are indicative of organisms with long established and well tested lifestyles. Herein is one of the major mysteries associated with these crop pathogens. The pathogenic microbe is frequently found only sporadically within natural ecosystems or may be apparently entirely absent. They appear to have evolved their current lifestyles in parallel with their hosts as these were domesticated by man.

Devastating disease epidemics appear to result from or be associated with extensive and intensive mono-cropping of vulnerable host genotypes. In this context the pathogens cause "iatrogenic diseases of cultivation". The consequence of this status for sustainable crop production demands environmentally balanced disease control achieved by using husbandry systems which are compatible with host growth and productivity and the maintenance of benign microbe populations which benefit the host. This means integrating husbandry, biological, chemical, genetical and legislative techniques into logical and coherent strategies. Acceptance of this approach reflects changing agronomic attitudes driven at least partially by the withdrawal of toxic chemical agents such as the sterilant methyl bromide (Martin 2003). Considerable re-appraisal of the importance of cultural controls and the needs for crop protection strategies using integrated systems aiming to stimulate host growth and reproduction, encourage soil health and repress pathogens result from this approach.

The Impact of Soil-Borne Pathogens on Crops

The term pathogen indicates the ability of a microbe to cause disease in a particular host (Anon 1973). Pathogenesis, the process of disease development, results from an intimate relationship between host plants and microbes. Through this process the pathogen derives energy for growth and reproduction without any benefits devolving to the host. Disease is, in consequence, a metabolic shift away from the normal functioning of the host resulting in weaker growth, reduced yields and ultimately host death. Primary problems for those who study soil-borne organisms are establishing the mechanisms of pathogenesis, relationships with crop losses and defining soil environments which enhance the pathogenic load around roots and in the soil more widely. These issues were succinctly summarised by Garrett in 1956

(Garrett 1956) in his researches into pathogenic 'Inoculum Potential' in the vicinity of infection courts and defined as "the energy of growth of a fungal parasite available infection of the host at the surface of the host organ to be affected". Later, in 1960 a broader definition was favoured by Dimond and Horsfall (1960): "the number of independent infections that are likely to occur in a given situation in a population of susceptible healthy tissues". This perspective embraces the actions of environment, pathogen virulence, host susceptibility and amount of inoculum (Baker and Snyder 1965). Garrett's (1956) view is still very valuable as a philosophical concept for the studies of plant infection especially by soil-borne microbes despite technical difficulties which still exist in its quantification.

Quantifying Losses

Measuring and evaluating crop losses caused by aerial and soil-borne microbes have gained precision progressively over the last half century. This is driven by the environmental and financial costs of chemical and genetic crop protection. Resultant reductions in pesticide-use have progressed hand-in-hand with increased accuracy in their application and efficacy in their modes of action. In 1971 crop loss models were defined by Chiarappa (Chiarappa 1971) as "mathematical methods used to describe the relationship between yield reduction and the intensity of harmful organisms". Disease intensity is defined as "the total amount of disease present". Techniques used to assess disease presence and intensity are reviewed in the sequence of their historical development by Chiarappa (1971), James (1974), Walker (1983) and Smith et al. (1984). The methods they described produced quantitative relationships between yield reduction and the intensity of harmful microbes as disease progress curves.

Pathogenic microbes are distributed sporadically throughout soil profiles. Understanding the spatial distribution of soil-borne pathogenic microbes and their effects on crop losses evolved slowly compared with knowledge of aerial pathogens. Soil-borne microbes are distributed in three dimensions. Their impacts are distributed unevenly across fields or crops and consequently crop stand and yield reductions vary widely. While James (1974) postulated "the higher the aggregation of missing plants, the higher the loss", Bardner and Fletcher (1974) indicated that yield compensation complicates that picture. There is "an ability of plants or plant parts to make up for the yield producing functions of other (damaged) plants or plant parts"; but "compensation is less effective if killed or injured plants are aggregated". The spatial distribution of pathogenic microbes, resultant patterns of diseased plants and the expression of damage as crop losses require careful and close examination (Campbell and Noe 1985).

Over the last 20 years substantial progress has been achieved integrating forms of crop management and developing precision methods for crop protection. These are based on the use of modelling, diagnosis, prediction and decision-making systems and summarised by Madden et al. (Madden et al. 2008). Frequently, establishing mean population densities does not offer sufficient understanding of the relationship between patterns of crop loss and inoculum distribution, hence more detailed localised sampling is needed (Hughes 1996). The concept of a pathogenic 'threshold' represents a dividing line between the alternative courses of action needed for slight, moderate or severe infestations (Hughes 1999). Sampling crops for pathogen incidence is one means by which judgements can be guided. For precision integrated pathogen management the objective of sampling is establishing the actual presence of a pathogen and measuring its spatial variation.

Measuring the content of pathogens in soil *in situ* is technically difficult as identified by Nicot et al. (1984). There are problems in relating samples with their original locations in fields and then acquiring subsamples for analysis. Regardless of the statistical methods used analysing the occurrence of pathogenic inoculum as revealed by laboratory-based testing of soil samples and relating results to the dispersion of the pathogen and/or its propagules is far from simple. Pathogens may be concentrated near the soil surface or randomly distributed throughout the soil horizons. Determining these patterns in a coherent manner is hugely costly in terms of analytical and technical time and not suited to routine crop or cultivar evaluation programmes. At even the most empirical levels where a series of soil cores are bulked together there is then no means of knowing how the pathogen propagules were distributed amongst the cores and even less in relating results to positioning in a field or fields.

Crop pathogens are frequently major causes of gross yield loss with prominent examples such as potato late blight (*Phytophthora infestans*) in Western Europe and rice blast (*Magnaporthe grisea*) in Asia. But huge losses of yield (biomass) in the field and in store are only part of the picture. Pathogens reduce crop quality and value in many direct and indirect ways. The scale of losses caused by soil-borne pathogens is still subject to approximation despite many years of research (e.g. Dixon 2009b). Improving the accuracy of estimates is beset with problems such as sampling errors and spatial variation in the distribution of diseased populations although geographic information systems (GIS) could help improve the latter's accuracy. Early studies tended to focus on the physiological basis for yield loss, macro- and micro-economics of losses due to single diseases, effects of multiple infections by varying pathogens and on the incorporation of crop loss data into management programmes. Zadoks (1987) identified three phases in the study of disease induced loses: exploratory (for example early studies of late blight epidemics, *P. infestans* (e.g. Beaumont and Staniland 1933), emergency (during and post-World War II studies estimating crop losses in staple foods such as cereals and potato, (e.g. Large 1945; Van der Plank 1963) and implementary (shifting to focusing on host growth and development which led to linking effects of pathogens such as reduced leaf area or reduced photosynthesis using several equations with disease incidence, plant stand reductions and yield losses as factors (Madden et al. 2008). Unfortunately, very little of this research has been applied to soil-borne pathogens. Savary and others (2006) list the types of decisions flowing from a knowledge of crop losses as: tactical short-term, during the season decisions (T); strategic short-term between seasons decisions (S) for example pre-planting for annual crops; strategic long-term decisions (D) for example the design of breeding programmes or development of a Integrated Pest Management (IPM) strategies; and very–long term decisions relating to research prioritization (P). Here threshold theory has been used (Zadoks 1987) and a damage threshold became defined as the level of injury (yield or quality loss) where the benefit of control just exceeds its cost. This approach also defines warning and action thresholds as injury levels where control is planned and implemented respectively, in a timely manner preventing the damage threshold from being reached. An example is the EPIPRE project (EPIdemiology, PREdiction and PREvention) developed in the Netherlands from the late-1970s onwards. By 1985 (Zadoks 1990) 70% of Dutch winter wheat farmers were involved and its use had spread to: Belgium, Switzerland, Sweden, France and the United Kingdom.

Short-term strategic decisions (S & T) related to soil-borne pathogens can involve: choice of crop and cultivar, tillage and genotype plus the timing of crop establishment, types of seed treatment, forms and rates of fertiliser, sowing density and spatial geometry at drilling or transplanting and subsequently the form and rate of the application of irrigation. Knowledge of the density of soil-borne propagules (inoculum potential) is used to determine whether a field is suited for a particular crop and whether use should be made of resistant or tolerant cultivars as with pathogens such as pea root rot (*Fusarium solani*) (Oyarzun 1993), *Verticillium* wilt of cotton (*V. albo-atrum* and *V. dahliae*) (Gutierrez et al. 1983), potato early dying disease (*V. albo-atrum* and *V. dahliae*) (Francl et al. 1987) or clubroot (*P. brassicae*) (Faggian and Strelkov 2009).

Long-term decision making (type 3 knowledge "D") permits disease risk zoning as with clubroot (*P. brassicae*) in Canadian canola (oil seed rape, *Brassica napus*) which may permit assessments of the importance of a pathogen, efficiency of its management and research prioritisation. The final category "P" may be applied for example, in prioritising pathogen importance in relation to climate change (Coakley et al. 1999; Garrett et al. 2006).

Concern is increasing at the development of toxins in crops pre- and post-harvest which have grave consequences for human and animal health. Currently, their impact is difficult to assess but of vital significance with, for example *Fusarium* spp. in temperate cereals (Paul et al. 2005a, b) and tropical and subtropical crops, aflatoxins from *Aspergillus flavus* in peanuts (Porter et al. 1984). The disease intensity–toxin relationship depends on fungal species, fungal strain or physiological race, host plant resistance, disease dynamics and environmental factors (Porter et al. 1984; Bai et al. 2001; Champeil et al. 2004; Sinha and Savard 1997).

Spatial aspects of yield loss are especially significant with soil-borne pathogens. Physiological compensation can occur between neighbouring organs (leaves, roots, tillers) and may mitigate disease damage (James 1974; Zadoks and Schein 1979). There are social and economic needs for knowledge of the losses to crop biomass and quality. This information permits the solving of more immediate practical problems which reduces: management costs, pesticide usage, other husbandry resource inputs (such as: irrigation, fertiliser, energy, labour) and improved maintenance of natural ecosystem services such as: soil biodiversity, topsoil and water conservation and with directly marketed crops (the fresh produce industry) the stabilisation of market and system performance. Conventionally, the dimensions of damage are described in terms of: biomass (M) or biomass per unit of production area ($M.L^{-2}$), and basic factors including labour time (T), energy (E) and variable inputs such as fertilisers, traction, and monetary and environmental issues (Ordish 1952). By this process financial costs are aligned with husbandry, environmental and biodiversity costs. Developing knowledge of an injury-damage-loss paradigm integrates the study of pathogens and their effects into the broader framework of ecological systems analysis. This concept has been recognised by Cheatham and others (2009) who encourage the recognition that defining successful disease management should reach beyond short-term crop yield evaluations. Disease has direct impacts on ecosystem services through the loss of individual plants or whole crops and indirectly through management activities such as the use of pesticides or selection of tillage methods. As Cheetham et al. (2009) observe, enhanced biodiversity could reduce disease risk by diluting the amount of susceptible tissue in the ecosystem alternatively it might heighten it by adding further plant species which aid the pathogen in the completion of its life cycle.

Evaluating Losses

An evaluation of the losses caused by soil pathogens permits balanced judgements of their costs. This type of analysis has been done with increasing precision for some temperate crops. Analyses with annual crops are reasonably straightforward. With perennial crops however, it is necessary to consider how long soil-borne pathogens may remain within the host roots, the impact that this will have on resistance mechanisms and the effects of subsequent waves of invasion. For example, Dixon et al. (1989) established with considerable precision the yield losses resulting from the invasion of lucerne (alfalfa, *Medicago sativa*) rootstocks by the soil-borne wilt inducing pathogen *V. albo-atrum*, over several cropping cycles each of which lasted up to 5 years. Lucerne is a perennial crop which may be retained for 5 years and beyond if yield is financially viable. Field studies assessed disease incidence and yield in two cultivars: cv. Europe which is high yielding and has a rapid rate of growth especially in years 2 and 3 after sowing but no perceptible resistance genes and cv. Vertus which is less high yielding in the early cropping years but possesses polygenic quantitative resistance. Clear positive correlation ($r = +\ 0.975$, $P < 0.001$) was obtained between disease symptoms in year 4 and those in year 5. This strongly indicated that this soil-borne pathogen is capable of perennating in the rootstocks of lucerne thereby boosting the *in planta* inoculum potential and adding substantially to infection in each subsequent season. Conversely, there was strong negative correlation ($r = -0.877$, $P < 0.001$) between symptoms in year 4 and yield produced in year 5. There is, therefore, a continuation of symptom effects between the seasons with increasing perennation of inoculum potential. This contention was strongly supported by highly significant negative correlation ($r = -0.881$, $P < 0.01$) between symptom severity and yield in year 5.

Estimates of yield loss in annual crops such as potato caused to *Verticillium* spp. have also been related to soil inoculum content. These suggested that 10–15 colony forming units (cfus) of *V. dahliae* per cubic centimetre of soil resulted in 10% yield loss (Francl et al. 1987; Nnodu and Harrison 1979). Similarly, with *V. dahliae* inducing wilt in cotton grown in California, USA, correlations were developed for the lint yields of the wilt tolerant cv. Acala GC-510 compared with the susceptible cv. Acala SJ-2 and expressed as the equation:

$$Y = 97.08 * X^{0.038}$$

Correlation coefficients calculated for the 3 years separately and for all 3 years cumulatively were 0.88, 0.95, 0.86 and 0.87, respectively. The fitted line predicts that with about three propagules per gram of soil both cultivars (wilt-tolerant and wilt-susceptible) should produce equivalent yields. At higher inoculum densities wilt-tolerant cultivars should have higher lint yields than susceptible ones (Paplomatas et al. 1992). This extends earlier work (Gutierrez et al. 1983) which identified a relationship between the concentration of soil-borne propagules of *V. dahliae* and yield reduction in cotton. Key host factors in this relationship were the earliness of symptom development and percentage of plants with foliar symptoms.

The availability of high-speed and -capacity computing provides technology that now allows sophisticated forecasting of disease risks using mathematical models of actual and potential pathogen infestation integrated with meteorological and financial data. This has been developed recently for *Sclerotinia sclerotiorum* the soil-borne cause of stem rot in oilseed rape and white rot diseases in an extensive range of other farm and horticultural crops. In northern Europe this pathogen is an increasing threat to overwintered oil seed rape crops. The forecasting model provides decision support for farmers when considering applying fungicidal sprays at flowering stage in late April and May (Koch et al. 2007). The crop microclimate is measured by four weather variables: air temperature, relative humidity, rainfall and sunshine duration. Minimum environmental conditions for infection from ascospores, following their release from soil level ascocarps growing from sclerotia resting on or in the soil surface is calculated from laboratory studies to be 7–11°C and 80–86% relative humidity and expressed as an index of 'infection hours' (inh). Disease incidence (DI) correlated significantly with 'inh' at post-growth stage 58 (late bud stage) ($r^2 = 0.42$, $P \leq 0.001$). Summating 'inh' from continuous infection periods over 23 h significantly improved the correlation ($r^2 = 0.82$, $P \leq 0.001$). A parallel GS (growth stage) model calculates the developmental stages of oil seed rape based on temperature within the canopy and commences model calculations at GS = 58. The forecasting system named 'SkleroPro' has a two tiered approach, firstly providing a regional assessment of risks assumed when 23 'inh' have accumulated after GS = 58. The next tier offers financially based recommendations tailored to specific fields. This uses the costs of spray chemicals, expected yield and price of rapeseed calculating the number of 'inh' corresponding with DI at the economic damage threshold (Inh_i). There is a decision to spray when Inh ≤ Inh_i. Using historical field data (1994–2004) the effect of agronomic practices on *Sclerotinia* stem rot may be evaluated. Using a 2-year rotation enhanced disease risk and lowered the infection threshold by 0.8, conversely, a 4-year rotational break raised it by a factor of 1.3. The number of plants per square metre, level of nitrogen fertilisation and soil management had no effect on DI. Evaluation of 'SklerPro' in practice using 76 historical (1994–2004) and 32 current field experiments in 2005 showed that the financially correct decisions were made in 70% and 81% of cases, respectively. Comparing this technique with the conventional use of routinely scheduled spraying programmes using predetermined application intervals and growth stages showed that it saved 39% and 81% of fungicides, thereby increasing the net returns to the grower by 23 and 45 ha^{-1}, respectively and substantially reducing the environmental impact of crop protection in the oil seed rape crops and more widely in areas surrounding them.

Horticultural or fresh produce crops present much more complex patterns of damage and loss compared with broadacre ones because of the intricacies and inter-relationships of their components of quality responsible for attracting the retail customer which are affected by pathogenic microbes (Dixon 1981, 1989). The value of these crops frequently rests solely on the quality of their visual appeal with flowers, fruit, pot plants, trees, shrubs and vegetables. Alternatively, value may lie in an estimation of future productivity as with transplanting material sold for growing-on or forcing. Some estimation of these effects is obtained when the pathogen is removed. Supplies of virus-tested *Narcissus* (*N. pseudonarcissus*) bulbs were developed using heat treatment which removed virus pathogens some of which have soil-borne nematode vectors from a range of popular commercial cultivars. This provided accurate assessments of the damage to yield and the quality losses which these pathogens caused. A large scale multiplication scheme for virus-tested *Narcissus* operated successfully in Kincardineshire, Scotland in the 1980s which integrated virus removal, rapid twin-scale repropagation of bulbs and subsequent protected cultivation to tonnage quantities, followed by farm field scale cropping, inspection and certification. In the cv. Carlton mean bulb yield was 9–20% greater in virus-tested stocks than that achieved in comparable visually healthy commercial stocks (Sutton et al. 1988). Virus-tested stocks flowered 3–4 days in advance of commercial stocks with in some years increased flower stalk length. These improvements substantially increased the financial value of the crop.

The land areas devoted to individual horticultural crops are relatively small and often widely dispersed around the countryside but of very high financial and frequently social values. This contrasts markedly with the broadacre agricultural crops where cropped land area is the yardstick of over-riding importance. The cash value of individual horticultural plants can be very high, for example with trees and shrubs grown for several seasons possibly propagated from seed for sale as flowering or bare root specimens. Such plants are very vulnerable to damage from soil-borne pathogens (such as *Phytophthora* spp.). In fresh produce grown for consumption slight blemishes will substantially diminish or totally destroy market-value. For example field lettuce carrying foliar lesions of downy mildew (*Bremia lactucae*) can be unsaleable. Resting oospores of *B. lactucae* perennate in soil as part of crop debris germinating when a susceptible host is replanted. Similarly, grey mould (*Botrytis cinerea*), may destroy the visual value of a huge range of vegetable, fruit and flower crops and perennates in crop debris from preceding crops and as resting sclerotia.

The intensity and continuity of cultivation for fresh produce crops is severely limited by soil-borne pathogens. These affect the intensities of time scales over which crops are grown particularly where there is an intention to use sequential plantings on an area of land over a single season. A build-up of white mould (*S. sclerotiorum*) in soil as resting sclerotia can prevent land from being rotationally cropped with carrots, dwarf beans, lettuce and parsnips for instance. There may be market demands dictating that crops mature at specific dates. Crop scheduling can be wrecked by pathogens. For example, downy mildew of pea (*Peronospora viciae*) may invade from soil-borne resting oospores into seedlings and eventually cause uneven ripening in the maturing crop. This disrupts the industrial-scale machine vining of the processing pea crop causing substantial losses since entire lorry loads which are sampled and found to have unacceptable tenderometer readings are rejected. In extreme cases entire crops may be rejected for factory freezing and downgraded to use as dried peas or as sources of animal feed protein with very much lower financial value. Outbreaks of pea wilt caused by the soil-borne *Fusarium oxysporum* f. sp. *pisi*, were at least partially responsible for accelerating the demise of 'pulling pea' (hand harvested) crop production in Essex during the 1960s and the crop was relocated to Lincolnshire and Norfolk as vining peas grown using industrial principles. Similarly red core disease, colloquially known as the 'Lanarkshire Disease' (*Phytophthora fragariae*) was largely responsible for the demise of strawberry cropping in the Scottish Clyde Valley and West Coast in the 1950s. The appearance of the variant *P. fragariae* var. *rubi* which infests raspberry limited the cropping of cultivars in Perthshire from the 1970s onwards.

Losses to the asset value of land caused by soil-borne pathogens are insidious and frequently un-appreciated. Land which becomes infested with a particularly intractable disease causing pathogen that is difficult and expensive to eradicate such as clubroot (*P. brassicae*), white mould (*S. sclerotiorum*), onion white rot (*Sclerotium cepivorum*) or violet root rot (*Helicobasidium purpureum*) loses market value. Using infested land for high value vegetable crops becomes impossible because yield, continuity and quality are impaired. Similar but as yet only crudely unquantified effects arise where rural and urban landscapes are damaged by the presence of soil-borne pathogens. The effects of pathogens such as *Phytophthora ramorum* (sudden oak death) attacking temperate trees and shrubs, *Armillaria mellea*, the cause of honey fungus disease or *Phytophthora cinnamomi* destroying a range of economically important crop trees and shrubs and wild jarra forests in Australia diminish the social value of landscapes which in turn adversely affects human health and welfare.

Mitigation by Manipulating Husbandry

Rotation

Crop rotation is one of the oldest cultural techniques for the mitigation of crop damage caused by soil-borne pathogens. Estimates suggest it has been used for at least 2,000 years with for example, triennial rotations practised during Medieval times (Maloy 1993). The English four-course rotation was developed in Norfolk during the early eighteenth century initiating the Agricultural Revolution and permitting increases in yield which fed the urban factory workers needed by the subsequent nineteenth century Industrial Revolution. Scientific explanations for the effectiveness of crop rotation began appearing in the late nineteenth and early twentieth centuries as research centres such as Rothamsted Experiment Station (now Rothamsted Research) commenced their studies. Until the 1930s, rotations of 4 years' duration were widespread throughout Great Britain (Weston 1944). Consequently, soil-borne pathogens such as: *Gaeumannomyces graminis* var. *tritici* (take-all), *Bipolaris sorokinia* (formerly *Cochliobolus sativus*, root rot) and *Tapezia yallundae* (formerly *Pseudocercosporella herpotrichoides*, eyespot) were of little significance. After World War II (1939–1945) traditional farm rotations were abandoned in Great Britain and Western Europe as industrial principles were applied to crop production with the aim of avoiding the food shortages which plagued populations in the first half of the twentieth century (Dixon and Margerison 2009). Comparisons between monocropping and attendant disease decline and rotation whereby the amount of soil-borne inoculum is steadily eroded by soil microbes (Shipton et al. 1973) have classically been made with clubroot disease (*P. brassicae*). This disease is a prime cause of loss in highly valuable field vegetable brassica crops where huge financial penalties result from monocropping and waiting for disease decline to emerge, consequently this strategy is virtually impossible to test in practice. Potential rotational alternatives (such as leek and onion) have lower market values and hence are not realistic solutions in practice. Now that this disease is making substantial inroads into the Canadian canola crop (Strelkov et al. 2008) and infested farms are banned by statute from cropping for 5 years it may at last be possible to evaluate the usefulness of rotation as a tool for the control of *P. brassicae*.

Resting spores of *P. brassicae* (clubroot) are long lived, reputedly surviving for 20 years or more and they are highly resistant to degradation, being able to survive passage through the highly acidic guts of grazing animals (Gibbs 1931); and their cell wall has a structure that is impervious to predation by soil microbes. Consequently, the resting spores retain viability in soil despite exposure to repeated seasons of adverse weather. Although rotations of 6 or more years are recommended (Dixon 1981), these are of limited use in the management of land infested with *P. brassicae* because of the characteristics of the consumer market for fresh *Brassica* vegetables. Swedish field studies indicated the resting spores have a half life of at least 3.6 years and as a result some spores may exist for at least 18 years in the absence of suitable hosts before spore populations are eroded to undetectable levels (Wallenhammar 1998). In an attempt to contain the current outbreak of clubroot disease in Canadian canola infested fields are legally prohibited from returning to this crop for 5 years (S. Strelkov, 2007 personal communication). Resting spore longevity perplexed early researchers such as Gibbs (1939), not least because of *bona fide* reports that cruciferous crops grown on land previously carrying permanent grassland leys[1] could become rife with clubroot disease after 1 or 2 years of arable cropping with brassicas, frequently swedes (*B. napus*) grown for stock feed. Temperature, moisture content and position in the soil profile will influence spore longevity (Monteith 1924; Fedorintschik 1935).

Soil pH apparently affects the rate of release of primary zoospores from the resting spores. Zoospore numbers increased in acidic compared with alkaline soils (Bochow 1961) but without much change in total numbers germinating. Spore dormancy and the need for an external germination stimulus are factors in the relationship between *P. brassicae* and the soil environment (e.g. Honig 1931). At the release of new generations of resting spores from decaying clubs only a few germinate immediately (Ogawa et al. 2001), forms of external stimulus (Ohi et al. 2003; Hata et al. 2002) are needed which initiate the process. The state of 'readiness for germination' in spores released from host roots was examined by a consecutive string of researchers (e.g. Humphrey 1892; Chupp 1917; Naumov 1925; Honig 1931; Colhoun 1958); generally they concluded that bacteria and other organisms disintegrate the diseased host tissues and 'condition' spores for more effective germination. But these secondary microbes are not essential for the germination process itself. Undefined mechanisms within the resting spore initiate germination and control its speed. It appears that these mechanisms within individual spores operate quite separately from others in the population since not all germinate in synchrony. Rotating *Brassica*[2] crops with *Allium* spp. such as leeks or bulb onions reduces the disease causing ability of *P. brassicae* (J. Robak, personal communication 2000). Resistant forms of *B. oleracea* and *B. rapa* were advocated by Yamagishi et al. (1986) as inducing the germination of resting spores of *P. brassicae* but inhibiting the full completion of the pathogen life cycle thereby diminishing soil inoculum content. A similar contention was proposed by Hsieh and Yang (1985). Perennial ryegrass was suggested by Rod and Robak (1994) as a non-brassica crop capable of reducing *P. brassicae* possibly because the pathogen may complete parts of its life cycle in the root hairs but does not progress any further.

Changing the crop rotationally reduces opportunities for root invasion by soil-borne pathogens, exposes their resting bodies to extended periods of predation and enzymatic breakdown from antagonistic flora and fauna and to adverse physical and chemical environments. Hence rotation suppresses inoculum development and reduces inoculum potential. For example, pycniospores and chlamydospores of

---

[1] ley = English agricultural term for grassland pasture providing grazing over an extended number of seasons.

[2] In this chapter the latin form '*Brassica*' is used botanically and vernacular form 'brassicas' is used horticulturally.

Phoma medicaginis var. *pinodella* (foot rot) survive on buried crop debris for up to 3 years (Carr and Smith 1988); rotations longer than this are key management strategies for avoiding damage to pea crops. By contrast *Fusarium* spp. often have extended host ranges and several are natural soil inhabitants surviving as competitive saprotrophs on crop debris or as various forms of resting spore (Fehrmann 1988a, b, c), so that growing several crops in rotation does not necessarily reduce the inoculum potential of *Fusarium* induced diseases (Gair et al. 1972). Similarly *V. dahliae* affects a very large number of hosts and there are reports that it perennates on non-host roots (Rowe and Powelson 2002). But as McKay (1926) demonstrated in some of the earliest studies of rotation there are benefits which diminish *Verticillium* wilt of potato (McKay 1926). Detailed field trials by Emmond and Ledingham (1972) (Table 1) amply supported this contention.

Valuing rotation as a husbandry tool demands careful analysis, none more so than with take-all disease of cereals caused by *G. graminis* var. *tritici* (anamorph *Phialophora*-like sp.). Disease severity increases during the first cropping year and in subsequent wheat crops even after periods of break crops (Fig. 1).

Disease in mature plants during July is greatest in third and fourth wheat cropping seasons, after that severity declines to a stable lesser intensity. Although this still exceeds that than found in first year wheat crops. This phenomenon is termed "take-all decline" and is attributed to the increasing populations of antagonistic microbes (Hornby 1992). In the decline period ascospores and phialospores of *G. graminis* var. *tritici* may develop, but are devoid of a role in field infection (Hornby 1981). But mycelium, in the declining saprotrophic phase, present on crop debris provides a source of viable inoculum for up to 3 years depending on soil type (Lemaire 1988). Even with such a relatively short survival time, rotations may be of limited use in the management of take-all if short (2–4 years) sequences of susceptible cereals form part of the cropping regime. Take-all susceptible cereals (wheat and barley) are the foundation of arable agriculture in Europe. Oats are resistant to *G. graminis* var. *tritici*, but their market is very limited. Cereals can be used to diminish soil-borne diseases in non-cereal crops. In wheat and/or barley however, following periods of continuous cropping of up to 4 years duration take-all severity increases. Some nitrogen fertilisers used in non-cereal crops may

Table 1 Percentage of potato stems infected with *Verticillium albo-atrum* wilt and black dot caused by *Colletotrichum coccodes* in samples from a series of crop rotations (Adapted from Emmond and Ledingham 1972)

	Stems infected (%)	
Rotation components	*V. albo-atrum*	*C. coccodes*
A. Continuous potatoes	75	20
B. Potatoes/sweetclover/hay/hay/fallow	5	15
C. Potatoes/potatoes/alfalfa/crested wheatgrass/hay/hay/hay/hay/fallow (sampled year 1 potatoes)	25	25
D. Sampled year 2 potatoes	75	25

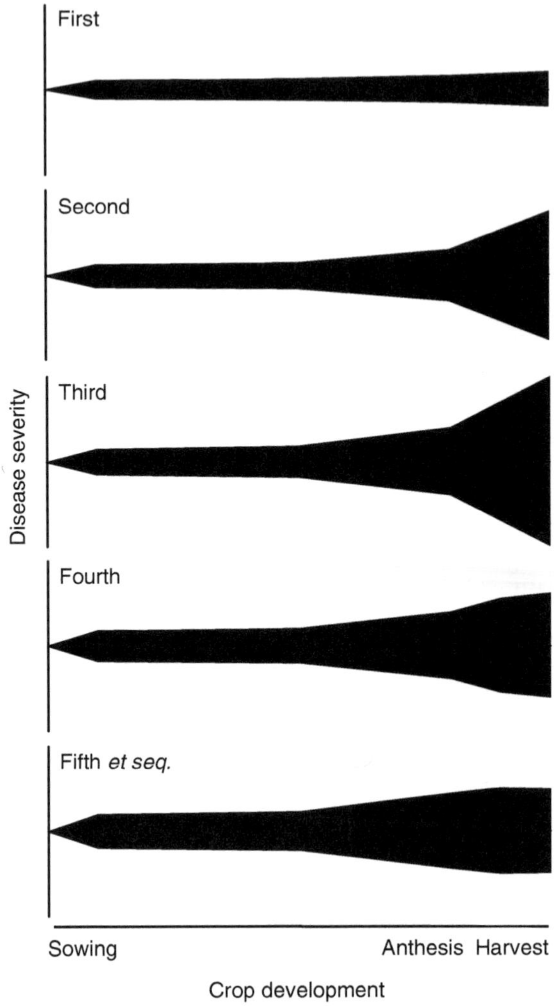

Fig. 1 Representative epidemiology of take-all disease in successive winter wheat crops under moderate to severe disease pressure

encourage *G. graminis* var. *tritici* by sustaining the saprotrophic survival phase. Take-all in cereals imposes significant agronomic constraints on cropping sequences (Hornby et al. 1998).

Soil Type

The effects of soil structure and texture on the biology of pathogenic microbes are well established at least in outline. Well structured freely draining soils normally harbour less plant pathogens. By contrast, badly abused and waterlogged soils tend to be associated with disease epidemics particularly damping-off, root, foot and collar rots. Recent research has highlighted the extent of this interaction. *Aspergillus flavus* the cause of aflatoxin contamination of cotton seed is a natural soil inhabitant. Aflatoxins are potent toxic and carcinogenic fungal metabolites that contaminate human and animal food and feed. Two strains of *A. flavus* are known; the S-strain which produces most toxin and the L-strain which generates lesser quantities (Jaime-Garcia and Cotty 2006). Incidence of S-strain is positively correlated with clay content and negatively correlated with sand content of soils. Studies in South Texas, USA related increased S-strain with successive cropping with cotton. There may be further complicating factors relating to the types of crop suited to the heavier soils and their abilities to retain a higher relative humidity within the canopy and encourage propagation of the fungus. A further example comes from Africa where in Cameroon andosols are suspected of suppressing cocoyam (*Xathosoma sagittifolium*) root rot caused by *Pythium myriotylum* compared with the effects of ferralsols. Andosols have higher contents of organic matter, calcium, potassium, magnesium and nitrogen which are negatively correlated with disease incidence (Adiobo et al. 2007). The clay and sand fractions were higher in ferrasols and correlated with disease severity. Suppressiveness is thought to relate to the organic matter content of andosols which improve soil structure, increase nutrient content, microbial biomass and activity. Particles of a larger size may contain a greater proportion of residual cellulose due to a smaller surface area. The larger size of particulate organic matter and therefore, comparatively smaller surface area will mean that the state of decomposition is less and that there may be smaller amounts of recalcitrant substrates.

The particle size distribution in soil has a significant bearing upon the volume of the interstitial spaces and hence the number of potentially habitable locations available for colonization by both microflora (Russell 1968; Van Elsas et al. 1986) and meiofaunal predators. For pathogens with motile zoospores the size of the interstitial spaces has important implications permitting or restricting the movement of these propagules through soil, the zoospores of *Phytophthora* spp. require water-filled pores of at least 40–60 μm in diameter before they can move through soil (Jackson 1975). Chemistry influences the interactions between particles and microorganisms with fluorescent pseudomonads being associated with particulate organic matter (Sivasithamparam et al. 1979a). Diseases too, can be influenced by soil texture, although for *G. graminis* var. *tritici* the effects are contradictory; coarse-textured soils of the Pacific North-west of the U.S.A. favour take-all (Cook et al. 1968) but silty soils in Brittany, France, are more favourable to the disease than are sandy soils (Lucas and Nignon 1987).

Take-all (*G. graminis* var. *tritici*) severity is partly dependent on certain soil types e.g. crops grown on fenland peats are at very high risk of take-all, but those on well-structured calcareous clay loams are at lowest risk. Soils with high organic matter content also favour the development of severe take-all as there is great potential for accumulation of nitrate due to mineralization of soil nitrogen, so fields of recently ploughed old pasture or old cleared wood land should be avoided (Hornby et al. 1998). Take-all severity in crops sown in fields known to be at risk can be reduced if firm seed beds, prepared from ploughed soil, are drilled at moderately low seeding rates and the crop provided with adequate amounts of nitrogen (especially as ammonium) and phosphate (Hornby et al. 1998). Another means of controlling take-all may be exploited where wheat is cropped continuously until take-all decline is achieved; decline can also be triggered more quickly by a severe outbreak of take-all. Soil in which take-all decline occurs could be described as suppressive (Simon and Sivasithamparam 1988). During the years preceding decline, however, the crops grown on that land will not make significant contributions to the farm's gross margin. Once take-all decline is established then its loss through planting of another type of crop, or possibly by liming (Simon et al. 1988), will produce a loss of both agronomic and financial investment (Hornby et al. 1998). The stability of take-all decline is also dependent on soil fertility, being unstable on soils of low fertility (Prew and McIntosh 1975), similarly the persistence of introduced antagonists is also strongly influenced by soil type (Capper and Higgins 1993).

Soil characteristics account for much of the variation in severity of soil-borne diseases found occurring between sites (Ball et al. 2005). Here it is necessary to distinguish between pathogen suppression and disease suppression. The former refers to the capacity of soil to limit the inoculum density of a pathogen and its saprophytic activity; the latter indicates a restriction in disease development even when the environment, the host and quantity of inoculum appear conducive. Disease suppression is related to the microbial characteristics of soil itself because this effect may be reduced or eliminated by soil sterilisation (Peters et al. 2003). Production of antibiotics, siderophores, nutrient competition, niche exclusion and an induction of systemic acquired host resistance (SAR) may affect the level of disease suppression. In turn these factors relate to total microbial activity in soil (Höper and Alabouvette 1996). Modifying soil conditions through the use of organic amendments, fertilisers and tillage practices, or the addition of microbial antagonists as biocontrol agents influence the degree of disease suppression. Pathogen suppression is suggested by Ball et al. (2005) to be a direct function of soil structure. An availability of oxygen and free-draining abilities are essential for the spread of many fungal pathogens and for the growth plant roots.

Soil structure affects the causation of soil-borne diseases. Otten and Gilligan (2006) clearly demonstrated that soil-borne fungal mycelia respond to soil structure. Where air filled pores are large then fungi form big masses of hyphae in an exploitive mode whereas where the pores are small and constrictive mycelia growth is limited to an explorative mode. They noted that even small changes in the physical soil environment resulted in "large and abrupt changes in fungal morphology". Soil compaction which is a function derived from the abuse of soil structure and texture encourages disease epidemics. These may result from primary infections developing from inoculum capable of saprophytic survival. The increase in the severity of soil-borne fungal diseases resulting from soil compaction (Höper and Alabouvette 1996) arises from an impact on the dynamics of microbial colony growth. The size of colony determines the distance over which a pathogen is able to make contact with hosts, the probability of infection is then determined by the density within the colony itself (Otten et al. 2001). Soil compaction alters the composition of root exudates and the structure of rhizosphere microbial communities; possibly such changes affect the ability of fungal pathogens to infect roots via for example, antibiosis or competitive interactions and hence the disease suppressivity of soils. Rotation and tillage practices both alter soil structure and then affect disease suppressivity and subsequent host–pathogen relationships.

Primary Cultivation: Tillage[3]

Physical disturbance of a soil is the primary effect of cultivation. Wetter, denser and cooler soil conditions typically resulting from reduced tillage systems increase the amounts of organic matter, encouraging greater microbial activity and biomass expansion, principally in the upper layers of the soil (Young and Ritz 2000). Disturbance of a soil results in the redistribution of pore networks and the resultant changed architecture permits the utilisation of newly exposed substrates and predation of freshly exposed organisms until an equilibrium is re-established or further disturbance maintains the dynamic nature of these processes.

The size of soil aggregates was generally found to be larger in no-till compared with tilled soil. In the upper-most 10 cm of soils, round pores were twice as abundant in no-till soils compared with tilled (Pagliani and De Nobili 1993). The tilled soils were characteristically platy at their surface and heterogeneous beneath that with numerous irregular pores and large compact aggregates separated by large planar pores. By contrast no-till soil had a more homogeneous subangular blocky structure. This is the reverse of the findings of Drees et al. (1994), hence the pedalogical effects of tillage and no-tillage regimes vary from site to site and the husbandry regimes employed. Such effects are microbiologically important since Otten et al. (2004) demonstrated that the spread of fungal pathogens through soil is not a random process. When spreading through soil and encountering a large continuous pore volume, fungal (in this case *Rhizoctonia solani*) spread is enhanced as it effectively by-passes the more tortuous smaller pathways through soil. Enhancing spread increases 'colonisation efficiency' which is the saprophytic equivalent of the pathozone (Gilligan 1985) and which describes the relationship between successful colonisation and the distance between inoculum and the target host(s).

Direct-seeding or no-till is defined as planting directly into the residues of previous crops without cultivation that mixes or stirs soil prior to planting. No-till is claimed to reduce soil erosion, improve soil structure, organic matter content and reduce fuel used by agricultural machinery (Paulitz 2006). No-tillage methods are widely used for cereal production in Australia, Argentina, Brazil and Canada but not in Europe nor in the Pacific Northwest of the USA. One limitation is that root diseases caused by species of *Fusarium*, *Pythium* or *Rhizoctonia* may increase with reduced tillage. This is a cause for alarm certainly in Europe because of the resultant increasing concentrations of fusarial toxins present in harvested grain. This situation is exacerbated by the removal from the market of several fungicides capable of controlling this pathogen. A 10-year study of conventional ploughing and no-tillage (Chervet et al. 2005) recognised the potential environmental advantages of reduced tillage but there remained a major human and animal health problem resulting from mycotoxin production in no-tilled winter cereal crops especially those which follow maize (corn). Issues of public health and safety which are

[3] Tillage is used here synonymously with cultivation.

consequential on changes to crop agronomy identify the need for an holistic approach carefully divorced from transitory and fashionable land-use policies.

The incorporation of crop debris into soil was once thought to affect the emergence of corn seedlings adversely of corn seedlings (McCalla and Duley 1949) and the injurious effects of phytotoxic substances extracted from wheat straw on germinating seeds and the growth of seedlings was demonstrated by Guenzi et al. (1967). Damage to seedlings or reductions in yield, following the incorporation of cereal residues, are most probably due to the increased activities of *Fusarium* spp. (Lipps and Deep 1991), *Pythium* spp. (Cook et al. 1990; Cook and Haglund 1991) and *R. solani* AG8 and *G. graminis* var. *tritici* (Cook and Haglund 1991). Yet in crops sown using direct-drilling techniques where crop residues are not buried by deep ploughing, the incidence of take-all can be reduced by 50%, even though inoculum levels are unaffected (Brooks and Dawson 1968). Lampkin (1990) noted that *G. graminis* var. *tritici* is not a major problem where direct drilling and conservation tillage practices are used and attributes this to the crop rotation and high microbial activity in the upper soil layers.

More generally, (Otten and Gilligan 2006) surveyed the effects of the use and the avoidance of tillage on disease incidence and found that conservation tillage can variously increase (54%), decrease (42%) or occasionally have nil effect (4%) on plant diseases. These authors note that unstable soil structure (as in silts) results in wetter, less well aerated and compacted soils in which pathogens such as *Phytophthora* and *Verticillium* spp. are encouraged while *R. solani* is discouraged (Alabouvette et al. 1996). But there is no simple relationship between soil conditions, pathogen vitality and disease incidence. Soil compaction reduced foot rot of winter wheat caused by *Microdochium nivale* and *Fusarium* spp. (Colbach et al. 1996) and sharp eyespot (*Rhizoctonia cerealis*) in winter wheat (Colbach et al. 1997). But had the reverse effect on common root rot of peas (*Fusarium solani* f. sp. *pisi*) (Fritz et al. 1995), crown and root rots in a crop rotation in spring barley (*Bipolaris* and *Fusarium* spp.) and soybean (Sturz and Carter 1995) and cereal root rot in humid climates (Sturz et al. 1997). Primary tillage using a chisel plough reduced root rot of peas (*Aphanomyces euteiches*) when associated with prior cropping with oats. Chisel ploughing retained much of the oat root and shoot tissue in the upper 15 cm of the soil profile whereas mouldboard ploughing deposited only 10% of these residues in that zone (Staricka et al. 1991). Prestes et al. (2002) noted that under crop rotation and nil-tillage husbandry as compared with monoculture and nil-tillage there was a reduced incidence of leaf blotches in wheat caused by *Drechslera* (*Pyrenophora*) *tritici-repentis*, *B. sorokiniana* and *Stagnospora* (*Phaeosphaeria*) *nodorum*. Direct drilling augments the build-up of organic carbon and microbial biomass in the surface layers of soil and raises the suppression of *G. graminis* var. *tritici* and *R. solani* (Pankhurst et al. 2002). Encouraging vesicular-arbuscular mycorrhizas (VAM) (see elsewhere in this book) by nil or limited tillage for low-input agriculture failed to increase maize growth or yield compared with conventional mouldboard ploughing systems (Galvez et al. 2001). Hence this fails to exploit the use of VAM for the reduction of soil-borne pathogens as frequently claimed.

Rhizoctonia spp. cause substantial yield losses in direct-seeded cereal crops compared with conventionally tilled ones. The strain *R. solani* AG-8 grew more rapidly through soil with a long-term history of direct seeding compared with tilled soils. This effect was ascribed to alterations in soil structure resulting from limited tillage whereby the residues left on the soil surface adversely affected soil moisture content (Schroeder and Paulitz 2008). No such effects were found with *R. oryzae*.

Disease incidence and severity were examined in oilseed rape in Germany under varying cultivation systems such as ploughing and drilling, cultivator and drilling and rotory cultivation (rotavation) and drilling. The incidence of *V. dahliae* was reduced with ploughing and drilling compared with cultivator or rotavator and drilling, but the incidence of black leg (*Leptosphaeria* [*Phoma*] *maculans*) was unaffected. Increasing nitrogen fertiliser applications had no effect on black leg but reducing nitrogen resulted in an increase in *S. sclerotiorum* (white mould) on aerial parts of the plants and pods at later growth stages. Yields and protein contents of the seed rose with increased nitrogen when associated with inversion husbandry (ploughing and drilling); oil content was reduced with that system (Sochting and Verreet 2004). Epidemics of early leaf spot (*Cercospora arachidicola*) on peanut were reduced in strip tilled or no tillage land compared with conventional full field cultivation in Florida, USA. The size of the disease reduction equalled that achieved by three to five applications of the fungicide, chlorothalonil (Cantonwine et al. 2007). Inoculum of *C. arachidicola* overwinters on crop debris from where primary infections spread onto the lower leaves. Strip tillage appears to reduce the inoculum potential which delays the build-up of epidemics and hence routine spraying is not required until later in the season.

Secondary Cultivation

A detailed 4-year study in southern England aimed to reduce the unit costs of growing consecutive wheat crops without reducing output, by focusing on the interactions between lower-cost establishment and the minimum of subsequent machinery passes through the crop (Knight 2003). On chalkland soil, minimum tillage was associated with the highest yields and biggest financial margins in all years. On medium loam, ploughing resulted in the highest yield in 1 year, but overall differences were small and direct sowing was the most cost-effective husbandry. No single technique was consistently better on heavy chalky clay-loam, but direct sowing was associated with lower yields in the second and third years, and minimum tillage produced the highest average financial margin over the three seasons. Direct sowing had the lowest energy costs for establishment, but also resulted in the lowest plant populations. Minimum tillage had the lowest total energy costs per tonne of grain harvested on both the light and heavy soils. Direct sowing resulted in less broad-leaved weeds compared with ploughing cultivation, but higher blackgrass (*Alopecurus myosuroides*) and brome-grass (*Bromus* spp.) populations. There was also a lower incidence of eyespot disease (*T. yallundae*) and incidence of *Septoria tritici* (*Mycosphaerella graminicola*) following direct sowing compared with after ploughing. The largest yield response to seed treatment was obtained in 1999, when it reduced take-all disease (*G. graminis* var. *tritici*) incidence by half. Despite having had the highest take-all disease levels in that first year, direct sowing showed the smallest average yield benefit on the light and medium soils. Reducing the number of machinery passes from 7 to 3 decreased yields at all sites in 2000. This was associated with inferior disease control, only one fungicide having been applied. There was no disadvantage associated with minimum pass cultivation in 2001, when disease pressure was lower. Where application and establishment costs were included, however, the minimum pass strategies were the most cost-effective. There were some interactions between the method of establishment and number of machinery passes. After ploughing, the 4-pass (with split applications of nitrogen fertiliser) was as cost-effective as the 3-pass system, whereas after direct sowing this treatment was no better than the 7-pass. At Biggleswade, Bedfordshire, the direct-sown blocks benefited more from the 7-pass than the other establishment methods. With consecutive wheat crops, minimum tillage or direct sowing can produce yields equal to or higher than ploughing on a range of soil types. The greatest benefit from ploughing is apparent on medium or heavy soils after a wet autumn, or where grass weeds are a problem. Adopting minimum pass husbandry can improve profitability, when the cost of each machinery movement is included in the accounts. For a single fungicide spray programme strategy to be effective, however, accurate timing of the application (at flag leaf) is vital. Low cost establishment does not dictate that a minimum pass approach will be less successful, and the combined benefits could range from £36 to £120 ha^{-1}. Although the single-sprays did not often result in significant yield penalties compared with the three-spray programme, they almost invariably were associated with inferior disease control, even in 2001. For example, at Andover, Hampshire, in 1999, the incidence of *S. tritici* on leaf 2 increased from 12% in the three-spray programme to 63–83% at the growth stage 32 (GS32) in single-spray treatments, but the associated penalty of yield losses was only between 1% and 6%. The absolute levels of disease were lowest, however in 2001, with almost all treatments recording less than 5% *S. tritici* on leaf 2. At Andover in 2000, the single-spray timing that was associated with the greatest control of *S. tritici* on leaf 2 (GS39–45) was also the one that gave the highest yields. Applications at GS32 gave the poorest disease control on leaf 2 in all 3 years at Andover, whereas at Biggleswade the GS39–45 single spray was equally ineffective. These findings help to explain why applications of fungicide at GS37 appeared to be the highest-yielding single spray timing at Biggleswade, whereas at Andover it was at GS39–45. At Biggleswade in 2000, control of brown rust (*Puccinia recondita*) was penalized more than the control of *S. tritici* by the change from a three-spray programme to the single-spray strategy.

In other broadacre crops, secondary tillage practices have a significant effect on the vertical distribution of residual stubble which carries disease. For example, oil seed rape crops and associated subsequent epidemics of soil and stubble-borne blackleg-canker caused by *Leptosphaeria* (*Phoma*) *maculans* (Schneider et al. 2006). Secondary cultivation and seed drilling raise previously buried infested stubble residues to the soil surface resulting in opportunities for the transfer of infection into autumn sown crops. Modelling tillage practices contributes towards formulating robust forms of sustainable disease control by integrating techniques to best advantage.

Soil Acidity and Alkalinity (Hydrogen Ion Content, pH)

Microbes in the group Plasmodiophorales are probably the pathogens most noted for their reaction to changing acidity and alkalinity in soil. The most vexed issue relating to clubroot disease (*P. brassicae*) is soil calcium content and the associated hydrogen–ion content (pH) of soil. Calcium emerges as a fundamental factor in the life cycles of both *P. brassicae* and its hosts. Datnoff and others (2007) summarised the involvement of calcium in host metabolism, physiology and signalling of many host–pathogen interactions indicating a relationship with expression of resistance. From the earliest studies of *P. brassicae* and clubroot onwards the disease was associated with acidic soils and claims that it was alleviated by the use of various forms of agricultural lime. Much of the work is, however, contradictory in terms of the forms of lime used, their sources, rates applied, date of application, recipient soil types and the measurement used to evaluate efficacy. It is now possible following 5 decades of research to conclude that clubroot disease incidence is not limited at pH 7.0 as is still claimed especially in much farm advisory and home gardening literature. As commented by Colhoun (1958): "results obtained by field experimentation show the difficulty encountered in determining the *exact upper limit* (our italics) of the soil pH at which infection can (*still*) occur". This begs the question as to whether there is an *exact upper limit*. Colhoun (1958) goes on to argue that "observations have been made without due attention to the variety of other factors which also influence infection" are of little if any value. He advocated the use of potted seedling tests which could be completed in 'controlled' conditions. As he also indicates pot tests have been undertaken at high soil moisture content but have failed to control spore load for example and they are much affected by seasonality. Glasshouse experiments running through a winter are far less acceptable because of the weaker host growth at low light levels and shorter day-lengths compared with those made in spring or early autumn, while summer-time experiments are likely to suffer from excessive lifts in air temperature. The chemical and physical forms and quantities of calcium used also affect the results and add further levels of variables to each experiment. Here again Colhoun (1958) reinforces, as with moisture and temperature, lessons from the classic studies of Samuel and Garrett (1945) related to the impact of spore load, inoculum potential and intensity. Theirs was one of the earliest scientific validations that the effects of pH and of calcium could be separated and quantified individually as factors influencing the environmental success of *P. brassicae*.

Subsequent to Colhoun (1958) practical studies indicated that the impact of the balance of nutrients in the soil is significant while the actual content of individual ions is still important. For example, Myers and Campbell (1985) suggested that clubroot disease expression depends on a balance between pH and the amounts of calcium and magnesium in the soil. While Dobson et al. (1983) concluded from their work using roughly and thoroughly mixed limed soils that if roots and spores occur within small pockets of low calcium and/or low pH, invasion is possible despite high overall soil calcium and pH estimations. Fletcher and others (1982) achieved greatest effects of clubroot disease with field applications of calcium carbonate and calcium nitrate which increased pH to 7.9 and 8.3, respectively. They also concluded that although pH was a major factor in reducing disease expression, some other factor than pH possibly the Ca^{2+} (calcium) ion itself was involved. Work conducted using controlled conditions has formed similar conclusions (Hamilton and Crête 1978). These results still however, beg the question of "where and when is *P. brassicae* influenced by the presence of calcium and by pH value?" There is a tendency to assume that these factors affect the microbe while in the soil but since *P. brassicae* spends most of its life cycle within the host it could be fair to suggest that calcium and pH also affect these environments. A role for calcium in the post-infection development of *P. brassicae* is supported by the demonstration that incorporation into roots is pH-dependent (Myers and Campbell 1985; Campbell and Greathead 1996) contended that *P. brassicae* is affected at more than one point in the life cycle between spore germination and the completion of resting spore formation in the cortical cells by pH and calcium concentration. Numerous detailed long-term experiments have confirmed this (Webster 1986; Dixon and Webster 1988; Webster and Dixon 1991a; Dixon and Page 1998; Page 2001). It is evident that the greatest impact of calcium is when it is present in the period between spore germination to post-penetration of root-hairs. The latter appears to be when root-hair infection has the biggest impact on subsequent gall formation. There may apparently be separate mechanisms since the periods 0–3 and 0–7 days post-penetration seem to be separated in the extent of their influence on subsequent disease development. The expression of effect seems to be cumulative since it took longer when a 30 mel^{-1} Ca^{2+} solution was used as compared with one containing 55 mel^{-1} Ca^{2+} in order to achieve similar final results. The host–pathogen response varies also with pH however, that is a separate factor. But it is worth recording here that calcium at pH 7.2 needed to be present by day 14 in order to suppress root-hair infection or alter the progress of galling. The pathogen may be affected by the calcium environment in the root-hair and this alters subsequent behaviour in the cortical cells. The work of Dixon and co-workers (Webster 1986; Dixon and Webster 1988; Webster and Dixon 1991a; Dixon and Page 1998; Page 2001) is supported by results of Donald and co-workers (Donald et al. 2004; Donald 2005) in Australia. Of major significance is the finding that high concentrations of calcium at pH 6.2 or 7.2 reduce total numbers of root-hair infections and the rate of maturation through plasmodial, sporangial and zoosporangial stages as compared with the controls. Raised concentrations of calcium completely inhibit the later stages of *P. brassicae* development in the root-hair even where high inoculum doses are applied. The calcium effect commences in the soil since Dixon and Page (1998) showed that the germination of resting spores, motility of zoospores and the composition of benign microbial flora around roots are altered.

High concentrations of calcium could possibly reduce flagellar action as Satir (1982) and Sleigh and Barlow (1982) reported that changes in calcium of the order of 10^{-6}–10^{-4} M affected the action of demembranated flagellae, whether this effect would hold for the flagellae of *P. brassicae* has yet to be determined.

Recently, Wallenhammar (1999) pointed to the uneven distribution of acidic and alkaline areas of soil in individual fields with pH ranging from 5.73 to 8.45 in localised patches. Mattsson (1995) identified that pH values of the subsoil are frequently more alkaline than the upper horizons in Sweden especially in the calcareous glacier clay region near Uppsala in eastern central Sweden. This modernises aspects of *Colhoun's Dilemma* related to pH. Earlier Palm (1958, 1963) had concluded that the effect of pH is not restricted solely to the establishment of *P. brassicae* as a parasite because the rate of gall proliferation was markedly suppressed by an alkaline condition of the medium after infection in the host tissues. It was suggested that changes in the soil reaction may have more drastic effects on gall development than on the number of infections by zoospores. The use of organic buffers to adjust pH and calcium content separately from each other and showed that at 10 mel^{-1} Ca^{2+} and a pH of above 7.1 reduced the numbers of primary zoosporangia in root-hairs thus inhibiting galling (Myers and Campbell 1985). Webster and Dixon (1991a) demonstrated that the effects of pH are independent of calcium concentration and found that alkaline pH reduced total root-hair infection number and retarded the maturation of plasmodia, sporangia and zoosporangia. The pH effect on the maturation of root-hair infections is activated by exposure to alkaline pH within 3 days of penetration. Prolonged exposure beyond 3 days gives no additional effect.

There may be a dual effect in that alkaline pH increases sensitivity of the host and/or *P. brassicae* to calcium effects as well as increasing the efficiency of calcium uptake. The effects of pH and calcium are remarkably similar but this does not necessarily mean they are one and the same as has been suggested by some workers. They may regulate the pathogenic potential of an inoculum quite separately. Since pH regulates the response to calcium, intracellular function may be modified in addition. A high concentration of H^+ ions in plant tissues is potentially antagonistic to calcium. Membrane permeability is lowered by both alkaline pH and by high calcium. This environment could affect the growth and reproduction of *P. brassicae* as it proliferates within the host root-hair and epidermal cells or within the cortical cells. Alkaline environments could influence primary and secondary invasions, cortical migration and cell hypertrophy. An involvement of Ca^{2+} ions in the growth and reproduction of *P. brassicae* ultimately leading to induced cell death or hypersensitivity is suggested by Takahashi et al. (2006). At the agronomic level promoting high alkalinity linked with continuous cropping is suggested by Shinoda et al. (2005) as a means of reducing the soil inoculum load.

In contrast to *P. brassicae* another Plasmodiophorales, *Streptomyces scabies,* the cause of common scab in potatoes is encouraged by alkaline soil conditions. Manipulating pH towards acidic values considerably reduces disease incidence. Adding sulphur to soil increases soil acidity and helped reduce the impact of *S. scabies* on the marketable yield of potato crops (Pavlista 2008).

Cavity spot disease of carrot (*Daucus carota*) has been another intractable problem where soil hydrogen ion and calcium content moderate incidence and severity. Carrots are rejected at grading with one or two visible lesions, and when disease incidence passes a relatively low threshold it becomes uneconomic to harvest crops (Hiltunen and White 2002). Early studies attributed cavity spot incidence to interactions with the presence and absence of calcium in soil classing the syndrome as a 'physiological disorder'. The first solid indication of the involvement of a pathogen was when three different fungicides with activity against Oomycete fungi all reduced disease. Very quickly thereafter, the causal agents *Pythium violae* and *P. sulcatum* were isolated from cavity spot lesions and Koch's postulates satisfied. Calcium carbonate is known to have significant effects on cavity spot, probably by inducing a soil microflora inhibitory to filamentous fungi. The management of agronomic aspects such as irrigation, soil cultivation, and the length of time for which crops are grown may all be used, while carrot cultivars with some level of field resistance may be beneficial. One of the most significant factors is disease avoidance by not selecting fields with high inoculum levels. One serology-based risk assessment test has been produced and commercialized, and molecular probes which could be the basis of more sensitive tests are available for both pathogens.

Investigations have been made of the relative importance of primary and secondary infections (auto- and allo-infections) in the development of a carrot cavity spot (CCS) epidemic caused by *Pythium* spp. Three cropping factors: fungicide application, soil moisture and planting density, were selected as the key variables affecting the disease syndrome (Suffert et al. 2008). Their effects on: (i) disease measurements at a specific time, (ii) the areas under the disease progress curves (AUDPCs) and (iii) a time-dependent parameter in a pathometric incidence-severity relationship, were studied. A deficit of soil moisture limited the movement of *Pythium* propagules towards host tissue, and thus reduced primary infections in the field; it also promoted the healing of lesions, limiting lesion expansion and the potential for allo-infections (6.8–7.5 lesions per root in irrigated plots compared with 2.4 lesions in non-irrigated plots). A negative relationship between the mean root-to-root distance and the rate of allo-infections was established in microcosms; a reduction in mean planting density was also effective in limiting CCS development (0.5, 1.6 and 2.0 lesions per root in microcosms containing 8, 16 and 31 roots, respectively). Free calcium in soil has been demonstrated by Heyman et al. (2007) as substantially contributing to reductions in the effects of *A. euteiches* a cause of root rot in peas. These authors established a negative correlation between calcium content and disease severity in Swedish soils. Part of this effect may result from an inhibition of zoospore release from oospores present in these soils.

Nutrient Status

Emphasis in the management of crop diseases is returning towards the concept of enhancing plant or crop health as compared with previous objectives of eradicating or diminishing pathogen populations. This approach was favoured until the advent of extensive numbers of synthetic agrochemicals in the 1970s, with little research interest from then until the late 1990s. Nutrient management is now increasingly gaining stature as a practical and environmentally safe approach (Adhilakshmi et al. 2008) towards crop health which is associated with the stimulation of beneficial soil microbes as opposed to exploiting directly fungitoxic effects. Mineral nutrition may also directly affect the ability of plants to resist harmful pathogens (and other soil inhabitants) by up-regulating genetic mechanisms. The literature on the effects of nitrogen influencing the outcome of both pest and pathogen invasion is very extensive but requires careful analysis, understanding of the factors involved and subsequent interpretation. For example, the effects of nitrogen (N) nutrition on pests of onions, carrots and cabbage were evaluated on organic and mineral soils in Ontario, Canada in 2000 and 2001 (Westerveld et al. 2002). The damage caused by onion thrips (*Thrips tabaci*) was lower on cabbage that received higher than recommended nitrogen. This effect may have been due to delayed maturity of plants that received high nitrogen levels. These field trials also suggested that leaf blight (*Cercospora carotae*) severity decreased with increasing nitrogen. In both instances it might be expected that an increase in succulence could predispose these crops to predation. Elsewhere the use of high levels of nitrogen fertilisation has been associated with increased disease severity, but the formulation of nitrogen used may be at least as important as the rates used (Huber and Watson 1974).

Arguments for the use of ammonium and nitrate based nutrition were reviewed by Barker and Mills (1980) and in great depth by Wild (1988). In general ammonium–nitrogen appears associated with increased disease severity more often than nitrate–nitrogen, but this is not always a hard and fast rule. Early research tended to be confused due failures to adjust the rates of nitrogen applied relative to molecular content, an array of formulations used and the huge diversity of environments, crops and host–pathogen combinations studied. A brief syntheses of results are provided by Marschner (1990) and Engelhard (1996), while earlier Huber (1978) considered the impact of pathogen infection on host nutritional metabolism. The concept that forms of nitrogen might influence the course of disease was pioneered at the University of Wisconsin from the 1930s onwards by J C Walker who showed for example, that vascular wilt of tomato (*V. albo-atrum*) increased with raised levels of nitrate–nitrogen (Walker et al. 1954), but cabbage yellows (*Fusarium oxysporum* f. sp. *conglutinans*), tomato wilt (*F. oxysporum* f. sp. *lycopersici*) and pea wilt (*F. oxysporum* f. sp. *pisi*) decreased as the concentration of nitrate-nitrogen increased (Gallegly and Walker 1949). While bacterial wilt and canker of tomato (*Ralstonia* (*Pseudomonas*) *solanacearum* and *Corynebacterium michiganense*) increased in severity concomitantly with increased nitrate-nitrogen. Nitrogen fertilisation, as ammonium but not nitrate, was thought to stimulate fluorescent pseudomonads within the rhizosphere (Smiley 1978). It is now apparent that plant nutrition must be considered as a component part of integrated control management for diseases (Dixon 2002, 2009c) in which combinations of nitrogen and calcium offer the most effective means of countering pathogenic microbes through balanced host nutrition.

In some instances, acid-forming fertilisers, such as ammonium sulphate or phosphate help to reduce disease incidence as with potato common scab (*S. scabies*) (Lazarovits et al. 2007). In other instances, a reverse effect is obtained, as with clubroot (*P. brassicae*) (Webster and Dixon 1991a). In general the literature on the influence of soil chemical constituents on common scab severity for example, is best described as inconclusive. Reviewing the effects of ten plant nutrients on scab Keinath and Loria (1989) concluded that some of this confusion could be resolved by examining interactions on the tuber surface between edaphic factors and disease intensity. The influence of pH is seen as the most consistent factor associated with potato common scab which is generally most serious with pH ranges of 5–8.

Long-term crop protection can be achieved by modification of the physico-chemical properties of the soil in favour of the development of an antagonistic microbial population. This can be achieved by the manipulation of some of the macro-nutrient and micro-nutrient balances within the soil. Disease suppression was not associated with the conduciveness of a soil to take-all, but rather to the supportiveness of a soil to biocontrol activity. Biocontrol activity was positively correlated with iron, nitrate-nitrogen, boron, copper, soluble magnesium and the percentage of clay and negatively correlated with soil pH and available phosphorus. In Western Australia take-all decline has been attributed to populations of *Trichoderma* spp. that suppress the pathogen's saprophytic as well as parasitic activity (Simon and Sivasithamparam 1989). Soil treatment with *T. koningii* the most abundant anatagonist in these suppressive soils reduced saprophytic growth of *G. graminis* var. *tritici* and increased survival of wheat seedlings by about 50%. In glasshouse experiments with natural soil, grain yield was increased about 10% in three out of six field trials. A strain of *T. koningii* originally isolated in Western Australia reduced take-all severity and increased yield as much as 65% when added to the seed furrow in field trials in Washington State, USA (Duffy et al. 1996). Both the take-all pathogen and indigenous antagonists are sensitive to soil pH and mineral nutrition. Microbially mediated take-all decline occurs after three wheat crops in slightly acidic soils but is delayed until the sixth or seventh crop when soil pH is elevated to 7.0 by liming (Cook and Baker 1983). Ammonium-nitrogen fertilisers reduce the rhizosphere pH of wheat and increase the proportion of rhizosphere bacteria that are antagonistic to *G. graminis* var. *tritici*; the reverse is true of nitrate–nitrogen (Smiley 1978). Similarly the take-all suppressive activity of indigenous populations of *Trichoderma* spp. is increased by soil acidification with ammonium fertilisers, but this effect is lost with liming (Simon and Sivasithamparam 1989). An adequate supply of micro nutrients such as copper, iron, manganese and zinc alleviates take-all and their increased availability under acidic conditions is due in part, to metabolic activity of certain indigenous microorganisms. In this study there was a negative interrelationship between boron, copper, iron, soluble magnesium, nitrate-nitrogen, percentage of clay and take-all severity in the presence of *T. koningii* on eight silt loams.

An exceptional diversity of mechanisms contribute to the biocontrol activity of *Trichoderma* spp. including the production of antibiotics, volatile organic compounds and lytic enzymes; mycoparasitism and competiveness in the rhizosphere, bulk soil or crop residues. While each of these has been proposed in different reports to be primarily responsible for biocontrol, the relative importance of individual mechanisms more than likely fluctuates depending on the strain, environmental conditions and pathosystem. Soil nitrate levels have been positively correlated with cellulase production and may favour competitiveness of the biocontrol agent with the pathogen. While nitrate fertilisers have been identified as increasing take-all, partially at least because they raise rhizosphere pH, which is favourable to the pathogen and unfavourable to the biocontrol agent (Simon and Sivasithamparam 1989). This contrasts with the findings that nitrate–nitrogen levels were positively correlated with *T. koningii* biocontrol activity. This suggests that pathogen suppression in the bulk soil with an acidic pH is less affected by nitrates (Smiley and Cook 1973) and may have been a major component of biocontrol during early plant growth and other factors may become important as biocontrol shifts from the infection court to the rhizosphere. Understanding which abiotic soil factors have most influence on the biocontrol activity of *T. koningii* and how these factors interact may provide clues to biocontrol mechanisms and their regulation in situ. This is especially important with regard to micro-elements such as iron, pH, boron, copper and soluble magnesium, which affect the production of anti-fungal compounds by *T. koningii* and competition with the pathogen. Some key soil factors favourable to fungal antagonists (acidic pH and high iron) contrast with those factors favourable to bacterial fluorescent pseudomonads (alkaline pH and low iron).

To solve the problem of decline in land productivity and spread of soil-borne disease such as *Fusarium* wilt (*Fusarium oxysporum* f. sp. *cubense*) in banana orchards caused by the degradation of land quality, some integrated measures including soil disinfection by application of modified lime-nitrogen (calcium cyanamide) fertiliser, decontamination of irrigation water, continued soil disinfection and field experiments were carried out in an abandoned banana orchard in China (ZhiYong and XiaoLin 2008). The results showed that the incidence of the disease and disease index of *Fusarium oxysporum* f. sp. *cubense* at key growth stages were significantly lower with treatment of integrated measures than those with the aim of using a single control treatment. The sterilization of irrigation water, to cut the route of the *Fusarium* wilt distribution, produced a significantly positive result. The disease incidence was 67.58% in the control treatment, and 13.75% in the integrated treatment. Harvest area increased from 32.42% in the control to 86.25% in the integrated treatment. Fruit yield per tree was 24.4 kg in the control and 26.9 kg in the integrated ones. The yield increased more than 10%. It could be concluded that integrated *Fusarium* wilt management might be the best method of control for this disease. More widely, benefits accruing from the use of calcium cyanamide are consistently greater than might be anticipated solely from the calcium (50%) and nitrogen (20%) content of this fertiliser. Use of this chemical is associated with improved soil fertility in terms of increased benign microbial activity. This apparently leads to a reduction in the pathogenic activity of several highly aggressive soil-borne microbes particularly those such as *Plasmodiophora brassicae*, the causal agent of clubroot disease in brassicas and *Sclerotinia sclerotiorum*, a cause of stem and root white rot in an extensive range of crops which are discouraged by alkaline pH values (Dixon 2009d).

High potassium levels were believed to reduce infection by some wilt inducing pathogens, such as *Fusarium* wilt of tomato (Walker and Foster 1946). But they were without effect on *Verticillium* and bacterial wilts of tomato (Walker et al. 1954; Gallegly and Walker 1949). For foot and root rots the influence of potassium is rarely noted, but high potassium, especially if associated with high nitrogen, reduced *Fusarium culmorum* on wheat. The reaction with *Gibberella zeae* (stalk rot) on maize is more complex and depends on nitrogen level; an increase in potassium reduced stalk rot when nitrogen was low, had no effect at medium nitrogen, and increased it at high nitrogen; with low potassium then nitrogen level had no effect. Root rots are generally reduced by high phosphate nutrition; in Canada the reappearance of browning root rot of cereals due to *Pythium* spp. was attributed to insufficient use of phosphate fertilisers. The relationship of calcium to disease severity is even more tantalising since there various side effects such as the action on pH, causing in turn imbalances of trace elements. Also related to calcium is the influence of sodium. Increased sodium in the nutrient solution increased the susceptibility of tomato to wilt (*Fusarium oxysporum* f. sp. *lycopersici*). This may be connected with the lowering of calcium induced by sodium since high concentrations of calcium are related to reductions of disease severity in tomatoes infected with *Verticillium albo-atrum*.

Disease management strategies, such as the application of fungicides (Bateman 1993) can also modify the severity and composition of the disease burden of a crop. Control of eyespot (*Tapezia yallundae*) can lead to significant increases in sharp eyespot (*R. cerealis*), possibly due to the loss of the exclusive capacity of the W-type pathotype of *T. yallundae* (Murray et al. 1998). Prior stem colonization by *Fusarium* spp., (especially *F. avenaceum*) is able to suppress development of lesions of *T. yallundae* caused by the more virulent W-pathotype. If, however, infection by *Fusarium* spp. occurs after that by *T. yallundae*, then foot rot severity may be increased (Bateman and Munnery 1995); possibly because the primary colonizer weakens or alters stems increasing their suitability for colonization by secondary colonizers (Bateman 1993). The incidence of *F. culmorum* has also been observed to increase at the same time as decline in *G. graminis* var. *tritici* severity (Vojinovic 1972). With regards to *Fusarium* spp. alone, supplementary nitrogen has been reported to increase the incidence of infection by up to 125% (Martin et al. 1991). Parry et al. (Parry et al. 1995) conclude, however, that the effect of nitrogen applications and *Fusarium* spp. are unclear. As with take-all it could be another example of regional variation due to prevailing climatic conditions, soil types and agronomic practices. The situations in America and the U.K. are very different and successful strategies imported from other countries often fail to give the same level of control elsewhere (Hornby et al. 1998).

Potassium phosphonate was found to have fertiliser properties which are linked to increasing plant growth and vigour and hence reductions in pathogen incidence for example with root rots caused by *Phytophthora palmivora* in pawpaw in Northern Queensland, Australia. This treatment was successfully integrated with the use of plastic sheeting and organic mulches and growing the plants on 0.75 m mounds (Vawdrey et al. 2004). Relationships between sugarcane rust (*Puccinia melanocephala*) and soil nutrients and pH have been suggested (Johnson et al. 2007). Phosphorus content correlates positively with rust incidence soil organic matter, sulphur and potassium appear to have similar effects whereas magnesium is negatively correlated with disease. Additives such as phosphonate (potassium phosphite) protect trees against infection, increase tree (coconut and durian) survival and yields and play an important role in integrated control systems (Guest 2002).

Increasing soil phosphorus level is suggested to accelerate plant growth and this may become more susceptible to this pathogen. High nutrient levels are associated with vigorous crop growth, denser canopies and earlier canopy closure creating an environment conducive to rust development. Results suggest there is a threshold value for nutrient concentration, below the threshold there is no stimulation and indeed there may be reductions. This is demonstrated with pH where at values of ≤6.0 reduced and >6.0 disease is stimulated.

Availability of micro-elements to microbes and plants is involved in suppressiveness towards several diseases. High concentrations of iron, manganese, copper and zinc were associated with suppressiveness to *Streptomyces*, *Aphanomyces*, *Gaeumannomyces* and *Thielaviopsis* spp. A high level of micro-nutrients should contribute to a more diversified microbial flora and hence a more complex and efficient antagonism to the pathogen. Micro-nutrients are also probably involved in host defence reactions. In the case of black root rot (*Thielaviopsis basicola*) of tobacco, iron appears to be a prerequisite for the antagonistic activity of fluorescent *Pseudomonas* spp. and the efficiency of the bacterium in controlling the pathogen. The suppressive effect of *Pseudomonas* results from the high production of hydrogen cyanide which is toxic to the pathogen. Cyanogenesis is regulated by the availability of iron to the bacterium. Diseases are suppressed in acid soils where iron, manganese, copper and zinc form water soluble salts that are readily available. In alkaline conditions these elements are insoluble. Thus adding micro-nutrients to alkaline soils does not necessarily lead to disease control. Soil factors influencing the availability of micro-nutrients include the following:

- Micro-nutrient content of soil minerals
- Age of the soil which affects the degree of alteration of soil minerals
- Leaching of microelements from the soil
- Redox potential

A low redox potential increases the availability of some micro-nutrients especially iron and manganese which form water soluble salts in the reduced state. The redox potential is low in fine-textured soils under conditions of high soil moisture and high microbial activity such as after the incorporation of large quantities of organic matter. The combination of high soil moisture or low redox potential and high level of micro-nutrients has been demonstrated to be effective in the suppressiveness of potato scab (which is suppressed in moist soils in which the availability of manganese and iron is high). By comparison fusarium wilt is enhanced by impeded drainage and high soil moisture that is conditions which increase the availability of iron for the pathogen.

The application of manganese has been associated with the stimulation of soil bacteria (Casida 1968) and in conjunction with nitrogen (ammonium) fertilisation, reductions in take-all severity (Brennan 1992a).The reduction of soil pH following ammonium fertilisation would favour manganese-reducing organisms and would increase the amount of available (to plants) manganese (Mn^{2+}). Plants deficient in manganese may have impaired abilities to synthesize phenolics and lignin within the root system and thus are more susceptible to physical penetration by *G. graminis* (Rengel et al. 1994). It is under conditions of both manganese and zinc deficiency that populations of fluorescent pseudomonads are increased, but within these populations, the incidence of both manganese-oxidizers and reducers is low (Rengel et al. 1996). Manganese can also have direct effects on *G. graminis* since in its reduced form manganese is toxic, but *G. graminis* and other rhizosphere microbes are able to oxidize manganese to its non-toxic and unavailable form (Mn^{4+}). The manganese oxidizing ability of *G. graminis* is partly dependent on the virulence of the pathogen, with virulence being correlated positively with manganese-oxidizing ability (Pedler et al. 1996). Manganese is one heavy metal upon which there has been a significant *G. graminis* var. *tritici*-based research effort, particularly in Australia (Rengel et al. 1994) and America (McCay-Buis et al. 1995). The effects of manganese are also dependent on plant health and nutrition, which is again a subject of considerable work in Australia by Brennan (1989, 1992b). Many mineral ions e.g. phosphorus, nitrogen, manganese, zinc, copper and chloride, have been shown to play significant roles in the responses of plants to attack by *G. graminis*. In the majority of this work the ion was applied in isolation, or perhaps in combination with nitrogen, and until a greater understanding of the inter-relationships within nutrient and host–parasite-wider microbial community interactions are obtained, greater exploitation of nutrient provision in the management of *G. graminis* will remain anecdotal.

The interactions between the inhibitory effects of fungal and bacterial antagonists and nutrients have been tested under *in vitro* conditions against the wilt pathogen of alfalfa (lucerne) *Fusarium oxysporum* f. sp. *medicaginis*, *Trichoderma harzianum* and *Pseudomonas fluorescens* (PI 5) and were found to be effective. Manganese sulphate at 500 and 750 ppm inhibited the mycelial growth of *F. oxysporum* f. sp. *medicaginis* under *in vitro* conditions. In pot culture studies, manganese sulphate at 12.5 mg kg^{-1} reduced the wilt incidence (23.33%). Combined application of manganese sulphate 12.5 mg kg^{-1} *T. harzianum* 1.25 mg kg^{-1} of soil significantly reduced the wilt incidence accompanied by improved plant growth and yield in pot culture. The mixture of manganese sulphate (25 kg ha^{-1}) *T. harzianum* (2.5 kg ha^{-1}) significantly reduced the wilt incidence when applied as a basal dose in the field conditions. The average mean of disease reduction was 62.42% over control.

Soil boron content has been associated with adveresely affecting the activities of *P. brassicae* since the 1930s (O'Brian and Dennis 1936) onwards. One of the first controlled studies was that of Palm (1963) who investigated the effect of boron on

P. brassicae in sand cultures and recorded maximum root-hair infection at a concentration of 0.3 mel^{-1} boron or less. He further demonstrated that in the absence of boron the inhibitory effect of calcium on root-hair infection is suppressed, he suggested that lime may fail to diminish clubroot disease in boron deficient soils. Extended research (Dixon and Wilson 1983, 1984a, b, 1985; Dixon 1983, 1984a, b, 1985) achieved significant reductions in disease index with sodium tetraborate applied to acidic granitic soils in 3 successive years of field studies. More recent studies showed that environments with elevated boron concentration there are significant effects both on the root-hair and cortical phases of *P. brassicae*. Throughout the *in planta* stages of the life cycle of *P. brassicae* boron has an impact on the microbe. There appears also to be a relationship with the quantity of boron in the plant which is moderated by uptake over time and space as determined by the size of the plant root system and its capacity to absorb boron. It is likely that there are interactions with other ions. For example, lime applications in the forms of calcium carbonate or oxide may alter the nutrient environment in soil to the detriment of *P. brassicae* and therefore, make the host–parasite association more affected by other factors such as boron. Alternatively, boron may have a primary effect because Webster and Dixon (1991b) found that the effects of boron interact with both the primary and secondary stages of development of *P. brassicae* ultimately affecting the intensity of symptom expression. The environment induced by boron in cells where membrane permeability and wall structure are altered may be to the detriment of *P. brassicae*. It could also make for conditions less conducive for nuclear division by the microbe. Quite possibly boron effects are distinct from those of calcium and pH. Dixon (1991) identified that boron affects the progress of *P. brassicae* by retarding the rate of sporangial maturation. The correlation of diminished intensity of disease expression and boron suppression of root-hair infection and gall formation appears related to host exposure. Long exposures to low concentrations seem to equate with the effects of shorter exposures to higher concentrations. Field and controlled laboratory studies (Craig and Dixon 1993a, b) identified that boron has a substantial effect on the ability of *P. brassicae* to invade root-hairs and establish colonisation in the field. Raising the boron content of the rhizosphere prior to the availability of a susceptible host to infested soil limited the subsequent ability of *P. brassicae* zoospores to penetrate, colonise root hairs and cause symptoms.

Water Management

High levels of soil moisture are often associated with increased disease severity. This may be due to the presence of increased moisture films through which motile spores move. Also, water-logged soil or compost will weaken and asphyxiate root systems, thereby giving portals for the entry of pathogenic organisms. Californian studies of the incidence of soil-borne *Sclerotinia minor* (lettuce drop) as influenced by type of irrigation and the form of tillage indicated that disease incidence remained significantly higher under furrow irrigation compared with subsurface drip and was significantly greater on autumn as opposed to spring crops (Wu and Subbarao 2003). With furrow irrigation the number of sclerotia at the end of the crop season increased significantly over that at the beginning of the year, but no significant changes were detected over-years. By contrast, the number of sclerotia within a single season did not increase significantly under subsurface drip irrigation, nor was the year-on-year accumulation of sclerotia significantly increased. The degree of aggregation of sclerotia increased significantly during a cropping season under furrow irrigation but not under subsurface drip irrigation. The conventional tillage after harvest under furrow irrigation decreased the degree of aggregation of sclerotia after each season but the distribution pattern of sclerotia under subsurface-drip irrigation changed slightly by association with minimum tillage. Spatial pattern analyses suggested that the aggregation of *S. minor* sclerotia occurred at a scale of not more than 1 m and the distribution of diseased lettuce plants was random at a scale larger than 1 m. The combination of fewer sclerotia produced by each crop and its unaltered distribution under subsurface drip irrigation and associated minimum tillage makes it a valuable cultural practice for lettuce drop disease management. Wet ground (Brenchley 1968) and areas where soil structure has deteriorated (Yarham 1981) have been associated with increased take-all (*G. graminis* var. *tritici*) severity, however it is more likely that the poor root growth in such areas predisposes plants to severe take-all (Gair et al. 1972; Parry 1990). Where soil moisture rises from 50% of maximum water holding capacity up to saturation (Dixon 1981, 1984b) clubroot disease (*P. brassicae*) severity escalates. Lange and Olson (1983) emphasised the dependence of zoosporic microbes on free water existing between the soil crumbs for the movement of zoospores. Free water is critically important for the formation, discharge and dispersal of zoospores and may influence the encystment and penetration processes at the root-hair surface. The distances travelled by soil-borne zoospores are relatively short, probably between 10 and 20 mm judging by information for *Olpidium brassicae* or *Synchitrium endobioticum* both relatives of *P. brassicae*. Invasion of root hairs occured up to 75 mm from the source of *P. brassicae* infection in soils where water mass movements were minimised (Watson 1967). Mathematical modelling by Yang et al. (2004) demonstrated a relationship between soil moisture and host invasion. Similarly a soil moisture deficit limits the movement of propagules of *Pythium violae*, the cause of cavity spot in carrots (Suffert et al. 2008) reducing primary infections in the field, promotes lesion healing, limits primary infections and subsequent spread. Integrating elements of husbandry such as harvest date, soil moisture content, sowing density and crop geometry linked with appropriate fungicide use offers means for controlling this previously intractable disease (Suffert et al. 2008).

Irrigation monitoring and scheduling are imperative as water becomes increasingly scarce producing a drive for greater water-use efficiency which in part at least aims to control those plant pathogens which are favoured by excessive application (Passioura 2002). Australian wheat yields have increased by improvements to health and nutrition of crops and effective water management ensuring that it is effectively turned into photosynthate used for grain filling. A steady annual yield increase of 1% can be achieved through plant breeding linked with greater understanding of the relationship of roots and soil microbes and knowledge of how roots utilise soil resources such as water (Passioura 2002). To sustain production of annual crops in Australian Mediterranean environments, for example, both agronomic and genetic options have been used. An analysis of the yield increases in wheat in Mediterranean climatic regions shows that there has generally been an increase in the yields over the past decades, albeit at a lower rate than in more temperate regions. Approximately half of this increase can be attributed to agronomic and half to genetic improvements. The agronomic improvements that have been utilised to sustain the increased yields include earlier planting to match crop growth more closely with rainfall distribution, use of fertilisers to increase early growth, minimum tillage to enable earlier planting and increase plant transpiration at the expense of soil evaporation, rotations to reduce weed control and disease incidence and use of herbicides, insecticides and fungicides to reduce losses from weeds, pests and pathogens. Genetic improvements include changing the phenological development for better matching with rainfall patterns, increased early vigour, deeper rooting, osmotic adjustment, increased transpiration efficiency and improved assimilate storage and remobilisation. Mediterranean environments that are subjected to annual terminal drought can be environmentally and economically sustainable but to maximise plant water-use efficiency while maintaining crop productivity requires an understanding of the interaction between genotypes, environment and management (Ashcroft et al., 2003).

Irrigation may enhance sustainable crop production but only if applications are made as part of an integrated management package. Water productivity (WP) expresses the value or benefit derived from the use of irrigation (Singh et al. 2006). Substantial variations appear between water productivities of varying crops such as wheat, rice and cotton, within crops each relative to the manner of use, prevailing husbandry practices such as the use of mulches and soil physical and chemical compositions. Soil moisture content affects the relationships between of *R. solani* and its antagonist *Trichoderma harzianum* (de Paula and Hau 2002). The antagonistic ability of *T. harzianum* was greater in soils held at intermediate moisture content as compared with very wet or very dry conditions. Soil water logging appeared to aid the establishment of a new pathogen (*Phytophthora botryose*) in plantations of *Elmerrillia tsiampacca* and *E. ovalis*, whereas previously this pathogen had been known on *Heavea braziliensis* (rubber) (de Kam and Sukmajaya 2002). *Aphanomyces euteiches* induced root-rot in pea causes serious crop loss (Allmaras et al. 2003) and the pathogen remains viable in soil for many years. The disease is favoured in poorly drained soils, especially when soil compaction caused by trafficking with heavy axle loaded vehicles impairs drainage. Soil water potentials (SWPs) of P >–5 kPa at the 15 cm depth were associate with pea crop failures especially where impaired drainage was in the upper 50 cm of fields. Soil–water retention and field observed SWP indicated <10% air-filled pores frequently occurred in the 15–30 cm layer, symptoms of anoxia were absent from the crops. Poor soil drainage influences the mode of infection. The virulence and inoculum potential (Ulacio et al. 1999) of *R. solani* propagules affecting rice plants were related to the period of time where they were exposed to anaerobic conditions.

Glasshouse and field experiments were conducted in Ibaraki, Japan, in 2001, studied the effects of soil moisture, pH and pathogen resting spore density on the effectiveness of the biological control of clubroot (*P. brassicae*) by a fungal endophyte (*Heteroconium chaetospira*) in Chinese cabbage cv. Shin-Riso (Narisawa et al. 2005). Conditions favouring disease development included low pH (5.5) and high soil moisture content (80%), with significant reductions in the disease observed at a higher pH (6.3 and 7.2) and lower soil moisture content (40% and 60%). In glasshouse tests, *H. chaetospira* effectively controlled clubroot (reducing the disease by 90–100%) at pathogen resting spore densities of 10^4 and 10^5 spores g^{-1} soil at all soil pHs tested (5.5, 6.3 and 7.2). When the resting spore density was 10^6 spores g^{-1} soil however, plants were severely infected, regardless of treatment, and *H. chaetospira* had no effect on the disease. At soil moisture content of 40%, disease occurrence was low, regardless of pathogen spore density, but the disease was significantly lower in *H. chaetospira* treated plants with a pathogen spore density of 10^5 spores g^{-1} soil. At 60% soil moisture content, *H. chaetospira* was significantly impaired the effectiveness of the pathogen (*P. brassicae*) at spore densities of 10^4 and 10^5 but not on 10^4 g^{-1} soil. At 80% soil moisture content, *H. chaetospira* had no effect on pathogen density. In situ soil moisture contents were constantly adjusted to relatively low to moderate values (pF 2.2–2.4 and pF 2.0–2.2, respectively) and high values (pF 1.6–1.8). Other environmental conditions, i.e. resting spore density and soil pH, were maintained at constant levels. Control plants (not treated with (*H. chaetospira*) showed uniformly high disease levels and proportions of diseased plants across all moisture treatments (disease index = 72–80, proportion of diseased plants = 85–97%). In the field, *H. chaetospira*-treated plants at low soil moisture (pF 2.2–2.4) had 68% disease reduction compared with the untreated controls and 49% reduction at moderate moisture (pF 2.0–2.2). *Heteroconium chaetospira* had no effect on the disease at high soil moisture (pF 1.6–1.8).

Sowing and Planting Times

Autumn sowing favours the development of take-all (*G. graminis* var. *tritici*) for climatic and temporal reasons (Hornby et al. 1998) and early infections cause the greatest degree of physiological problems. Ninety per cent of total nitrogen and phosphorus content of mature cereals is absorbed before the plant has achieved 25% of its final dry weight (Ayres 1984), therefore the rate and volume of soil explored in the early growth-stages has long-term implications for the subsequent performance of the plant. This is of particular importance to the uptake of phosphorus, which in combination with potassium increases cell wall thickness and tissue hardiness thus providing additional mechanical resistance to pests and pathogens (Maloy 1993). Though following infection with *G. graminis* var. *tritici* phosphorus uptake may not be impaired as uptake of phosphorus can occur along the entire length of the root (Fitt and Hornby 1978; Hornby and Fitt 1982) and transport is unaffected by the presence of well-developed lesions on seminal roots (Clarkson et al. 1975). A reduction in the volume of soil colonized by the root system may limit phosphorus uptake because it diffuses slowly through soil and does not move more than the width of a root hair, so zones of depletion develop in the vicinity of roots (Harper 1977).

One of the most dramatic examples of the changing interaction between sowing date and disease incidence is the soil-borne microbe *P. brassicae* and clubroot in oil seed rape (*B. napus*). Previously, this pathogen was held in check in the British crop because it was a predominantly sown in the autumn. This meant that it was drilled in late August to early September into cooling soil. The seed germinated and produced rosette structured plants by November which formed their components of yield before growth recommenced in mid to late February. The pathogen was inactive in the cold winter soils remaining as dormant resting spores. Consequently, the crop grew away and yielded in early summer with little damage from *P. brassicae* in contradistinction to the spring drilled crops of Continental Europe (France, Germany and Scandinavia) which succumbed to clubroot as both host and pathogen developed as the soils were warming. Now the British crop is being sown in late July to early August and soils are retaining heat and moisture through the increasingly milder autumn periods as a consequence *P. brassicae* remains actively causing damage throughout the year. Greater soil moisture content in the autumn and winter because of increased rainfall has only served to offer the pathogen additionally improved opportunities for spread and multiplication (Dixon 2006). Cabbage and other vegetable cole crops which are grown in the colder seasons traditionally suffer less from *Plasmodiophora brassicae*. The pathogen is less active in cold soil whereas summer cabbage and cauliflower are at high risk from it.

Disease may be avoided by early or late sowing or planting so that the crop and its pathogen are out of phase with each other and an epidemic cannot develop. A traditional method for reducing *Erysiphe cruciferarum* (powdery mildew) infection on brassicas is to sow late so that the foliage is immature and cannot support the pathogen during the main period when the crop is at risk. A heavy price is paid, however, in terms of yield. Mid- and late-season Brussels sprouts in Bedfordshire and the Vale of Evesham are less badly attacked by *E. cruciferarum*, perhaps because the environmental conditions are such that the pathogen is at a disadvantage compared with early season crops grown for processing in Lincolnshire (Dixon 1981).

Soil Solarisation

Solarisation is the practice whereby soil is covered with polyethylene sheeting capturing heat and raising the temperature substantially such that the surface layers are sterilised. Obviously, such techniques are only feasible in countries where intense sunshine is converted rapidly to solar gain such as parts of southern Europe, Israel, states such as California, USA and parts of Australia. In these areas solarisation has been used successfully in order to remove pathogens such as *Verticillium* and *Fusarium* spp. The difficulty lies in that most other microbes are also killed leaving soil populations disbalanced and vulnerable to rapid recolonisation by the more aggressive forms which are usually pathogenic. In which case if severe disease epidemics are to be avoided following solarisation the process must be linked in with other measures encouraging the re-entry of more beneficial forms. Columns of moist sandy loam soil inoculated with several soil-borne pathogens were inserted into fields and then subjected to solarisation by covering with transparent polyethylene sheets (Porter and Merriman 1983). Temperatures achieved were within the range 38–55°C, this was sufficient to eradicate: *V. dahliae*, *S. cepivorum* and *S. minor*. Less success was achieved with *F. oxysporum*, *P. irregulare* and *P. brassicae*. The soil was penetrated by high temperatures to 6–11 cm with sufficient effect to destroy most propagules to the point where they could not be subsequently detected. Mechanisms for this effect were examined by Horiuchi et al. (1982). They suggest that temperatures in excess of 45°C were sufficient to eradicate *P. brassicae* and the effect was enhanced by the presence of organic matter or the fertiliser calcium cyanamide. Their laboratory experiments suggested that a relationship between temperature and the time of exposure could exist since extending treatment at lower temperatures achieved similar effects to shorter periods of more intense treatment.

Solarisation was examined by Stevens et al. (2003) who found that soils amended with high nitrogen organic fertiliser (chicken manure) contained significantly increased soil rhizobacteria such as *Bacillus* spp. and fluorescent pseudomonas in the rhizospere, rhizoplane and root interior of sweet potatoes and possibly suppressed soil pathogenic organisms such as Fusarium root rot (*Fusarium solani*) and Java black rot (*Diplodia tubericola*). Soil solarization offers a means for killing teliospores of the soil-borne fungal wheat pathogen *Tilletia indica* (Peterson et al. 2008). A rapid decline in teliospore viability occurred at all treatment depths over 38 days. Initial viability rates of 43%, 71%, and 82% germination were reduced to 0.1%, 7.7%, and 0.2% after 38 days (across all depths) over 3 years – 2003 to 2006. Mean daily maximum soil temperatures at 5 and 20 cm under clear plastic in 2003, 2005, and 2006 were 67°C, 53°C and 60°C and 43°C, 38°C, and 43°C, respectively. Soil solarisation was evaluated for the control of *Phytophthora* root rot of raspberry (*P. rubi*) (Pinkerton et al. 2009). Cumulative time with soil temperatures >29°C at 30 cm soil depth in solarised plots exceeded 200 h. Results indicated that soil solarisation can be an effective component of integrated management of root rot in the Pacific Northwest, especially when combined with using raised beds and gypsum (calcium sulphate) amendments.

Combining soil solarisation with the addition of organic amendments derived from the cruciferous plant wild rocket (*Diplotaxis tenuifolia*) or thyme (*Thymus vulgaris*) raised the control of the wilt *Fusarium oxysporum* f. sp. *radicis-lycopersici* (Klein et al. 2007). Adding lucerne (alfalfa) leaves to soil increased the content of the nematode-trapping fungus (*Dactylellina candidum*) and microbivorus nematodes (Jaffee 2006), but the inter-relationships between these events remains obscure. Soil solarization in combination with the introduction of biocontrol agents was evaluated as a potential disease management strategy for tomato damping-off caused by *Pythium* spp. (Jayaraj and Radhakrishnan 2008). A rifampicin resistant *Pseudomonas fluorescens* strain (PfT-8) and a carbendazim resistant *Trichoderma harzianum* strain (ThM-1) were introduced into soil following solarization. Tomato seeds were planted into treated field plots. Damping-off incidence was significantly reduced in solarised plots compared with controls. Soil inoculation of biocontrol agents into solarised plots resulted in the highest suppression of damping-off incidence (PfT-8 up to 92%; ThM-1 up to 83%), and increase in plant biomass (PfT-8 up to 66%; ThM-1 up to 48%) when compared with un-solarised control plots. Rhizosphere populations of introduced biocontrol agents gradually increased (PfT-8 up to 102% and ThM-1 up to 84%) in solarised soils when compared with unsolarized controls. The population of *Pythium* spp. in rhizosphere soil was reduced up to 55% in solarised plots; whereas, application of bicontrol agents to solarised soils reduced the rhizosphere population of *Pythium* spp. by 86% and 82% in *P. fluorescens* and *T. harzianum* applied plots respectively.

Cover, Trap and Biofumigation Cropping

Cover crops are used primarily as a means of protecting uncropped land from erosion but they may have an effect on the survival of soil-borne pathogens. The use of cover crops such as hairy vetch (*Vicia villosa*) and rye (*Secale cereale*) in combination with nil tillage husbandry produced beneficial effects by reducing the incidence of blight (*Plectrosporium tabacinum*) and the need for fungicidal sprays on pumpkin crops (Everts 2002). Antifungal effects of intercrops, e.g. peppermint and *iso*thiocyanate production by brassica crops like mustard have been examined. Use of crops to stimulate pathogen germination and then exterminate the propagules produced (so called trap crops) has long been sought with little really practically effective effects. The phasing-out of methyl bromide brings the need to develop a scientific basis for these effects into particularly sharp focus (Matthiessen and Kirkegaard 2006). Biofumigation is the beneficial use of brassica green manures that release *iso*thiocyanates chemically similar to methyl *iso*thiocyanate, the active agent from the synthetic fumigant metam sodium, which is used as a substitute for methyl bromide in some systems. A systematic approach to research into biofumigation, specifically aimed at overcoming a long history of empiricism, has started to see significant advances in both basic and applied knowledge. A key development has been achievement of maximal biofumigation potential through greatly enhanced release of appropriate *iso*thiocyanates into soil. These advances have led to commercial adoption, demonstrating that biofumigation, when applied to appropriate production systems, can have efficacy and offer cost savings. Despite this success, biofumigation is not yet seen as being sufficiently powerful or practical in its implementation to be a realistic alternative to methyl bromide.

Enhanced microbial biodegradation is a cryptic phenomenon that can enhance or diminish the efficacy of soil-applied pesticides, including *iso*thiocyanates and most other currently available methyl bromide substitutes. This has been aided for some particularly vulnerable environments by recent clarification of key risk factors associated with soil type, soil pH, and calcium content. The hydrolysis of glucosinolates contained within the tissue of *Brassica juncea* releases volatile *iso*thiocyanates and these have been shown to inhibit mycelial growth of *G. graminis* and *R. solani* but not *B. sorokiniana* (Angus et al. 1994; Kirkegaard et al. 1996). The toxicity of aliphatic *iso*thiocyanates to fungal pathogens is correlated inversely with side chain length and of the pathogens tested by (Sarwar et al. 1998), *G. graminis* was the most sensitive and *R. solani* and *Fusarium* spp. were more tolerant. Fluorescent pseudomonads are also able to produce a related compound: hydrogen cyanide (Défago et al. 1990), which may be the active component of the volatile substances produced by some of the isolates in this study.

In Germany, where a high value is set on the maintenance of good soil conditions (H. Toxopeus 1972, personal communication), considerable use is made of green manuring crops such as mustard and radish. These have anti-fungal properties due to their mustard oil content which reduce the populations of soil-borne pathogens. *Brassica juncea* cv. Vitasso grown as an intercrop for strawberry cv. Elsanta provided biofumigation. This resulted in a significant reduction in the inoculum potential of *V. dahliae* microsclerotia (Steffek et al. 2006). Soil-borne organisms vary greatly in their response to brassicaceous soil amendments. For example, while the *Citrus* nematode was consistently suppressed the effects on *Fusarium* spp. and weed survival were variable. Zasada et al. (2003) contend that consistency of effect will only come from an analytical understanding of the active principles involved.

Combining cover crops with the incorporation of compost increased soil quality in irrigated, intensive lettuce and green broccoli (calabrese) production in the Salinas Valley, California, USA (Jackson et al. 2003). There were beneficial effects increasing soil microbial biomass, increasing total soil carbon and nitrogen, reducing surface bulk density and decreasing the potential for groundwater pollution as a result of nitrate leaching below the root zone. Yields were comparable to traditional husbandry and occasionally resulted in fewer weeds and lower lettuce corky root disease (*Sphingomonas suberifaciens*). Although surface minimum tillage reduced yields there was less nitrate leaching below the root zone. The use of conventional tillage, cover crops, and compost produced high vegetable yields and acceptable net economic returns over a 2-year period, but broccoli was more profitable than lettuce under this regime. Cover cropping has potential benefits especially where producers are in transition between conventional and organic systems (Baysal et al. 2008). Comparisons of husbandry-business systems included: tilled fallowing, mixed-species hay production, low-intensity vegetables and intensive vegetable production using high tunnels. Biologically most advantageous was mixed-species hay production where percentage carbon and total phosphorus in soil increased by two- and fivefold and potassium by between sixfold and 12-fold. Damping-off (*Pythium ultimum* and *R. solani*) in soybean and tomato was reduced by 2–30% in pot tests and in the field. The suppressiveness of soil was enhanced by cropping with mixed species hay systems.

Organic Additives

The overall effects of mulching with organic materials are beneficial (Quarles 2008), they conserve water and reduce the need for pesticide use. They also reduce soil compaction and erosion; buffer soil against temperature extremes; reduce plant diseases, insect pests and weeds; and encourage beneficial insects. Green manuring is a traditional method for improving soil fertility and eradicating weeds, pests and pathogens using uncomposted plant materials grown and incorporated in situ. For example, it formed an integral part of the 'Scarlett System' developed in the Scottish Lothians for land restoration which eradicated perennial weeds and enhanced soil fertility in combination with the use of the nitrogenous fertiliser calcium cyanamide which stimulates the beneficial microbial flora and is only leached-out slowly through soil (Scarlett 1937). Soil amended with organic additives such as green manures, composts and animal manures suppresses some soil-borne microbes (Millard and Taylor 1927; Sanford 1926) with other benefits including less surface crusting, increased water infiltration and increased activity by benign microbes. Organic matter additions are thought to suppress the cause of potato early dying syndrome (*V. dahliae* and *V. albo-atrum*) indirectly through increases in soil microbial activity or by release of toxic compounds (Rowe and Powelson 2002). More recently green manuring (Wiggins and Kinkel 2005), of buckwheat, canola or fallow linked with three crop sequences: alfalfa (lucerne)–potato, corn–potato and potato–potato has been evaluated. When compared with fallow treatments potato tubers grown in buckwheat treated soil had significantly lower *Verticillium* (*V. dahliae*) wilt ratings and tubers grown in buckwheat or canola treated soil had greater yields. Potatoes grown in soil planted to corn or alfalfa (lucerne) the previous year had significantly lower *Verticillium* wilt and potato scab (*S. scabies*) ratings as well as higher yields than potatoes grown in soil previously planted to that crop. Other *Streptomyces* spp. taken from soils collected from green-manure treated plots tended to have greater *in vitro* pathogen inhibitory activity than *Streptomyces* spp. from fallow-treated plots. *Streptomyces* pathogen inhibitory activity was frequently negatively correlated with plant disease and positively correlated with potato yield. These results offer opportunities for the management of soil-borne *Streptomyces* content with a view to reducing the impact of soil pathogens such as *Verticillium*. A potato-scab suppressive soil developed in Minnesota, USA, following long-term potato monoculture and had greater densities of *Streptomyces*, greater proportions of antagonistic *Streptomyces* spp. and *Streptomyces* isolated from these soils had greater mean inhibition zone sizes compared with nearby conducive soil (Lorang et al. 1989 and L. L. Kinkel unpublished data cited in Wiggins and Kinkel 2005).

Composts affect soil fertility and plant health beneficially or detrimentally depending on their quality. Their organic substance and nutrient content varies greatly the origin of materials was the major factor influencing these values (Fuchs et al. 2008). While the majority of composts protected cucumber plants against *Pythium ultimum*, only a few suppressed *R. solani* infecting basil. The incidence of damping-off caused by *R. solani* was significantly and consistently suppressed in soils amended with residues of clover, peanut or *Brassica rapa* spp. *rapifera* cv. Saori (Kasuya et al. 2006). Tests using antibacterial antibiotics indicated that a viable microbial community is needed. That contention is supported by Mazzola (2007) and their studies of apple replant disorder caused by *R. solani* AG-5. A composite of *Brassica napus* and *B. rapa* seed meal produced the highest disease suppression. Suppression was a result of the proliferation of *Streptomyces* spp. not the glucosinolate content of the meal.

A general or long-term effect of organic soil amendments is the enhancement of microbial activity including biocontrol agents which enhances natural disease suppressive conditions (Abbasi et al. 2009). Additionally short term effects may be derived from some amendments such as fish emulsion which these authors found capable of substantially reducing the viability of microsclerotia of *V. dahliae*, effects attributed to the presence of organic aliphatic acids such as glycolic, acetic, formic, *n*-butyric and propionic. Damping-off caused by *P. ultimum* damaging cucumber declined over time following the addition of fish emulsion as a result of the erosion of inoculum concentrations.

Research which investigated the relationship between propagule numbers and genetic diversity of *Trichoderma* species and Southern blight of tomato caused by soil-borne plant pathogen *Sclerotium rolfsii* in soils with long-term organic, sustainable, and conventional farms is reported by Liu et al. (2008). Dilution plating was used to quantify the propagule numbers of *Trichoderma*, denaturing gradient gel electrophoresis (DGGE) and DNA sequence analysis were used to identify *Trichoderma* species, and glasshouse assays were used to test for soil suppressiveness to Southern blight. The propagule numbers of *Trichoderma* tended to be higher in soils from conventional farms. There was no clear separation for the propagule numbers of *Trichoderma* based on different management systems using canonical correspondence analysis (CCA). There was, however, general separation for total microbial communities based on organic and conventional management systems using CCA. That suggests that the differences in soil suppressiveness to disease from organic, sustainable, and conventional farms is due to the effects of the total microbial diversity but not directly due to the *Trichoderma* populations in each farming system. The propagule numbers of soil *Trichoderma* did not significantly correlate with the disease suppressiveness, although individual species such as *Trichoderma harzianum* were shown to be related to disease suppressiveness. Moreover, several *Trichoderma* species were found in the soil tested based on DGGE and DNA sequence analysis. *Trichoderma hamatum*, *T. harzianum*, *Trichoderma virens*, and *Trichoderma erinaecem* were the most abundant species in tested soil.

Control of *Allium* white rot caused by *Sclerotium cepivorum* (Coventry et al. 2006), was developed through the incorporation of onion waste compost into soil. This reduced the viability of sclerotia of *S. cepivorum* and the incidence of white rot disease in the field and was as effective as using the fungicide tebuconazole. Two mechanisms of suppression were suggested: firstly, a reduction in the soil population of viable sclerotia which may result from the action of volatile sulphur compounds present in the composted onion waste, and secondly, prevention of infection of onion plants from sclerotia. *Allium* white rot attacks the root system resulting in either death before harvest or rotting of stored bulbs. The pathogen survives in soil as sclerotia which may remain dormant in the absence of a host for more than 20 years. Sclerotia are stimulated into germination by thiols and sulphides released by soil microorganisms from alk(en)yl cysteine sulphoxides secreted by root of *Allium* spp.

Greenwaste compost has proven broadacre activity against *G. graminis* var. *tritici* (Tilston et al. 2005) and based on experiments performed under protected conditions, activity against many soil-borne pathogens of cultivated crops. Pathogens suppressed by greenwaste compost include *Sclerotium cepivorum*, *Fusarium oxysporum* f. sp. *dianthi* and *Phoma medicaginis* var. *pinodella* (Pitt et al. 1998; Tilston et al. 2002). Composts provide a source of nutrients and when applied in conjunction with biological control agents can increase the efficacy of introduced microorganism, especially this is done in a non-competitive environment (Deacon and Berry 1993). Other organic soil amendments also stimulate the activity of soil microbes and suppress the survival and activity of *Phytophthora* spp. Natural products such as clove oil, neem oil, pepper extract with mustard oil, cassia extract and synthetic cinnamon oil were tested for the control of *Phytophthora nicotianae* affecting periwinkle (*Catharanthus roseus*). With the exception of neem oil all these substances appeared to reduce disease expression (Bowers and Locke 2004). Claims have been made that fish oil emulsion added to soil will reduce soil-borne pathogen damage.

Among the most potent organic extracts are those from the various seaweeds which inhabit marine continental shelves around the world. In Europe the prime source is from *Ascophyllum nodosum* harvested from Scandinavian and British waters. Such extracts may alter the mode of activity of microorganisms as to whether they operate as parasites, mutualists or as saprophytes and whether they enter states of imposed passivity or inherent dormancy. This may lead to these organic extracts, when placed in the soil environment either altering directly or indirectly, characteristics such as: root colonisation and penetration, competition and microbiostasis and antibiosis. The growth *in vitro* of *P. putida* and the abilities to produce siderophores in the presence and absence of organic extracts are illustrated in Figs. 2 and 3 (Dixon and Walsh 1998).

The production of siderophores, which are low molecular weight iron binding compounds, has been associated with the fungal pathogen suppressive properties of several fluorescent *Pseudomonas* spp. (Leong and Winkelmann 1998). It is suggested that the iron sequestering abilities of siderophores may deprive a pathogen of available iron (Kloepper and Schroth 1981). Hence, stimulation of siderophore production by the presence of organic extracts correlates with increased soil suppressive properties. Increasing biological activity by benign organisms in the soil may be correlated with rising production of extra-cellular enzymes. These degrade organic materials within the soil providing mutualists such as *P. putida* with energy supplies which in turn aids their capacity to inhibit the activities of pathogenic organisms. Samples of compost were treated with organic extracts derived from seaweed and this increased activities of the enzymes, amylase, cellulase (β-1–4-glucanase) and glucanase (β-1–3-glucanase) detected within 48 h of incubation at 25°C (Fig. 4).

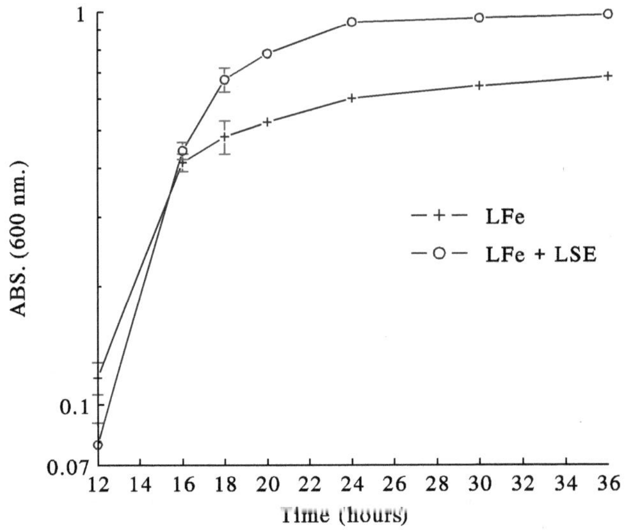

Fig. 2 The effect of organic extracts on the *in vitro* growth of *Pseudomanas putida* Key: LFe = low iron content medium; LSE = liquid seaweed extract (Dixon and Walsh 1998)

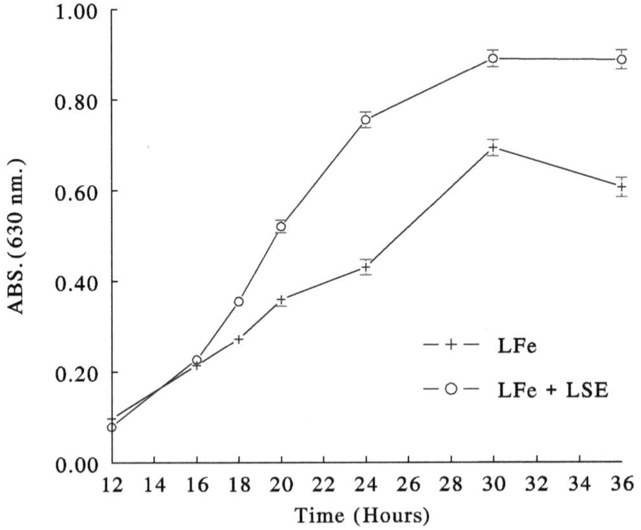

Fig. 3 The effects of organic extract on the *in vitro* production of siderophores by*Pseudomonas putida* Key: LFe = low iron content medium; LSE = liquid seaweed extract (Dixon and Walsh 1998)

In an extended study Abbasi et al. (2006) showed that reductions in potato scab (*S. scabies*) and *V. dahliae* wilt of potatoes using organic extracts were unaffected by soil type but varied in efficacy from season to season and was ineffective at high inoculum pressures. Agricultural wastes such as rice straw, rice hull, groundnut

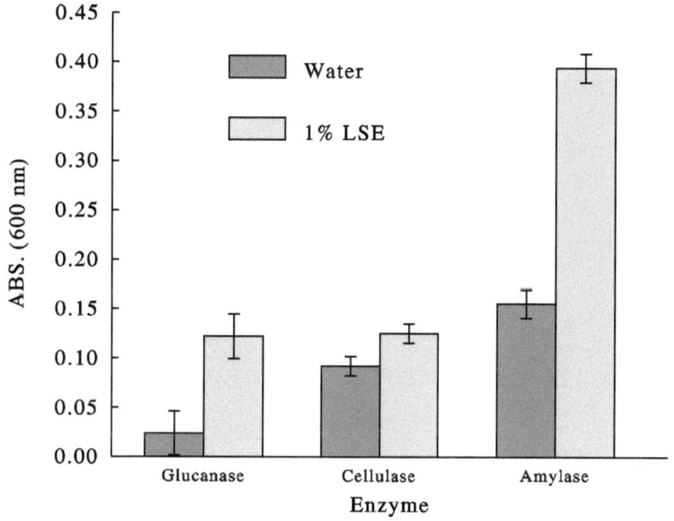

Fig. 4 Extra-cellular enzymes produced by *Pseudomonas putida* in compost in response to the addition of organic extracts Key: LSE = liquid seaweed extract (Dixon and Walsh 1998)

husk, maize cob, bagasse, rape seed pomace, castor seed pomace, tree bark, mushroom growth medium waste and shrimp shell powder are reviewed by Jennwen (2005) for their use in land reclamation and in the production of horticultural crops. Their effects include improving soil fertility, increasing soil organic matter and in some instances reducing the incidence of plant diseases. The latter is attributed to a combination of several effects including direct impact on pathogenic microbes and improving the vigour of host plants. Amending soils with composts increased beneficial microbial (actinomycetes, bacteria and fungi) activity especially where chitin (crab shell) was added (Escuadra and Amemiya 2008) and raised soil suppressiveness to pathogenic organisms. Alkaline stablised biosolids produced from dried sludge, cement kiln dust and unspecified fillers are claimed to reduce the impact of several soil-borne pathogens by changing pH to alkaline values, adding nutrients and improving structure (Ben-Yephet et al. 2005). Frequently this requires the use of several factors as with tan spot of cereals caused by *Pyrenophora tritici-repentis* which was encouraged by nil-tillage, reduced nitrogen fertilisation, continuous cropping with wheat or rotation with maize. Integrated use of nitrogen up to 100 kg ha^{-1} N, conventional tillage and the use of fungicides resulted in the highest yields at El Batan, Mexico (Duveiller et al. 2002).

Crop husbandry practices may lead to reduced or improved soil structure and concomitantly increased or reduced soil-borne plant diseases (Bailey and Lazarovits 2003). Agricultural practices such as incorporating organic amendments and managing the type and quantity of crop residue directly affect plant health and crop productivity. Soil management practices such as tillage, rotation, and stubble (residue) burning affect the quantities and qualities of organic matter returning into the soil. This influences pathogen viability and distribution, nutrient availability, and the release of biologically active substances from both crop residues and soil microorganisms as illustrated by a model system of *B. sorokinia* on the development of common root rot in cereals (Bailey and Lazarovits 2003). The application of organic amendments, manures and composts rich in nitrogen, may reduce soil-borne diseases as a result of releasing allelochemicals by microbial decomposition. The form of nitrogen released can be crucially important when seeking to achieve this effect. Developing disease suppressive soils by introducing organic amendments and crop residue management takes several seasons but benefits accumulate as soil health and structure improve.

Characteristics of Soil Microbial Communities

Inoculum Potential

Balanced in the equation of soil factors and husbandry techniques which support sustainable crop growth are the general traits of microbes which give them advantages as pathogens. Pathogens can only dominate the soil environment and cause disease when they are present in sufficiently large quantities that they are able to breakdown the genetic barriers possessed by crop hosts and overcome edaphic biological, chemical and physical soil environments which are adverse to them. Raising the microbial diversity of soils is associated for example, with increased suppressiveness towards soil-borne pathogens. Grassland soils for example, possess higher microbial diversity compared with arable ones and are able to diminish the effects of *R. solani* (van Elsas et al. 2002). A cardinal characterisitic for successful pathogenesis by soil-borne organisms is the inoculum potential of a pathogen. This concept embraces for the pathogen all of those factors which cumulatively lead to successful invasion, colonisation and pathogenesis. Where inoculum potential is high one or possibly a few disease-causing microbes form a substantial proportion of the total soil microflora population and incite ill-health in the host. Despite this concept being conceived by Garrett in 1956 (Garrett 1956) it remains one of the corner stones of our understanding of host and pathogen biology in the soil. The concept of inoculum potential applies to both pathogens and to the benign microbes which antagonise them and hence may have value in evaluating biocontrol agents.

The biodiversity of benign soil microbes which are antagonisitic to pathogenic forms is critical in order to maintain soil health and quality (Garbeva et al. 2004). Diversity is affected by crop type, soil type and their management as discussed earlier in this chapter. Characterising the diversity of benign microbes is frequently defeated by sheer scale of soil inhabitants. In the top layers of soil even with only moderately fertility there are 10^9 bacterial cells per gram most of which resist culturing. Less than 5% of soil microbial biomass has been cultured and taxonomically characterised (Torsvik and Ovreas 2002). The relationship between the size of microbial biomass and the capacity of a soil to suppress pathogens was demonstrated by van Os and van Ginkel (2001). A *Pythium* sp. was suppressed by high microbial biomass which reduced the impact on bulbous *Iris*. Mechanisms suggested which underpined suppression included nutrient competition, commensalism, microbial antagonism, parasitism and systemic induced resistance. Host effects which lead to suppressivity have been ascribed to continuous monocropping leading to: disease decline, cover cropping, compost application, tillage and rotation. Adding organic amendments such as manure, compost and cover crops especially when combined with biocontrol agents such as *Trichoderma* or *Gliocladium*, limit pathogenesis as seen with reductions of sheath blight in rice (Sreenivasaprajad and Johnson 2001). But the position is far from clear as for example, where minimum or nil-tillage are associated with reduced pathogen impact in some instances while in others crop damage increases. This lack of clarity is hardly surprising since there is little evidence with which to understand the difference between pathogenic and antagonisitic microbes and comprehend the factors that engender either pathogenicity towards crop plants or antagonism towards the pathogens. Stukenbrock and McDonald (2008) suggested that pathogens emerge as a result of cultivation where a homogeneous host population and a conducive environment coincide. If this is accepted then it may be reasonable to assert that the contrary state of suppressiveness also results from the appropriate combination of microbial populations and soil environmental factors. Understanding the development of suppressive soils may be obtained from firstly following the development of suppressiveness over time, secondly examining naturally occurring suppressive soils and thirdly using soil treatments which encourage suppressiveness (Borneman and Becker 2007).

The horizontal and vertical distribution of *Aphanomyces* spp. causing root rot disease of field peas is reported by Moussart et al. (2009) and provides clues in understanding the distribution patterns of pathogen propagules and the onset of disease. Measurements of inoculum potential indicated a horizontal distribution of propagules among several foci in the field. These foci differed in size and disease intensity. A significant relationship was established between disease severity on maturing pea plants and total soil inoculum potential. In the vertical distribution of inoculum, maximum values were found at 10–40 cm but the pathogen could be detected to a depth of 60 cm. Inoculum potential was least at depths of 0–10 and 50–60 cm. It appears in this case that the top soil horizon and the lowest one possessed features which vitiated against the pathogen. Sandwiched between these horizons the pathogen had an advantage. Similarly the development of arbuscular mycorrhizal fungi and their inoculum potential in soil varied with different hosts for colonisation such as on prairie little bluestem grass and wild rye. Total soil nitrogen correlated with host root mass and negatively with the root colonisation (Anderson 2008). Stimulating vigorous host root growth apparently discouraged the pathogen. In another example intercropping apparently aided the inoculum potential of beneficial microbes such arbuscular mycorrhizas in tropical agroforestry with trees such as *Albizzia procera* (safed siris), *Ambilica officinalis* (aonla) and *Eucalyptus teritciornis* using black gram and wheat or green gram and mustard (Kumar et al. 2007). Knowledge of inoculum concentration is equally important as that for the environment. Thus studies of soil inoculum densities and *Verticillium* wilt (*V. dahliae*) in *Acer platanoides* and *Catalpha bignonioides* indicated that 5% diseased plants occurred at 1–2 microsclerotia per gram of soil (Gould 2003). Bio-assay techniques used to establish concentrations of propagules of pathogens in soil have tended to remain empirical and lack analytical precision. Recently, however, improvements have been achieved using enzyme-based and PCR-based systems as reviewed by Faggian and Strelkov (2009) for *P. brassicae*, the cause of clubroot in brassicas. Methods for the bioassay of soil suppressiveness to *Fusarium* wilts, evaluating inoculum potential and the impact of soil fungistasis are reviewed by Alabouvette et al. (2005).

Disease management may be either largely empirically driven by experience and practice or it may designed around an epidemiological framework in which the mechanisms responsible for an epidemic are identified and the knowledge gained is used to target, improve and deploy methods which reduce yield losses (Gilligan 2002). This requires an understanding of the processes controlling the invasion and persistence of inoculum and subsequent disease development and expansion. The spatial structure of susceptible crops affects the spread of diseases across a landscape. When susceptible crops are rotated the connectivity between contiguous fields of susceptible crops changes and transmission between neighbouring fields alters. Spread by soil-borne fungi such as *R. solani* is classed by Bailey et al. (2000) as invasive or non-invasive (finite) when moving from one site to another, as studied using the colonisation of poppy seeds from agar plugs as a model system. Invasive spread did not depend on the furthest extent of mycelia growth evident in the tails of colonisation, but was instead associated with threshold distances which were different for low and high nutrient sites. As the probability of colonisation increases above a threshold value so the likelihood of invasive spread increased. Conversely, below that threshold finite growth was more common. In nature the colony size, nutritional state and availability of organic matter are likely to be very variable. Small differences in soil physical properties, in particular the continuity of air-filled pore space will have a profound effect on the growth and expansion of a single propagule or from a colony concentrated round a nutrient source. The colonisation profile is affected by soil physical properties and this character is analogous to the pathozone profile for the growth or movement of fungal parasites during primary and secondary infection. In primary infection the fungus grows saprophytically from a source of previously colonised host material towards a susceptible host. The analogies between parasitic and saprotrophic invasion can be extended by considering saprotrophy by soil microbes as a special case in the general epidemic processes in which colonisation replaces infection. From this there can be a general interpretation of the dynamics and spread of pathogens and saprophytes, parasites spread horizontally through a host population (such as roots) and saprophytes move through a discrete series of nutrient sites. Unlike saprophytes the parasitic fungus could experience a switch from finite to invasive growth. At low root densities the pathogen is constrained by the availability of host tissue as this changes there is a switch to rapid invasive spread. This could explain phenomena such as sudden increases in the rate of infection during the spread of take-all (*G. graminis* var. *tritici*) on wheat as the epidemic switches from primary to secondary infection in response to changes in the density of roots (Bailey and Gilligan 1999).

Entry and Colonisation

Root colonisation is an important first stage in infection by pathogens. Early research by Hiltner in 1904 (Hiltner 1904) suggested the "rhizosphere effect" whereby microbes are attracted to nutrient supplies exuded by roots. This has been reviewed by Bais et al. (2006) who contend that in addition to providing a carbon-rich environment the roots initiate cross-talk signalling with soil microbes which in turn produce messages initiating colonisation. Motility is an important trait of both pathogens and beneficial microbes enabling participation in this cross-talk (De Weert et al. 2002; Lugtenberg et al. 2001). Chemical attraction (chemotaxis) is involved in cross-talk between microbes and root (Bais et al. 2004a). Electrical potentials resulting from electrogenic ion transport at the root surface, attracts swimming zoospores of oomycete organisms to the root surface (van West et al. 2002). Some rhizobacteria create suppressive soils which inhibit the activities of plant pathogenic microbes. Mechanisms for this form of biocontrol include competition for nutrients as supplied in root exudates, niche exclusion, induced host systemic resistance (ISR) and the production of antifungal metabolites. The best characterised biocontrol agents at the molecular level are species of *Pseudomonas* which form antifungal metabolites such as phenazines, pyrrolnitrin, 2,4-diacetylphloroglucinol (DAPG), pyoluteorin and several cyclic lipopeptides (as discussed elsewhere in this book). Stimulation of acquired resistance to cucumber mosaic virus (CMV) by a strain of *Bacillus pumilis* in *Arabidopsis thaliana* resulted from adapting a signalling pathway for virus protection (Ryu et al. 2004). *Bacillus subtilis* strains, on the other hand, assist plants to avoid the Gram-negative pathogen *Pseudomonas syringae* pv. *tomato* by forming a protective biofilm on the roots of *A. thaliana* and producing an antimicrobial cyclic lipopeptide, surfactin (Bais et al. 2004b). Root exudates substantially increase microbial activity in the rhizosphere (Ohi et al. 2003). Their role in pathogenesis caused by root invading bacteria and fungi has yet to be fully appreciated. This deficiency is the result of inadequate technology in the methods of analysis. It is however, recognised that there is constant "underground warfare" alleviated for host plants by production of phytoalexins and defense proteins. Sequencing the genomes of crop plants reveals that they are rich sources of antimicrobial indole, terpenoid, benzioxazinone and flavanoid / *iso*flavanoid natural products. This chemical diversity could yield direct forms of sustainable control of pathogenic microbes and/or indicate indirect means via the stimulation of their secretion (Bais et al. 2006).

The rhizosheath is the soil adhering directly to the roots and the rhizosphere is the zone of soil directly under the influence of the roots, the pathozone is the region of soil surrounding a root, seed or hypocotyls within which fungal propagules must exist if they are to have any chance of infecting the host (Gilligan 1990). The ability of a fungus to invade was not determined by the furthest extent of fungal growth but instead by the distribution and density of fungal biomass as captured by the pathozone (Otten and Gilligan 2006). The invasiveness of fungal spread can be predicted from fungal colony morphology together with the geometry of the root or host population, and small changes in fungal morphology or the distribution of susceptible hosts (both of which can be mediated by soil physical conditions) and can make a fungus switch between invasive and non-invasive spread (Otten and Gilligan 2006). The evolution of infection efficiency (Gilligan 1985; Gilligan and Bailey 1997) is summarised by a non-linear model for the changing probability of infection with distance of inoculum of, for example, *R. solani* from a host in which certain parameters vary with time. Thus the probability of infection *P*(*r*, *t*), depends on distance (*r*) and time (*t*):

$$P(r, t) = f(r|\theta(t))$$

where θ (t) are the time varying parameters.

The rhizosphere is the region of soil influenced most by the presence of roots (Lynch 1990). The pathozone forms the central environment where propagules of plant pathogens must lie if they are to have finite chances of infecting host roots (Gilligan 1985). Both pathozone and rhizosphere are closely related. Dimensions of the pathozone are dictated principally by the properties of the root and surrounding soil. The major components responsible for the size and shape of the pathozone also depend on the properties of the inoculum (Gilligan and Bailey 1997). The pathozone is characterised not only by the furthest extent from which infection can occur but also by a decrease in the probability of infection as the distance between inoculum and host increases (Gilligan 1985). The probability of infection equates with infection efficiency relating the spatial distribution of propagules and roots. The pathozone changes with time as inoculum germinates and grows or moves towards the host, within this the susceptibility of the host may also change. These dynamics affect the rate of spread of an epidemic from an initial inoculum source and could also affect, and be affected by, the presence of propagules of biocontrol organisms such as *T. viride*. Bailey and Gilligan (1997) showed that this biological control organism reduced the extent of pathozone influence and the infection efficiency of *R. solani* infecting radish seedlings. From this relationship the progress of disease could be predicted in a population of hosts. There are two interacting processes involving the density of mycelium that reaches the host and the net receptiveness of the host surface which increases during the period of inoculum challenge (Gilligan and Bailey 1997). High mycelial density may be affected by mechanical disturbance arising from host germination and root growth.

Invasion and persistence by pathogen in host populations are defined by Gilligan and van den Bosch (2008) in straightforward terms. Invasion they identify as "the introduction and subsequent increase of the pathogen in a host population" while persistence requires the host population to contain sufficient susceptible individuals such that pathogenesis continues for a prolonged period. Modifying these characteristics by forms of control limits the scale and extent of pathogen epidemics. The form of an epidemic results from the integration of the factors: inoculum, host, environment and time, environment includes aerial and edaphic meteorological factors and those of the soil itself such as pH, texture, bulk density and chemistry. The choice then falls to

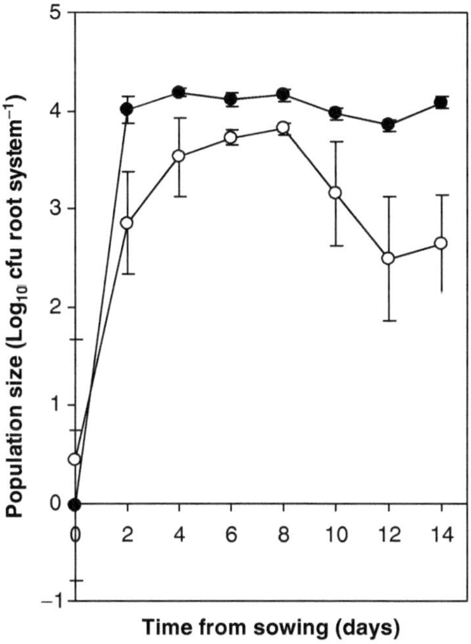

Fig. 5 Fluorescent pseudomonad colonization of winter wheat (cv. Hereward) seedlings grown in mature greenwaste compost artificially infected (*closed symbols*) or uninfected (*open symbols*) with *Gaeumannomyces graminis* var. *tritici*. (After E.L. Tilston, 2000)

which forms of control will yield the most sustainable result in terms of the environment and crop production. The transition from inoculum to epidemic is driven by fungal (pathogen) growth, the infection dynamics of a single host and invasive spread of the disease-causing microbe through a population of hosts (Otten and Gilligan 2006). Like soil, the rhizosphere and more specifically the rhizoplane, has microsites where the microbial residents are concentrated. These microsites are often a function of the architectural complexity of roots and provide an array of potential niches for antagonists (Burdon 1987). The number of microsites may limit the size of rhizosphere populations and account for the colonisation plateau observed in Fig. 5.

Pathogenesis

The invasion and disruption of root tissues following infection of a root system by a pathogen has a profound effect on the efficiency of the supply of water, ions and growth regulatory substances and on the rate of root growth (Ayres 1984). The position and depth of lesions on wheat roots dictate the emergent physiological effects (Fitt and Hornby 1978). More severe take-all (*G. graminis* var. *tritici*) for example, may develop following infection of newly formed, undifferentiated tissues than after infection of differentiated tissues where subsequent vascular colonization can by limited by the presence of an endodermis (Deacon and Henry 1980), particularly if the lesions are close to the shoot (Ayres 1984). Penetration to the stele effectively excises the distal portions of roots (Manners and Myers 1981), though this may not necessarily translate into reduced growth rates (Andrews and Newman 1968) as there is a state of functional equilibrium in the partition of resources between the roots and the shoot (Brouwer 1977). Low levels of infection can stimulate production of nodal roots (and account for increases in root-to-shoot ratios), with the result of at least off-setting and possibly over-compensating for those lost due to penetrating infections (Asher 1972), though the reverse may also be true (Hornby et al. 1998). The regenerative capacity depends on the growth stage (Seidel et al. 1981) and of agricultural cereals, barley has greater regenerative capacity than wheat and this is the reason why wheat is affected more severely by take-all (*G. graminis* var. *tritici*) than barley (Manners and Myers 1981). Cereal plants are most vulnerable to infection and it is most severe in the early season, especially where there are high levels of soluble nitrogen available in the soil and a high population density (Papendick and Cook 1974). Initially crop growth is vigorous so there is increased rate of water uptake and its rapid depletion consequently water-stress takes effect earlier, predisposing the plants to attack by *G. graminis* var. *tritici*. The low water content restricts the activities of other microflora, especially bacteria so levels of antagonism by *Pseudomonas* spp., for example, are often low. In addition, the area of the root–soil interface will decrease as the dehydrating roots shrinks away from the soil particles as its volume decreases (Weatherly 1979). Disease incidences arising from primary infection (soil-borne inoculum, auto-infection) and secondary infection (root-to-root passage, allo-infection) are distinguished by Suffert (2007) in modelling *Pythium violae* (a cause of cavity spot in carrot) resulting in the confirmation that this disease syndrome is polycyclic.

Necrotic host tissues, such as diseased roots, are known to support a diverse array of saprophytic organisms and fluorescent pseudomonads often colonize and dominate in the community arising in take-all (*G. graminis* var. *tritici*) lesions (Wilkinson et al. 1982; Barnett et al. 1999). Within colonized *G. graminis* lesions, it is suggested that antagonists prevent pathogen outgrowth from the lesion and limit subsequent autoinfection (Baker and Paulitz 1996) and infection by other pathogens resulting in further tissue destruction (Leath et al. 1989). The bacteria present within the rhizosphere are less fastidious with their requirements for pre-formed organic compounds, vitamins and growth factors than those present in non-rhizosphere soil (Brown 1975). The quantitative and qualitative composition of sugars, organic acid, amino acids, glycosides, nucleotides and their bases, enzymes, vitamins, indole-derivatives, phenolic substances and other substances within root exudates changes during plant growth and root development, and in response to foliar applications of chemicals (Macura 1968) and environmental and climatic factors (Brown 1975). With increasing root age cell debris augments the diversity of nutrients within the rhizosphere. The amount of organic material released by roots can be considerable with 1–2% of the carbon translocated to the roots being released into the soil as water-soluble exudate (0.2–0.4%) and 0.8–1.6% as insoluble mucilaginous material (Brown 1975).

If the substances actively secreted by roots, the lysates of root material and gases released from the roots, referred to collectively as rhizodeposits (Whipps 1990) are included in such calculations then 12–40% of photosynthate enters the soil in this manner (Whipps and Lynch 1985). As well as stimulating antagonistic organisms and other metabolic types of bacteria such as carbohydrate fermenters, cellulose decomposers, ammonifiers and denitrifiers; pathogens are attracted towards and stimulated by root exudates (Burdon and Shattock 1980).

The rhizosphere therefore, presents a suitable microhabitat in which events relating to disease suppression could occur (Troxler et al. 1997). In suppressive soils, the zone of suppression around roots extends for up to 1.5 mm from the rhizoplane (Wildermuth et al. 1985), but the rhizosphere can influence soil up to 2 cm from the rhizoplane (Brown 1975). For rhizosphere organisms to effect disease suppression, they must be in place before arrival of the pathogen within the infection court (Campbell 1989a; Harris et al. 1997). Like roots, germinating seeds exude a rich supply of nutrients, which also support a diverse microflora (Brown 1975). Microorganisms of the spermatosphere originate from the testa, and the wider soil, and have the potential to participate in root colonization. A few bacteria originating from the testa migrate to and persist within the rhizosphere where colonists are drawn from often dormant propagules in the soil (Macura 1968). The difference between spermatosphere and rhizosphere microbial communities probably being determined by quantitative and qualitative differences in the exudates from each organ and the poor persistence and unimpressive *in vivo* antagonistic ability probably means that the contribution to suppression by the spermatosphere community is minor.

In field-grown wheat the seminal roots harbour small, sparse populations of bacteria located mainly at the junctions of cortical cells. Whereas in old nodal roots large aggregates of bacteria colonize the troughs created by the extensive longitudinal corrugations and intercellular spaces in the cortex (Darbyshire and Greaves 1971) and may even be found in the xylem (Troxler et al. 1997). The colonization of furrows on roots has been attributed to increased exudation at cell junctions but these sites may also offer protection against desiccation and dislocation due to abrasion (Mosse 1975). Generally nodal roots exert a greater rhizosphere effect on antagonistic microorganisms and support larger populations than seminal roots (Sivasithamparam and Parker 1979b). The difference between nodal and seminal roots may have a profound influence on the prospects of biological control of take-all (*G. graminis* var. *tritici*) in Britain; where prolonged protection of the nodal root system of winter wheat during the spring and early summer may be of greater importance than protection of seminal roots (Hornby et al. 1998).

To suppress, displace or pre-empt a pathogen in the invasion of an host the antagonist must be a specialized and aggressive colonist of host tissue. Furthermore an antagonist must be able to use most of the nutrients needed by the pathogen and preferably the antagonist should have some means of inhibiting the growth of the pathogen. Pathogens and their antagonists inhabit ecological niches with a high degree of equivalence and the only way in which they differ is that antagonists lack the ability to initiate the flow and transport of food out from the host. Biocontrol agents can inhibit the growth of pathogens by the following modes of action: pre-emptive competitive exclusion, antagonism by antibiosis, interference, parasitism or predation, induction of host resistance and transmission of hypovirulence genes (see elsewhere in this book).

Characterising Sustainability Through Suppressive Soils

Much has already been said in this chapter about the occurrence of suppressive soils. The key objective in developing microbially sustainable soils is the enhancement of pathogen suppressivity. A healthy soil is defined by van Bruggen and Termorshuizen (2003) as a stable soil system with high levels of biological diversity and activity, internal nutrient cycling and resilience to disturbance. This implies these authors suggest that microbial fluctuations after a disturbance would dampen more quickly in a healthy than in a chronically damaged and biologically impoverished soil. Soil may be disturbed by, for example, adding a nutrient source, tillage or drying and rewetting. Disturbance results in the numbers of heterotrophic bacteria and of individual species starting to oscillate both in time and space. Oscillations appear as moving waves along the path of an active nutrient source such as a root tip. The phase and period for different trophic groups and species of bacteria may be shifted indicating that oscillation occurs. Analyses of the subsequent populations in oscillations confirm that there is a cyclic succession in microbial communities. Microbial diversity oscillates in the opposite direction to microbial population size. In healthy soil the amplitude of these oscillations will be small but the background levels of microbial diversity and activity are high so that soil-borne diseases will face more competitors and antagonists. Soil-borne pathogens and antagonists alike will fluctuate in time and space as a result of growing plant roots and other disturbances and the periods and phases of the oscillations may vary. As a result biological control by members of a single trophic group or species may never be complete as pathogens will encounter varying populations of the antagonistic agent at the root surface. A mixture of different trophic groups occurs at subsequent locations along a root. Alternatively regular addition of organic matter to the soil may increase background levels of microbial activity, increase nutrient cycling, lower the concentrations of easily available nutrient sources, increase microbial diversity and enhance natural disease suppression.

The antithesis of avoiding land conducive to disease is the intentional use of suppressive soils. In suppressive soils disease incidence remains low, despite opportunities in which the pathogen could have established an inoculum potential of economic importance, under favourable climatic conditions and in the presence of a susceptible host. Suppressive soils have been identified for *T. yallundae* (Baker 1970), *R. solani* of cereals (Wiseman et al. 1996), *G. graminis* var. *tritici* (Cook and Rovira 1976), *P. brassicae* (Worku and Gerhardson 1996) and *Fusarium* wilt of melon (Alabouvette 1986). Not all suppressive soils are pathogen specific; soil from a wheat field at Roseworthy Agricultural College in southern Australia was suppressive to cereal pathogens other than *G. graminis* var. *tritici* and included *R. solani*, *B. sorokiniana* and to a lesser degree *F. culmorum* (Wildermuth 1982).

The presence of an active community of antagonistic microflora and microfauna is usually proposed as the *raison d'être* for suppressive soils, and microbial isolates taken from these soils are often used for further research, e.g. *Pseudomonas fluorescens* strain CHAθ (Steffek et al. 2006).

Disease suppression is inhibition of the pathogen when it is growing parasitically in or on the host (Alabouvette et al., 1996) while pathogen suppression is inhibition of the pathogen in its saprotrophic growth and, or survival phase(s) in the soil (Cook and Baker 1983; Alabouvette et al., 1996). *Gaeumannomyces graminis* var. *tritici* has four types of saprotrophic existence within soil viz.: (Pretty 1990) survival on crop residues, (Houghton 1996) growth in soil, (Britton 1990) colonization of organic residues and ectotrophic growth on the host root preceding infection (Simon and Sivasithamparam 1989). It is difficult to measure true pathogen suppression, particularly for *G. graminis* var. *tritici* where spores probably do not perform an active role in the field-based infection process and because of the absence of a satisfactory selective isolation medium for use with soil; although immunological or molecular methods may eventually overcome this.

Pathogen suppression may be specific or general. Specific suppression is that which develops only in the presence of a virulent pathogen (Gerlagh 1968), or is due to the activities of specific groups or populations of antagonists. The definition of specific pathogen suppression may be refined further by the inclusion of a clause relating to transferable suppression. Transferable suppression is that which can be moved to a conducive medium within a small amount (1% w/w) of test soil. This implies that the active factor is biological in nature and the amount of soil introduced is too small to change the nutritional status of the conducive medium (Shipton et al. 1973) and can provide a confirmatory test for a suppressive soil. Greater amounts of soil have been used to demonstrate transferable suppression, but there is the risk of obtaining an erroneous result if the nutritional status of the conducive medium is less than that of test soil (Simon and Sivasithamparam 1989). A general form of pathogen suppression arises from the activity of unspecified (generalised) soil microbiota which can be manipulated by practices such as crop rotation and cultivation (Gerlagh 1968). General suppression is not dependent upon characterized antagonistic abilities but operates by virtue of wider activities such as spatial exclusion. Wider microbial communities have effects that contribute to the phenomenon of general suppression and the two mechanisms (specific and general) can be complementary rather than opposed, e.g. the Fusarium wilt-suppressive soils of France which have been described as showing both general and specific suppression (Alabouvette 1986). Pathogen suppression and disease suppression, however, are not complementary within ecological niches. Simon and Sivasithamparam (1989) define pathogen suppression as that occurring outside the rhizosphere, while disease suppression occurs within the rhizosphere. For microbially-based disease suppression it is important that the antagonistic organism is present within the rhizosphere before arrival of the pathogen. *Gaeumannomyces graminis* var. *tritici* has runner (ectotrophic) hyphae that colonize the root surface and later penetrate the root growing within the intracellular spaces of the cortex and stele interface, where they could be protected from the antagonistic activities of other microorganisms.

Naturally occurring disease-suppressive soils have been documented in a variety of cropping systems, and in many instances the biological attributes contributing to suppressiveness have been identified. While these studies have often yielded an understanding of operative mechanisms leading to the suppressive state, significant difficulty has been encountered in the transfer of this knowledge into achieving effective field-level disease control. Early efforts focused on the inundative application of individual or mixtures of microbial strains recovered from these systems and known to function in specific soil suppressiveness. The introduction of biological agents into non-native soil ecosystems however, typically yielded inconsistent levels of disease control (Mazzola 2007). Of late, greater emphasis has been placed on manipulation of the cropping system to manage resident beneficial rhizosphere microorganisms as a means for suppressing soil-borne plant pathogens. One such strategy is the cropping of specific plant species or genotypes or the application of soil amendments with the goal of selectively enhancing disease-suppressive rhizobacteria communities. This approach has been utilized in a system attempting to employ biological elements resident to orchard ecosystems as a means to control the biologically complex phenomenon termed apple replant disease. Control of the fungal pathogen *R. solani* AG-5 was achieved by cropping with wheat in apple orchard soils prior to re-planting the site with apple cultivars. Disease control was elicited in a wheat cultivar-specific manner and functioned through transformation of the fluorescent pseudomonad population colonizing the rhizosphere of apple. Wheat cultivars that induced disease suppression enhanced populations of specific fluorescent pseudomonad genotypes with antagonistic activity toward *R. solani* AG-5, but cultivars that did not elicit a disease-suppressive soil did not modify the antagonistic capacity of this bacterial community.

It has been suggested that fluorescent pseudomonads are unlikely to make significant contributions to the suppression of soil-borne pathogens (Sivasithamparam et al. 1979a; Tuitert et al. 1998). Both groups of authors justify this view with evidence of low population densities of fluorescent pseudomonads within soil and compost. Sivasithamparam et al. (Sivasithamparam et al. 1979a) do acknowledge however, that at certain microsites there may be intense antagonistic activity by fluorescent pseudomonads. These microsites are highly dispersed in soil and it has been suggested that the inhabiting colonies of soil bacteria are often small, comprising of fewer than ten cells (Gray et al. 1968). Antibiotics are organic compounds of low molecular weight, produced by microbes and at low concentrations have deleterious effects on growth and, or other metabolic activities of other microorganisms (Fravel 1988) and may be the cause of the inhibition zones. Under *in vitro* conditions fluorescent pseudomonads have been shown to produce a range of phenolic polypeptide antibiotics, e.g. pyoluteorin and pyrrolnitrin, phloroglucinols, e.g. 2,4-diacetylphloroglucinol and monoacetylphloroglucinol (Howell and Stipanovic 1980), and phenazines, e.g. phenazine-1-carboxylic acid and 2-hydroxyphenazine-1-carboxylic acid (Thomashow and Weller 1988). The ability to synthesize antibiotics does not ensure efficacy as a biocontrol agent. Non-antibiotic producing mutants may exert suppression equal to that of their wild type (Fravel 1988); suggesting that general environmental competence and, or other activities may be limiting or augmentative (Campbell 1989b; Ryder and Rovira 1993). Exposure to ecological stresses is probably necessary for the conservation of antibiotic production, as isolates maintained on culture media of high nutrient value often lose the ability to produce antibiotics over time (Bruehl et al. 1969). The production and activity of antibiotics in soil has been subject to debate for many years, with evidence that antibiotics are frequently bound to clay and organic matter, which reduces their effective concentrations. But in turn this may serve to increase the concentration within microsites (Fravel 1988). With newer techniques evidence is emerging that antibiotic molecules such as, phloroglucinol are produced *in vivo* by rhizosphere colonizing fluorescent pseudomonads and that they do have activity against pathogens such as *G. graminis* in the field (Raaijmakers et al. 1999).

The lack of continuation of the suppression is due to poor longer-term persistence of antagonistic microbes and the failure for the initial protection to establish a permanent presence under less suitable physico-chemical conditions (Capper and Higgins 1993). Limiting physico-chemical conditions could arise in the summer when soil water potential declines and restricts the growth of antagonists; *G. graminis* var. *tritici* is able to grow over a wide range of water potentials and would therefore dominate (Campbell and Clor 1985). The problems of known biological control agents namely longevity of suppression and of applying sufficient inoculum are common for all externally applied agents.

Forms of crop husbandry such as the addition of soil amendments have a clear capacity to enhance disease suppression. This may as Mazzola (2004) notes differ from the manner by which suppressivity operates in natural soils. Crop plants themselves will affect the forms of suppressivity by altering the characteristics, activity and composition of the soil microbial community. Opportunities for comparative studies of microbial communities have increased significantly with the advent of molecular diagnostic and analytical tools (Rondon et al. 1999). All soils, with the possible exception of the most badly degraded, have some capacity to suppress pathogenic microbes as identified by the rapidity and intensity of disease development in plants exposed to them in sterilised and non-sterile states. Such simple comparisons identify the presence of generalised suppressivity in soils resulting from their natural microbial soil flora. Raising this status from generalised to specific and obtaining the suppression of particular pathogens requires the development of viable disease management strategies (van Elsas et al. 2002) which integrate a panoply of agronomic, biologic, genetic and chemical techniques.

It should be apparent that soil management practices can affect the population dynamics of soil microbial communities. Cultural practices can be adequately combined to benefit natural populations of microorganisms that may have a role in biological control (*Trichoderma* spp., and *Gliocladium* spp.), thus for example contributing to the management of peanut fungal soil-borne diseases in a sustainable manner within ecological boundaries (Vargas et al. 2008). During six agricultural cycles, rhizosphere soil samples were taken from a field subjected to crop rotation (soybean, peanut, and maize), peanut being under two tillage systems (no till, reduced tillage) with the aim of quantifying populations of soil microorganisms. The incidence of diseases caused by soil-borne fungi in peanut was determined at harvest. The highest amount of *Trichoderma* spp., and *Gliocladium* spp. were recorded when maize was the preceding crop. Regarding tillage systems, the populations of these groups of microorganisms were higher in peanut under no tillage than under reduced tillage. Under these conditions, the lowest incidence of peanut blight (*Sclerotinia minor*) and root rot (strains of *F. solani*) was observed, suggesting a possible natural control of peanut soil-borne pathogens. The quantification of *Trichoderma* spp., and *Gliocladium* spp. was used as a tool to explore the impacts of different management systems on microbial groups that may be involved in the biological control of soil-borne diseases, with the aim of combining those practices that improve native populations of possible beneficial microorganisms. This manipulation can provide sustainable management strategies in the control of soil-borne diseases, avoiding the use of artificial inoculations of microorganisms, and reducing agrochemical application.

Integrative Pathogen Management

Plant pathologists have traditionally viewed soil as a hostile environment, harbouring microbial opponents that have adverse effects on plant health (Raaijmakers et al., 2010). Increasing the efficiency of crop disease control is an essential prerequisite towards the provision of adequate food supplies for the world's burgeoning population. Currently, at least 15–20% of all crops are lost to the activities of pathogens either in the field or during preparation, processing and presentation to the ultimate consumer. Unfortunately, conventional agriculture has had major environmental impacts, in particular with respect to soil degradation. Soil structure, fertility, microbial and faunal biodiversity have declined, and root diseases are common unless genetic resistance, soil fumigation and, or seed treatments are used (van Bruggen and Termorshuizen 2003). Primarily for environmental reasons and increasing demands for safe and healthy food from the public, farmers must switch to more sustainable systems. This chapter has identified from the perspectives of crop husbandry and pathogen ecology means by which diseases can be diminished and yields increased sustainably. Essentially this uses the somewhat wider view adopted by those microbiologists who approach the soil ecologically and see as it populated by a mixture of microbes fitted to a variety of niches with some even capable of exploiting more than one lifestyle. Approaching disease control with this vantage point accepts the emerging concept that sees the need to increase soil quality and health and approaches pathogens as components of the total soil biological diversity. Disease is seen as resulting from a disturbance to the balance between functional groups in soil. An experimental platform for this concept is beginning to emerge. Molecular methods based on polymerase chain reaction (PCR) of DNA extracted from soil allow rapid assessment of genetic diversity and increasingly will be used to measure functional diversity as well taxonomic placing. This enables the relationships between diversity and disease suppression to be characterised. Suppressiveness of soils to pathogenic organisms is biological in nature, although modified by abiotic factors and is of two types. General suppression depends on overall diversity and activity of the soil biota and acts against a broad range of pathogens. Specific suppression is due to particular antagonists or functional groups and acts against single pathogens. Studies on the effects of management practices on pathogen suppression are still limited in scope, and are often difficult to interpret because most practices have direct effects on pathogen populations as well as on suppressiveness. Continuous cropping of a plant species selects for microflora adapted to its rhizosphere, which may suppress the activities of some pathogens while increasing those of others. Rotations and reduced tillage should increase microbial diversity and increase suppressiveness but evidence for these causal links is hard to find. Treatments that increase soil organic matter such as, residue retention and application of manure or compost may increase general suppression, and certain types of manure may also increase specific suppression.

Soil biodiversity, pathogen suppression and management practices are yet to be fully synthesised into models that have general applicability. This means that management thresholds remain speculative in most instances (Alabouvette et al. 2004). Soil agroecosystems can be modified through rotation, conservation tillage, nutritional management and other agronomic measures to improve disease suppression by enhancing the antibiosis abilities of endophytic and root zone bacteria (Peters et al. 2003). Recognition is emerging that soil suppressiveness is a quantitative property which may be enhanced by agricultural practice as equally it may be destroyed by adverse husbandry systems. Sustainable control of soil-borne pathogens demands an integrated approach in which husbandry forms the foundation of a systematic approach over which other factors are added. The extent to which this pyramid is developed is moderated by the inoculum potential of the soil-borne pathogen(s) and the capacity of benign (beneficial) microbes to exert natural mechanisms of control. Integration is becoming the key to sustainable pathogen control. Sieling et al. (2007) report on extensive studies of wheat grown continuously and in rotations with break crops such as oil seed rape, nitrogen fertilisation and the use of seed treatment with the chemical fluquinconazole for control of take-all (*G. graminis* var. *tritici*). The preceding crop whether it is wheat or a rotational component change the inoculum density of *G. graminis* var. *tritici* and soil structure, weed infestation and population composition, soil nutrient status and the amount and composition of crop residues returning to the soil. This is the type of multifactorial analysis which is required if reliable integrated systems are to be made financially attractive to farmers.

Undoubtedly the intensification of agricultural practices will continue driven by diminishing land available for cropping and an almost exponential increase in demand for food. This, as experience of the last century of agricultural expansion, has demonstrated encourages pests and pathogens to thrive. It is now understood that soil cultivation techniques interact with species antagonistic to crops in different ways and hence the intensity of tillage adopted (from full inversion to no-till) may result in greater or lesser pest or pathogen pressure depending upon the organisms involved (Leake 2003). Selecting one husbandry factor for change such as tillage alone cannot provide the solution. The requirement is for integration of the full range of methods including cultural, biological, genetical, mechanical and where appropriate chemical methods.

The objectives of research must be to evaluate the impact of organic, sustainable, and conventional management strategies in growers' fields on soil physical, chemical, and biological factors including soil microbial species and functional diversity and their effect on plant pathogens. This type of approach has been adopted with *Sclerotium rolfsii*, causal agent of Southern blight in the USA. Soils from ten field locations including conventional, organic and sustainable farms were sampled and assayed for disease suppressiveness in glasshouse assays, and for soil quality indicators. Soils from organic and sustainable farms were more suppressive to Southern blight than soils from conventional farms (Liu et al. 2007). Soils from organic farms had improved soil chemical factors and higher levels of extractable and microbial biomass carbon and nitrogen and net mineralizable nitrogen. In addition, soil microbial respiration was higher in soils from organic than sustainable or conventional farms, indicating that microbial activity was greater in these soils. Populations of fungi and thermophiles were significantly higher in soils from organic and sustainable than conventional fields. The diversity of bacterial functional communities was also greater in soils from organic farms, while species diversity was similar. Soils from organic and sustainable farms had improved soil health as indicated by a number of soil physical, chemical and biological factors and reduced disease.

A clear example of the manner by which cropping systems have a substantial effect on the microbial competence of soils has been shown by Alvarez-Solis and Anzuetto-Martinez (2004). The effects of agricultural intensification on the numbers of microbes, microbial respiration, organic carbon mineralization, mycorrhizal inoculum potential and biological nitrogen fixation and their relationships with soil fertility were assessed in Mexican maize cropping systems using long-fallow, short-fallow, pasture-cultivation rotations and annual continuous cropping. Microbial respiration was 38% lower in plots with corn under annual continuous cultivation or pasture-cultivation rotation than in long or short fallow, irrespective of their condition in cultivation (corn) or fallow (tree and shrub vegetation). Organic carbon mineralization was 31% and 43% lower in plots with corn under annual continuous cultivation and pasture cultivation rotation than in cultivated plots of long and short fallow, respectively. Mycorrhizal colonization of corn was 3.1 times higher in grassland than in cultivated soils with long fallow. The decrease of microbial respiration was related to the decline of soil organic reserves, the increase in soil acidity and decrease of basic cations. These results indicate the importance of the periodic addition of organic and mineral amendments that should return the soil organic reserves and basic cationic reserves to improve heterotrophic microbial activity in these acid soils with intensive agricultural use. Also results suggest the importance of grassland to maintain mycorrhizal inoculum potential in soils.

Currently, since the adoption of organic systems correlates with substantially reduced yields the prime demand is for research to develop means for utilising its environmental advantages while increasing productivity. Sustainable control changes the environment that is encountered by the pathogen either inside or outside the host, thereby providing conditions conducive to host growth but adverse to those of the pathogen. Some microbes are antagonistic to the growth of pathogens while others enhance the growth of crops by benign activities. Antagonism between microbes utilizes factors such as competition, antibiosis and antagonisms. Relatively recently it has been realised that biological control, for example, is most effective when mixed populations of microbes are used by comparison with the previously prevailing approach of using single strains taken from a specific species. Breeding for genetic resistance is probably the most sustainable form of disease control. It aims to produce plants which are either unaffected by or withstand the pathogen per se. Developing new higher yielding and pathogen resistant cultivars is one of the major factors whereby crop yields can be raised to match rising human demand. The requirement now is not only to produce pathogen resistant cultivars but to understand the ecological implications of their use on soil microbial ecosystems. Chemical control which prevents either invasion, colonisation of sporulation of a microbe by destroying the pathogen either immediately before or after penetration into the host still has a continuing role to play. The agrochemical industry has matched society's demands for environmental care with the continuing need for increased production by designing molecules which possess greater efficiency in their modes of action, enhanced sprayer operation and much reduced volumes. Successful production of the large areas of broadacre crops demanded by the world's population will only be feasible if this industrial capacity continues in the long-term. The demand now is that as with resistant cultivars the soil ecological implications of using pesticide molecules is understood in advance of their application in practice. But the nub of this Chapter demonstrates diseases in agriculture are often iatrogenic, so therefore it is possible to alter management practices to reduce disease severity and incidence. Husbanding soil health recognizes that economic management of diseases is not achieved by the using one single procedure but results from the accumulation of multiple strategies which cumulatively constrain diseases below economic thresholds (Katon 1981). The aim of integrated pathogen management is to prevent diseases reaching economic thresholds rather than through curative actions once economic losses might occur. Integrated pathogen management systems exploit the benefits of organic agriculture within essentially conventional systems in a way that will enable agricultural production to be extended beyond present levels with less reliance upon external inputs (Anon 1997). But it will not be possible to exclude using external inputs. Integrated (Sustainable) Crop Management (ICM) unifies all of the field processes and procedures with the objective of growing top quality, healthy, visually attractive crops. Husbandry management starts with site selection, avoiding land which because of aspect or for edaphic reasons may be conducive to weed, pest or pathogen problems. The aim is then to diminish the inoculum potential of organisms that compete with crops for resources by use agronomic knowledge and expertise. The one key resource which is missing in the drive for sustainable crop production is agronomic knowledge and expertise. Around the world governments have neglected to sustain their educational capacities for the continuing supply of personnel capable of viewing soil microbiology and sustainable crop production as a continuum from the laboratory bench to the farmers' fields (as identified by Dixon (2007) in a Report of a Survey of Members Qualifications, Careers and Employers Needs prepared for the British Society for Plant Pathology and available on "www.bspp.org.uk" (accessed 2010) and in the United States (Martyn 2009). This critical deficiency is probably the biggest single impediment to achieving a sufficiency of nutritious and attractive food which satisfies the needs and aspirations of the burgeoning world population.

References

Abbasi PA, Conn KL, Lazarovits G (2006) Effect of fish emulsion used as a preplanting soil amendment on Verticillium wilt, scab, and tuber yield of potato. Can J Plant Pathol 28:509–518

Abbasi PA, Lazarovits G, Jabaji-Hare S (2009) Detection of high concentrations of organic acids in fish emulsion and their role in pathogen or disease suppression. Phytopathology 99:274–281

Adhilakshmi M, Karthikeyan M, Alice D (2008) Effect of combination of bio-agents and mineral nutrients for the management of alfalfa wilt pathogen *Fusarium oxysporum* f. sp. *medicaginis*. Arch Phytopathol Plant Prot 41:514–525

Adiobo A, Oumar O, Perneel M, Zok S, Hofte M (2007) Variation in Pythium induced cocoyam root rot severity in response to soil type. Soil Biol Biochem 39:2915–2925

Alabouvette C (1986) Fusarium-wilt suppressive soils from the Châteaurenard region: a review of a 10-year study. Agronomie 6:273–284

Alabouvette C, Höper H, Lemanceau P, Steinberg C (1996) Soil suppressiveness to disease induced by soil borne pathogens. In: Bollag JM (ed) Soil biochemistry. Marcel Dekker, New York, pp 371–413

Alabouvette C, Backhouse D, Steinberg C, Donovan NJM, Edel-Hermann V, Burgess LW (2004) Microbial diversity in soil – effects on crop health. In: Schjonning P, Elmholt S, Christensen BT (eds) Managing soil quality: challenges in modern agriculture. CABI Publishing, Wallingford, pp 121–138

Alabouvette C, Raaijmakers J, de Boer W, Notz R, Défago G, Steinberg C, Lemanceau P (2005) Concepts and methods to assess the phytosanitary quality of soils. In: Bloem J, Hopkins DW, Benedetti A (eds) Microbiological methods for assessing soil quality. CABI Publishing, Wallingford, pp 257–269

Allmaras RR, Fritz VA, Pfleger FL, Copeland SM (2003) Impaired internal drainage and *Aphanomyces euteiches* root rot of pea caused by soil compaction in a fine-textured soil. Soil Tillage Res 70:41–52

Alvarez-Solis JD, de Anzuetto-Martinez MD (2004) Soil microbial activity under different corn cropping systems in the Highlands of Chiapas, Mexico. Agrociencia (Montecillo) 38: 13–22

Anderson RC (2008) Growth and arbuscular mycorrhizal fungal colonisation of two prairie grasses grown in soil from restorations of three ages. Restor Ecol 16:650–656

Andrews RE, Newman EI (1968) The influence of root pruning on the growth and transpiration of wheat under different soil moisture conditions. New Phytol 67:617–630

Angus JF, Gardner PA, Kirkegaard JA, Desmarchelier JM (1994) Biofumigation: Isothiocyanates released from Brassica roots inhibit growth of the take-all fungus. Plant Soil 162:107–112

Anon (1973) A guide to the use of terms in plant pathology (prepared by the Terminology Sub-committee of the Federation of British Plant Pathologists, now known as The British Society for Plant Pathology). Phytopathological Papers No. 17. Kew: Commonwealth Agricultural Bureau and Commonwealth Mycological Institute

Anon (1997) Diseae suppression on the links. Biocycle October: 62–65.

Ashcroft WJ, Qassim A, Aleemullah M, Hickey MJ, Hoogers R (2003) Irrigation best management practices – the Australian experience. Acta Hortic 613:147–150

Asher MJC (1972) Effects of *Ophiobolus graminis* infection on the growth of wheat and barley. Ann Appl Biol 70:215–223

Ayres PG (1984) Effects of infection on root growth and function; consequences for plant nutrient and water relations. In: Wood RKS, Jellis GJ (eds) Plant diseases: infection, damage and loss. Blackwell Scientific, Oxford, pp 105–118

Badri D, Vivanco JM (2009) Regulation and function of root exudates. Plant Cell Environ 32:666–681

Bai GH, Plattner R, Desjardins AE, Kolb F (2001) Resistance to Fusarium head blight and deoxynivalenol accumulation in wheat. Plant Breed 120:1–6

Bailey DJ, Gilligan CA (1997) Biological control of pathozone behaviour and disease dynamics of *Rhizoctonia solani* by *Trichoderma viride*. New Phytol 136:359–367

Bailey DJ, Gilligan CA (1999) Dynamics of primary and secondary infection in take-all epidemics. Phytopathology 89:84–91

Bailey KL, Lazarovits G (2003) Suppressing soil-borne diseases with residue management and organic amendments. (Special issue: Soil agroecosystems: impacts of management on soil health and crop diseases). Soil Tillage Res 72:169–180

Bailey DJ, Otten W, Gilligan CA (2000) Saprotrophic invasion by the soil-borne fungal plant pathogen *Rhizoctonia solani* and percolation thresholds. New Phytol 146:535–544

Bais HP, Park SW, Weir TL, Callaway RM, Vivanco JM (2004a) How plants communicate using underground information superhighway. Trends Plant Sci 9:26–32

Bais HP, Fall R, Vivanco JM (2004b) Biocontrol of *Bacillus subtilis* against infection of Arabidopsis root by *Pseudomonas syringae* is facilitated by biofilm formation and surfactant production. Plant Physiol 134:307–319

Bais HP, Weir TL, Perry LG, Gilroy S, Vivanco JM (2006) The role of root exudates in rhizosphere interactions with plants and other organisms. Annu Rev Plant Biol 57:233–266

Baker KF (1970) Selective killing of soil microorganisms by aerated steam. In: Tousson T, Bega RV, Nelson PE (eds) Root diseases and soil-borne pathogens. University of California Press, Berkley, CA, pp 234–239

Baker R, Paulitz TC (1996) Theoretical basis of pathogen suppression by antibiosis in the rhizosphere. In: Hall R (ed) Principles and practice of managing soilborne plant pathogens. American Phytopathological Society, St Paul, MN, pp 50–79

Baker KF, Snyder WC (eds) (1965) Ecology of soil-borne plant pathogens. John Murray, London

Ball BC, Bingham I, Rees RM, Watson CA, Litterick A (2005) The role of crop rotations in determining soil structure and crop growth conditions. Can J Soil Sci 85:557–577

Bardner R, Fletcher KE (1974) Insect infestations and their effects on the growth and yield of field crops: a review. Bull Entomol Res 64:141–160

Barker AV, Mills HA (1980) Ammonium and nitrate nutrition of horticultural crops. Hortic Rev 2:395–423

Barnett SJ, Singleton I, Ryder M (1999) Spatial variation in populations of *Pseudomonas corrugata* 2140 and pseudomonads on take-all diseased and healthy root systems of wheat. Soil Biol Biochem 31:633–636

Bateman GL (1993) Development of disease symptoms and fungal pathogens on shoot bases in continuous winter wheat, and effects of fungicides. Plant Pathol 42:595–608

Bateman GL, Munnery IC (1995) Development of eyespot and brown foot rot in wheat plants inoculated with mixtures of *Pseudocercosporella herpotrichoides* types and Fusarium species. J Phytopathol 143:147–151

Baysal F, Bemnitez M-S, Kleinhenz MD, Miller SD, McSpadden-Gardener BB (2008) Field management effects on damping-off and early season vigour of crops in a transitional organic cropping system. Phytopathology 98:562–570

Beaumont A, Staniland LN (1933) Annual Report of Seale Hayne Agricultural College for 1932; Newton Abbot: Seale Hayne Agricultural College
Ben-Yephet Y, Tsror L, Reuiven M, Gips A, Bar Z, Einstein A, Turjeman Y, Fine P (2005) Effect of ecosoil and NH_4 in controlling soilborne pathogens. Acta Hortic 698:115–121
Bochow H (1961) Über die Beeinflessung von *Plasmodiophora brassicae* Wor. durch Mineralsalze. Phytopathol Z 40:420–421
Borneman J, Becker JO (2007) Identifying microorganisms involved in specific pathogen suppression in soil. Annu Rev Phytopathol 45:153–172
Bowers JH, Locke JC (2004) Effect of formulated plant extracts and oils on population density of *Phytophthora nicotianae* in soil and control of Phytophthora blight in the green house. Plant Dis 88:11–16
Brenchley GH (1968) Aerial photography for the study of plant diseases. Annu Rev Phytopathol 6:1–22
Brennan RF (1989) Effect of nitrogen and phosphorus deficiency in wheat on the infection of roots by *Gaeumannomyces graminis* var. *tritici*. Aust J Agric Res 40:489–495
Brennan RF (1992a) The role of manganese and nitrogen nutrition in the susceptibility of wheat plants to take-all in Western Australia. Fertil Res 31:35–41
Brennan RF (1992b) Effect of ammonium chloride, ammonium sulphate and sodium nitrate on take-all and grain yield of wheat in Southwestern Australia. J Plant Nutr 15:2039–2051
Britton D (1990) Recent changes and current trends. In: Britton D (ed) Agriculture in Britain: changing pressures and policies. CABI, Wallingford, pp 1–33
Brooks DH, Dawson MG (1968) Influence of direct-drilling of winter wheat on incidence of take-all and eyespot. Ann Appl Biol 61:57–64
Brouwer R (1977) Root functioning. In: Landsberg JJ, Cutting CV (eds) Environmental effects on crop physiology. Academic, London, pp 229–245
Brown ME (1975) Rhizosphere micro-organisms – opportunists, bandits or benefactors. In: Walker N (ed) Soil microbiology – a critical review. Butterworth, London, pp 21–38
Bruehl GW, Millar RL, Cunfer B (1969) Significance of antibiotic production by *Cephalosporium gramineum* to its saprophytic survival. Can J Plant Sci 49:235–246
Burdon JJ (1987) Diseases and plant population biology. Cambridge University Press, Cambridge
Burdon JJ, Shattock RC (1980) Disease in plant communities. Appl Biol 5:145–219
Campbell R (1989a) The use of microbial inoculants in the biological control of plant diseases. In: Campbell R, Macdonald RM (eds) Microbial inoculation of crop plants. Special Publications of the Society for General Microbiology, vol 25. IRL Press, Oxford, pp 67–77
Campbell R (1989b) Biological control of microbial plant pathogens. Cambridge University Press, Cambridge
Campbell R, Clor A (1985) Soil moisture affects the interaction between *Gaeumannomyces graminis* var. *tritici* and antagonistic bacteria. Soil Biol Biochem 17:441–446
Campbell RN, Greathead AS (1996) Control of clubroot of crucifers by liming. In: Engelhard AW (ed) Soil borne pathogens: management of diseases with macro – and microelements. American Phytopathological Society, St Paul, MN, pp 90–101
Campbell CL, Noe JP (1985) The spatial analysis of soil borne pathogens and root diseases. Annu Rev Phytopathol 23:129–148
Cantonwine EG, Culbreath AK, Stevenson KL (2007) Characterization of early leaf spot suppression by strip tillage in peanut. Phytopathology 97:187–194
Capper AL, Higgins KP (1993) Application of *Pseudomonas fluorescens* isolates to wheat as potential biological control agents against take-all. Plant Pathol 42:560–567
Carr AJH, Smith IM (1988) *Phoma medicaginis* Malbr. & Roum. In: Smith IM, Dunez J, Lelliot RA, Phillips DH, Archer SA (eds) European handbook of plant diseases. Blackwell Scientific, Oxford, p 389

Casida LE (1968) Report of the discussion session on 'the growth of bacteria in soil'. In: Gray TRG, Parkinson D (eds) The ecology of soil bacteria. Liverpool University Press, Liverpool, pp 623–631

Champeil A, Doré T, Fourbet JF (2004) Fusarium head blight: epidemiological origin of the effects of cultural practices on head blight attacks and the production of mycotoxins by Fusarium in wheat grains. Plant Sci 166:1389–415

Cheatham MR, Rouse MN, Esker PD, Ignacio S, Pradel W, Raymundo R, Sparks AH, Forbes GA, Gordon TR, Garrett KA (2009) Beyond yield: plant disease in the context of ecosystem services. Phytopathology 99:1228–1236

Chervet A, Ramseier L, Sturny WG, Tschannen S (2005) No-tillage versus conventional tillage: ten years of comparison. Agrarfirschung 12:184–189

Chiarappa L (ed) (1971) Crop loss assessment methods. Commonwealth Agricultural Bureaux, Sloough

Chupp C (1917) Studies on clubroot of cruciferous plants. In: Bulletin of Cornell Agricultural Experimental Station, No. 387, pp 419–452

Clarkson DT, Drew MC, Ferguson IB, Sanderson J (1975) The effect of take-all fungus, *Gaeumannomyces graminis*, on the transport of ions by wheat plants. Physiol Plant Pathol 6:75–84

Coakley SM, Scherm H, Chakraborty S (1999) Climate change and plant disease management. Annu Rev Phytopathol 37:399–426

Colbach N, Maurin N, Huet P (1996) Influence of cropping system on foot rot of winter wheat in France. Crop Prot 15:295–305

Colbach N, Duby C, Cavelier A, Meynard JM (1997) Influence of cropping systems on foot and root rot diseases of winter wheat: fitting of a statistical model. Eur J Agron 6:61–77

Colhoun J (1958) Club root of crucifers caused by *Plasmodiophora brassicae* Woron. A monograph. Pytopathological paper No. 3. Commonwealth Mycological Institute, Kew

Cook RJ, Baker KF (1983) The nature and practice of biological control of plant pathogens. American Phytopathological Society, St Paul, MN

Cook RJ, Haglund WA (1991) Wheat yield depression associated with conservation tillage caused by root pathogens in the soil and not phytotoxins from the straw. Soil Biol Biochem 23:1125–1132

Cook RJ, Rovira AD (1976) The role of bacteria in the biological control of *Gaeumannomyces graminis* by suppressive soils. Soil Biol Biochem 8:269–273

Cook RJ, Huber DM, Powelson RL, Bruehl GW (1968) Occurrence of take-all in wheat in the Pacific Northwest. Plant Dis Rep 52:716–718

Cook RJ, Chamswarng C, Tang WH (1990) Influence of wheat chaff and tillage on Pythium populations in soil and Pythium damage to wheat. Soil Biol Biochem 22:939–947

Coventry E, Noble R, Mead A, Marin FR, Perez JA, Whipps JM (2006) Allium white rot suppression with composts and *Trichoderma viride* in relation to sclerotia viability. Phytopathology 96:1009–1020

Craig MA, Dixon GR (1993a) The influence of boron on the growth of *Plasmodiophora brassicae* in summer cabbage. In: Proceedings crop protection in Northern Britain, Scottish Crop Research Institute, Dundee, pp 277–282.

Craig MA, Dixon GR (1993b) Alterations to the growth of *Plasmodiophora brassicae* Wor. (Clubroot) associated with boron. In: Proceedings 6th international congress of plant pathology, National Council for Research Canada, Montreal, Canada, poster 3.4.12

Darbyshire JF, Greaves MP (1971) The invasion of pea roots, *Pisum sativum* L. by soil microorganisms *Acanthamoeba palestinensis* (Reich) and *Pseudomonas* sp. Soil Biol Biochem 3:151–155

Datnoff LE, Elmer WH, Huber DM (2007) Mineral nutrition and plant disease. American Phytopathological Society, St Paul, MN

de Kam M, Sukmajaya D (2002) A serious disease of seedlings of Elmerrillia caused by a Phytophthora species and measures for control. FAO Regional Office for Asia and the Pacific 2002; Publication No. 30, pp 59–62 de Paula TJ, Hau B (2002) Effect of soil moisture on the survival of *Rhizoctonia solani* and *Trichoderma harzianum*. Bulletin of the International Organisation for Biological and Integrated Control of Noxious Animals and Plants (OIBC/OILB), West Palaearctic Regional Section (WPRS/SROP), Dijon, France, 25, pp 295–298

De Weert S, Vermeiren H, Mulders IH, Kuipert I, Hendricks N, Bloemberg GV (2002) Flagella-driven chemotaxis towards exudates components is an important trait for tomato root colonisation by *Pseudomonas fluorescens*. Mol Plant-Microbe Interact 15:1173–1180

Deacon JW, Berry LA (1993) Biocontrol of soil-borne plant pathogens: concepts and their application. Pest Sci 37:417–426

Deacon JW, Henry CM (1980) Age of wheat and barley roots and infection by *Gaeumannomyces graminis* var. *tritici*. Soil Biol Biochem 12:113–118

Défago G, Berling CH, Burger U, Haas D, Kahr G, Keel C, Voisard C, Wirthner P, Wüthrich B (1990) Suppression of black root rot of tobacco and other root diseases by strains of *Pseudomonas fluorescens*: potential applications and mechanisms. In: Hornby D (ed) Biological control of soil-borne plant pathogens. CABI, Wallingford, pp 93–108

Dimond AE, Horsfall JG (1960) Plant pathology – an advanced treatise, vol 3. Academic, London

Dixon GR (1981) Vegetable crop diseases. Macmillan, London

Dixon GR (1983) Vegetable husbandry, research investigations and field trials. Horticulture Division, Aberdeen School of Agriculture (University of Aberdeen Department of Agriculture and North of Scotland College of Agriculture), Aberdeen

Dixon GR (1984a) Vegetable husbandry, research investigations and field trials. Horticulture Division, Aberdeen School of Agriculture (University of Aberdeen Department of Agriculture and North of Scotland College of Agriculture), Aberdeen

Dixon GR (1984b) Galls caused by fungi and bacteria. In: Wood RKS, Jellis GJ (eds) Plant diseases: infection, damage and loss. Blackwell Scientific, Oxford, pp 189–197

Dixon GR (1985) Vegetable husbandry, research investigations and field trials. Horticulture Division, Aberdeen School of Agriculture (University of Aberdeen Department of Agriculture and North of Scotland College of Agriculture), Aberdeen

Dixon GR (1989) Cost/benefit analyses in horticultural fruit and vegetable crops. In: Hosie G, Davis RP (eds) The use of cost/benefit analyses in crop protection. British Crop Protection Council, Farnham, pp 28–32

Dixon GR (1991) Primary and secondary stages of *Plasmodiophora brassicae* (clubroot) as affected by metallic cations and pH. In: Beemster ABR, Bollen GJ, Gerlagh M, Ruissen MA, Schippers B, Tempel A (eds) Biotic interactions and soil – borne diseases, no 23 in Developments in agricultural and managed forestry ecology. Elsevier, Amsterdam, pp 381–386

Dixon GR (2002) Interactions of soil nutrient environment, pathogenesis and host resistance. Plant Prot Sci 38:87–94, special edition no 1

Dixon GR (2006) The biology of *Plasmodiophora brassicae* – a review of recent advances. Acta Hortic 706:271–282

Dixon GR (2007) Report on a Skills Audit undertaken for the British Society for Plant Pathology (BSPP). www.bspp.org.uk accessed 2010

Dixon GR (2009a) *Plasmodiophora brassicae* in its environment. In: Dixon GR (ed) *Plasmodiophora brassicae* (Clubroot): a plant pathogen that alters host growth and productivity. J Plant Growth Regul 28:212–228

Dixon GR (2009b) The occurrence and economic impact of *Plasmodiophora brassicae* and Clubroot disease. In: Dixon GR (ed.) *Plasmodiophora brassicae* (Clubroot): a plant pathogen that alters host growth and productivity. J Plant Growth Regul 28:194–202

Dixon GR (2009c) Husbandry – the sustainable means of controlling soil borne pathogens: a synoptic review. Acta Hortic 817:233–242

Dixon GR (2009d) Calcium cyanamide – a century of integrated control. In: Report of plasmodiphorids and related organisms workshop, Turin, Italy 23 August 2009. Plant Prot Sci 45:25–38

Dixon GR, Margerison C (2009) Biodiversity and land-use: co-operation and conflict. Biologist 34:202–206

Dixon GR, Page LV (1998) Calcium and nitrogen elicting alterations to growth and reproduction of *Plasmodiophora brassicae* (clubroot). Acta Hortic 459:343–349

Dixon GR, Walsh UF (1998) Suppression of plant pathogens by organic extracts – a review. Acta Hortic 469:383–390

Dixon GR, Webster MA (1988) Antagonistic effects of boron, calcium and pH on pathogenesis caused by *Plasmodiophora brassicae* (clubroot) – a review of recent work. Crop Res (Hortic Res) 28:83–95

Dixon GR, Wilson F (1983) Evaluation of calcium cyanamide for control of *Plasmodiophora brassicae* (clubroot). Tests of agrochemicals and cultivars no 4. Ann Appl Biol 102:50–51

Dixon GR, Wilson F (1984a) Field evaluation of chemicals for control of clubroot (*Plasmodiophora brassicae*). In: Proceedings of the crop protection in Northern Britain conference, Dundee, pp 400–405

Dixon GR, Wilson F (1984b) Field evaluation of WL 105305 (NK483) for control of clubroot (*Plasmodiophora brassicae*). Tests of Agrochemicals and Cultivars no. 5. Ann Appl Biol 104:34–35

Dixon GR, Wilson F (1985) Fungicides applied to propagation blocks and the field to control *Plasmodiophora brassicae*, clubroot on cabbage. Tests of Agrochemicals and Cultivars no 6. Ann Appl Biol 106:44–45

Dixon GR, Kershaw CD, Hunter EA (1989) Crop yields from lucerne (*Medicago sativa* L.) cultivars displaying gradations in resistance to wilt (*Verticillium albo-atrum*). J Agric Sci (Cambridge) 112:387–394

Dobson RL, Gabrielson RL, Baker AS, Bennett L (1983) Effects of lime particle size and distribution and fertiliser formulation on clubroot disease caused by *Plasmodiophora brassicae*. Plant Dis 67:50–52

Donald EC (2005) The influence of abiotic factors and host plant physiology on the survival and pathology of *Plasmodiophora brassicae* of vegetable brassicas. unpublished Ph.D. thesis, University of Melbourne, Melbourne

Donald EC, Lawrence JM, Porter IJ (2004) Influence of particle size and application method on the efficacy of calcium cyanmaide for control of clubroot of vegetable brassicas. Crop Prot 23:297–303

Drees LR, Karathanasis AD, Wilding LP, Blevins RL (1994) Micromorphological characteristics of long-term no-till and conventionally tilled soils. Soil Sci Soc Am J 58:508–517

Duffy BK, Simon A, Weller DM (1996) Combination of *Trichoderma koningii* with fluorescent pseudomonads for control of take-all on wheat. Phytopathology 86:188–194

Duveiller E, Garcia A, Fischer RA, Budhathoki C (2002) Effect of crop rotation, tillage and residue management on tan spot in the subhumid tropical highlands. In: Proceedings of the 4th international wheat tan spot and spot blotch workshop, Bemidji, Minnesota, MN, pp. 86–92, 21–24 July 2002. Agricultural Experiment Station, North Dakota State University, Fargo, 2003

Emmond GS, Ledingham RJ (1972) Effects of crop rotation on some soilborne pathogens of potato. Can J Plant Sci 52:605–611

Engelhard AW (ed) (1996) Soil borne pathogens: management of diseases with macro- and microelements. American Phytopathological Society, St Paul, MN

Escuadra GME, Amemiya Y (2008) Suppression of Fusarium wilt of spinach with compost amendments. J Gen Plant Pathol 74:267–274

Everts KL (2002) Reduced fungicide applications and host resistance for managing three diseases in pumpkin grown on a no till cover crop. Plant Dis 86:1134–1141

Faggian R, Strelkov SE (2009) Detection and measurement of *Plasmodiophora brassicae*. In: Dixon GR (ed.) *Plasmodiophora brassicae* (Clubroot): a plant pathogen that alters host growth and productivity. J Plant Growth Regul 28:282–288

Falik O, Reides P, Gersani M, Novoplansky A (2003) Self/non-self discrimination in roots. J Ecol 91:525–531

Fedorintschik NS (1935) Life history of club – root of cabbage (*Plasmodiophora brassicae* Woron.). (In Russian) Summary Science Research Wk. Institute of Plant Protection Leningrad, pp 69–70. (abstract in Rev Appl Mycol 16:10)

Fehrmann H (1988a) *Fusarium culmorum* (W. G. Sm.) Sacc. In: Smith IM, Dunez J, Lelliot RA, Phillips DH, Archer SA (eds) European Handbook of Plant Diseases. Blackwell Scientific, Oxford, pp 287–289
Fehrmann H (1988b) *Gibberella avenacea* R.J. Cooke. In: Smith IM, Dunez J, Lelliot RA, Phillips DH, Archer SA (eds) European handbook of plant diseases. Oxford, Blackwell Scientific, pp 277–278
Fehrmann H (1988c) *Monographella nivalis* (Schaffnit) E. Muller & v. Arx. In: Smith IM, Dunez J, Lelliot RA, Phillips DH, Archer SA (eds) European handbook of plant diseases. Oxford, Blackwell Scientific, pp 335–337
Fitt BDL, Hornby D (1978) Effects of root-infecting fungi on wheat transport processes and growth. Physiol Plant Pathol 13:335–346
Fletcher JT, Hims MJ, Archer FC, Brown A (1982) Effects of adding calcium and sodium salts to field soils on the incidence of clubroot. Ann Appl Biol 100:245–251
Francl LJ, Madden LV, Rowe RC, Riedel RM (1987) Potato yield loss prediction and discrimination using preplant population densities of *Verticillium dahliae* and *Pratylenchus penetrans*. Phytopathology 77:579–584
Fravel DR (1988) Role of antibiosis in the biocontrol of plant diseases. Annu Rev Phytopathol 26:75–91
Fritz VA, Allmaras RR, Pfleger FL, Davis DW (1995) Oat residue and soil compaction influences on common root-rot (*Aphanomyces euteiches*) of peas in a fine-textured soil. Plant Soil 266:235–244
Fuchs JG, Berner A, Mayer J, Smidt E, Schleiss K (2008) Influence of compost and digestates on plant growth and health: potentials and limits. Compost and digestate: sustainability, benefits, impacts for the environment and for plant production. In: Proceedings of the International Congress CODIS 2008, Solothurn, Switzerland, 27–29 February, 2008, pp 101–110. Research Institute of Organic Agriculture (FiBL), Frick, Switzerland
Gair R, Jenkins JEE, Lester E (1972) Cereal pests and diseases. Farming Press, Ipswich
Gallegly ME, Walker JC (1949) Plant nutrition in elation to disease development. V. bacterial wilt of tomato. Am J Bot 36:613–623
Galvez L, Douds DD, Drinkwater LE, Wagoner P (2001) Effect of tillage and farming system upon VAM fungus populations and mycorrhizas and nutrient uptake of maize. Plant Soil 228:299–308
Garbeva P, van Veen JA, van Elsas JD (2004) Microbial diversity in soil: selection of microbial populations by plant and soil type and implications for disease suppressiveness. Annu Rev Phytopathol 42:243–270
Garrett DS (1956) Biology of root invading fungi. Cambridge University Press, Cambridge
Garrett KA, Dendy SP, Frank EE, Rouse MN, Travers SE (2006) The impact of global climate change on plant diseases. Annu Rev Phytopathol 44:489–509
Gerlagh M (1968) Introduction of *Ophiobolus graminis* into new polders and its decline. Neth J Plant Pathol 74:1–97
Gibbs JG (1931) Dissemination of club root in the dung of farm stock. N Z J Agric 42:193–198
Gibbs JG (1939) Factors influencing the control of club root. N Z J Sci Technol 20A:409–412
Gilligan CA (1985) Probability models for host infection by soil-borne fungi. Phytopathology 75:61–67
Gilligan CA (1990) Mathematical models of infection. In: Lynch JM (ed) The rhizosphere. Wiley, Chichester, pp 207–232
Gilligan CA (2002) An epidemiological framework for disease management. Adv Bot Res 38:1–64
Gilligan CA, Bailey DJ (1997) Components of pathozone behaviour. New Phytol 136:342–358
Gilligan CA, van den Bosch F (2008) Epidemiological models for invasion and persistence of pathogens. Annu Rev Phytopathol 46:385–418
Gould JC (2003) Verticillium wilt in trees: detection, prediction and disease management. Wageningen University, Wageningen
Gray TRG, Baxby P, Hill IR, Goodfellow M (1968) Direct observation of bacteria in soil. In: Gray TRG, Parkinson D (eds) The ecology of soil bacteria. Liverpool University Press, Liverpool, pp 171–192

Guenzi WD, McCalla TM, Norstadt FA (1967) Presence and persistence of phytotoxic substances in wheat, oat, corn and sorghum residues. Agron J 59:163–165

Guest D (2002) Integrated management of Phytophthora diseases in tropical horticulture. Acta Hortic 575:447–451

Gutierrez AP, DeVay JE, Pullman GS, Friebertshauser GE (1983) A model of Verticillium wilt in relation to cotton growth and development. Phytopathology 73:89–95

Hamilton HA, Crête R (1978) Influence of soil moisture, soil pH and liming source on the incidence of clubroot, the germination and growth of cabbage produced in mineral and organic soils under controlled conditions. Can J Plant Sci 58:45–53

Harper JL (1977) Population biology of plants. Academic, London

Harris AR, Siwek K, Wiseman BM (1997) Interactions between damping-off fungi, antagonists and Capsicum seedlings. Appl Soil Ecol 6:251–263

Hata S, Sumi Y, Ohi M (2002) Dry powder and extract of *Posidonia australis* Hook. F., a species of seagrass, stimulate the germination of the pathogen *Plasmodiophora brassicae* and control clubroot of Chinese cabbage. J Jpn Soc Hortic Sci 71:197–202

Heyman F, Lindahl B, Persson L, Wikstrom M, Stenlid J (2007) Calcium concentrations of soil affect suppressiveness against Aphanomyces root rot of pea. Soil Biol Biochem 39: 2222–2229

Hiltner L (1904) Uber neure Erfahrungen und problem auf dem gebeit der bodenbackteriologie und unter besonderer berucksichtigung der grundungung und brache. Ark Deut Landwirsch Ges 98:59–78

Hiltunen LH, White JG (2002) Cavity spot of carrot (*Daucus carota*). Ann Appl Biol 141:201–223

Hodge AS (2009) Root decisions. Plant Cell Environ 32:628–640

Honig F (1931) Der Kohlkropfereger (*Plasmodiophora brassicae* Wor.). Gartenbauwiss 5:116–225

Höper H, Alabouvette C (1996) Importance of physical and chemical soil properties in the suppressiveness of soils to plant diseases. Eur J Soil Biol 32:41–48

Horiuchi S, Hori M, Takashi S, Shimizu K (1982). Factors responsible for the development of clubroot-suppressing effect in soil solarisation. In: The Bulletin of the Chigoku National Agricultural Experiment Station Series E (Environment Division) 20: 25–48

Hornby D (1981) Inoculum. In: Asher MJC, Shipton PJ (eds) Biology and control of take-all. Academic, London, pp 271–293

Hornby D (1992) New information about take-all decline and its relevance to research on the control of take-all by biological control agents. In: Tjamos EC, Papavizas GC, Cook RJ (eds) Biological control of plant diseases: progress and challenges for the future. Plenum Press, London, pp 95–98

Hornby D, Fitt BDL (1982) Effects of root-infecting fungi on structure and function of cereal roots. In: Ayres PG (ed) Effects of disease on the physiology of the growing plant. Cambridge University Press, Cambridge, pp 101–130

Hornby D, Bateman GL, Gutteridge RJ, Lucas P, Osbourn AE, Ward E, Yarham DJ (1998) Take-all disease of cereals, a regional perspective. CABI, Wallingford

Houghton J (1996) Royal Commission on Environmental Pollution, 19th Report. Sustainable use of soil. HMSO, London

Howell CR, Stipanovic RD (1980) Suppression of *Pythium ultimum*-induced damping-off of cotton seedlings by *Pseudomonas fluorescens* and its antibiotic pyoluteorin. Phytopathology 70:712–715

Hsieh WH, Yang CH (1985) Investigations on the clubroot and the susceptibility of crucifers in Taiwan. Plant Prot Bull (Taiwan ROC) 27:3–10

Huber DM (1978) Disturbed mineral nutrition. In: Horsfall JG, Cowling EB (eds) Plant disease – an advanced treatise, vol. 3. Academic, London, pp 163–181

Huber DM, Watson RD (1974) Nitrogen form and plant disease. Annu Rev Phytopathol 12:139–165

Hughes G (1996) Incorporating spatial pattern of harmful organisms in crop loss models. Crop Prot 15:407–421

Hughes G (1999) Sampling for decision making in crop loss assessment and pest management: introduction. (In: Symposium of sampling for decision making in crop loss assessment and pest management – the American Phytopathological Society meeting with the Entomological Society of America November 11 1998, Las Vegas, NV). Phytopathology 89:1080–1083

Humphrey JE (1892) Report on plant diseases etc. with observations in the field and in the vegetation house. Public Document Massachusetts Agricultural Experimental Station, no 33, 218–248 (Abstract in Bot Zbl 307–309)

Jackson RM (1975) Soil Fungi. In: Walker N (ed) Soil microbiology – a critical review. Butterworth, London, pp 165–180

Jackson LE, Ramirez I, Yokota R, Fennimore SA, Koike ST, Henderson DM, Chaney WE, Klonsky KM (2003) Scientists, growers assess trade-offs in use of tillage, cover crops and compost. Calif Agric 57:48–54

Jaffee BA (2006) Interactions among a soil organic amendments, nematodes and the nematode-trapping fungus *Dactylellina candidum*. Phytopathology 96:1388–1396

Jaime-Garcia R, Cotty PJ (2006) Spatial relationships of soil texture and crop rotation to *Aspergillus flavus* community structure in South Texas. Phytopathology 96:599–607

James WC (1974) Assessment of plant diseases and losses. Annu Rev Phytopathol 12:27–48

Jayaraj J, Radhakrishnan NV (2008) Enhanced activity of introduced biocontrol agents in solarized soils and its implications on the integrated control of tomato damping-off caused by *Pythium* spp. Plant Soil 304:189–197

Jennwen H (2005) Extension bulletin: management of crop diseases with agricultural wastes. Food and Fertiliser Technology Centre for the Asian and Pacific Region, Tapei

Johnson RM, Grisham MP, Richard EP (2007) Relationship between sugarcane rust severity and soil properties in Louisiana. Phytopathology 97:748–755

Kasuya M, Olivier AR, Ota Y, Tojo M, Honjo H, Fukui R (2006) Induction of soil suppressiveness by incorporation of dried plant residues into soil. Phytopathology 96:1372–1379

Katon J (1981) Solar heating (solarization of soil for control of soilborne pests). Annu Rev Phytopathol 19:211–236

Keinath APLR, Loria R (1989) Management of common scab of potato with plant nutrients. In: Engelhard AW (ed) Soilborne plant pathogens: management of disease with macro- and microelements. American Phytopathological Society, St Paul, MN, pp 152–166

Kirkegaard JA, Wong PTW, Desmarchelier JM (1996) In vitro suppression of fungal root pathogens of cereals by Brassica tissues. Plant Pathol 45:593–603

Klein E, Kataan J, Austerweil M, Gamliel A (2007) Controlled laboratory system to study soil solarisation and organic amendment effects on plant pathogens. Phytopathology 97:1476–1483

Kloepper JW, Schroth MN (1981) Plant growth-promoting rhizobacteria and plant growth under gnotobiotic conditions. Phytopathology 71:642–644

Knight SM (2003) Effects of establishment technique and number of management passes on winter wheat production costs. Home Grown Cereals Authority (HGCA) Report no. 311. Home Grown Cereals Authority, London

Koch S, Dunker S, Kleinhenz B, Röhrig M, von Tiedeman A (2007) A crop-loss related forecasting model for Sclerotinia stem rot in winter oilseed rape. Phytopathology 97:1186–1194

Kumar A, Hashmni S, Shukla A, Chaturvedi OP (2007) Effect of intercropping on inoculum potential of arbuscular mycorrhizal fungi in agroforestry systems. Ann Arid Zone 46:31–36

Lampkin N (1990) Organic farming. Farming Press, Ipswich

Lange L, Olson LW (1983) The fungal zoospore – its structure and biological significance. In: Buczacki ST (ed) Zoosporic plant pathogens – a modern perspective. Academic, London, pp 1–42

Large EC (1945) Field trials of copper fungicides for the control of potato blight. 1. Foliage protection and yield. Ann Appl Biol 32:319–329

Lazarovits G, Hill J, Patterson G, Conn KL, Crump NS (2007) Edaphic soil levels of mineral nutrients, pH, organic matter and cationic exchange capacity in the geocaulosphere associated with potato common scab. Phytopathology 97:1071–1082; quoting: Shiga H, Suzuki K (2004) Effect of soil acidity, organic soil amendment and green manure on potato scab. International Potato Scab Symposium, Sapporo, Japan

Leake AR (2003) Integrated pest management for conservation agriculture. In: García-Torres L, Benites J, Martínez-Vilela A, Holgado-Cabrera A (eds) Conservation agriculture: environment, farmers experiences, innovations, socio-economy, policy. Kluwer, Dordrecht, The Netherlands, pp 271–279

Leath KT, Lukezic FL, Pennypacker BW, Kendall WA, Levine RG, Hill RR (1989) Interaction of *Fusarium avenaceum* and *Pseudomonas viridiflava* in root rot of red clover. Phytopathology 79:436–440

Lemaire JM (1988) *Gaeumannomyces graminis* (Sacc.) v. Arx & H. Oliver. In: Smith IM, Dunez J, Lelliot RA, Phillips DH, Archer SA (eds) European handbook of plant diseases. Oxford, Blackwell Scientific, pp 323–324

Leong SA, Winkelmann G (1998) Molecular biology of iron transport in fungi. In: Sigel A, Sigel H (eds) Metal ions in biological systems. Marcel Dekker, New York, pp 147–185

Lipps PE, Deep IW (1991) Influence of tillage and crop rotation on yield, stalk rot and recovery of *Fusarium* and *Trichoderma* spp. from corn. Plant Dis 75:828–833

Liu B, Tu C, Hu SJ, Gumpertz M, Ristaino JB (2007) Effect of organic, sustainable, and conventional management strategies in grower fields on soil physical, chemical, and biological factors and the incidence of Southern blight. Appl Soil Ecol 37:202–214

Liu B, Glenn D, Buckley K (2008) Trichoderma communities in soils from organic, sustainable, and conventional farms, and their relation with Southern blight of tomato. Soil Biol Biochem 40:1124–1136

Lorang JM, Anderson NA, Lauer FL, Wildung DK (1989) Disease decline in a Minnesota potato scab plot. Am Potato J 66:531

Lucas P, Nignon M (1987) Importance of soil type and physicochemical characteristics on relations between a wheat variety (*Triticum aestivum* L. cv. Rescler) and two virulent or hypovirulent strains of *Gaeumannomyces graminis* (Sacc.) Von arx & Oliver var. *tritici* Walker. Plant Soil 97:105–117

Lugtenberg BJ, Dekkers L, Bloemberg GV (2001) Molecular determinants of rhizosphere colonisation by Pseudomonas. Annu Rev Phytopathol 39:461–490

Lynch JM (1990) Microbial metabolites. In: Lynch JM (ed) The rhizosphere. Wiley, Chichester, pp 177–206

Macura J (1968) Physiological studies of rhizosphere bacteria. In: Gray TRG, Parkinson D (eds) The ecology of soil bacteria. Liverpool University Press, Liverpool, pp 379–395

Madden LV, Hughes G, van den Bosch F (2008) The study of plant disease epidemics. American Phytopathological Society, St Paul, MN

Maloy OC (1993) Plant disease control. Wiley, New York

Manners JG, Myers A (1981) Effects on host growth and physiology. In: Asher MJC, Shipton PJ (eds) Biology and control of take-all. Academic, London, pp 237–248

Marschner H (1990) Relationship between mineral nutrition and plant diseases and pests. In: Marschner H (ed) Mineral nutrition of higher plants. Academic, London, pp 369–390

Martin FN (2003) Development of alternative strategies for management of soil borne pathogens currently controlled with methyl bromide. Annu Rev Phytopathol 41:325–350

Martin RA, Macleod JA, Caldwell C (1991) Influences of production inputs on incidence of infection by Fusarium species on cereal seed. Plant Dis 75:784–788

Martyn RD (2009) Where will the next Norman Borlaugh come from? A U.S. perspective of plant pathology education and research. Plant Prot Sci 45:125–139

Matthiessen JN, Kirkegaard JA (2006) Biofumigation and enhanced biodegradation: opportunity and challenge in soilborne pest and disease management. Crit Rev Plant Sci 25:235–265

Mattsson L (1995) Skördevariationer inom enskilda fält. Storlek och tänkbara orsaker. Swedish University of Agricultural Sciences, Department of Soil Sciences, Division of Soil Fertility. Report 196

Mazzola M (2004) Assessment and management of soil microbial community structure for disease suppression. Annu Rev Phytopathol 42:35–59

Mazzola M (2007) Manipulation of rhizosphere bacterial communities to induce suppressive soils. J Nematol 39:213–220

McCalla TM, Duley FC (1949) Stubble-mulch studies. III. Influence of soil microorganisms and crop residues on germination, growth and direction of root growth of corn seedlings. Soil Sci Soc Am Proc 14:196–199

McCay-Buis TS, Huber DM, Graham RD, Philips JD, Miskin KE (1995) Manganese seed content and take-all of cereals. J Plant Nutr 18:1711–1721

McKay MB (1926) Potato wilt and its control. In: Oregon Agricultural Experiment Station Bulletin No. 22, pp 1–23

Millard WA, Taylor CB (1927) Antagonism of microorganisms as the controlling factor in the inhibition of scab by green manuring. Ann Appl Biol 14:202–215

Monteith J (1924) Relation of soil temperature and soil moisture to infection by *Plasmodiophora brassicae*. J Agric Res 28:549–561

Mosse B (1975) A microbiologist's view of root anatomy. In: Walker N (ed) Soil microbiology – a critical review. Butterworth, London, pp 39–66

Moussart A, Wicker E, le Delliou B, Abelard JM, Esnault R, Lemarchand E, Rouault F, le Guennou F, Pilet-Nayel ML, Baranger A, Rouxel F, Tivoli B (2009) Spatial distribution of *Aphanomyces euteiches* inoculums in a naturally infested pea field. Eur J Plant Pathol 123:153–158

Murray TD, Parry DW, Cattlin ND (1998) A colour handbook of diseases of small grain cereals. Manson Publishing, London

Myers DF, Campbell RN (1985) Lime and the control of clubroot of crucifers: effects of pH, calcium, magnesium and their interaction. Phytopathology 75:670–673

Narisawa K, Shimura M, Usuki F, Fukuhara S, Hashiba T (2005) Effects of pathogen density, soil moisture, and soil pH on biological control of clubroot in Chinese cabbage by *Heteroconium chaetospira*. Plant Dis 89:285–290

Naumov NA (1925) Contribution to the study of club root of cabbage. II. (In Russian). Morb Plant 14:49–73

Nicot PC, Rouse DI, Vandell BS (1984) Comparison of statistical methods for studying spatial patterns of soilborne plant pathogens in the field. Phytopathology 74:1399–1402

Nnodu EC, Harrison MD (1979) The relationship between *Verticillium albo-atrum* inoculum density and potato yield. Am Potato J 56:11–25

O'Brian DG, Dennis RWG (1936) Further information relating to control of rain in Sweden. Scott J Agric 19:1–18

Ogawa S, Takahashi H, Hayakawa T, Ito K, Mitsui T, Hori H, Kiso A (2001) Enhancement of germination of spores from obligatory plant pathogen, *Plasmodiophora brassicae* which causes clubroot disease. Bull Fac Agric Niigata Univ, Jpn 54:35–43

Ohi M, Kitamura T, Hata S (2003) Stimulation by caffeic acid, coumalic acid, and corilagin of the germination of resting spores of the clubroot pathogen *Plasmodiophora brassicae*. Biosci Biotechnol Biochem 67:170–173

Ordish G (1952) Untaken harvest. Constable, London

Otten W, Gilligan CA (2006) Soil structure and soil-borne diseases: using epidemiological concepts to scale from fungal spread to plant epidemics. Eur J Soil Sci 57:26–37

Otten W, Hall D, Harris K, Ritz K, Young IM, Gilligan CA (2001) Soil physics, fungal epidemiology and the spread of *Rhizoctonia solani*. New Phytol 151:202–215

Otten W, Harris K, Young IM, Ritz K, Gilligan CA (2004) Preferential spread of the pathogenic fungus *Rhizoctonia solani* through structured soil. Soil Biol Biochem 36:203–210

Oyarzun PJ (1993) Bioassay to asses root rot in pea and effect on root rot on yield. Neth J Plant Pathol 99:61–75

Page LV (2001) Studies of components for a potential integrated control system for *Plasmodiophora brassicae*. unpublished Ph.D. thesis, University of Strathclyde, Glasgow

Pagliani M, De Nobili M (1993) Relationships between soil porosity, root development and soil enzyme activity in cultivated soils. Geoderma 56:243–256

Palm ET (1958) Effect of mineral nutrition on invasiveness of *Plasmodiophora bnrassicae* Wor. and the development of club root. Diss Abstr 19:425–426

Palm ET (1963) Effect of mineral nutrition on the invasion and response of turnip tissue to *Plasmodiophora brassicae* Wor. Contrib Boyce Thompson Inst 22:91–112

Pankhurst CE, McDonald HJ, Hawke BG, Kirkby CA (2002) Effect of tillage and stubble management on chemical and microbiological properties and the development of suppression towards cereal root disease in soils from two sites in NSW, Australia. Soil Biol Biochem 34:833–840

Papendick RI, Cook RJ (1974) Plant water stress and development of Fusarium foot rot in wheat subjected to different cultural practices. Phytopathology 64:358–363

Paplomatas EJ, Bassett DM, Broome JC, DeVay JE (1992) Incidence of Verticillium wilt and yield losses to cotton cultivars (*Gossypium hirsutum*) based on soil incoculum density of *Verticillium dahliae*. Phytopathology 82:1417–1420

Parry DW (1990) Plant pathology in agriculture. Cambridge University Press, Cambridge

Parry DW, Jenkinson P, McLeod L (1995) Fusarium ear blight (scab) in small grain cereals – a review. Plant Pathol 44:207–238

Passioura JB (2002) Environmental biology and crop improvement. Funct Plant Biol 29:537–546

Paul PA, El-Allaf SM, Lipps PE, Madden LV (2005a) Relationships between incidence and severity of Fusarium head blight on winter wheat in Ohio. Phytopathology 95:1049–1060

Paul PA, Lipps PE, Madden LV (2005b) Relationship of visual estimates of Fusarium head blight intensity and deoxynivalenol accumulation in harvested wheat grains: a meta-analysis. Phytopathology 95:1225–1236

Paulitz TC (2006) Low input no-till cereal production in the Pacific Northwest of the US: the challenges of root diseases. Eur J Plant Pathol 115:271–281

Pavlista AD (2008) Sulfur and marketable yield of potato. In: Jez J (ed) Sulfur: a missing link between soils, crops and nutrition. Agronomy Monograph No. 50. American Society of Agronomy, Madison, WI, pp 171–182

Pedler JF, Webb MJ, Buchhorn SC, Graham RD (1996) Manganese-oxidizing ability of isolates of the take-all fungus is correlated with virulence. Biol Fertil Soils 22:272–278

Peters RD, Sturz AV, Carter MR, Sanderson JB (2003) Developing disease-suppressive soils through crop rotation and tillage management practices. Soil Tillage Res 72:181–192

Peterson GL, Kosta KL, Glenn DL, Phillips JG (2008) Utilization of soil solarization for eliminating viable *Tilletia indica* teliospores from Arizona wheat fields. Plant Dis 92:1604–1610

Pinkerton JN, Bristow PR, Windom GE, Walters TW (2009) Soil solarization as a component of an integrated program for control of raspberry root rot. Plant Dis 93:452–458

Pitt D, Tilston EL, Groenhof AC (1998) Recycled organic materials (ROM) in the control of plant disease. Acta Hortic 469:391–403

Porter IJ, Merriman PR (1983) Effects of solarisation of soil on nematode and fungal pathogens at two sites in Victoria. Soil Biol Biochem 15:39–44

Porter DM, Smith DH, Rodríguez-Kábana R (eds) (1984) Compendium of peanut diseases. American Phytopathological Society, St. Paul, MN

Prestes AM, Santos HP, Reis EM (2002) Effect of cultural practices on the incidence of leaf blotches of wheat. Pesqui Agropec Bras 37:791–797

Pretty J (1990) Sustainable agriculture in the Middle Ages: the English manor. Agric Hist Rev 38:1–19

Prew RD, McIntosh AH (1975) Effects of benomyl and other fungicides on take-all, eyespot and sharp eyespot diseases of winter wheat. Plant Pathol 24:67–71

Quarles W (2008) Mulch optimization in integrated pest management, vol. 30. IPM Practitioner Bio Integral Research Centre, Berkeley, CA, pp 1–10

Raaijmakers JM, Bonsall RF, Weller DM (1999) Effect of population density of Pseudomonas fluorescens on production of 2, 4-diacetylphloroglucinol in the rhizosphere of wheat. Phytopathology 89:470–475

Raaijmakers JM, Paulitz TC, Steinberg C, Alabouvette A, Moenne-Loccoz Y (2010) The rhizosophere: a playground and battlefield for soilborne pathogens and beneficial microorganisms.

Plant & Soil 321(1–2):341–361, published as Development in Plant and Soil Sciences no 104 Rhizosphere: Achievements and Challenges. edits: Y. Dessaux, P. Hinsinger & P. Lemanceau, Springer, The Netherlands.

Rengel Z, Pedler JF, Graham RD (1994) Control of Mn status in plants and rhizosphere: genetic aspects of host and pathogen effects in the wheat take-all interaction. In: Manthey JA, Crowley DE, Luster DG (eds) Biochemistry of metal micronutrients in the rhizosphere. CRC Press, Boca Raton, FL, pp 125–145

Rengel Z, Gutteridge R, Hirsch P, Hornby D (1996) Plant genotype, micronutrient fertilization and take-all infection influence bacterial populations in the rhizosphere of wheat. Plant Soil 183:269–277

Rod J, Robak J (1994) The effect of some crops on inactivation of clubroot (*Plasmodiophora brassicae*) in soils. Ochr Rostl 30:267–272

Rondon MR, Goodman RM, Handelsman J (1999) The earth's bounty: assessing and accessing soil microbial diversity. Trends Biotechnol 17:403–409

Rowe RC, Powelson ML (2002) Potato early dying: management challenges in a changing production environment. Plant Dis 86:1184–1193

Russell EW (1968) Discussion: the environment of soil bacteria. In: Gray TRG, Parkinson D (eds) The ecology of soil bacteria. Liverpool University Press, Liverpool, pp 89–94

Ryder MH, Rovira AD (1993) Biological control of take-all of glasshouse-grown wheat using strains of *Pseudomonas corrugata* isolated from wheat field soil. Soil Biol Biochem 25:311–320

Ryu CM, Farag MA, Hu CH, Reddy MS, Kloepper JW, Pare PW (2004) Bacterial volatiles induce systemic resistance in Arabidopsis. Plant Physiol 134:1017–1026

Samuel G, Garrett SD (1945) The infected root hair count for estimating the activity of *Plasmodiophora brassicae* Woron. in the soil. Ann Appl Biol 32:96–101

Sanford GB (1926) Some factors affecting the pathogenicity of *Actinomyces scabies*. Phytopathology 16:525–547

Sarwar M, Kirkegaard JA, Wong PTW, Desmarchelier JM (1998) Biofumigation potential of brassicas III. In vitro toxicity of isothiocyanates to soil-borne fungal pathogens. Plant Soil 201:103–112

Satir P (1982) Mechanisms and controls of microtubule sliding in cilia. In: Amos WB, Duckett JG (eds) Prokaryotic and eukaryotic flagella (Symposium of the society for experimental biology, symposium no. XXXV). Cambridge University Press, Cambridge, pp 179–201

Savary S, Teng PS, Willocquet L, Nutter FW (2006) Quantification and modelling of crop losses: a review of purposes. Annu Rev Phytopathol 44:89–112

Scarlett RL (1937) Cleaning dirty land to produce soil fertility. Scott J Agric 20:41–45

Schneider O, Roger-Estrade J, Aubertot JN, Dore T (2006) Effect of seeders and tillage equipment on vertical distribution of oilseed rape stubble. Soil Tillage Res 85:115–122

Schroeder KL, Paulitz TC (2008) Effect of inoculum density and soil tillage on the development and severity of Rhizoctonia root rot. Phytopathology 98:304–314

Seidel D, Spaar D, Wächter V (1981) Einfluss eines Befalls durch *Gaeumannomyces graminis* var. *tritici* Walk. auf die Ertragskomponenten und das Wurzelwachstum in Abhängigkeit von den Entwicklungsstadien der Weizenpflanze. Arch Phytopathol Pflanzenschutz 17:95–104

Shinoda H, Murakami K, Gotto I (2005) Effect of amelioration of soil acidity and continuous cropping of cruciferous vegetables on the incidence of clubroot disease and resting spore density in soil. Jpn J Soil Sci Plant Nutr 76:891–896

Shipton PJ, Cook RJ, Sitton JW (1973) Occurrence and transfer of a biological factor in soil that suppresses take-all of wheat in eastern Washington. Phytopathology 63:511–517

Sieling K, Ubben K, Christen O (2007) Effects of preceding crop, sowing date, N fertilization and fluquinconazole seed treatment on wheat growth, grain yield and take-all. J Plant Dis Prot 114:213–220

Simon A, Sivasithamparam K (1988) Crop rotation and biological suppression of *Gaeumannomyces graminis* var. *tritici* in soil. Mycol Res 91:279–286

Simon A, Sivasithamparam K (1989) Pathogen-suppression: a case study in biological suppression of *Gaeumannomyces graminis* var. *tritici* in soil. Soil Biol Biochem 21:331–337

Simon A, Sivasithamparam K, Macnish GC (1988) Effect of application to soil of nitrogenous fertilizers and lime on biological suppression of *Gaeumannomyces graminis* var. *tritici*. Mycol Res 91:287–294
Singh E, van Dam JC, Feddes RA (2006) Water productivity analysis of irrigated crops in Sirsa district, India. Agric Water Manage 82:253–278
Sinha RC, Savard ME (1997) Concentration deoxynivalenol in single kerels and various tissues of wheat heads. Can J Plant Pathol 19:8–12
Sivasithamparam K, Parker CA, Edwards CS (1979a) Bacterial antagonists to the take-all fungus and fluorescent pseudomonads in the rhizosphere of wheat. Soil Biol Biochem 11:161–165
Sivasithamparam K, Parker CA, Edwards CS (1979b) Rhizosphere micro-organisms of seminal and nodal roots of wheat grown in pots. Soil Biol Biochem 11:155–160
Sleigh MA, Barlow DI (1982) How are different ciliary beat patterns produced? In: Amos WB, Duckett JG (eds) Prokaryotic and eukaryotic flagella. Cambridge University Press, Cambridge, pp 139–157 (Symposium of the Society for Experimental Biology, symposium no XXXV)
Smiley RW (1978) Antagonists of *Gaeumannomyces graminis* from the rhizoplane of wheat in soils fertilized with ammonium- or nitrate-nitrogen. Soil Biol Biochem 10:169–174
Smiley RW, Cook RJ (1973) Relationship between take-all of wheat and rhizosphere pH in soils fertilized with ammonium - vs nitrate-nitrogen. Phytopathology 63:882–890
Smith IM, Chiarappa L, van der Graf NA (1984) World crop losses: an overview. In: Wood RKS, Jellis GJ (eds) Plant diseases, infection, damage and loss. Blackwell Scientific, Oxford, pp 213–223
Sochting HP, Verreet JA (2004) Effects of different cultivation systems (soil management and nitrogen fertilisation) on the epidemics of fungal diseases in oilseed rape (*Brassica napus*). Z Pflanzenkr Pflanzenschutz 111:1–29
Sreenivasaprajad S, Johnson R (2001) Major fungal diseases of rice: recent advances. Kluwer Academic Publishers, Dordrecht
Staricka JA, Allmaras RR, Nelson WW (1991) Spatial variation of crop residue incorporated by tillage. Soil Sci Soc Am J 55:1668–1674
Steffek R, Spornberger A, Altenburger J (2006) Detection of microsclerotia of *Verticillium dahliae* in soil samples and prospects to reduce the inoculum potential of the fungus in the soil. Agric Conspectus Sci (Zagreb, Croatia) 71:145–148
Stevens C, Khan VA, Rodriguez-Kabana R, Ploper LD, Backman PA, Collins DJ, Brown JE, Wilson MA, Igwegbe ECK (2003) Integration of soil solarisation with chemical, biological and cultural control for the management of soilborne diseases of vegetables. Plant Soil 253:493–506
Strelkov SE, Manoli VP, Hwang SF, Howard RJ, Manoli AV, Zhou Q, Holtz M, Yang Y (2008) Incidence of clubroot on canola in Alberta in 2007. Can Plant Dis Surv 88:101–102
Stukenbrock EH, McDonald BA (2008) The origins of plant pathogens in agro-ecosystems. Annu Rev Phytopathol 46:75–100
Sturz AV, Carter MR (1995) Conservation tillage systems, fungal complexes and disease development in soybean and barley rhizospheres in Prince Edward Island. Soil Tillage Res 34:225–238
Sturz AV, Carter MR, Johnston HW (1997) A review of plant disease, pathogen interactions and microbial anatagonism under conservation tillage in temperate humid agriculture. Soil Tillage Res 41:169–189
Suffert F (2007) Modelling of the kinetics of cavity spot caused by a complex of pathogenic agents of the genus Pythium dominated by *Pythium violae*. Can J Plant Pathol 29:41–55
Suffert F, Delalande D, Prunier M, Andrivon D (2008) Modulation of primary and secondary infections in epidemics of carrot cavity spot through agronomic management practices. Plant Pathol 57:109–121
Sutton MW, Robinson DL, Dixon GR, Wilson F (1988) A comparison of growth and yield by virus-tested and visually healthy commercial Narcissus stocks from different localities. J Hortic Sci 63:479–487
Takahashi H, Ishikawa T, Kaido M, Takita K, Hayakawa T, Okazaki K, Itoh K, Mitsui T, Hori H (2006) *Plasmodiophora brassicae* induced cell death and medium alkalisation in clubroot-resistant cultured roots of *Brassica rapa*. J Phytopathol 154:156–162

Thomashow LS, Weller DM (1988) Role of a phenazine antibiotic from *Pseudomonas fluorescens* in biological control of *Gaeumannomyces graminis* var. *tritici*. J Bacteriol 170:3499–3508
Tilston EL (2000) Phytoprotective properties of composted recycled organic matter against soil-borne plant pathogens. Ph.D. thesis, University of Exeter
Tilston EL, Pitt D, Groenhof AC (2002) Composted recycled organic matter suppresses soil-borne diseases of field crops. New Phytol 154:731–740
Tilston EL, Pitt D, Fuller MP, Groenhof AC (2005) Compost increases yield and decreases take-all severity in winter wheat. Field Crops Res 94:176–188
Torsvik V, Ovreas L (2002) Microbial diversity and function in soil: from genes to ecosystems. Curr Opin Microbiol 5:240–245
Troxler J, Berling C-H, Moënne-Loccoz Y, Keel C, Défago G (1997) Interactions between the biocontrol agent *Pseudomonas fluorescens* CHA0 and *Thielaviopsis basicola* in tobacco roots observed by immunofluorescence microscopy. Plant Pathol 46:62–71
Tuitert G, Szczech M, Bollen G (1998) Suppression of *Rhizoctonia solani* in potting mixes amended with compost made from organic household waste. Phytopathology 88:764–773
Ulacio D, Nass H, Pineda J (1999) Viability of *Rhizoctonia solani* AG1-1A in flooding conditions. II. Effects of propagules in rice plants. Bioagro 11:61–68
van Bruggen AHC, Termorshuizen AJ (2003) Integrated approaches to root disease management in organic farming systems. Australas Plant Pathol 32:141–156
Van der Plank JE (1963) Plant diseases: epidemics and control. Academic, London
Van Elsas JD, Dijkstra AF, Govaert JM, van Veen JA (1986) Survival of *Pseudomonas fluorescens* and *Bacillus subtilis* introduced into two soils of different texture in field microplots. FEMS Microbiol Ecol 38:151–160
van Elsas JD, Garbeva P, Salles J (2002) Effects of agronomical measures on the microbial diversity of soils as related to the suppression of soil-borne plant pathogens. Biodegradation 13:29–40
van Os GJ, van Ginkel JH (2001) Suppression of Pythium root rot in bulbous Iris in relation to biomass and activity of the soil microflora. Soil Biol Biochem 33:1447–1454
van West P, Morris BM, Reid B, Appiah AA, Osborne MC, Campbell TA, Shepherd SJ, Gow NAR (2002) Oomycete plant pathogens use electrical fields to target roots. Mol Plant-Microbe Interact 15:790–798
Vargas GS, Meriles JM, Haro R, Casini C, March GJ (2008) Crop rotation and tillage systems as a proactive strategy in the control of peanut fungal soilborne diseases. Biocontrol 53:685–698
Vawdrey LL, Grice KE, Peterson RA, De Faveri J (2004) The use of metalaxyl and potassium phosphonate, mounds, and organic and plastic mulches for the management of Phytophthora root rot of papaya in far northern Queensland. Australas Plant Pathol 33:103–107
Vojinovic ZO (1972) Antagonists from soil and rhizosphere to phytopathogens. Final technical report. Institute of Soil Science, Beograd. Cited in: Wildermuth GB (1982) Soil suppressive to *Gaeumannomyces graminis* var. *tritici*: effects on other fungi. Soil Biol Biochem 14:561–567
Walker PT (1983) Crop diseases: the need to quantify the effects of pests, diseases and weeds on agricultural production. Agric Ecosyst Environ 9:119–158
Walker JC, Foster RE (1946) Plant nutrition in relation to disease development. III. Fusarium wilt of tomato. Am J Bot 33:259–264
Walker JC, Gallegly ME, Bloom JR, Scheffer RP (1954) Relation of plant nutrition to disease development. viii. Verticillium wilt of tomato. Am J Bot 41:760–762
Wallenhammar AC (1998) Observations on yield loss from *Plasmodiophora brassicae* infections in spring oilseed rape. Z Pflkrankh Pflschutz 105:1–7
Wallenhammar A-C (1999) Monitoring and control of *Plasmodiophora brassicae* in spring oil-seed brassica crops. Acta Universitatis Agriculturae Sueciae, Agraria 183. Ph.D. thesis, Swedish University of Agricultural Sciences, Uppsala
Watson AG (1967) The movement of *Plasmodiophora brassicae* in soil. Phytopathology 57:608
Weatherly PE (1979) The hydraulic resistance of the soil-root interface – a cause of water stress in plants. In: Harley JL, Scott-Russell R (eds) The soil-root interface. Academic, London, pp 275–286

Webster MA (1986) pH and nutritional effects on infection by *Plasmodiophora brassicae* Wor. and on clubroot symptoms. unpublished Ph.D. thesis, University of Aberdeen, Aberdeen
Webster MA, Dixon GR (1991a) Calcium, pH and inoculum concentration influencing colonization by *Plasmodiophora brassicae*. Mycol Res 95:64–73
Webster MA, Dixon GR (1991b) Boron, pH and inoculum concentration influencing colonization by *Plasmodiophora brassicae*. Mycol Res 95:74–79
Westerveld SM, McDonald MR, Scott-Dupree CD, McKeown AW (2002) The effect of nitrogen on insect and disease pests of onions, carrots, and cabbage. J Veg Crop Prod 8:87–99
Weston WARD (1944) Take-all or whiteheads of wheat and barley. Agriculture (London) 51:226–227
Whipps JM (1990) Carbon economy. In: Lynch JM (ed) The rhizosphere. Wiley, Chichester, pp 59–97
Whipps J, Lynch JM (1985) Energy losses by the plant in rhizodeposition. In: Fuller KW, Gallon JR (eds) Plant products and the new technology, Annual proceedings of the phytochemical society of Europe, vol 26. Clarendon, Oxford, pp 59–71
Wiggins BE, Kinkel LL (2005) Green manures and crop sequences influence potato diseases and pathogen inhibitory activity of indigenous Streptomyces. Phytopathology 95:178–185
Wild A (1988) Plant nutrients in soil – nitrogen. In: Wild A (ed) Russell's soil conditions and plant growth. Longman Scientific & Technical, London, pp 652–694
Wildermuth GB (1982) Soil suppressive to *Gaeumannomyces graminis* var. *tritici*: effects on other fungi. Soil Biol Biochem 14:561–567
Wildermuth GB, Rovira AD, Warcup JH (1985) Mechanism and site of suppression of *Gaeumannomyces graminis* var. *tritici* in soil. Mycol Res 84:3–10
Wilkinson HT, Weller DM, Alldredge JR (1982) Enhanced biological control of wheat take-all when inhibitory Pseudomonas strains are introduced on inoculum or seed as opposed to directly into soil. Phytopathology 72:978–949
Wiseman BM, Neate SM, Keller KO, Smith SE (1996) Suppression of *Rhizoctonia solani* anastomosis group 8 in Australia and its biological nature. Soil Biol Biochem 28:727–732
Worku Y, Gerhardson B (1996) Suppressiveness to clubroot, pea root rot and Fusarium wilt in Swedish soils. J Phytopathol 144:143–146
Wu BM, Subbarao KV (2003) Effects of irrigation and tillage on the temporal and spatial dynamics of *Sclerotonia minor* sclerotia and lettuce drop incidence. Phytopathology 93:1572–1580
Yamagishi H, Yoshikawa H, Ashizawa M, Hida K-i, Yui S (1986) Effects of resistant plants as a catch crop on the reduction of resting spores of clubroot (*Plasmodiophora brassicae* Worn.) in soil. J Jpn Soc Hortic 54:460–466
Yang MY, Yanf JL, Sun DW, Yan WZ (2004) Effects of soil moisture on the incidence of clubroot. SW China J Agric Sci 17:482–483
Yarham DJ (1981) Practical aspects of epidemiology and control. In: Asher MJC, Shipton PJ (eds) Biology and control of take-all. Academic, London, pp 353–384
Young IM, Ritz K (2000) Tillage, habitat space and function of soil microbes. Soil Tillage Res 53:1229–1232
Zadoks JC (1987) Rationale and concepts of crop loss assessment for improving pest management and crop protection. In: Teng PS (ed) Crop loss assessment and pest management. American Phytopathological Society, St Paul, MN, pp 1–5
Zadoks JC (1990) Management of wheat diseases in NorthWest Europe. Can J Plant Pathol 12:117–122
Zadoks JC, Schein RD (1979) Epidemiology and plant disease management. Oxford University Press, Oxford
Zasada IA, Ferris H, Elmore CL, Roncoroni JA, MacDonald JD, Bolkan LR, Yakabe LE (2003) Field applications of brassicaceous amendments for control of soil borne pests and pathogens. Plant Health Prog November:1–8
ZhiYong D, XiaoLin F (2008) Results of banana Fusarium wilt control and banana reproduction in degraded orchard by use of modified lime nitrogen. J Fruit Sci 25:373–377

Chapter 7
The Impact of Land-Use Practices on Soil Microbes

Emma L. Tilston, Tom Sizmur, Geoffrey R. Dixon, Wilfred Otten, and Jim A. Harris

Introduction

The extensive use of land resources for food production, fibre for construction, wood pulp for paper, removal for extractive industries, sealing for urban and industrial development and as a receiver (either deliberate or accidental) of polluting substances has wrought huge changes in the chemistry, structure and biology of soils, away from their natural state.

These pressures are likely to continue, if not increase in the coming century due to the ineluctable increase of human population; in the order of an additional two to four billion in the next half century (Cohen 2003). All of these people will require food, shelter, goods and services, as well as anticipating an increase in material wealth through the development of their nation states.

E.L. Tilston (✉)
Scottish Universities Environmental Research Centre, Scottish Enterprise Technology Park, East Kilbride, G75 0QF, UK
e-mail: e.tilston@suerc.gla.ac.uk

T. Sizmur
Department of Soil Science, University of Reading, Whiteknights, Berkshire, RG6 6DW, UK
e-mail: t.p.sizmur@reading.ac.uk

G.R. Dixon
Centre for Horticulture & Landscape, School of Biological Sciences, Whiteknights, University of Reading, Berkshire, RG6 6AS, UK
e-mail: geoffrdixon@btinternet.com

W. Otten
SIMBIOS Centre, University of Abertay, Bell Street, Dundee, DD1 1HG, Scotland, UK
e-mail: w.otten@abertay.ac.uk

J.A. Harris
Natural Resources Department, School of Applied Sciences, Cranfield University, Cranfield, Bedfordshire, MK43 0AL, UK
e-mail: j.a.harris@cranfield.ac.uk

G.R. Dixon and E.L. Tilston (eds.), *Soil Microbiology and Sustainable Crop Production*,
DOI 10.1007/978-90-481-9479-7_7,

The central challenge this offers is how we can continue to maintain soil health whilst managing the trade-offs between optimization of agricultural production and the other ecosystem goods and services provided by plant-soil systems (Kibblewhite et al. 2008). These authors indicate how the various flows of a range of ecosystem goods and services are dependant upon the activities of the soil biological community (see Fig. 1).

The effects of this increased pressure on soil resources will be manifest in a number of areas, notably hydrology and soil organic matter. Work by Bellamy et al. (Bellamy et al. 2005) has demonstrated substantial losses of soil organic matter, in some cases up to 50% in a 25 year period, probably driven by agricultural intensification with an element of climate change. The management of these carbon stocks to mitigate, and adapt to, climate change and increased pressures for food production, will be essential to maintain a civilized society (Lal 2007). Even before this there were calls for institutional changes to increase focus on soil biology, rather than traditional approaches based on the manipulation of chemistry and physics of soils (Sherwood and Uphof 2000).

Measuring the Soil Microbial Community

In order to detect the effects that differing land uses and management strategies undoubtedly have on soil microbial communities it is necessary to have techniques for measuring their size, composition and function. There are two principal ways of organising the way in which this can be done and we can adopt a typology based on:

- Size – how much living microbial biomass there is
- Composition – what that biomass comprises of in terms of different "taxonomic" groups
- Activity – at what rates the microbial community processes materials and energy
- Physical arrangement of the community – using a variety of visualisation techniques

Or we can look at the levels of organisation deriving from the evolutionary history of soil organisms

- Genotype – the blueprint of the community
- Phenotype – the expressed parts
- Functional capabilities – the potential for the extant community to process substrates

For a fuller review of the methodologies available see Harris and Steer (2003).

Size

This is the amount living biomass and is usually expressed as biomass-C and may be determined in a number of ways, directly and indirectly. Fumigation of soils

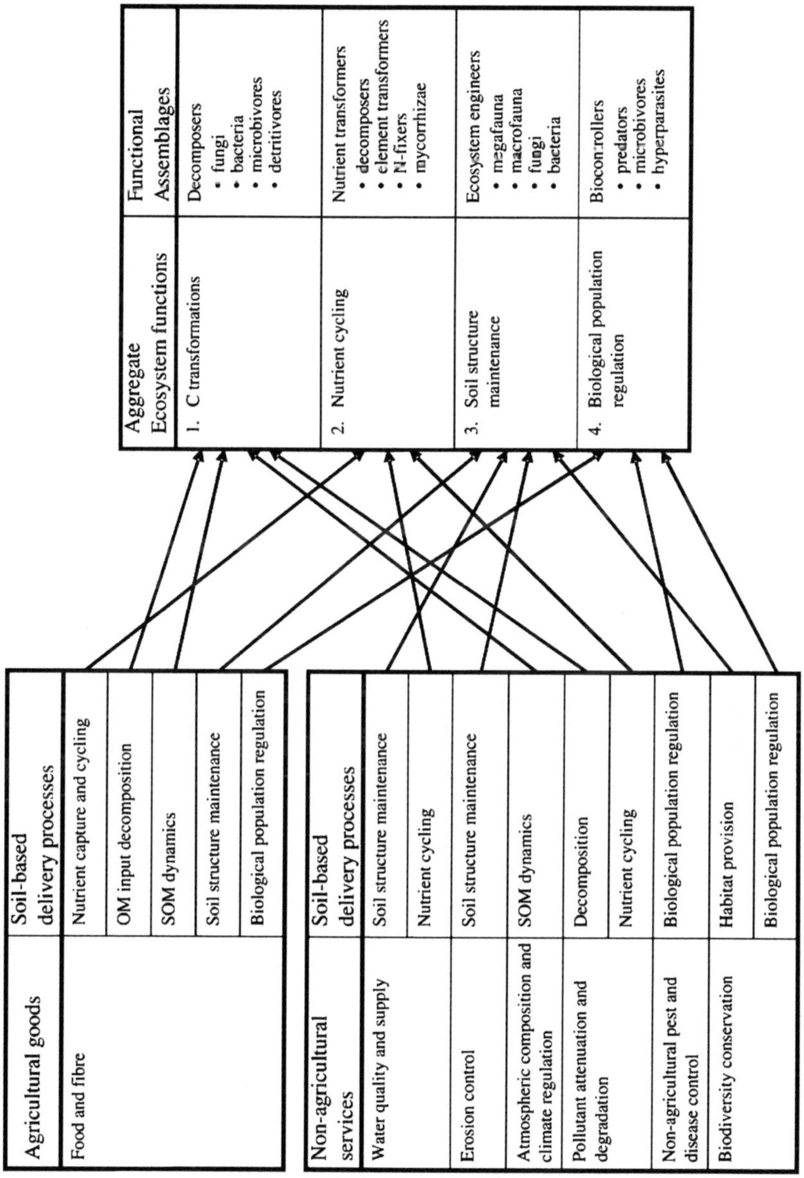

Fig. 1 Relationships between the activities of the soil biological community and ecosystem services (From Kibblewhite et al. 2008). Reproduced with permission from the Royal Society

with chloroform followed by extraction and determination of soluble carbon, in comparison with a non-fumigated control is commonly used (Bloem et al. 2005), as is determination of the soils adenosine tri-phosphate content (ATP), e.g. (Bentham et al. 1992). This latter method is less commonly used, but has the advantage of being exquisitely sensitive to low concentrations of biomass, and being absolutely dependant on the presence of living cytoplasm within a viable cell – it is not likely to produce artefacts as it is metabolised almost instantaneously on cell death.

Composition

The composition of communities has been commonly used in the sense of species numbers and their abundance. This is not easy to determine, as the community is incredibly diverse, with estimates of 10,000 species of soil bacteria alone (Torsvik et al. 1990), with in excess of 5,000 species in a single gram. Traditional methods of culturing "plate-counts" seriously underestimate this diversity, with as little as 0.1% of soil microbial species being culturable (Ritz 2007). Modern approaches rely on biochemical methods such as the extraction of phospholipid fatty acid profiles (Peacock et al. 2001a; Steer and Harris 2000), which provide a community "finger-print" which can be used to ascertain the effects of management, pollution, ecosystem health, and plant growth; or molecular and approaches which indicate the genotypic diversity of the microorganisms in a sample (Torsvik and Ovreas 2002).

Activity

Some measurements are carried out *in situ* and are particularly useful when considering changes over short time spans. However, most measurements on restored sites are those carried out on samples returned to the laboratory. Common measurements include enzymatic assays and respirometry. Enzyme assays are particularly versatile and can indicate both non-specific (e.g. dehydrogenase – DHA) and specific activity within the soil microbial community. Assays for phosphatase, sulphatase and urease are of particular interest because of their importance to phosphorus, sulphur and nitrogen cycling (Speir and Ross 2002).

Physical Arrangement

Determining the physical arrangement of the microbial community, particularly with respect to organic matter resources and soil physical architecture, can be very rewarding in terms of elucidating the recovery of biotic-abiotic linkages and structural stability. It is, however, difficult and time consuming but continues to be very useful in a research context.

There is a gap between the scale at which we observe and manage soil ecosystems and the scale at which processes involving microorganisms take place, and it is becoming increasingly clear that the emergent behaviour of soils can only be fully understood if we account for the underlying heterogeneity. Much of our understanding of spatial heterogeneity in soils relies heavily on destructive sampling and the concept of soil aggregates. These techniques exert physical forces in order to produce and break aggregates. Whilst having brought great insight, such approaches do ignore much of the spatial heterogeneity which may be key to soil functioning. Mineralogical and biological thin sections of soils revealed the first clear quantitative data of the enormous heterogeneity of soils at microbial scales (Tippkötter et al. 1986). These methods still provide the highest spatial resolution to study the distribution of soil microorganisms (Nunan et al. 2002), and when combined with catalysed reporter deposition and fluorescence *in situ* hybridization (CARD-FISH), enable phylogenetic staining of microorganisms *in situ* (Eickhorst and Tippkötter 2008). Advancements of microbial techniques to the 3-D environment where bacteria can be active in relatively thin films adhering to surfaces are however slow.

In contrast, the ability to visualize and quantify the physical heterogeneity at microscopic scales has advanced tremendously in recent years with the development and application of neutron scanning (Schaap et al. 2008), nuclear magnetic resonance (Pohlmeier et al. 2009) and X-ray tomography (Carminati et al. 2008; Deurer et al. 2009). This has brought about a revolution in the way soil can be characterized with a shift in emphasis from destructive analysis of the solid phase to non-destructive characterization of the geometry and connectivity of the pore volume. Carminati et al. (2008) demonstrated the intimate relationship between the distribution of soil water and the micro structure using synchrotron X-ray tomography. The distribution of water within a physical structure determines the connectivity of physical habitats and contributes to the great microbial diversity in soils (Or et al. 2007). Nowadays, micron scale resolutions can be achieved for small samples, but larger samples (plant-root microcosms) can also be readily scanned at high spatial resolutions (Fig. 2). This provides great insight into the complex space within which water, air and resources move, and microorganisms live and interact.

In addition to the physical heterogeneity, soils are chemically heterogeneous, with, for example, the distribution of carbon intimately linked to soil structure. Conventional chemical analyses often take place after homogenizing relatively large soil samples, and even studies with small samples of soils are effectively bulk analyses (Wan et al. 2007). Recently, microscopic and micro-spectroscopic analyses are beginning to address this. Examples include the use of stable carbon isotope ratio measurements by laser combustion with a typical resolution of a few hundred micrometres, soft X-ray techniques such as STXM (soft tissue X-ray microscopy) and NEXAFS (near edge X-ray absorption fine structure) spectroscopy which enable the study of soil organic carbon distributions at a spatial resolution of 30 nm (Wan et al. 2007). To date, these techniques are restricted to small samples (few millimetres in diameter) and require access to synchrotron facilities, but more accessible micro computed-tomography (CT)/micro X-ray fluorescence (XRF) scanners for non-destructive 3-D

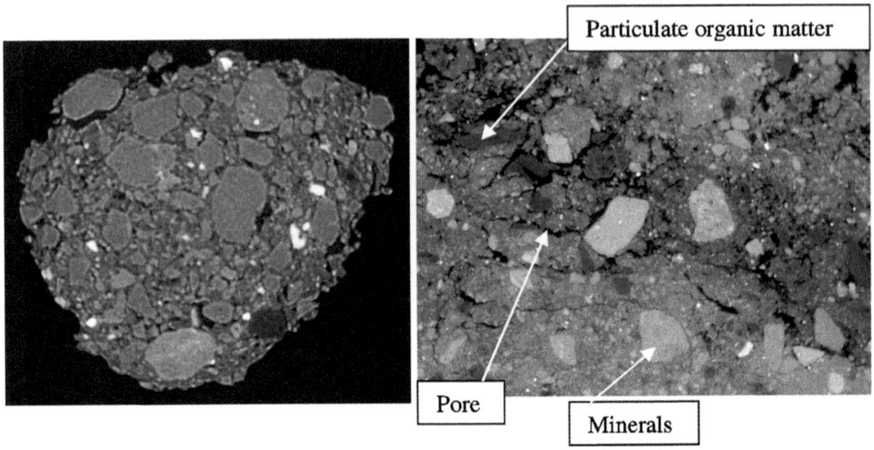

Fig. 2 Visualization of the internal structure of soils: Transect through a 3-D volume showing the internal structure of an aggregate (approx. 2 mm in diameter) scanned at the Advance Photon Source (APS, Chicago, Dr. M. Rivers) with a spatial resolution of [illegible] μm (*left*), and the internal structure of a soil core (approx. 4 cm) from the top layer of an arable field after ploughing (30 μm resolution). Pores are visible as dark areas (black), particulate organic matter as dark grey, and denser minerals are presented by lighter grey scale values.
Images courtesy of the SIMBIOS Centre of the University of Abertay Dundee (P. Dello Sterpaio, D. Grinev, R. Pajor and W. Otten). The image on the left was performed at GeoSoilEnviroCARS (Sector 13), Advanced Photon Source (APS), Argonne National Laboratory with Dr M. Rivers.
GeoSoilEnviroCARS is supported by the National Science Foundation - Earth Sciences (EAR-0622171) and Department of Energy-Geosciences (DE-FG02-94ER14466). Use of the Advanced Photon Source was supported by the U. S. Department of Energy, Office of Science, Office of Basic Energy Sciences, under Contract No. DE-AC02-06CH11357

chemical analysis are making rapid progress (Sasov et al. 2008), and will soon deliver major advancement in our understanding of the chemical heterogeneity of soils.

Future efforts should address how to integrate these novel developments and how to assess its impact on larger scale processes. Such a holistic approach to soil systems is currently not exploited to its full potential as the technological advances are predominantly made within separate disciplines. Some difficulties may be overcome readily via an integrated experimental design whereas others are advanced via modelling and statistical approaches. This offers challenges for soil scientists as well great opportunities to deepen our understanding of soil microhabitats and deliver novel insight into soil ecosystem functioning.

Genotype

This approach looks at the soil microbial community in two ways: firstly, by determination of gene sequences identified with particular groups or traits; secondly by considering the soil microbial "genome" as a whole – here it is possible to examine differences in soil communities caused by differing management regimes, vegetation types, and external stressors and perturbations by determination of diversity based on sequences reflective of the fungal or bacterial components of the community. Examples of this approach include terminal restriction fragment length polymorphism (T-RFLP) analysis, and have proven powerful methods of analysis.

Analysis of particular functional groups in this way also remains as a strong presence in the field – particularly examination of groups associated with functions related to nitrogen cycling, such as ammonium oxidising bacteria.

Phenotype

The determination of the soil microbial community "phenotype" has produced a large body of work in the field. One approach stands out as a powerful, reliable and discriminatory tool – that of phospholipid fatty acid (PLFA) analysis. Phenotypic profiling based upon PLFA analysis is used to provide a "fingerprint" of the extant community composition. Since this analysis is based upon the membrane composition of soil organisms, it reflects the status of living and intact cells. Typically, an adaptation of a well-established method for extraction of neutral, glyco- and polar lipids with a single phase mixture of chloroform, methanol and water, followed by fractionation according to polarity, by adsorption to silica and selective elution and derivatisation to the fatty acid methyl esters (FAMEs) prior to analysis by gas chromatography/mass spectroscopy (GC-MS). This also allows the determination of stress/storage lipids, and, critically, the ratio of fungal-to-bacterial biomass (Zelles 1999). It is one of the principal methods identified by Ritz et al. (2009) for providing clear discrimination between land-use impacts.

Functional Capability

Recent attempts have been made to determine the ability of the soil microbial community to process a range of carbon substrates. One common approach has been the use of the BIOLOG MicroPlate™ bacterial identification system; a microtitre plate system with 95 carbon substrates – these are inoculated with a soil suspension and substrate utilisation is detected by a colour change. One drawback of this approach is that the culturable fraction of the microbial community is favoured, which may only represent less than 1% of the soil microbial community (Preston-Mafham et al. 2002). Degens and Harris (1997) have also developed a catabolic community profiling technique in which the substrate is taken to the community, and the response measured over the first 4 h, before any cell division is likely to have occurred and therefore is more likely to represent the response of the in situ community. This methodology is quite labour intensive and more recently Campbell et al. (2003) have developed a simpler method to carry out this assessment, which is capable of discrimination and interpretation when assessing the impact of land-use and management practice (Ritz et al. 2009).

Conventional Agriculture

Conventional agriculture operates on a simple principle of excluding all but the crop of interest from a given area. Therefore the microbial community characteristics reflect that focus on production. As there are clear changes in the microbial community structure during the course of succession (Harris 2009), it follows that any management intervention, in terms of land use change, are likely to lead to significant changes away from that successional trajectory. The major pressures brought to bear on the composition of the soil microbial community are:

- Cultivation – disruption of fungal networks are significant and long-lasting
- Fertilisation – altering nutrient balance in the soil impacts on the composition of the microbial biomass
- Pesticides – affect microbial community directly by acting as food sources, or indirectly as they impact on other components of soil food webs and plant communities
- Veterinary medicines – in livestock production these compounds have a significant effect on the soil microbial community antibiotics

Cultivation

Repeated destruction of stable soil structure by intensive cultivation disrupts fungal networks, and prevents their re-establishment. Alguacil et al. (2008) examined the effects of tillage practices on arbuscular mycorrhizal fungal (AMF) diversity in subtropical crops and found that the agricultural systems which included tillage had significantly lower AMF diversity than non-tilled reference sites. This effect on AMF is highly significant as they have been demonstrated to have a central role in the maintenance of plant health and soil fertility (Jeffries et al. 2003) with impacts on phosphorus-uptake, nutrient uptake generally, water relations, and above ground plant productivity. These factors are substituted with specific management intervention strategies in conventional agriculture, but the potential for the use of AMF in sustainable systems is promising, so their continued impoverishment in agricultural systems is problematical if we are to secure effective sustainable farming systems in the future (Gosling et al. 2006).

Fertilisation

There is an extensive literature on the impact of agricultural inputs on the soil microbial community (Bünemann et al. 2006). Although direct effects of fertilisers are sometimes difficult to clearly identify, addition of nitrogen in almost any form affects the carbon-to-nitrogen ratio and increases the degradability of soil organic matter stocks. Also, an indirect effect of nitrogen addition is that of acidification, which tends to result in decreased biomass over long periods of time, unless countered with applications of lime (which in itself may not be a sustainable practice in the long-term).

Nitrogen management is of central importance in productive systems and there has been serious effort to develop chemical interventions to inhibit nitrification – in this sense these amount to a pesticide intervention. These have met with variable success (Subbarao et al. 2006) and there is renewed interest in suppressing nitrification using biological control methods, such as using plants known to suppress nitrogen losses in this way (Subbarao et al. 2007).

Pesticides

The direct effects of herbicides on the soil microbial community would appear to be minor. There is, however, evidence for impacts of insecticides and fungicides (as one might expect), as both target important soil groups, with knock-on effects for other components of the soil food web. Bending et al. (2007) investigated the impacts of a range of fungicides on soil biology, using molecular techniques, and demonstrated no measurable impact on bacterial community structure, but in certain fungicide treatments losses in both fungi and protozoa.

Livestock Management: Pastures

Pastures would, on the face of it, appear not to suffer from the disturbances associated with aggressive cultivation management. There are disturbances, however, engendered by the impact of animals on soil structure (Palmer et al. 2006). Wakelin et al. (2009) have studied the interacting effects of pasture type, liming, phosphorus-fertilisation and grazing pressure (and sampling data) on soil microbial community structure, to identify the most sensitive components, on pasture systems in Australia. Using molecular tools (denaturing gradient gel electrophoresis, DGGE and T-RFLP) they found that liming had the biggest effect (mirrored on studies of pastures in the Scottish Borders; Pawlett et al. 2009), followed by time during growing season; fungal phylotype richness responded most markedly – capacity for denitrification hardly at all. This underlines the important general observation that microbial responses to land-use management at the community level are complex.

Sustainable Agriculture

In sustainable agriculture, the emphasis is on the management of natural resources so as to maximize production within the constraints of minimizing synthetic inputs sourced off-farm. By definition, therefore, the ecosystem services of soil microbial communities are valued more than they are in conventional agriculture and it is believed that sustainable management practices have positive impacts on microbial biomass, community structure and function.

Minimal or No Tillage and Organic Agriculture

Minimal cultivation systems probably offer the best prospect for the development of sustainable production systems based in part on biological control of pests and pathogens, and biological promotion of fertility (Kennedy 1999). There is some tentative evidence from a Canadian study that tree based intercropping systems can increase microbial diversity, but these are early results and more work needs to be carried out to confirm this (Lacombe et al. 2009).

The recovery of biodiversity does occur quickly as a result of switching from conventional to no-till/min-till or organic agriculture, but takes some time for the complexity of communities to develop – probably due to the time it takes to build organic matter stocks, rather than a lack of appropriate propagules being available in the soil or arriving from other sources (Simmons and Coleman 2008). Adl et al. (2006) demonstrated that even after 26 years in no-tillage management communities of micro-arthropods, nematodes and protozoa had not fully developed the richness and complexity found in natural communities. All these practices have been found to result in increased microbial biomass, and a general increase in fungal contribution to this biomass, leading to improved soil aggregation and stabilisation of soil organic matter (Six et al. 2006). There is also probably a role for the management of soil microbial communities specifically for the suppression of soil-borne pathogens in agricultural systems (Watt et al. 2006; Janvier et al. 2007).

Soil Amendments: Biochar

There has been increasing interest recently in the prospects for employing pyrolysis-derived organic materials, colloquially known as charcoal or "biochar", as amendments for degraded soil systems to improve soil nutrient holding and structural characteristics (Marris 2006). This suggestion arises from studies in the lowland humid tropics of "Terra Preta" areas found in South America, which reflect aboriginal practices of returning charcoal to cleared forest areas to provide for agricultural production on a small scale (typically circa 20 ha) on soils ostensibly too infertile to support sustainable agriculture (Glaser et al. 2001), see elsewhere in this Book. This has lead to various proposals for the widespread adoption of the use of this approach to not only support sustainable agriculture in humid regions, but more universally as a method for sequestering atmospheric carbon (Lehmann et al. 2006).

The dynamics of organic matter in systems where biochar has been added are poorly understood, if at all. Several studies suggest, however, that the addition of non-native pyrolysed materials may actually result in the accelerated decom- position of native soil organic matter. Wardle et al. (2008) have recently demonstrated that addition of fire-derived charcoal to forest humus resulted in a significant loss of native organic matter over a 10 year period. It is known that addition of charred plant materials to soil accelerates breakdown of simple carbohydrates (Hamer et al. 2004), and that microbial community activity and metabolic efficiency increases linearly with the addition of charcoal to soils (Steiner et al. 2008). Several potential mechanisms for these observations are possible – including the increased retention of soluble forms of nitrogen by high cation-exchange-capacity of pyrolised materials facilitating the accelerated decomposition of native organic matter by overcoming recalcitrance of the native materials. This suggests that adding pyrolysed organic materials to soils may result in a net loss, not gain, in soil carbon, and irrevocably change soil organic matter dynamics, but such effects may be modulated by the soil type and its carbon status.

Animal Manures, Sewage Sludge and Other Organic Wastes

The application of animal manures, sewage sludges and composts have the potential not only to increase soil organic matter stocks, but also to increase the size (Anderson and Domsch 1989; Guerrero et al. 2007), activity (Bolton et al. 1985) and diversity (Hassink et al. 1991) of the soil microbial community. Such enhancements to the soil microbial community may have benefits for plant productivity through increased nutrient cycling rates (Doran et al. 1987; Kirchner et al. 1993) and reducing the risk of soil erosion through improved soil structure. Promotion of aggregation (Angers et al. 1992) and the allied properties macroporosity and pore continuity (Bronick and Lal 2005), also reduce nutrient losses through leaching (Elliott and Coleman 1988).

Phospholipid fatty acid profiling of soils from long-term agricultural field trials (10–100+ years) has proved particularly useful in describing the changes in microbial community structure developing in soils amended with inorganic nitrogen and farmyard manure. In agreement with other biological and chemical parameters (Kandeler et al. 1999), the relative responses are typically no amendment < inorganic nitrogen < farm yard manure (Peacock et al. 2001b; Wander et al. 1995; Böhme et al. 2005; Toyota and Kuninaga 2006). The abundance of biomarkers for Gram-negative bacteria are normally increased in soils receiving application of farmyard manures (Peacock et al. 2001b; Böhme et al. 2005; Bossio et al. 1998), but decreased after application of inorganic nitrogen (Peacock et al. 2001b; Böhme et al. 2005). Applications of inorganic nitrogen (Peacock et al. 2001b) or compost (Bossio et al. 1998), are however associated with increased abundance of Gram-positive biomarkers. Sewage sludge applications elicit responses comparable with those for farmyard manure (Petersen et al. 2003).

One disadvantage of the extensive and repeated application of organic amendments, including farmyard manure, is the accumulation of heavy metals and other xenobiotic compounds within soil (DEFRA 2009; Hamscher et al. 2005; Stoob et al. 2006). Antibiotics, copper and zinc are commonly added to livestock feed as growth promoters, to increase the supply of trace elements or for medicinal purposes (DEFRA 2009). Without changes to upstream animal husbandry and waste management, greater use of manures and allied materials within a more sustainable form of conventional agriculture has the potential to produce substantial, long-lived changes to soil microbial community composition through the introduction of metals and antibiotics/other veterinary medicines (Abaye et al. 2005). Elevated levels of metals in soils decrease soil microbial biomass, activity and diversity, and change the structure of the microbial community (Bååth 1989). These changes probably occur at lower concentrations than those at which toxic effects to plants or animals are detected (Giller et al. 1998). As concentrations are built up slowly over time it is likely that the changes develop from stress responses rather than profound disturbance responses. The ratio of cyclopropyl fatty acids to their metabolic pre-cursors is commonly used to indicate stress within microbial communities (Bossio and Scow 1998) and has been used to indicate long-term metabolic stress associated with sewage sludge application (Petersen et al. 2003). Applications of farmyard manure, on the other hand, tend to decrease stress indicators suggesting alleviation of sub-optimal nutritional status (Bossio et al. 1998; Bossio and Scow 1998). Sulphonamides are broad-range antibiotics commonly used in European agriculture, they inhibit dihydropteroate synthesis in the folic acid pathway (O'Neil et al. 2001) reducing bacterial reproduction. They enter the soil either by direct deposition via livestock, or indirectly through the spreading of manure or slurry and it would appear that the presence of manure increases the tolerance of soil microbial communities to sulphonamides antibiotics (Demoling et al. 2009; Hammesfahr et al. 2008).

Use of Plant Growth Promoting Rhizobacteria (PGRP)

The importance of bacterial symbionts to legumes has long been known, but wider use of plant growth promoting bacteria has become more commonplace. There are now several PGPR formulations available which are used commercially in agricultural production (Lucy et al. 2004). Direct inoculation of soils with microbial preparations has also been shown to increase the efficiency of fertiliser uptake (Adesemoye and Kloepper 2009). They are also being used in forest re-establishment and phytoremediation of contaminated soils.

Impact of Civil Engineering, Mineral Extraction and Other Forms of Extreme Disturbance

In response to changes in policy and economic activity, it is not unusual for land formerly subject to industrial uses e.g. coal spoil tips to be restored to agricultural uses such as pasture and biofuel production (Davies 1999). The terrestrial civil engineering programmes for mineral extraction and landscape re-instatement inevitably lead to significant degradation of the soil ecosystem. Principally, there are large shifts in the functional and phenotypic characteristics of the soil microbial community, which are consistent and predictable for the type of degradative activity applied. Similarly, as the system recovers (or not as the case may be) during the course of reclamation and restoration programmes there are significant shifts in the community structure towards the type of profile found in the desired target system. Harris (2009) has suggested that the close relationship between the successional state of an ecosystem, the fungal-to-bacterial ratio and size of the microbial community in bulk soil offers a template for assessing the degree of degradation and the extent of ecosystems affected by civil engineering-based activities (Fig. 3).

The handling and re-instatement of soils and soil forming materials is usually one of the principal activities of the land reclamation process. There is little consideration given to the soil as a living material, rather it is treated as an engineering material with some unfortunately recalcitrant characteristics. What is meant by restoration and reclamation is central to the assessment of the success or otherwise of such programmes, and still the subject of much debate (van Diggelen et al. 2001). Harris (2003) has reviewed the use of soil microbial community measurements to determine the "state" or "quality" of soils in restoration programmes,

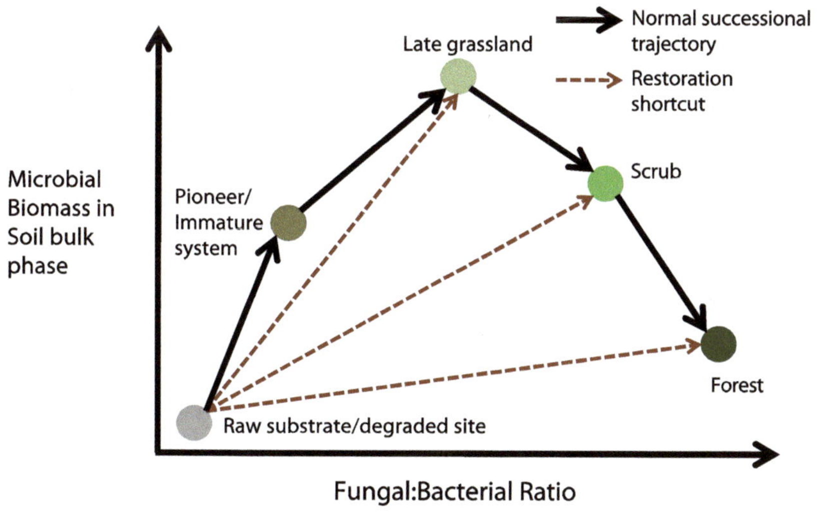

Fig. 3 Relationship between ecosystem successional state and microbial community size and composition. Copyright: JA Harris. From (Harris 2009)

related to the physico-chemical condition of soil, and their role in facilitating and assessing restoration success (Harris 2009).

There is a plethora of methods now available for the rapid and reliable assay of the soil microbial community (Ritz et al. 2009). These allow for assessment on a time scale fast enough to make management interventions to enhance or correct the trajectory of soil recovery to a condition being targeted to secure the target state. This offers the prospect of the development of decision support tools encompassing the full range of variables associated with soil handling and re-establishment of the soil-plant system. Eventually this may lead to a reduction in the number of restoration schemes subject to failure.

Changes During Disturbance

Impacts on the soil microbial community during movement and trafficking operations are significant (Harris et al. 1989). They are not confined to the mineral extraction and construction industries: Frey and co-workers (Frey et al. 2009) examined the impact of logging operations on soil bacterial community structure, with particular reference to the effects of compaction. Using analysis of soil microbial biomass-carbon and T-RFLP community genotypic profiling, they detected a decrease in microbial biomass, and shifts in the bacterial community composition. They ascribed these changes to the decreases in soil porosity (up to 17%) and large decreases in pores of >50 μm, which were associated with higher water-logging and restricted gas exchange.

Changes During Topsoil Storage

There are large and rapid changes in the soil community as a result of the handling, and movement of soils during civil engineering activities, with these damaging effects made worse by storage for long periods in stockpiles (Harris et al. 1993; Stahl et al. 2002). Bacterial numbers initially increase as a result of the large amounts of nutrient made available by the death of larger organisms (most significantly the fungi and invertebrates, such as earthworms) caused by the shear and compaction forces applied during the soil lifting and store construction process. Following this the bacterial component of the soil biomass declines as both nutrients and oxygen are exhausted at depth in the stockpile, particularly in fine textured soils, until only anaerobic metabolism persists.

Invertebrates, particularly earthworms, are severely affected by handling procedures. They are liable to be crushed and unable to find physical refuges from the large compaction and shearing forces being imparted to the soil by the heavy earth-moving equipment, leading to large scale declines in their populations (Scullion 1994). Eventually the soil microbial biomass in stockpiles declines to values less than

5% of the undisturbed values, and, depending on texture, the microbial biomass in less than 10% of the volume of stockpiled soil recovers to pre-disturbance values (Harris and Birch 1990a). This recovery is further destroyed, however, when the soil is re-instated at the end of coaling operations.

Recovery After Disturbance

Ruzek et al. (2001) demonstrated clear relationships between time since restoration and increases in soil microbial biomass, in reclaimed sites in the Czech Republic and Germany. Ruzek and co-workers (2001) indicated that this was related to both organic matter content as a starting point in new reclamations, and the textural characteristics of the soils reclaimed, and developed an algorithm which could be applied to different soil types, with modifications appropriate to regions with differing soil resources.

Total biomass is not the only characteristic which changes with time. Yin et al. (2000), showed that the total number of bacterial species on a site recovering from mining activity increased fairly quickly around pioneer vegetation, but the proportion of these species which were active (as determined by a radio-label incorporation method) took much longer to increase, and at no point matched the undisturbed reference forest site This offers a potentially very sensitive approach for determining the efficacy of treatment strategies for restoring function to systems.

Relationship Between Soil Microbial Community and Other Parameters

Of great interest to those attempting to restore function to soils after civil engineering activities is the interaction between the microbial community and soil structural characteristics.

Edgerton and co-workers (Edgerton et al. 1995) investigated the recovery of soil structural characteristics and the size of the microbial biomass in a number of soils re-instated after opencast mine restoration (Fig. 4). This relationship is linear for the sites shown in Fig. 3, after which time the relationship becomes log-linear. This development of structural stability, principally in this case the ability to retain structure when subject to wetting and shaking (quite vigorously) is essential if the soil is to function and redevelop drainage channels. Otherwise the re-instated soil will remain compacted and liable to shed rather than absorbing water – causing off-site pollution. The microbial communities in such degraded soils will never recover function and will be droughty in the summer and water-logged in winter. This process takes longer than the standard 5-year post-mining recovery period normally allowed. This also illustrates that measurements of the microbial community can

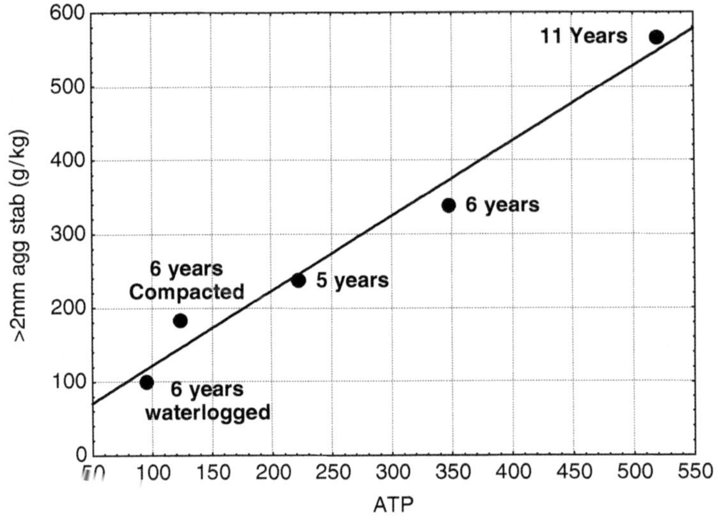

Fig. 4 Relationship between water stable aggregates (>2 mm g kg^{-1}) and microbial biomass (ATP ng g^{-1}) at five restored sites (Redrawn from Edgerton et al. 1995). Reproduced with permission of Elsevier

often be used as a surrogate for the recovery of other soil characteristics and functions, with assessments rapidly and reliably available.

Effects of Management

The importance of the type of management employed in restoring sites are illustrated by the work on restored species rich meadows in the English Midlands around Nottingham, subject to soil re-instatement after opencast coal mine operations (Harris and Birch 1990b). There were two undisturbed meadows, classified as "wet" and "dry" and two sites re-instated some 5 years previously, and two 10 years previously. Two management regimes had been applied – grazing or cutting and leaving the aftermath on the field. Determinations of soil activity by soil enzyme measurements (dehydrogenase activity which can be used as a surrogate for respiration) were made at 0–5 cm intervals to a depth of 30 cm. A clear result emerges from this study – cutting and leaving the aftermath was superior to grazing in encouraging microbial activity leading to a more rapid recovery of the activity profile. What is also clear, however, is that these soils will take in excess of 20 years of this type of careful management before they begin to approach the undisturbed reference sites with regards to microbial community activity. This gradual development of soil catabolic capabilities and respiration dependant on management practice has been found in other studies, e.g. (Chodak et al. 2009).

The use of other biomarkers is becoming routine in the determination of ecosystem status with respect to land use change – the lipids are the major group used for this purpose, and have found widespread application (Zelles 1999). Mummey and co-workers (Mummey et al. 2002) investigated the application of one group of lipid biomarkers, the fatty acid methyl esters (FAMEs). Lipids were extracted and derivatized to FAMEs, from samples taken from a surface mine reclamations of different ages, and an undisturbed reference site. They were clearly able to distinguish between the different stages of reclamation, and identified a trend towards the undisturbed reference condition. They suggest that the ratio of fungal-to-bacterial biomarkers is a useful indicator of reclamation progress, and this is a view supported by other authors (Harris 2003, 2009).

Quite simple measurements of soil microbial lipid characteristics can yield useful information. Figure 5 shows the effects of varying intensities of military traffic/training on the soil microbial biomass, as determined by PLFAs (Peacock et al. 2001a). All levels of traffic caused some decrease in biomass, but heaviest traffic (i.e. tank training) caused a decrease by an order of magnitude. In an area which had been remediated (i.e. 10 years after replanting with trees) the microbial biomass had returned to the levels found in light and moderately trafficked areas but note that the variability here was very high indicating the patchy nature of the recovery.

Bentham et al. (1992) investigated the effects of time and management by comparing reclaimed opencast coal mining sites restored to either grassland or woodland with a number of reference sites, including flood meadow, chalk grassland, rough grassland and woodland. The size and activity of the soil microbial communities were measured by adenosine triphosphate determination (microbial biomass), ergosterol determination (fungal biomass); and, dehydrogenase activity (microbial activity). Multivariate statistical analysis showed that the site restored to grassland clustered with a "rough" (improved) grassland; and, the restored

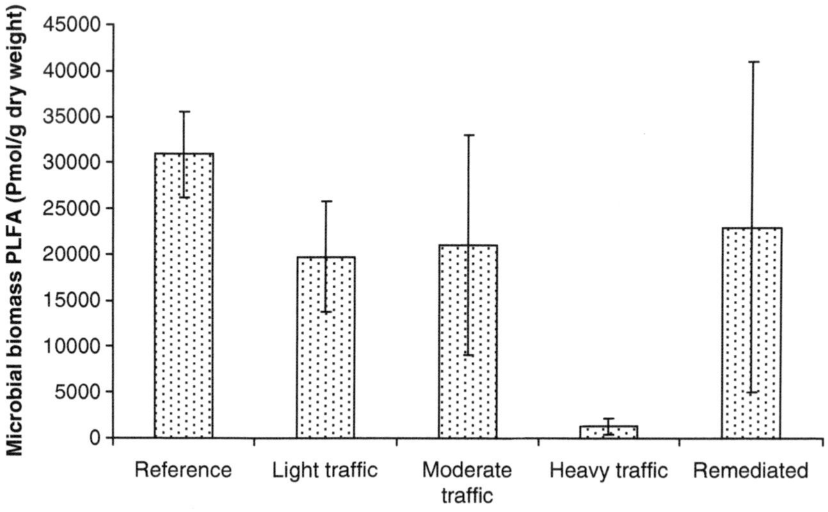

Fig. 5 The effect of military traffic and remediation on soil microbial biomass as determined by PLFA (pmol g^{-1}) (Redrawn from Peacock et al. 2001a). Reproduced with permission of Elsevier

woodlands are "moving away" from the 5 year grassland restoration in the direction of woodland reference sites. A more comprehensive database would facilitate interpretation of data of this type, allowing us to identify the effects of management interventions.

Concluding Remarks

It is clear that there are profound changes in the soil microbial community in response to the land uses currently practiced. The evidence comes form a wide variety of sources and needs to be brought together with physico-chemical and pedological data and knowledge in a unified model or "theory of soil", a notion first suggested by McBratney over a decade ago (McBratney 1998). This is a major collaborative project for society, without which sustaining both food supply and environment will be inefficient and costly.

We have, however, a framework for putting this work in the context of society as a whole in the form of the concept of ecosystem services. Barrios (2007) has reviewed the role that soil biota plays in underpinning the delivery of ecosystem goods and services, and particularly its importance in promoting productivity. He identifies research gaps and opportunities, in particular:

- Integration of spatial variability research in soil ecology and a focus on "hot spots" of biological activity
- Using a selective functional group approach to study soil biota and function
- Combining new and existing methodological approaches to that link the temporal-spatial dynamics of soil organisms to the delivery of specific services
- How these relationships might be directly or indirectly manipulated to enhance particular functions (this will result in trade-offs of function)
- How remote sensing of vegetation condition and composition could inform understanding of soil microbiological communities at a landscape scale; and
- Developing land quality monitoring systems that inform land users of ecosystem service delivery to aid policy and decision making

With respect to this latter point Ritz and co-workers (Ritz et al. 2009) have carried out a sophisticated analysis of the plethora of methods available (184 and counting) used to determine the size, composition and activity of the soil biological community in the context of informing national soil quality and ecosystem assessment schemes. The results of this work are very promising, and there is a real prospect of deploying a handful of techniques to capture the entirety of ecosystem level responses to land-use and climate change, better to inform the policy and decision making as called for by Barrios (2007). This fits neatly in the concept of sustainable agriculture, as it provides a context for determining the impact of different farming practices, and a readily available toolkit of techniques with which to assess their success or failure.

References

Abaye DA, Lawlor K, Hirsch PJ et al (2005) Changes in the microbial community of an arable soil caused by long-term metal contamination. Eur J Soil Sci 56:93–102

Adesemoye AO, Kloepper JW (2009) Plant-microbes interactions in enhanced fertilizer-use efficiency. Appl Microbiol Biotechnol 85:1–12

Adl S, Coleman DC, Read F (2006) Slow recovery of soil biodiversity in sandy loam soils of Georgia after 25 years of no-tillage management. Agric Ecosys Environ 114:323–334

Alguacil MM, Lumini E, Roldan A et al (2008) The impact of tillage practices on arbuscular mycorrhizal fungal diversity in subtropical crops. Ecol Appl 18:527–536

Anderson TH, Domsch KH (1989) Ratios of microbial biomass carbon to total organic carbon in arable soils. Soil Biol Biochem 21:471–479

Angers DA, Peasant A, Vigneux J (1992) Early cropping-induced changes in soil aggregation, organic matter, and microbial biomass. Soil Sci Soc Am J 56:115–119

Bååth E (1989) Effects of heavy metals in soil on microbial processes and populations (a review). Water Air Soil Pollut 47:335–379

Barrios E (2007) Soil biota, ecosystem services and land productivity. Ecol Econ 64:269–285

Bellamy PH, Loveland PJ, Bradley RI et al (2005) Carbon losses from all soils across England and Wales 1978–2003. Nature 437:245–248

Bending G, Rodriguez-Cruz M, Lincoln SD (2007) Fungicide impacts on microbial communities in soils with contrasting management histories. Chemosphere 69:82–88

Bentham H, Harris JA, Birch P et al (1992) Habitat classification and soil restoration assessment using analysis of soil microbiological and physico-chemical characteristics. J Appl Ecol 29:711–718

Bloem J, Hopkins DW, Benedetti A (eds) (2005) Microbiological methods for assessing soil quality. CABI, Wallingford, UK

Böhme L, Langer U, Böhme F (2005) Microbial biomass, enzyme activities and microbial community structure in two European long-term field experiments. Agric Ecosyst Environ 109:141–152

Bolton HJ, Elliott LF, Papendick RI (1985) Soil microbial biomass and selected soil enzyme activities: effect of fertilization and cropping practices. Soil Biol Biochem 17:297–302

Bossio DA, Scow KM (1998) Impacts of carbon and flooding on soil microbial communities: phospholipid fatty acid profiles and substrate utilisation patterns. Microb Ecol 35:265–278

Bossio DA, Scow KM, Gunapala N et al (1998) Determinants of soil microbial communities: effects of agricultural management, season, and soil type on phospholipid fatty acid profiles. Microb Ecol 36:1–12

Bronick CJ, Lal R (2005) Manuring and rotation effects on soil organic carbon concentration for different aggregate size fractions on two soils in northeastern Ohio, USA. Soil Till Res 81:239–252

Bünemann EK, Schwenke GD, Van Zweiten L (2006) Impact of agricultural inputs on soil organisms – a review. Aust J Soil Res 44:379–406

Campbell CD, Chapman SJ, Cameron CM et al (2003) A rapid microtiter plate method to measure carbon dioxide evolved from carbon substrate amendments so as to determine the physiological profiles of soil microbial communities by using whole soil. Appl Environ Microbiol 69:3593–3599

Carminati A, Kaestner A, Lehmann P et al (2008) Unsaturated water flow across soil aggregate contacts. Adv Water Resour 31:1221–1232

Chodak M, Pietrzykowski M, Niklinska M (2009) Development of microbial properties in a chronosequence of sandy mine spoils. Appl Soil Ecol 41:259–268

Cohen JE (2003) Human population: the next century. Science 302:1172–1175

Davies CS (1999) Derelict land. In: Alexander DE, Fairbridge RW (eds) Encyclopaedia of environmental science. Springer, New York

DEFRA (2009) A code of good agricultural practice for farmers, growers and land managers. Department for Environment, Food and Rural Affairs and The Stationery Office, Norwich, UK

Degens B, Harris JA (1997) Development of a physiological approach to measuring the catabolic diversity of soil microbial communities. Soil Biol Biochem 29:1309–1320

Demoling LA, Bååth E, Greve G et al (2009) Effects of sulfamethoxazole on soil microbial communities after adding substrate. Soil Biol Biochem 41:840–848

Deurer M, Grinev D, Young IM et al (2009) The impact of soil carbon management on soil macropore structure: a comparison of two apple orchard systems in New Zealand. Eur J Soil Sci 60:945–955

Doran JW, Fraser DG, Culick MN et al (1987) Influence of alternative and conventional agricultural management on soil microbial processes and nitrogen availability. Am J Alternat Agr 2:99–109

Edgerton DL, Harris JA, Birch P et al (1995) Linear relationship between aggregate stability and microbial biomass in three restored soils. Soil Biol Biochem 27:1499–1501

Eickhorst T, Tippkötter R (2008) Improved detection of soil microorganisms using fluorescence in situ hybridization (FISH) and catalyzed reporter deposition (CARD-FISH). Soil Biol Biochem 40:1883–1891

Elliott ET, Coleman DC (1988) Let the soil work for us. Ecol Bull 39:23–32

Frey B, Kremer J, Rudt A et al (2009) Compaction of forest soils with heavy logging machinery affects soil bacterial community structure. Eur J Soil Biol 45:312–320

Giller KE, Witter E, McGrath SP (1998) Toxicity of heavy metals to microorganisms and microbial processes in agricultural soils: a review. Soil Biol Biochem 30:1389–1414

Glaser B, Haumaier L, Guggenberger G et al (2001) The "Terra Preta" phenomenon: a model for sustainable agriculture in the humid tropics. Naturwissenschaften 88:37–41

Gosling P, Hodge A, Goodlass G et al (2006) Arbuscular mycorrhizal fungi and organic farming. Agric Ecosyst Environ 113:17–35

Guerrero C, Moral R, Gomez I et al (2007) Microbial biomass and activity of an agricultural soil amended with the solid phase of pig slurries. Bioresour Technol 98:3259–3264

Hamer U, Marschner B, Brodowski S et al (2004) Interactive priming of black carbon and glucose mineralisation. Org Geochem 35:823–830

Hammesfahr U, Heuer H, Manzke B et al (2008) Impact of the antibiotic sulfadiazine and pig manure on the microbial community structure in agricultural soils. Soil Biol Biochem 40:1583–1591

Hamscher G, Pawelzick HT, Höper H et al (2005) Different behaviour of tetracyclines and sulfonamides in sandy soils after repeated fertilization with liquid manure. Environ Toxicol Chem 24:861–868

Harris JA (2003) Measurements of the soil microbial community for estimating the success of restoration. Eur J Soil Sci 54:801–808

Harris JA (2009) Soil microbial communities and restoration ecology: facilitators or followers? Science 325:573–574

Harris JA, Birch P (1990a) The effects of heavy civil engineering and stockpiling on the soil microbial community. In: Howsam P (ed) Micro-organisms in civil engineering, Proceedings of the conference, FEMS symposium No. 59. London, pp 274–286

Harris JA, Birch P (1990b) Application of the principles of microbial ecology to the assessment of surface mine reclamation. In: Skousen J, Sencindiver J, Samuel D (eds) Proceedings of the 1990 mining and reclamation conference. American Society of Surface Mining and Reclamation, West Virginia, pp 111–120

Harris JA, Steer J (2003) Modern methods for estimating soil microbial biomass and diversity: an integrated approach. In: Sasek V (ed) The utilization of bioremediation to reduce soil contamination: problems and solutions, NATO CCMS Symposium. Kluwer, Dordrecht, pp 29–48

Harris JA, Birch P, Short KC (1989) Changes in the microbial community and physico-chemical characteristics of topsoils stockpiled during opencast mining. Soil Use Manage 5:161–168

Harris JA, Birch P, Short KC (1993) Changes in the microbial community during the construction and subsequent storage of soil stockpiles: a strategist theory interpretation. Restor Ecol 1:88–100

Hassink J, Voshaar JH, Nijhus EH et al (1991) Dynamics of microbial populations of a reclaimed-polder soil under a conventional and a reduced-input farming system. Soil Biol Biochem 23:515–524

Janvier C, Villeneuve F, Alabouvette C et al (2007) Soil health through soil disease suppression: which strategy from descriptors to indicators? Soil Biol Biochem 39:1–23

Jeffries P, Gianinazzi S, Perotto S et al (2003) The contribution of arbuscular mycorrhizal fungi in maintenance of plant health and soil fertility. Biol Fertil Soils 37:1–16

Kandeler E, Stemmer M, Klimanek E-M (1999) Response of soil microbial biomass, urease and xylanase within particle size fractions to long-term soil management. Soil Biol Biochem 31:261–273

Kennedy AC (1999) Bacterial diversity in agroecosystems. Agric Ecosyst Environ 74:65–76

Kibblewhite MG, Ritz K, Swift MJ (2008) Soil health in agricultural systems. Philos T Roy Soc B-Biol Sci 363:685–701

Kirchner MJ, Wollum AG, King LD (1993) Soil microbial populations and activities in reduced chemical input agroecosystems. Soil Sci Soc Am J 57:1289–1295

Lacombe S, Bradley RL, Hamel C et al (2009) Do tree-based intercropping systems increase the diversity and stability of soil microbial communities? Agric Ecosyst Environ 131:25–31

Lal R (2007) Soil science and the carbon civilization. Soil Sci Soc Am J 71:1425–1437

Lehmann J, Gaunt J, Rondon M (2006) Bio-char sequestration in terrestrial ecosystems—a review. Mitigat Adapt Strateg Global Change 11:403–427

Lucy M, Reed E, Glick BR (2004) Applications of free living plant growth promoting rhizobacteria. Antonie Leeuwenhoek 86:1–25

Marris E (2006) Black is the new green. Nature 442:624–626

McBratney AB (1998) Some considerations on methods for spatially aggregating and disaggregating soil information. Nutr Cycl Agroecosyst 50(1–3):51–62

Mummey DL, Stahl PD, Buyer JS (2002) Microbial biomarkers as an indicator of ecosystem recovery following surface mine reclamation. Appl Soil Ecol 21:251–259

Nunan N, Wu K, Young IM et al (2002) In situ spatial patterns of soil bacterial populations, mapped at multiple scales, in an arable soil. Microb Ecol 44:296–305

O'Neil MJ, Smith A, Heckelmann PE (eds) (2001) The Merck Index, 13th edn. Merck, Whitehouse Station, NJ

Or D, Smets BF, Wraith JM et al (2007) Physical constraints affecting bacterial habitats and activity in unsaturated porous media – a review. Adv Water Resour 30:1505–1527

Palmer RC, Burton RGO, Hannam JA et al (2006) Comparison of soil structural conditions in Tone and Parrett catchments during winter periods 2002–03 and 2005–06. Environment Agency Contract Report

Pawlett M, Hopkins DW, Moffett BF et al (2009) The effect of earthworms and liming on soil microbial communities. Biol Fertil Soils 45:361–369

Peacock AD, McNaughton SJ, Cantu JM et al (2001a) Soil microbial biomass and community composition along an anthropogenic disturbance gradient within a long-leaf pine habitat. Ecol Indic 1:113–121

Peacock AD, Mullen MD, Ringelberg DB et al (2001b) Soil microbial community responses to dairy manure or ammonium nitrate applications. Soil Biol Biochem 33:1011–1019

Petersen SO, Henriksen K, Mortensen GK et al (2003) Recycling of sewage sludge and household compost to arable land: fate and effects of organic contaminants, and impact on soil fertility. Soil Till Res 72:139–152

Pohlmeier A, Haber-Pohlmeier S, Stapf S (2009) A fast field cycling nuclear magnetic resonance relaxometry study of natural soils. Vadose Zone J 8:735–742

Preston-Mafham J, Boddy L, Randerson PF (2002) Analysis of microbial community functional diversity using sole-carbon-source utilisation profiles – a critique. FEMS Microb Ecol 42:1–14

Ritz K (2007) The Plate Debate: cultivable communities have no utility in contemporary environmental microbial ecology. FEMS Microbiol Ecol 60:358–362

Ritz K, Black HIJ, Campbell CD et al (2009) Selecting biological indicators for monitoring soils: a framework for balancing scientific and technical opinion to assist policy development. Ecol Indic 9:1212–1221

Ruzek L, Vorisek K, Sixta J (2001) Microbial biomass-C in reclaimed soil of the Rhineland (Germany) and the north Bohemian lignite mining areas (Czech Republic): measured and predicted values. Restor Ecol 9:370–377

Sasov A, Liu X, Rushmer D (2008) Compact micro-CT/micro-XRF system for non-destructive 3D analysis of internal chemical composition. In: Luysberg M, Tillmann K, Weirich T (eds) Instrumentation and methods, vol 1, EMC 2008. Springer, Berlin, pp 705–706. doi:10.1007/978-3-540-85156-1_353

Schaap JD, Lehmann P, Kaestner A et al (2008) Measuring the effect of structural connectivity on the water dynamics in heterogeneous porous media using speedy neutron tomography. Adv Water Resour 31:1233–1241

Scullion J (1994) Restoring farmland after coal: the Bryngwyn project. British Coal Opencast, Mansfield, UK

Sherwood S, Uphof N (2000) Soil health: research, practice and policy for a more regenerative agriculture. Appl Soil Ecol 15:85–97

Simmons BL, Coleman DC (2008) Microbial community response to transition from conventional to conservation tillage. Appl Soil Ecol 40:518–528

Six J, Frey SD, Thiet RK et al (2006) Bacterial and fungal contributions to carbon sequestration in agroecosystems. Soil Sci Soc Am J 70:555–569

Speir TW, Ross DJ (2002) Hydrolytic enzyme activities to assess soil degradation and recovery. In: Burns RG, Dick RP (eds) Enzymes in the environment: activity, ecology and application. Marcel Dekker, New York, pp 407–432

Stahl PD, Perryman BL, Sharmasarkar S et al (2002) Topsoil stockpiling versus exposure to traffic: a case study on in situ uranium wellfields. Restor Ecol 10:129–137

Steer J, Harris JA (2000) Shifts in the microbial community in rhizosphere and non-rhizosphere soils during the growth of *Agrostis stolonifera*. Soil Biol Biochem 32:869–878

Steiner C, Das KC, Garcia M et al (2008) Charcoal and smoke extract stimulate the soil microbial community in a highly weathered xanthic Ferralsol. Pedobiologia 51:359–366

Stoob K, Singer HP, Stettler S et al (2006) Exhaustive extraction of sulfonamide antibiotics from aged agricultural soils using pressurized liquid extraction. J Chromatgr A 1128:1–9

Subbarao GV, Ito O, Sahrawat KL et al (2006) Scope and strategies for regulation of nitrification in agricultural systems – challenges and opportunities. Crit Rev Plant Sci 25:303–335

Subbarao GV, Wang HY, Ito O et al (2007) NH_4^+ triggers the synthesis and release of biological nitrification inhibition compounds in *Brachiaria humidicola* roots. Plant Soil 290:245–257

Tippkötter R, Ritz K, Darbyshire JF (1986) The preparation of soil thin-sections for biological studies. J Soil Sci 37:681–690

Torsvik V, Ovreas L (2002) Microbial diversity and function in soil: from genes to ecosystems. Curr Opin Microbiol 5:240–245

Torsvik V, Sørheim R, Goksøyr J (1996) Total bacterial diversity in soil and sediment communities—a review. J Ind Microbiol 17:170–178

Toyota K, Kuninaga S (2006) Comparison of soil microbial community between soils amended with or without farmyard manure. Appl Soil Ecol 33:39–48

van Diggelen R, Grootjans AP, Harris JA (2001) Ecological restoration: state of the art or state of the science? Restor Ecol 9:115–118

Wakelin SA, Gregg AL, Simpson RJ et al (2009) Pasture management clearly affects soil microbial community structure and N-cycling bacteria. Pedobiologia 52:237–251

Wan J, Tyliszczak T, Tokunaga TK (2007) Organic carbon distribution, speciation, and elemental correlations within soil micro aggregates: applications of STXM and NEXAFS spectroscopy. Geochim Cosmochim Acta 71:5439–5449

Wander MM, Hedrick DS, Kaufman D et al (1995) The functional significance of the microbial biomass in organic and conventionally managed soils. Plant Soil 170:87–97

Wardle DA, Nilsson M-C, Zackrisson O (2008) Fire-derived charcoal causes loss of forest humus. Science 320:629

Watt M, Kirkegaard JA, Passioura JB (2006) Rhizosphere biology and crop productivity – a review. Aust J Soil Res 44:299–317

Yin B, Crowley D, Sparovek G et al (2000) Bacterial functional redundancy along a soil reclamation gradient. Appl Environ Microbiol 66:4361–4365

Zelles L (1999) Fatty acid patterns of phospholipids and lipopolysaccharides in the characterization of microbial communities in soil: a review. Biol Fertil Soils 29:111–129

Chapter 8
The Effects of Plant Breeding on Soil Microbes

Petra Marschner and Zed Rengel

Introduction

Breeding wheat in the twentieth century, for example has been successful on many accounts, resulting in increased yields and grain quality (and contributing to the "Green Revolution"). One of the consequences of breeding for increased yield and partitioning of biomass into grain (raised harvest index, i.e. the grain-to-straw biomass ratio) has however been a decrease in the root-to-shoot biomass ratio (Rengel 2005). This had significant implications for crop nutrition, since the smaller root system can only support the relatively larger shoots when provided with large quantities of inorganic fertilizers. What remains unclear, and needs to be one of priorities for future research, is whether altered root-to-shoot ratios influence root exudation and therefore the structure of microbial communities in the rhizosphere.

Soil microbes can play an important role in growth and nutrient uptake by plants, as well as modifying the tolerance of plants to biotic and abiotic stresses; so are worthy of stronger consideration in plant breeding. Those microorganisms colonizing the roots as symbionts and in the rhizosphere are particularly important in plant–soil relations and some of the observed effects of breeding, e.g. enhanced nutrient acquisition, probably result not only from changes in plant physiology, root morphology and function alone but also from changes in the nature of the interactions with soil microorganisms.

Physiological or morphological changes induced by breeding can affect microorganisms in the rhizosphere and bulk soil both directly and indirectly (Table 1), thereby potentially affecting microbial community structure and activity as well as their effect on plants. In this review, we will focus on the effect of plant breeding

P. Marschner (✉)
Soils, School of Agriculture, Food & Wine, The University of Adelaide, Adelaide, South Australia 5005, Australia
e-mail: petra.marschner@adelaide.edu.au

Z. Rengel
School of Earth and Environment, The University of Western Australia, Crawley, Western Australia 6009, Australia
e-mail: zed.rengel@uwa.edu.au

G.R. Dixon and E.L. Tilston (eds.), *Soil Microbiology and Sustainable Crop Production*,
DOI 10.1007/978-90-481-9479-7_8,

Table 1 Direct and indirect effects of plant breeding on soil microorganisms

Direct	Indirect
Root exudates (amount and composition)	Nutrient and water uptake (competition with microbes, water availability)
Root morphology	Changes in nutrient availability
Composition of residues (including genes from other organisms)	Soil temperature (more or less shading)
Susceptibility/tolerance to infection	Amount of residues

on soil microorganisms, particularly symbionts and rhizosphere microorganisms. We discuss how the changes induced by plant breeding could affect microbial diversity and function as well as nutrient cycling in ways that are usually not considered by plant breeders.

Breeding for Modified Nutrient Uptake

Associative and Symbiotic Nitrogen Fixation

Under certain conditions, biological nitrogen fixation via associative bacteria may contribute up to 50 kg/ha or 30% of nitrogen needs of wheat and up to 150 kg N ha-*1* year-*1* in sugar cane fields in Brazil (Rengel 2002). Brazilian cereal and sugar cane genotypes may have inadvertently been selected for efficient symbiosis with associative diazotrophs (Dobereiner 1997), thus significantly decreasing or completely eliminating the need for inorganic nitrogen fertilization of these crops.

Genotypic differences in capacity to support associative nitrogen fixation have been found among cereal species and genotypes (see Rengel 2002). Rice genotypes show more than one order of magnitude difference in the percentage of nitrogen derived from air (Ndfa) by associative fixation (Shrestha and Ladha 1996). Some maize genotypes had similar yield when inoculated with *Azospirillum* and when artificially fertilized with 100 kg N ha-*1* (Garcia de Salamone et al. 1997).

The symbiosis between legumes and *Rhizobium* is highly specific; inoculation with a given *Rhizobium* genotype may increase nitrogen fixation strongly in one cultivar while having little effect on that process in another cultivar (Hardarson and Atkins 2003). The higher capacity of certain *Rhizobium* genotypes to form successful symbioses with a particular plant genotype increases their survival and competitive abilities in soil, hence affecting the composition the rhizobial community in the soil. Increased nitrogen fixation, either from associative or symbiotic bacteria, will have indirect effects on soil microorganisms via the higher nitrogen content of the resultant plant residues. These effects will occur mainly after harvest or decay of the plants but may also happen during plant growth when roots die and are decomposed. Effects can include increased activity and/or abundance of microorganisms involved in nitrogen mineralization, increased microbial activity more generally because of the higher nitrogen supply, stronger decomposition of native soil organic matter (priming effect) (Kuzyakov et al. 2000), increased abundance of microbial species with high nitrogen requirement and reduced nitrogen fixation by free-living bacteria (see chapters elsewhere in this Book). This may affect microbial community composition, and in the longer term, nutrient cycling itself. However, to our knowledge, there are no studies which consider the effects of enhanced nitrogen fixation on soil microorganisms and nutrient cycling. Long-term field experiments would be required in order to assess this. Nevertheless, some of the effects induced by the growth of plant genotypes associated with high nitrogen fixation could be similar to those of high increased availability induced by high doses of inorganic fertilizer which have been shown to change the bacterial colonization of roots (Marschner et al. 1999), microbial community composition (Marschner et al. 2004), growth and activity of bacteria (Soederberg and Baath 2004; Alarcon-Gutierrez et al. 2008) and bacteria-feeding nematodes (Griffiths et al. 1992) and increase or decrease the density of diazotrophs (Azaizeh et al. 1996). However, the effect of plant residues with higher nitrogen content may differ from that resulting from inorganic fertilizer because they also supply carbon. Since soil microorganisms are primarily limited by carbon availability (Kemmitt et al. 2008; Hoyle et al. 2008), changes induced by high nitrogen content in residues are likely to be greater than those caused by the simple addition of inorganic fertilizer.

Arbuscular Mycorrhiza

Arbuscular mycorrhizal (AM) fungi are less host-specific than rhizobia, but their rate of colonization and other benefits to the host are plant (host) genotype-dependent. Zhu et al. (2001) showed that mycorrhizal responsiveness in terms of phosphorus uptake was lower in modern wheat genotypes than in older cultivars, suggesting that current plant breeding practices have selected for genotypes that are less dependent on mycorrhiza for their phosphorus uptake (Zhu et al. 2001). Likewise, Xavier and Germida (1998) found that wheat genotypes differ in the capacity to sustain AM colonization. Yield responses varied from zero to both increased or decreased values. In durum wheat genotypes, the benefits from the AM symbiosis were not proportional to the extent of the root colonization because the genotypes differed in degree to which they were dependent on mycorrhiza (Al-Karaki and Al-Raddad 1997). Similar results were obtained for barley genotypes varying in their phosphorus use efficiency (Baon et al. 1993).

Differences in the extent of root colonization of plant genotypes may affect the survival of mycorrhizal fungi. If host roots are poorly colonized, the fungus may not derive sufficient energy to sporulate and reproduce. Consequently, the colonization potential of the soil may be reduced with resultant decreases in colonization of following crops. Differences in AM colonization can also affect other soil microorganisms either directly via the fungus or indirectly through effects on host plants. Arbuscular mycorrhizal colonization decreases root exudation in some cases (Dixon et al. 1989; Graham et al. 1981; Marschner et al. 1997) although in others there may be no effect (Azaizeh et al. 1995). Mycorrhizal colonization may affect exudate composition (Marschner et al. 1997; Po and Cumming 1997) and the carbohydrate metabolism of roots (Buwalda and Goh 1982; Shachar-Hill et al. 1995). Mycorrhizal fungi themselves may release exudates which selectively influence the microorganisms of the rhizosphere. Thus, differential colonization affects exudates composition and hence has an influence on other soil microorganisms. Indeed, previous studies have shown that AM colonization itself and the extent of colonization affect the structure of microbial communities in the rhizosphere (Amoralazcano et al. 1998; Andrade et al. 1997; Marschner et al. 2001; Marschner and Baumann 2003; Meyer and Linderman 1986; Posta et al. 1994; Wamberg et al. 2003; Secilia and Bagyaraj 1987). Additionally, bacterial community structure in the rhizosphere is affected by interactions between host genotypes and mycorrhizal fungal species (Marschner and Timonen 2004). Colonization by AM can affect nematode infection of roots (Cooper and Grandison 1986; Elsen et al. 2001), which is likely to affect the community structure of the parasite, in turn, this could have a cascading effect on the types, density and interactions between microbes.

Phosphorus Efficiency

Plant genotypes differ in phosphorus-use efficiency, i.e. capacity to grow with low phosphorus supply. Differential phosphorus-use efficiency can be due to differences in uptake and/or internal utilisation. Phosphorus-use efficiency has been studied extensively in cereals, particularly wheat (e.g. Zhu et al. 2001; Osborne and Rengel 2002a; b; Marschner et al. 2006) and legumes (Bolland et al. 1999; Horst et al. 2001; Lynch and Brown 2001) but relatively little is known about differences in phosphorus-use efficiency in *Brassica* species or in vegetable and oil seed rape genotypes (Greenwood et al. 2005). Although most studies have focussed on plant growth and yield, a number have investigated the effect of phosphorus-use efficiency on soil microorganisms.

Comparisons of modern and old wheat cultivars, Zhu et al. (2001) showed that the greater the phosphorus utilisation efficiency of the non-mycorrhizal plants, the smaller the growth response of the plant to mycorrhizal colonization. Moreover, despite similar AM colonization rates, old cultivars were more responsive to mycorrhizal colonization than modern cultivars.

In a comparison between two wheat cultivars (P-efficient cv. Goldmark and P-inefficient cv. Janz), Marschner et al. (2006) found that genotypic differences in phosphorus efficiency is explained by plant traits, including root growth, phosphorus uptake and utilization and rhizosphere properties such as acid phosphatase activity and concentrations of available and microbial phosphorus. The genotypes differed in their associated microbial community composition, which might indicate that a differential capacity of the genotypes for growth in soil with low phosphorus availability is, at least partially, due to differences in microbial community composition in the rhizosphere. Interestingly, there were no differences in mycorrhizal colonization between the genotypes. A positive correlation between microbial phosphorus in the rhizosphere and shoot dry weight, phosphorus uptake per plant and available phosphorus in the rhizosphere suggested the importance of this source for plant uptake. These findings supported the previous hypotheses (Oberson et al. 2001; Seeling and Zasoski 1993) that an active microbial biomass with a high turnover rate provides a slow-release, sustained source of available phosphorus. However, while phosphorus-efficient cv. Goldmark appeared to compete effectively with rhizosphere microorganisms for phosphorus and might have taken advantage of the microbial pool as a source, this was not the case for phosphorus-inefficient cv. Janz. Root exudates of cv. Goldmark may favour rapid microbial turnover and/or stimulate the growth of rhizosphere microorganisms with high phosphorus mobilization capacities and low requirements. The correlation between microbial community composition and acid phosphatase activity indicates that different microbial species release varying amounts of phosphatase, and/or that microbial species differ in their capacities to decompose the phosphatase enzyme. Moreover, microbial community composition was correlated with phosphorus availability in the rhizosphere. This suggests that microbial species differ in phosphorus uptake, requirement and/or mobilization capacity (Banik and Dey 1983). Thus in wheat, phosphorus-efficient genotypes appear to have rhizosphere microbial communities that mobilize this element but also have high turnover rates, releasing it for subsequent plant uptake. By contrast, in the rhizosphere of phosphorus-inefficient genotypes, rhizosphere microbial communities also mobilize this element, but due the low turnover rate, it remains locked-up in the microbial biomass.

Interestingly, microbial biomass phosphorus in the rhizosphere was not correlated with plant uptake in three brassicas (two canola, *Brassica napus* genotypes and a mustard, *Sinapis* sp.) (Marschner et al. 2007), which differed in phosphorus efficiency. In the brassicas, shoot uptake was positively correlated with root length as well as phosphorus availability and phosphatase activity in the rhizosphere. The genotypes also differed in microbial community composition in the rhizosphere, but community composition did not appear to be related with a differential capacity for growth in phosphorus-limiting conditions. Thus in brassicas, traits such as root length and the capacity to maintain high phosphorus availability in the rhizosphere by mobilization via carboxylate or phosphatase release explains the observed differences in plant growth and phosphorus uptake. It cannot be excluded however, that some of the observed differences in mobilization between the genotypes were due not only to exudates released by the roots but also to differential capacities of the rhizosphere microorganisms to mobilize phosphorus.

Some plants, such as grasses, are unable to mobilize sufficient phosphorus from phytate (inositol hexaphosphate) under sterile conditions (Richardson et al. 2000; Hayes et al. 2000a), because they release only very small amounts of phytase into the rhizosphere (Hayes et al. 2000b). Inoculation with a soil suspension (containing microorganisms) restores the capacity to utilize phytate (Hayes et al. 2000a; Richardson et al. 2001). Moreover, some soil microorganisms, particularly fungi can utilize phytate *in vitro* (Richardson and Hadobas 1997; George et al. 2007). To increase the capacity of grasses to grow in soils with low phosphorus availability, fungal phytase genes have been introduced into wheat and these plants are able to utilize phytate under sterile conditions (George et al. 2005). There are also wheat genotypes with an over-expression of genes involved in citrate or malate synthesis and release from roots (Delhaize et al. 1993; Pellet et al. 1995). It was suggested that engineering plants with increased synthesis and exudation of organic acid anions might increase their capacity for growth in phosphorus-limiting conditions (Lopez-Bucio et al. 2000), but this has yet to be demonstrated (Delhaize et al. 2001).

Plant genotypes that increase phosphorus availability in the rhizosphere would affect soil microorganisms by preventing deficiency. If the higher phosphorus availability is associated with higher carbon and nitrogen availability, this may lead to increased growth and activity by rhizosphere microorganisms, and could favour microbial species that have high growth rates under conditions of phosphorus sufficiency. On the other hand, slower growing microbial species with high phosphorus mobilization capacities would be out-competed. If plants with high inherent capacities for mobilizing phosphorus are grown repeatedly in a soil, this might lead to an overall decrease in mobilization capacity of the soil microbial community. It remains to be experimentally verified however if this would really be the case.

Manganese Efficiency and Availability

Yield of cereals on calcareous soils is frequently limited by manganese deficiency caused by low availability, rather than low total manganese content in soil (Rengel 2000). Compared to inefficient genotypes, manganese-efficient genotypes take up more of the element from soils with limited availability, but the physiological mechanisms underlying efficiency are poorly understood (Rengel 2000, 2001a).

As a result of root uptake and poor mobility, the concentration of manganese in the rhizosphere is lower than in the bulk soil. Such depletion is more prominent in the rhizosphere of manganese-efficient than inefficient *Triticum aestivum* genotypes (Marschner et al. 2003). In another study however, the genotypic difference in efficiency between *Triticum aestivum* and *T. durum* genotypes was not due to differences in chemical mobilization of manganese in the rhizosphere (Sadana et al. 2002).

Reduction and oxidation of manganese by microorganisms are important components of manganese cycling in soil. Whereas reduction increases manganese availability, oxidation decreases it. Interestingly, manganese reducers appear to be more abundant in the rhizosphere of some efficient compared with inefficient *Triticum aestivum* genotypes (Rengel et al. 1998).

Differential manganese efficiency in cereals might also affect their susceptibility to Take-all (*Gaeumannomyces graminis* var. *tritici*) (Rengel 2001b). It has been shown that increasing fertilization of manganese decreases Take-all infection in wheat and this is explained by two factors: (i) increased defence reaction by roots to Take-all infection, and (ii) inhibition of the growth of the take-all fungus by high manganese concentrations in soil (Marschner et al. 1991). The increased defence reaction in plants that have adequate manganese status is related to its role in the synthesis of phenolic compounds and lignin which form a mechanical barrier in roots preventing penetration by the pathogen into cells, thereby inhibiting or limiting infection. The rate of lignin accumulation in the whole root system was lower in manganese-efficient than in inefficient genotypes (Rengel et al. 1994). In that study, manganese fertilization did not affect lignin concentration in the whole root system but still decreased Take-all infection. These apparent contradictions could be due to localized high concentrations of lignin around infection points rather than an increased lignin concentration in the entire root system.

Differential manganese availability in the rhizosphere of manganese-efficient and inefficient plant genotypes may affect microbial community composition (Marschner et al. 2003); high availability could favour microbial species with a high manganese requirement, and decrease the density of microbial species sensitive to high concentrations. A higher lignin concentration of the roots may reduce root colonization by beneficial microorganisms such as AM fungi or pathogens and could also affect root exudation; and, consequently microbial community composition and nutrient cycling. Moreover, higher lignin concentration may reduce decomposition rates of roots and favour microorganisms with capacities to decompose lignin such as fungi and actinobacteria.

Other Plant Breeding Effects

Breeding can affect soil microorganisms by changing the chemical composition of the plants. Here we discuss the effect of enhanced glucosinolate concentration and the introduction of the Bt genes. Lastly, we briefly mention an example of breeding for enhanced tolerance to *Rhizoctonia* induced disease and how this may affect soil microorganisms.

Glucosinolates in Brassicas

Brassicas contain glucosinolates (GSL), which are hydrolyzed by the enzyme myrosinase into glucose, sulphate and biocidal products such as *iso*thiocyanates, nitriles and ionic thiocyanate (Brown and Morra 1997; Kirkegaard and Sarwar 1998a). Among these metabolites, *iso*thiocyanates are considered to be the most toxic to organisms (Kirkegaard and Sarwar 1998a). The toxicity of *iso*thiocyanates is due to irreversible reaction with sulphhydryl groups, amine groups and disulphide bonds of proteins, which can lead to enzyme inactivation (Kirkegaard and Sarwar 1998a). Glucosinolates and myrosinase are spatially separated in intact cells, therefore *iso*thiocyanates are produced only when cell integrity is disrupted (Kirkegaard and Sarwar 1998a).

The formation of biocidal metabolites from GSLs is used in 'biofumigation', a suppression of soil-borne pests and pathogens after incorporating green manure or harvest residues from canola (*B. napus* L.) or other high glucosinolate brassicas (Kirkegaard and Sarwar 1998b; Kirkegaard and Matthiessen 1999). The dynamics of GSL hydrolysis after incorporation of plant material into soil has been studied intensively (for an overview see (Kirkegaard et al. 1999; Rumberger and Marschner 2003). *Iso*thiocyanates are degraded in soil within a few days and are adsorbed by organic matter (Gimsing and Kirkegaard 2006; Motisi et al. 2009). Despite the rapid hydrolysis of glucosinolates and high rates of *iso*thoicyanate degradation in soil, addition of residues high in GSL may inhibit growth of microorgansims for up to 2 weeks after addition to soil. This suggests that the inhibitory effect of residues is due to the release of *iso*thiocyanates during the first days following incorporation, but that other mechanisms are likely to contribute to lasting persistence (Scott and Knudsen 1999). Even longer lasting effects of incorporation of residues with high GSL concentration are reported by Smith et al. (2004), where incorporation of brassica residues caused changes in the rhizosphere microbial community composition of the following wheat crop. Daily addition of the *iso*thiocyanate 2-phenylethyl*iso*thiocyanate (PEITC) for 5 days affected both bacterial and eukaryotic community structure (Rumberger and Marschner 2003). On the other hand, incorporation of high GSL brassica residues did not affect the substrate-utilization pattern of bacteria in the rhizosphere of pea although it inhibited the growth of the seedlings (Scott and Knudsen 1999).

In canola roots GSLs are highly localised, the highest concentrations were measured in two cell layers just below the outermost periderm layer, being up to 100-fold higher than published concentrations for whole roots; by contrast, primary tissues contained negligible GSL concentrations (McCully et al. 2008). McCully et al. (2008) hypothesised that release of GSLs is a normal developmental process during the secondary thickening of stems with senescence of surface cells leading to release of GSLs and their biocidal hydrolysates into the rhizosphere of intact roots. These authors did not detect myrosin idioblasts close to the root surface and suggested that GSLs released during plant growth are hydrolyzed by myrosinase in the rhizosphere, ensuring a continuous localized source of biotoxic hydrolysates in the long-lived components of the root system.

In agreement with the findings of McCully et al. (2008), Rumberger and Marschner (2003) found in a pot study with different brassica genotypes, that the concentration of PEITC in the rhizosphere of living canola roots was greater in first order laterals than in those of the second order. The PEITC concentration detected in the rhizosphere was in the range that had been shown to affect microbial community structure when added directly to the soil. Indeed, there was a significant correlation between the PEITC concentration in the rhizosphere and the community structure of the active fraction of eukaryotes and bacteria.

In a field experiment with several canola cultivars, PEITC concentrations in the rhizosphere ranged between 0 and 12,000 pmol g^{-1} and this was affected by plant growth stage and environmental conditions (Rumberger and Marschner 2004). Maximal PEITC concentrations were found at the boot stage in the winter-grown cultivars and at flowering in the spring cultivars. PEITC concentrations in the rhizosphere decreased towards maturity, which agrees with the low root GSL concentration in mature canola (Kirkegaard et al. 2001b; Potter et al. 2000). Surprisingly, there was no difference in PEITC concentrations in the rhizosphere between cultivars with high and low seed GSL concentrations, suggesting that seed concentration is a poor indicator for release of *iso*thiocynates from roots. As in the pot experiment (Rumberger and Marschner 2003), bacterial community composition in the rhizosphere was affected by PEITC concentration. Both pot and field experiment showed, however, that other factors such as soil moisture, plant developmental stage and dry matter had a greater effect on the microbial community composition in the rhizosphere, suggesting a limited effect of PEITC released by roots on microbial community composition in the rhizosphere. Moreover, *iso*thiocyanate release by roots is highly variable. It varies with plant developmental stage as described above and is also related to growth conditions, with adverse growing conditions such as low light and temperature or pathogen attack decreasing *iso*thiocyanate release (Kirkegaard et al. 2001a). Potter et al. (2000) showed that root GSL concentration and thus probably also *iso*thiocyanate release into the rhizosphere can differ widely between otherwise phenotypically similar plants of the same cultivar.

Thus, brassicas with high GSL concentrations can affect soil microorganisms during plant growth by release of GSLs from roots and their hydrolysis into toxic *iso*thiocyanates in the rhizosphere and following incorporation of GSL-rich residues into the soil. In biofumigation, the aim of the latter is a decrease of pathogenic microorganisms, however a long-lasting effect, particularly with repeated cultivation of GSL-rich cultivars, on microbial community composition and nutrient cycling cannot be ignored.

Transgenic Plants with Increased Resistance Against Damage by Pathogens or Insects

In order to increase resistance against bacterial pathogens, transgenic potato plants containing lysozyme, which degrades the murein in bacterial cell walls, were produced. *Bacillus subtilis* cells are killed on contact with roots of transgenic potato (Heuer et al. 1999) and the bacterial community composition in the rhizosphere of transgenic potato assessed by molecular techniques differs from that of the wild type (Lottmann et al. 1999). On the other hand, the number of culturable total aerobic bacteria or phytohormone-producing bacteria in the rhizosphere did not differ between transgenic and wild-type potatoes (Flores et al. 2005).

The insecticidal Cry protein from *Bacillus thuringiensis* (*Bt*) has been introduced in crops such as: cotton, maize, potato, rice and tobacco in order to decrease insect damage (Stotzky 2005; Saxena and Stotzky 2000). These crops synthesise Cry proteins in all tissues, and it has been shown that these proteins are released by roots of transgenic maize (Icoz and Stotzky 2008a; Saxena and Stotzky 2002). The question therefore arises about the fate of Cry1Ab in soils and if it can affect soil microbial populations.

There are conflicting results regarding the uptake of Cry proteins by plants. Saxena and Stotzky (2002) found that the protein is not taken up from soil or in hydroponic culture by: carrot, maize, radish, or turnip. The Cry protein was recently detected however, by Western blot in shoots of a range of plant species after its addition or that of residues of transgenic maize into soil (Fiorito et al. 2008).

Persistence of Cry protein in soil is increased by the presence of clay minerals such as kaolinite and montmorillonite, probably by adsorption (Saxena and Stotzky 2002). Fiorito et al. (2008) showed that when they are bound to clay, the Cry proteins were not utilized for growth by soil microorganisms, whereas the microorganisms readily utilized the free proteins as sources of carbon and energy. Bound Cry proteins remain active however and therefore kill larvae of soil insects (Stotzky 2005), which via a cascading effect through the soil food web, could indirectly affect soil microorganisms.

Two to 3 weeks after addition of transgenic maize tissue to soil, the Cry3Bb1 protein could no longer be detected, suggesting a low persistence in soil (Saxena and Stotzky 2002) however, the persistence of different Cry proteins may vary. After 4 years of consecutive planting of a number of transgenic maize cultivars (some expressing Cry1Ab1, others Cry3Bb1) and incorporating the residues, Cry1Ab1 could be detected in the soil whereas Cry3Bb1 could not (Saxena and Stotzky 2002). In that field study, there were no consistent statistically significant differences in the density of different groups of microorganisms and the activities of various enzymes in comparison of soils planted with *Bt* and non-*Bt* maize, suggesting that although the proteins may persist in soil, there is little effect on soil microorganisms even after 4 years of continuous cultivation.

The decomposition rate of residues of transgenic *Bt* plants was lower than that of their near-*iso*genic non-*Bt* plant counterparts (Stotzky 2005), which could be due to the higher lignin content of the residues of the transgenic *Bt* plants. The density of culturable bacteria and fungi and the activity of a range of enzymes involved in residue decomposition were not significantly different between soils amended with residues of *Bt* or non-*Bt* maize. The discrepancy between decreased decomposition rate on one side and no apparent effect on cuturable bacteria and fungi could be due to the fact that less than 5% of soil microorganisms are culturable (Bakken 1985). Thus, the decreased decomposition of the residues of *Bt* maize is probably due to reduced microbial activity or changes in community composition of the non-culturable microorganisms which represent the vast majority of microorganisms in soil.

In a review, Icoz and Stotzky (2008b) noted few or no toxic effects of Cry proteins on: collembola, earthworms, mites, nematodes, protozoa and woodlice and the activity of various enzymes in soil. Although some significant effects of *Bt* plants on microbial communities in soil have been reported, they were transient and not related to the presence of the Cry proteins and appeared to be plant genotype and site-specific (soil type and climate).

Thus, the effect of transgenic plants containing Cry proteins on soil microorganisms remains unclear. Some negative effects have been reported however, such as the decreased decomposition of residues of *Bt* maize and the persistence of certain

Cry proteins in soil where they retain their insecticidal activity and could therefore affect soil microorganisms indirectly via cascading effects within the soil food web. More studies, particularly long-term field studies, are required to identify the plant genotype-soil-climate combinations in which effects on soil microorganisms in terms of community structure and activity occur and to identify the mechanisms that lead to these effects.

Control of Rhizoctonia

Pathogenic fungi in the genus *Rhizoctonia* (such as *R. solani,* or *R. cerealis*) affect many cultivated plants causing a range diseases, e.g. sheath blight in rice (Lu et al. 2009), sharp eye spot in wheat (Nagendran et al. 2009), damping-off in sugar beet (Bokmeyer et al. 2009) and brown patch in tall fescue (Zhang et al. 2009). Chemical suppression of *Rhizoctonia* is relatively poorly effective, albeit new chemical products show promise (Gallou et al. 2009). By contrast, biocontrol agents (such as *Trichoderma* spp.) were found to be very effective, e.g. *Rhizoctonia solani* can be controlled with *Trichoderma harzianum* (Ruocco et al. 2009). Biocontrol agents as well as *Rhizoctonia* and other pathogenic fungi can exude antimicrobial compounds into the surrounding soil (Vleesschauwer et al. 2009), potentially altering the composition of microbial communities. The effects of pathogenic fungi or biocontrol agents on soil and rhizosphere microflora are poorly understood. Using biocontrol agents against one pathogenic fungus may increase plant susceptibility to others, as shown by using *Serratia plymuthica* to increase resistance of rice plants to blast disease caused by *Magnaporthe oryzae*, which resulted in plants becoming more sensitive to *Rhizoctonia solani* (Vleesschauwer et al. 2009).

Intensive breeding efforts aiming to increase the resistance of cultivated plants to *Rhizoctonia* via traditional as well as transgenic technologies (Nagendran et al. 2009) may result in an exacerbation of antimicrobial effects in the rhizospheres of new cultivars, which would affect not only the targeted phytopathogenic fungi but may have significant (and potentially undesirable) effects on the microbial communities in the rhizosphere. Hence, increasing attention needs to be given to the effects of plant breeding in changing the structure of the rhizosphere microbial communities as breeders strive for increased release and activity of antimicrobial compounds in the rhizosphere.

Conclusions

Some plant breeding effects on soil microbes are more obvious than others e.g. raising the tolerance of genotypes to pathogens will decrease their growth and survival in the soil or legume genotypes favouring selected *Rhizobium* genotypes because they can only be colonized by this particular strain. As discussed in this chapter however, plant breeding may radically change microbial community composition and/or activity (Table 2, Fig. 1) in often unpredictable ways. Changes in microbial community *composition* may or may not be associated with changes in function, but changes in microbial *activity* are likely to change function (nutrient release or immobilization, release of enzymes); the changes may be transient and the result (e.g. net mobilization or net immobilization) is often unknown. However, the high spatial and temporal variability of soil microorganisms and their activity makes it unlikely that a particular plant genotype will have a consistent effect on microbial community composition and activity.

Soil microorganisms and their activity are not affected only by plants, plant residues and root exudates but also by environmental factors such as climate and soil properties (e.g. moisture, temperature and texture). Thus, a plant genotype may have a strong effect on soil microorganisms in one set of conditions, but not in others.

Because of this apparent variability of effect, it is essential to increase the knowledge about specific plant genotype-microbe interactions so as to enhance our understanding of long-term effects specific genotypes might have on the structure and function of microbial communities. Such knowledge will guard against selecting crop genotypes that might result in losses to the diversity and/or functionality of soil microbial communities. This is particularly important because microbial activity and community composition in turn affect plant growth and nutrient uptake. In parallel, the long-term negative effects of a plant genotype on soil microorganisms could be counterproductive to the undoubted benefits of plant breeding as a route towards sustainable crop production.

Table 2 Potential changes in soil microorganisms induced by plant breeding (see text for references)

Effect of plant breeding	Microbial community composition	Microbial biomass	Potential effects
Increased N and P availability	Species with high N, P demand ↑ Species with low N, P demand ↓ Species mobilizing P, N_2 fixers ↓	Increase in microbial C, N, P Increased activity	Microbial biomass turnover? Priming effect? Loss of Species? Diversity? Functionality? Food web structure?
Inhibitory compounds e.g. *iso*thiocynates, antibiotics	Tolerant species ↑ Susceptible species ↓	Biomass ↓ Activity ↓	
Higher lignin concentration in residues	Lignin-degrading species ↑ Species unable to degrade lignin ↓	Biomass ↓ Activity ↓	

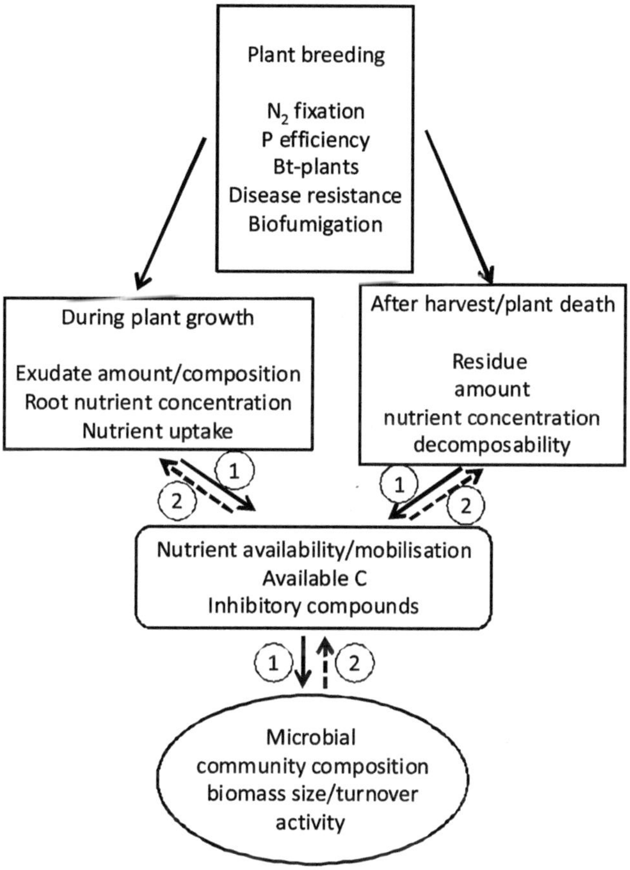

Fig. 1 Schematic diagram of effects of plant breeding on soil microorganisms during plant growth and after harvest/plant death, 1 indicates effects of plant breeding on soil microorganisms, 2 indicates effects of soil microorganisms on plant growth and residue decomposition

References

Alarcon-Gutierrez E, Couchaud B, Augur C, Calvert V, Criquet S (2008) Effects of nitrogen availability on microbial activities, densities and functional diversities involved in the degradation of a Mediterranean evergreen oak litter (*Quercus ilex* L.). Soil Biol Biochem 40:1654–1661

Al-Karaki GN, Al-Raddad A (1997) Drought stress and VA mycorrhizal fungi effects on growth and nutrient uptake of two wheat genotypes differing in drought resistance. Crop Res 13:245–257

Amoralazcano E, Vazquez MM, Azcon R (1998) Response of nitrogen-transforming microorganisms to arbuscular mycorrhizal fungi. Biol Fertil Soils 27:65–70

Andrade G, Mihara KL, Linderman RC, Bethlenfalvay GJ (1997) Bacteria from the rhizosphere and hyphosphere soils of different arbuscular mycorrhizal fungi. Plant Soil 192:71–79

Azaizeh HA, Marschner H, Roemheld V, Wittenmayer L (1995) Effects of a vesicular mycorrhizal fungus and other soil microorganisms on growth, mineral nutrient acquisition and root exudation of soil-grown maize plants. Mycorrhiza 5:321–327

Azaizeh HA, Neumann G, Marschner H (1996) Effects of thiamine and nitrogen fertilizer form on the number of N_2-fixing and total bacteria in the rhizosphere of maize plants. Z Pflanz Bodenkunde 159:1–6

Bakken LR (1985) Separation and purification of bacteria from soil. Appl Environ Microbiol 49:1482–1487

Banik S, Dey BK (1983) Phosphate-solubilizing potentiality of the microorganisms capable of utilizing al phosphate as a sole phosphorus source. Zentralbl Mikrobiol 138:17–23

Baon JB, Smith SE, Alston AM (1993) Mycorrhizal responses of barley cultivars differing in P efficiency. Plant Soil 157:97–105

Bokmeyer JM, Bonos SA, Meyer WA (2009) Broad-sense heritability and stability analysis of brown patch resistance in tall fescue. Hort Sci 44:289–292

Bolland MDA, Siddique KHM, Loss SP, Baker MJ (1999) Comparing responses of grain legumes, wheat and canola to applications of superphosphate. Nutr Cycl Agroecosys 53:157–175

Brown PD, Morra MJ (1997) Control of soil-borne plant pests using glucosinolate-containing plants. Adv Agron 61:167–231

Buwalda JG, Goh KM (1982) Host-fungus competition for carbon as a cause of growth depressions in vesicular-arbuscular mycorrhizal ryegrass. Soil Biol Biochem 14:103–106

Cooper KM, Grandison GS (1986) Interaction of vesicular-arbuscular mycorrhizal fungi and root-knot nematode on cultivars of tomato and white clover susceptible to *Meloidogyne hapta*. Ann Appl Biol 108:555–565

Delhaize E, Ryan PR, Randall PJ (1993) Aluminum tolerance in wheat L.).II. Aluminum- stimulated excretion of malic acid from root apices. Plant Physiol 103:695–702

Delhaize E, Hebb DM, Ryan PR (2001) Expression of a *Pseudomonas aeruginosa* citrate synthase gene in tobacco is not associated with either enhanced citrate accumulation or efflux. Plant Physiol 125:2059–2067

Dixon RK, Garrett HE, Cox GS (1989) Boron fertilization, vesicular-arbuscular mycorrhizal colonization and growth of *Citrus jambhiri* Lush. J Plant Nutr 12:687–700

Dobereiner J (1997) Biological nitrogen fixation in the tropics: social and economic contributions. Soil Biol Biochem 29:771–774

Elsen A, Declerck S, De Waele D (2001) Effects of *Glomus intraradices* on the reproduction of the burrowing nematode (*Radopholus similis*) in dixenic culture. Mycorrhiza 11:49–51

Fiorito TM, Icoz I, Stotzky G (2008) Adsorption and binding of the transgenic plant proteins, human serum albumin, beta-glucuronidase, and Cry3Bb1, on montmorillonite and kaolinite: microbial utilization and enzymatic activity of free and clay-bound proteins. Appl Clay Sci 39:142–150

Flores S, Saxena D, Stotzky G (2005) Transgenic Bt plants decompose less in soil than non-Bt plants. Soil Biol Biochem 37:1073–1082

Gallou A, Cranenbrouck S, Declerck S (2009) *Trichoderma harzianum* elicits defence response genes in roots of potato plantlets challenged by *Rhizoctonia solani*. Eur J Plant Pathol 124:219–230

Garcia de Salamone IE, Dobereiner J, Urquiaga S, Boddey RM (1997) Biological nitrogen fixation in *Azospirillum* strain-maize genotype associations as evaluated by the ^{15}N isotope dilution technique. Biol Fertil Soils 23:249–256

George TS, Richardson AE, Simpson RJ (2005) Behaviour of plant-derived extracellular phytase upon addition to soil. Soil Biol Biochem 37:977–988

George TS, Simpson RJ, Gregory PJ, Richardson AE (2007) Differential interaction of *Aspergillus niger* and *Peniophora lycii* phytases with soil particles affects the hydrolysis of inositol phosphates. Soil Biol Biochem 39:793–803

Gimsing AL, Kirkegaard JA (2006) Glucosinolate and isothiocyanate concentration in soil following incorporation of Brassica biofumigants. Soil Biol Biochem 38:2255–2264

Graham JH, Leonard RT, Menge JA (1981) Membrane-mediated decrease of root exudation responsible for phosphorus inhibition of vesicular-arbuscular mycorrhiza formation. Plant Physiol 68:548–552

Greenwood DJ, Stellacci AM, Meacham MC, Broadley MR, White PJ (2005) Phosphorus response components in different *Brassica oleracea* genotypes are reproducible in different environments. Crop Sci 45:1728–1735

Griffiths BS, Welschen R, Van Arendonk JJCM, Lambers H (1992) The effect of nitrate-nitrogen supply on bacteria and bacterial feeding fauna in the rhizosphere of different grass species. Oecologia 91:253–259

Hardarson G, Atkins C (2003) Optimizing biological N_2 fixation by legumes in farming systems. Plant Soil 252:41–54

Hayes JE, Simpson RJ, Richardson AE (2000a) The growth and phosphorus utilisation of plants in sterile media when supplied with inositol hexaphosphate, glucose-1-phosphate or inorganic phosphate. Plant Soil 220:165–174

Hayes JE, Richardson AE, Simpson RJ (2000b) Components of organic phosphorus in soil extracts that are hydrolysed by phytase and acid phosphatase. Biol Fertil Soils 32:279–286

Heuer H, Hartung K, Wieland G, Kramer I, Smalla K (1999) Polynucleotide probes that target hypervariable region of 16S rRNA genes to identify bacterial isolates corresponding to bands of community fingerprints. Appl Environ Microbiol 65:1045–1049

Horst WJ, Kahm M, Jibrin JM, Chude VO (2001) Agronomic measures for increasing P availability to crops. Plant Soil 237:211–223

Hoyle FC, Murphy DV, Brookes PC (2008) Microbial response to the addition of glucose in low-fertility soils. Biol Fertil Soils 44:571–579

Icoz I, Stotzky G (2008a) Cry3Bb1 protein from *Bacillus thuringiensis* in root exudates and biomass of transgenic corn does not persist in soil. Transgen Res 17:609–620

Icoz I, Stotzky G (2008b) Fate and effects of insect-resistant Bt crops in soil ecosystems. Soil Biol Biochem 40:559–586

Kemmitt SJ, Lanyon CV, Waite IS, Wen Q, Addiscott TM, Bird NRA, O'Donnell AG, Brookes PC (2008) Mineralization of native soil organic matter is not regulated by the size, activity, or composition of the soil microbial biomass – a new perspective. Soil Biol Biochem 40:61–73

Kirkegaard JA, Matthiessen JN (1999) Biofumigation research - beyond empiricism. In: Magarey RC (ed) Proceedings of 1st Australian soil-borne disease symposium. Watson Ferguson, Brisbane, Australia

Kirkegaard JA, Sarwar M (1998a) Biofumigation potential of brassicas. I. Variation in glucosinolate profiles of diverse field-grown brassicas. Plant Soil 201:71–89

Kirkegaard JA, Sarwar M (1998b) Biofumigation potential of brassicas. II. Effect of environment and ontogeny on glucosinolate production and implications for screening. Plant Soil 201:91–101

Kirkegaard JA, Matthiessen JN, Wong PTW, Mead A, Sarwar M, Smith BJ (1999) Exploiting the biofumigation potential of brassicas in farming systems. In: Proceedings of 10th international rapeseed congress, Canberra, Australia

Kirkegaard JA, Sarwar M, Wong PTW, Mead A, Howe G, Newell M (2001a) Field studies on the biofumigation of take-all by Brassica break crops. Aust J Agric Res 51:445–456

Kirkegaard JA, Smith BJ, Morra MJ (2001b) Biofumigation: soil-borne pest and disease suppression by Brassica roots. In: Proceedings of the 6th symposium of the international society of root research, Nagoya, Japan, pp 416–417

Kuzyakov Y, Friedel JK, Stahr K (2000) Review of mechanisms and quantification of priming effects. Soil Biol Biochem 32:1485–1498

Lopez-Bucio J, De la Vega OM, Guevara-Garcia A, Herrera-Estrella L (2000) Enhanced phosphorus uptake in transgenic tobacco plants that overproduce citrate. Nature Biotech 18:450–453

Lottmann J, Heuer H, Smalla K, Berg G (1999) Influence of transgenic T4-lysozyme-producing potato plants on potentially beneficial plant-associated bacteria. FEMS Microb Ecol 29:365–377

Lu Y, Zhang Z, Ren L, Liu B, Liao Y, Xu H, Du L, Ma H, Ren Z, Jing J, Xin Z (2009) Molecular analyses on Rs-AFP2 transgenic wheat plants and their resistance to *Rhizoctonia cerealis*. Acta Agron Sin 35:640–646

Lynch JM, Brown KM (2001) Top soil foraging – an architectural adaptation of plants to low phosphorus availability. Plant Soil 237:225–237

Marschner P, Baumann K (2003) Changes in bacterial community structure induced by mycorrhizal colonisation in split-root maize. Plant Soil 251:279–289

Marschner P, Timonen S (2004) Interactions between plant species and mycorrhizal colonization on the bacterial community composition in the rhizosphere. Appl Soil Ecol 28:23–36

Marschner P, Ascher JS, Graham RD (1991) Effect of manganese-reducing rhizosphere bacteria on the growth of *Gaeumannomyces graminis* var. *tritici* and on manganese uptake by wheat (*Triticum aestivum* L.). Biol Fertil Soils 12:33–38

Marschner P, Crowley DE, Higashi RM (1997) Root exudation and physiological status of a root-colonizing fluorescent pseudomonad in mycorrhizal and non-mycorrhizal pepper (*Capsicum annuum* L.). Plant Soil 189:11–20

Marschner P, Gerendas J, Sattelmacher B (1999) Effect of N concentration and N source on root colonization by *Pseudomonas fluorescens* 2-79RLI. Plant Soil 215:135–141

Marschner P, Yang CH, Lieberei R, Crowley DE (2001) Soil and plant specific effects on bacterial community composition in the rhizosphere. Soil Biol Biochem 33:1437–1445

Marschner P, Fu QL, Rengel Z (2003) Manganese availability and microbial populations in the rhizosphere of wheat genotypes differing in tolerance to Mn deficiency. J Plant Nutr Soil Sci 166:712–718

Marschner P, Crowley DE, Yang CH (2004) Development of specific rhizosphere bacterial communities in relation to plant species, nutrition and soil type. Plant Soil 261:199–208

Marschner P, Solaiman MZ, Rengel Z (2006) Rhizosphere properties of *Poaceae* genotypes under P-limiting conditions. Plant Soil 283:11–24

Marschner P, Solaiman MZ, Rengel Z (2007) Brassica genotypes differ in growth, phosphorus uptake and rhizosphere properties under P-limiting conditions. Soil Biol Biochem 39: 87–98

McCully ME, Miller C, Sprague SJ, Huang CX, Kirkegaard JA (2008) Distribution of glucosinolates and sulphur-rich cells in roots of field-grown canola (*Brassica napus*). New Phytol 180: 193–205

Meyer JR, Linderman RG (1986) Selective influence on populations of rhizosphere or rhizoplane bacteria and actinomycetes by mycorrhizas formed by *Glomus fasciculatum*. Soil Biol Biochem 18:191–196

Motisi N, Montfort F, Dore T, Romillac N, Lucas P (2009) Duration of control of two soilborne pathogens following incorporation of above- and below-ground residues of *Brassica juncea* into soil. Plant Pathol 58:470–478

Nagendran S, Hammerschmidt R, McGrath JM (2009) Identification of sugar beet germplasm EL51 as a source of resistance to post-emergence *Rhizoctonia* damping-off. Eur J Plant Pathol 123:461–471

Oberson A, Friesen DK, Rao IM, Buehler S, Frossard E (2001) Phosphorus transformations in an oxisol under contrasting land-use systems: the role of the soil microbial biomass. Plant Soil 237:197–201

Osborne L, Rengel Z (2002a) Genotypic differences in wheat for uptake and utilisation of P from iron phosphate. Aust J Agric Res 53:837–844

Osborne L, Rengel Z (2002b) Growth and P uptake by wheat genotypes supplied with phytate as only P source. Aust J Agric Res 53:845–850

Pellet DM, Grunes DL, Kochian LV (1995) Organic acid exudation as an Al-tolerance mechanism in maize (*Zea mays* L.). Planta 196:788–795

Po C, Cumming JR (1997) Mycorrhizal fungi alter the organic acid exudation profile of red clover rhizospheres. In: Flores HE, Lynch JP, Eissenstat D (eds) Radical biology: advances and perspectives on the function of plant roots. Am Soc Plant Physiol, pp 517–519

Posta K, Marschner H, Roemheld V (1994) Manganese reduction in the rhizosphere of mycorrhizal and non- mycorrhizal maize. Mycorrhiza 5:119–124

Potter MJ, Vanstone VA, Davies KA, Rathjen AJ (2000) Breeding to increase the concentration of 2-phenylethyl-glucosinolate in the roots of Brassica napus. J Chem Ecol 26:1811–1820

Rengel Z (2000) Uptake and transport of manganese in plants. In: Sigel A, Sigel H (eds) Metal ions in biological systems. Marcel Dekker, New York, pp 57–87

Rengel Z (2001a) Genotypic differences in micronutrient use efficiency in crops. Comm Soil Sci Plant Anal 32:1163–1186

Rengel Z (2001b) The role of micronutrient fertilization in managing soil-borne plant diseases. In: Singh K, Mori S, Welch RM (eds) Perspectives on the micronutrient nutrition of crops. Scientific Publishers, Jodhpur, India, pp 53–69

Rengel Z (2002) Breeding for better symbiosis. Plant Soil 245:147–162

Rengel Z (2005) Breeding crops for adaptation to environments with low nutrient availability. In: Ashraf M, Harris PJC (eds) Abiotic stresses: plant resistance through breeding and molecular approaches. The Haworth Press, New York, pp 239–276

Rengel Z, Graham RD, Pedler JF (1994) Time-course of biosynthesis of phenolics and lignin in roots of wheat genotypes differing in manganese efficiency and resistance to take-all fungus. Ann Bot 74:471–477

Rengel Z, Ross G, Hirsch P (1998) Plant genotype and micronutrient status influence colonization of wheat roots by soil bacteria. J Plant Nutr 21:99–113

Richardson AE, Hadobas PA (1997) Soil isolates of *Pseudomonas* spp. that utilize inositol phosphates. Can J Microbiol 43:509–516

Richardson AE, Hadobas PA, Hayes JE (2000) Acid phosphomonoesterase and phytase activities of wheat (*Triticum aestivum* L.) roots and utilization of organic phosphorus substrates by seedlings grown in sterile culture. Plant Cell Environ 23:397–405

Richardson AE, Hadobas PA, Hayes JE, O'Hara CP, Simpson RJ (2001) Utilisation of phosphorus by pasture plants supplied with myo-inositol hexaphosphate is enhance by the presence of soil micro-organisms. Plant Soil 229:47–56

Rumberger A, Marschner P (2003) 2-Phenylethylisothiocyanate concentration and microbial community composition in the rhizosphere of canola. Soil Biol Biochem 35:445–452

Rumberger A, Marschner P (2004) 2-Phenylethylisothiocyanate concentration and bacterial community composition in the rhizosphere of field-grown canola. Funct Plant Biol 31:623–631

Ruocco M, Lanzuise S, Vinale F, Marra R, Turra D, Woo SL, Lorito M (2009) Identification of a new biocontrol gene in *Trichoderma atroviride*: the role of an ABC transporter membrane pump in the interaction with different plant-pathogenic fungi. Mol Plant Microbe Interact 22:291–301

Sadana US, Kusum L, Claassen N (2002) Manganese efficiency of wheat cultivars as related to root growth and internal manganese requirement. J Plant Nutr 25:2677–2688

Saxena D, Stotzky G (2000) Insecticidal toxin from *Bacillus thuringiensis* is released from roots of transgenic Bt corn in vitro and in situ. FEMS Microb Ecol 33:35–39

Saxena D, Stotzky G (2002) Bt toxin is not taken up from soil or hydroponic culture by corn, carrot, radish, or turnip. Plant Soil 239:165–172

Scott JS, Knudsen GR (1999) Soil amendment effects of rape (*Brassica napus*) residues on pea rhizosphere bacteria. Soil Biol Biochem 31:1435–1441

Secilia J, Bagyaraj DJ (1987) Bacteria and actinomycetes associated with pot cultures of vesicular-arbuscular mycorrhizas. Can J Microbiol 33:1069–1073

Seeling B, Zasoski RJ (1993) Microbial effects in maintaining organic and inorganic solution phosphorus concentrations in a grassland topsoil. Plant Soil 148:277–284

Shachar-Hill Y, Pfeffer PE, Douds D, Osman SF, Doner LW, Ratcliffe RG (1995) Partitioning of intermediary carbon metabolism in vesicual- arbuscular mycorrhizal leek. Plant Physiol 108:7–15

Shrestha RK, Ladha JK (1996) Genotypic variation in promotion of rice dinitrogen fixation as determined by nitrogen-15 dilution. Soil Sci Am J 60:1815–1821

Smith B, Kirkegaard JA, Howe GN (2004) Impacts of Brassica break-crops on soil biology and yield of the following wheat crop. Aust J Agric Res 55:1–11

Soederberg KH, Baath E (2004) The influence of nitrogen fertilisation on bacterial activity in the rhizosphere of barley. Soil Biol Biochem 36:195–198

Stotzky G (2005) Persistence and biological activity in soil of the insecticidal proteins from *Bacillus thuringiensis*, especially from transgenic plants. Plant Soil 266:77–89

Vleesschauwer DD, Chernin L, Hofte MM (2009) Differential effectiveness of *Serratia plymuthica* IC1270-induced systemic resistance against hemibiotrophic and necrotrophic leaf pathogens in rice. BMC Plant Biol 9:10–15

Wamberg C, Christensen S, Jacobsen I, Mueller AK, Sorensen SJ (2003) The mycorrhizal fungus (*Glomus intraradices*) affects microbial activity in the rhizosphere of pea plants (*Pisum sativum*). Soil Biol Biochem 35:1349–1357

Xavier LJC, Germida JJ (1998) Response of spring wheat cultivars to *Glomus clarum* NT4 in a P-deficient soil containing arbuscular mycorrhizal fungi. Can J Soil Sci 78:481–484

Zhang CQ, Liu YH, Ma XY, Feng Z, Ma ZH (2009) Characterization of sensitivity of Rhizoctonia solani, causing rice sheath blight, to mepronil and boscalid. Crop Protect 28:381–386

Zhu YG, Smith SE, Baritt AR, Smith FA (2001) Phosphorus (P) efficiencies and mycorrhizal responsiveness of old and modern wheat cultivars. Plant Soil 237:249–255

Chapter 9
Utilizing Soil Microbes for Biocontrol

Alison Stewart, Michael Brownbridge, Robert A. Hill, and Trevor A. Jackson

Introduction

This review focuses on the potential for microbial biological control of soil-borne pests and pathogens. The range of crops affected by key soil-borne pests and pathogens, and opportunities for biocontrol using composts, organic soil amendments and/or augmentation with selected bioactive microbes are discussed. Selected examples of biological control successes, constraints and likely future developments are also explored.

The Role of Soil Microbes in Biocontrol

The benefits of adding organic soil amendments, especially composts and mulches, to agricultural cropping systems has long been recognised. These amendments have been shown to increase soil fertility and suppress a range of soil-borne plant pests and pathogens. Additional benefits include water retention, increased soil organic matter and enhanced soil aeration. There have been many research studies on successful compost-related suppression of plant pests and diseases, particularly plant collar and root rots, although some studies using organic amendments have resulted in increased disease incidence (Hoitink and Boehm 1999; Hoitink et al. 1996a). A key factor in disease suppression is the

A. Stewart(✉) and R.A. Hill
Bio-protection Research Centre, Lincoln University, P.O. Box 84, Lincoln University, Lincoln 7647, Canterbury, New Zealand
e-mail: alison.stewart@lincoln.ac.nz; robert.hill@lincoln.ac.nz

M. Brownbridge
Vineland Research and innovation Center, P.O. Box 8000, Vineland Station, ON LOR 2EO, Canada
e-mail: michael.brownbridge@vinelandresearch.com

T.A. Jackson
AgResearch, Private Bag 4749, Christchurch 8140, New Zealand
e-mail: trevor.jackson@agresearch.co.nz

G.R. Dixon and E.L. Tilston (eds.), *Soil Microbiology and Sustainable Crop Production*, DOI 10.1007/978-90-481-9479-7_9, © Springer Science+Business Media B.V. 2010

degree of decomposition of organic matter in compost, which determines the numbers and kinds of miroorganisms responsible for disease suppression (Hoitink and Grebus 1994; Hoitink et al. 1996b). In this context, compost is defined as organic matter that has undergone controlled aerobic biological transformation, involving a heating cycle, resulting in pasteurisation and maturation. Mulching is similar to the natural process of organic matter decomposition on the forest floor and is defined as any organic material, composted or raw, that is placed on the soil surface (see other Chapters in this Book).

Disease Suppression Using Composts

Under natural conditions, healthy soils are rich in microbes, many of which can suppress plant diseases. Provision of organic amendments to the soil, by ploughing-in green manures or through the application of composts, can further promote disease suppression. A variety of disease suppressive effects from composts have been described and these effects comprise a combination of physiochemical and biological characteristics. One mechanism is thought to be a result of toxic compounds released from decomposing plant material (Gimsing and Kirkegaard 2006). However, disease suppression reported in the literature is mostly attributed to the biological activity of microorganisms, described as beneficial microorganisms or biological control agents. Species reported to act as biocontrol agents in compost-amended substrates include bacteria such as *Bacillus* spp., *Enterobacter* spp., *Flavobacterium balustinum* and *Pseudomonas* spp., and fungi such as *Penicillium* spp., *Gliocladium virens*, and several *Trichoderma* spp. (Litterick et al. 2004). These beneficial microorganisms can be released from the compost or the compost may provide nutrients that stimulate the proliferation of antagonistic bacteria and fungi in the rhizosphere (Noble and Coventry 2005; Green et al. 2006). Four main mechanisms of disease suppression have been described: competition for nutrients (between the microorganisms from compost and soil pathogens); antibiotic and enzyme production by beneficial microorganisms; parasitism and predation by beneficial microorganisms of pathogens; and enhanced plant disease resistance (both induced systemic resistance and systemic acquired resistance). In addition, more vigorous growth on and better use of decaying organic matter by beneficial microorganisms results in microbiostasis through the repression of pathogen spore germination and growth. Several or all these different mechanisms of disease suppression may occur simultaneously through the activity of one or more beneficial microorganisms present in disease suppressive composts. For example, several different *Trichoderma* spp. can compete with pathogens for nutrients and space, exhibit antibiosis, parasitize the pathogen and elicit induced plant resistance (Harman et al. 2004; Hill et al. 1999).

Natural disease suppression in some composts is due to the succession of antagonistic microbes that flourish in the organic material. These suppressive effects have been observed in a wide range of crops and soil-borne diseases, including: *Phytophthora* root rot in avocado (Costa et al. 1996); *Pythium* in cotton and turfgrass (Craft and Nelson 1996; McKellar and Nelson 2003); snow moulds in turf grass (Boulter et al. 2002); *Fusarium*, *Pythium* and *Rhizoctonia* in tomato and cucumber (Kannangara et al. 2004; Trillas et al. 2006), and many others in cereal, fodder, vegetable and horticulture crops (Litterick et al. 2004; Noble and Coventry 2005).

Composts can be pre-inoculated with beneficial microorganisms prior to application. Use of inoculated composts not only promotes disease suppression, but utilises an inexpensive waste product, adds value to the composted product, and additionally provides nutrients to the soil. Nakasaki et al. (1998) reported successful inoculation and growth of *Bacillus subtilis* in composted grass-clippings. When applied to turf grass, the *Bacillus*-augmented compost significantly reduced the severity of *Rhizoctonia* large-patch disease. While it is generally accepted that general pathogen suppression is provided by a consortia of microbes, Nakasaki et al. (1998) demonstrated the potential to inoculate compost with a specific bacterium to control a specific plant pathogen. *Bacillus*-compost preparations applied against *Phytophthora* blight of bell pepper provided a biocontrol efficacy of 60% and increased yield by 200% through disease suppression and growth promotion compared with untreated controls (Jiang et al. 2006). This study highlights the dual action of some suppressive microbes as plant growth promoters. Recent research suggests that *Trichoderma*-amended compost can reduce the incidence of plant pathogens. Trillas et al. (2006) showed suppression of *R. solani* in cucumber seedlings using compost inoculated with *T. asperellum*. A significant reduction in bacterial leaf spot of radish, lettuce and tomato has also been observed using bark compost inoculated with *T. hamatum* 392 (Aldahmani et al. 2005).

Pest Suppression Using Compost

Composts applied as a mulch not only suppress plant diseases and help protect and conserve agro-ecosystem health and quality, but also suppress insects, weeds and nematodes. Control of arthropod pests through mulch application occurs by increasing densities and diversity of generalist predators in the soil and on plants, by reducing environmental suitability for herbivore population growth (Brust 1994), or through biofumigation, the release of toxic compounds from decomposing plant material into the soil (Gimsing and Kirkegaard 2006). Most research concerning the inoculation of compost/growing media with microbes has focused on the suppression of insect pests in container-grown greenhouse and nursery ornamentals. Two major insect pests, black vine weevil, *Otiorhynchus sulcatus*, and western flower thrips, *Frankliniella occidentalis*, have been targeted in particular. The soil-dwelling stages of both these insect species are susceptible to entomopathogenic fungi, which possess several desirable traits that make them suitable for development as compost amendments. Use of the fungus *Metarhizium anisopliae* has been extensively studied for black vine weevil control in containerised production, and has provided efficient control of the pest on a range of plant types (Moorhouse et al. 1992a, 1993a, b; Bruck 2005, 2006). The fungus is effective in soil and a range of soil-less potting media (including peat, bark, coir and peat-blends incorporating composted green waste) whether applied as a drench or pre-mixed into the medium as conidia or granules, although pre-mixing provides a more homogenous distribution of infective material throughout the medium (Bruck 2006; Moorhouse et al. 1992b, 1993c; Kepler and Bruck 2006; Shah et al. 2007). A strain of *M. anisopliae* active against black vine weevil was shown to persist at infective levels for >4 months in all media and considerably longer (approximately 1 year) in peat- and bark-based media (Bruck 2005, 2006; Moorhouse et al. 1993c). Factors affecting pest and disease suppression by composts include the raw materials used, the composting process, conditions and microbial colonisation, the moisture content of the compost, compost stability/maturity, compost nutrient content, timing of compost application and rate of application and the longevity of disease suppression. The use of bioinoculated composts or composted mulches offers potential in many cropping systems such as orchards, soft or vine fruits, e.g., kiwifruits, strawberries, grapes. In addition to improving overall soil health and providing favourable conditions for insect infection to occur and disease-suppressive organisms to flourish, composts can be seeded with insect-active or antagonistic microbes and, thereby, provide an effective means of managing pest and disease problems that would complement environmentally sustainable production systems.

Suppressive Soils

Disease severity or incidence remains low in suppressive soils, in spite of the presence of the pathogen, a susceptible host plant, and climatic conditions favourable for disease development. This suppressiveness is attributable to microorganisms and has been divided into two main categories: general suppression, which is associated with the total amount of microbial activity in relation to the pathogen; and specific suppression, which is attributable to the activity of one or a few inhibitory microorganisms (Hoitink and Boehm 1999; Cook and Baker 1983).

Take-All Suppression

Take-all is an important root disease of grasses and cereal crops, especially wheat and barley, worldwide and is caused by the fungus *Gaeumannomyces graminis* var. *tritici* (Hornby et al. 1998; Asher and Shipton 1981). Significant yield reductions can occur after two successive crops in the same field as a result of take-all disease (MacSpadden Gardener and Weller 2001). However, in wheat cropping areas where monoculture is practiced, a severe outbreak of take-all followed by 4–6 years of wheat or barley monoculture can result in a natural disease suppression called take-all decline (MacSpadden Gardener and Weller 2001). Take-all decline is thought to be due to specific suppression of the disease with research showing that disease suppression is transferrable and can multiply (Cook 2003, 2007). Indeed, field plots inoculated with take-all produced good disease symptoms. However, plots treated at tillering with "starter" soil (0.5% by volume) from soils with 12 years wheat monoculture produced plants with healthy looking roots and stems with natural biological control being demonstrated as the cause of this take-all decline (Cook 2007). Specifically, strains of the rhizosphere-inhabiting bacterium *Pseudomonas fluorescens* that produce the broad spectrum antibiotic 2,4-diacetylphoroglucinol (DAPG) are responsible (Cook 2007). The genes associated with this antibiotic synthesis have been identified allowing the molecular detection and quantification of these bacterial strains in take-all suppressive soils (Weller et al. 2002). The threshold for *P. fluorescens* DAPG producing strains to cause take-all suppression was shown to be 10^5 CFU g^{-1} root (Weller et al. 2002). The elimination of these bacterial strains from the soil has been shown to stop take-all suppression, but it can be restored by their reintroduction (Cook 2007). DAPG-producing strains of *P. fluorescens* have also been shown to suppress soil-borne diseases in different host/pathogen systems (Cook 2007). In New Zealand wheat soils, particular species of actinomycetes, fungi and bacteria (including *P. fluorescens*) were associated with take-all suppressive soils (general and specific) (Chng 2009).

Fusarium Suppressive Soils

Soil suppressiveness to *Fusarium*, in most instances, has been attributed to biotic characteristics, although soil physiochemical factors are also involved (Amir and Alabouvette 1993; Hoper et al. 1995). Suppressiveness is lost through sterilisation and may be restored by the addition of a little untreated suppressive soil (Whipps 1997), while the addition of organic amendments and plant residues resulted in suppression of diseases caused by *Fusarium oxysporum* in field soils (Lodha 1995). A variety of different mechanisms have been demonstrated in this suppression, including general suppression (Serra-Wittling et al. 1996), specific antagonism (Trillas-Gay et al. 1986), propagule lysis (Sequeira 1962) and induced systemic resistance (Pharand et al. 2002; Larkin et al. 1996). For example, conifer litter on the forest floor induced germination and subsequent destruction of macroconidia and chlamydospores of *F. oxysporum* (Toussoun et al. 1969). Competition for substrate in the soil, often by non-pathogenic fusaria (Toyota et al. 1996), and for root infection sites (Olivain and Alabouvette 1999) contributes to suppressiveness in some soil, while antibiosis has been implicated in suppression of *Fusarium* wilt in peas (Weller et al. 2002).

Augmentation with Selected Bioactives

The soil can be augmented with selected microbes to bring about the biological control of soil-borne pests and pathogens. There are numerous comprehensive reviews on specialised topics describing the biocontrol activities of different microorganisms against plant pests and pathogens (Van Driesche and Bellows 1996; Vincent et al. 2007; Jaronski 2007). In this review section, we highlight specific examples of successful augmentation for biocontrol of insect pests, plant pathogenic nematodes and plant microbial pathogens.

Insect Pests

Entomopathogenic Fungi

Fungi are the predominant pathogens of insects and play a significant role in the natural regulation of soil-dwelling pests. Among all of the disease-causing microbes of insects, fungi are unique in their ability to infect their hosts through the external cuticle. Several hypocrealean entomopathogenic species are important constituents of the soil biota and appear to be ubiquitous inhabitants of soils worldwide, having been recovered from a diverse array of geographic, climatic and agro-ecological zones. Species such as *B. bassiana* and *M. anisopliae* are commonly found in both cultivated and undisturbed soils, although their natural distribution does appear to be linked to habitat (Meyling and Eilenberg 2006) and soil populations are influenced by agricultural practices (Jaronski 2007; Meyling and Eilenberg 2007).

Fungi have been widely evaluated as control agents for a variety of noxious arthropods of agricultural and horticultural importance. They have many desirable traits including; minimal risk posed to beneficial non-target species such as earthworms and Collembola, which are key service-providers in the soil ecosystem (Brownbridge and Glare 2007; O'Callaghan and Brownbridge 2009), and host-specificity, which enhances their potential role in integrated pest management (IPM). Furthermore, these beneficial fungi can persist in the soil for several years with new 'flushes' of inoculum provided following the successful infection and colonization of a susceptible host. This leads to high, localized concentrations of infective conidia and greater opportunities for insect infection to occur. Long-term survival of entomopathogenic fungi in soil does appear to be reliant upon their access to susceptible hosts, as they are generally considered poor saprophytes in the competitive soil environment (Jaronski 2007; Keller et al. 2003). For those species with relatively narrow host-spectra, this can limit their natural occurrence and longevity in treated soils (Meyling and Eilenberg 2007; Keller et al. 2003).

A diverse range of subterranean pests have been targeted for biocontrol. The majority of these initiatives have sought to use fungi as inundative control agents, an approach which is not reliant upon fungi recycling within the pest population. The primary candidate organisms have been *M. anisopliae*, *B. bassiana*, and *B. brongniartii* and several commercial products and proto-products have been developed (de Faria and Wraight 2007). Here, we will focus on biocontrol initiatives targeting insects, which cause economic losses through direct feeding damage to roots or root crops, although examples of initiatives against other important species with a soil-dwelling phase in their life cycle are included.

Fungi have been widely tested against root-feeding Coleoptera in particular, including pests of agricultural and horticultural importance. *Beauveria brongniartii* is used in grassland regions of Austria, Italy and Switzerland to control the European cockchafer, *Melolontha melolontha* (Keller et al. 2003; Keller 2000). Formulated on colonized barley kernels, the fungus is applied to the soil (pastures, grasslands, forests, orchards, etc.) via modified seed drills, and suppresses cockchafer populations for several years. In trials to assess effects of *B. brogniartii* on non-target organisms, the fungus was considered low-risk for a predatory carabid, *Poecilus versicolor*, which was exposed directly to different pathogen formulations and to mycosed cockchafer cadavers (Traugott et al. 2005). Thus, while active against *M. melolontha*, the pathogen does not appear to affect coleopteran natural enemies in the targeted ecosystem. The red-headed cockchafer (*Adoryphorus coulani*), a destructive root-feeding pest of pasture and field crops in Australia that was recently introduced into New Zealand, has been successfully controlled with a strain of *M. anisopliae* specifically selected for its activity at cooler temperatures and a commercial product (ChaferGuard®; Becker Underwood Pty Ltd, Australia) was developed (Rath et al. 1995). The fungus is produced on rice and formulated on the grains for application to soil using a seed drill at the time of pasture renovation (Rath et al. 1995; Rath 2002). In field trials, the pathogen regulated cockchafer populations in treated soils for more than 3 years, delivering a 23% increase in pasture productivity with no apparent impact on non-target invertebrates (Rath et al. 1995). Control of the New Zealand grass grub, *Costelytra zealandica*, may also be possible with *M. anisopliae* or *B. bassiana*, as larvae are hosts to these species and natural epizootics are periodically observed (Glare et al. 1994; Townsend et al. 1995). However, insect infection by a selected *M. anisopliae* isolate was inhibited at 15°C, indicating the necessity of matching a candidate strain to the prevailing environmental conditions in which it will be used (Glare et al. 1994). Research has demonstrated the virulence of an exotic *B. bassiana* isolate for clover root weevil, *Sitona lepidus*, a pest that has a severe impact on the productivity of New Zealand white clover-ryegrass pastures (Gerard 2001). This *B. bassiana* isolate is easily mass-produced on rice and persists in treated soils with farm-scale field trials on a prototype grain-based formulation drilled into clover root weevil-infested pasture significantly reducing larval and pupal populations without impacting non-target species (O'Callaghan and Brownbridge 2009; Brownbridge et al. 2006). *Metarhizium anisopliae* and *B. bassiana* are frequent natural pathogens of the black vine weevil, *O. sulcatus*, and *M. anisopliae* has provided excellent control of this pest, as discussed previously.

The larval stages of several dipteran species also cause direct economic damage through their feeding activity on root- and other field-crops. Few chemical control options are available for these pests and concerns over resistance, pesticide residues and environmental impacts have prompted the search for more benign alternatives, including entomopathogenic fungi. Sugarbeet root maggot, *Tetanops myopaeformis*, is a major problem for North American sugarbeet growers. *Metarhizium anisopliae* successfully reduced feeding damage and increased yield under low to moderate root maggot pressure in infested sugarbeet fields (Campbell et al. 2000).

The effectiveness of *M. anisopliae* treatments was further demonstrated in trials devised to evaluate effects of formulation and application rate on efficacy (Campbell et al. 2006). Although higher yields were obtained under an intensive insecticide program, particularly in areas where root maggot infestation levels were high, the fungus was seen as being complementary to an insecticide regime in such situations and integration of the two strategies could reduce overall pesticide use (Campbell et al. 2006). Larvae and adults of the onion maggot, *Delia antiqua*, are also susceptible to *M. anisopliae* and promising isolates have been identified in laboratory trials (Davidson and Chandler 2005). These strains were also pathogenic to cabbage maggot, *D. radicum*, a primary pest of brassicas in northern temperate regions (Chandler and Davidson 2005). Drench applications of the fungus reduced survival of cabbage maggot larvae and pupae by up to 90% and adult emergence by up to 92% in glasshouse trials. However, when applied to infested cauliflowers in the field, the fungus did not control *D. radicum* populations, although 30% of the larval cadavers recovered from treated plots were infected (Chandler and Davidson 2005). Advances in formulation and delivery techniques were considered critical to improving the efficacy and cost-effectiveness of this control approach in all of the above examples (Jaronski 2007; Campbell et al. 2006; Chandler and Davidson 2005).

Several pest species cause no crop damage during the soil phase of their life cycle. These stages are often targeted with insecticides that could be replaced with fungal treatments. Ekesi et al. (2005) showed that fungi were effective against soil-dwelling stages of three fruit-fly species in this pest complex. In field trials, emergence of adult fruit flies were significantly reduced over an extended period in soils treated with *M. anisopliae* formulations with this treatment providing superior control to diazinon. Furthermore, *M. anisopliae* had no adverse effects on the non-target fruit-fly parasitoids (Ekesi et al. 2005) and shows great promise for use in IPM strategies for this pest (Ekesi et al. 2007).

Western flower thrips (*F. occidentalis*), a pest of global significance on a wide range of economically important crops, largely pupate off-plant in the soil. Although western flower thrips are susceptible to *B. bassiana* and *I. fumosorosea* in peat and other plant-growing media, *M. anisopliae* has consistently provided the best levels of control (Ansari et al. 2007, 2008; Butt and Brownbridge 1997). Indeed, under greenhouse conditions *M. anisopliae* was effective when applied either as a drench or pre-mixed into the medium as dry conidia, and was more efficacious than either fipronil or impidacloprid treatments(Ansari et al. 2007, 2008).

Social insects are often considered poor targets for microbial control owing to their hygienic behaviour. However, opportunities exist to use fungi against subterranean termites. Isolates of *B. bassiana*, *M. anisopliae* and *Paecilomyces* appear to be non-repellent and capable of infecting and being transmitted among termites, causing rapid mortality in exposed individuals (Milner et al. 1998; Sun et al. 2003; Wright et al. 2002, 2005). Direct inoculation of termite mounds in Australia with *M. anisopliae* successfully controlled termite populations (Milner et al. 1998), while in Kenya, application of *M. anisopliae* at maize planting significantly reduced termite lodging and increased grain yield (Maniania et al. 2002).

Entomopathogenic Bacteria

The soil-dwelling larvae of the Scarabaeidae often feed on plant roots, causing significant losses in economically important crops worldwide. These larvae have proven especially difficult to control with all but the most toxic and persistent of insecticides. However, they are hosts to a wide range of pathogenic bacteria, some of which have been used in augmentative biological control (Jackson and Glare 1992). Milky disease, named from the milky colouration that develops after colonisation of the insect haemolymph by strains of the bacterium, *Paenibacillus popilliae*, is unique to scarab larvae. Although milky disease bacteria are frequently found in scarab populations, the disease rarely controls insect populations under natural conditions. Resistant spores, which can survive for many years in the soil, are released from the host after death. To increase the impact of this pathogen, milky disease bacteria have been cultured and introduced into turf for control of Japanese beetle (*Popillia japonica*) in the United States since the 1940s (Fleming 1968). Formulations of *P. popilliae* have been commercialised, but use of the bacterium has been limited by the high cost of *in vivo* production and the slow progression of disease. It has been suggested that the bacterium would be best used as an inoculative agent within wider pest management programmes rather than as a biopesticide to provide immediate control (Redmond and Potter 1995).

Strains of the non-spore forming bacteria, *Serratia* spp., cause amber disease in the New Zealand grass grub, *C. zealandica* (Jackson et al. 2001). Pathogenic strains contain a specific plasmid (Hurst et al. 2000), and ingestion brings about a cessation of feeding, clearance of the gut and eventual death of the larval host. Cultivation of pastures for cropping and resowing generally kills grass grub and eliminates pathogenic strains of bacteria, leaving new pastures vulnerable to pest attack. This provides an opportunity for biological control, where artificially produced bacteria can be applied to healthy grass grub populations to promote epizootics of disease and prevent pasture damage. Strains of *S. entomophila* have been commercialised as the products Invade™ and Bioshield™ and successfully used for grass grub biological control for more than a decade (Jackson and Klein 2006). For best effect, the bacterium is applied to a healthy grass grub population in young pastures. Recycling of the disease through the pest population provides a reservoir of pathogenic bacteria that prevent damaging outbreaks of the pest.

Antibiosis/Toxins

Insecticidal toxins are produced by most entomopathogenic fungi during pathogenesis. After successfully penetrating the insect cuticle, the fungi enter the haemocoel where they have to overcome insect immune responses in order to colonize and kill the host. Generally, these toxins are bioactive secondary metabolites secreted during growth inside the insect. The toxins can have a range of effects, and they are considered important pathogenicity determinants (Strasser et al. 2000). Toxic (and other) effects of metabolites produced by *M. anisopliae* (destruxins), *B. bassiana* (e.g., oosporein, beauvericin, bassianolide, oxalic acid) and *Tolypocladium* spp. (efrapeptins) have been well documented in insects and other arthropod pests (Vey et al. 2001; Charnley 2003; Kirkland et al. 2005; Quesada-Moraga et al. 2006; Zimmermann 2007a, b). Attempts to increase the potency of *M. anisopliae* through genetic manipulation have recently focused on production of isolates to express scorpion neurotoxin during colonization of the insect hemolymph (Wang and St Leger 2007).

The colonisation of plants by endophytic fungi can provide protection from insect predation via *in planta* production of metabolites by the endophyte. These metabolites can be directly toxic to the pest, inhibit insect development, or deter feeding activity. These associations tend to be beneficial to plant survival and include provision of protection against herbivorous insects and plant-parasitic nematodes (Vega et al. 2008). Many economically important turf and forage grasses are naturally infected with endophytes, and breeding programmes have provided lines that contain selected strains. For example, *Neotyphodium uncinatum* produces loline alkaloids in meadow fescue (*Festuca pratensis*), which are expressed throughout the plant (Bush et al. 1993). The alkaloids have bioactivity against a range of insects conferring resistance to foliar and root aphids, while deterring larval grass grub feeding and development. Murphy et al. (1993) observed such effects on scarab grubs in tall fescue (*F. arundinacea*), while Clement et al. (2005) found that *Neotyphodium*-infected wild barley induced a range of responses to different insect pest species.

Some entomopathogenic fungi are also endophytic (Vega et al. 2008). *Beauveria bassiana* for example, is an endophyte in a diverse assortment of host plants. In all cases, plants have been protected against insect damage yet mycosed insects have not been recovered. This evidence suggests that the suppressive effects are a result of feeding deterrence or antibiosis due to fungal metabolites produced in plant tissues, or direct toxic effects following consumption of these metabolites or vegetative mycelia. To date, evaluations have concentrated on effects against foliar-feeding pests. Effects against root-feeding insect pests have not been determined, but could provide a novel means of protecting crop plants.

Although soil-inhabiting insects appear to be resistant to most strains of *Bacillus thuringiensis* (Bt), the organism is the source of a wide and expanding range of Bt toxins. Resistance to Bt toxins may have evolved through close association of larvae with bacterial spores in the soil. However, one isolate, the Buibui strain, has been shown to be effective against some scarab pests, including Japanese beetle (Alm et al. 1997). It has been suggested that a protein/spore mix is necessary to induce mortality suggesting a role for the live cells beyond that of the toxin alone. Strains of *B. thuringiensis* serovar. *israelensis* are now being produced for control of tipulid larvae in pastures suggesting that, as a wider range of strains is discovered, there will be more uses for this bacterium.

The Tc toxin complex is another important toxin group that has been discovered associated with soil bacteria. This complex is found in strains of the nematode-associated bacteria *Xenorhabdus* and *Photorhabdus* spp., as well as soil bacteria from the Enterobacteriaceae. While many strains of Tc-producing bacteria are weakly pathogenic to insects, a novel bacterium, *Yersinia entomophaga*, has a wide insect host range and is capable of killing late instars of several soil dwelling pest species (Hurst et al. 2007). Bacterial toxins applied to the soil face similar problems to many chemicals by inactivation through adsorption or microbial activity. These problems may be overcome by formulation of the toxins into resistant granules and baits (Brownbridge et al. 2006).

Plant Parasitic Nematodes

Plant-parasitic nematodes cause yield losses worth billions of dollars each year. Currently, crop rotations with resistant or non-susceptible cultivars, combined with applications of nematicides, are used to regulate these pests. Nematicides can provide short-term gains in productivity. However, due to the ability of nematode populations to build up rapidly, the benefits are limited in both space and time. The ban on the soil sterilant methyl bromide has provided additional impetus to search for alternative, sustainable management options.

Nematodes are regulated by a diverse array of natural enemies including fungi and bacteria, as well as invertebrate predators. Considerable research has been done to evaluate different microbial control agents, yet few commercial products have resulted and none of these are in widespread use. There are many reasons for this, but the primary difficulties have been related to development of efficient mass production systems for the innoculative or innundative use of these organisms, limited host range, poor understanding of host/pathogen interactions, and lack of good ecological data to develop products that deliver consistent levels of suppression (Dong and Zhang 2006).

Microbial biocontrol agents of plant-parasitic nematodes appear to have a cosmopolitan distribution in soils worldwide. Their prevalence is generally correlated with high host nematode populations, but cropping sequence, tillage practices and ability to survive as a saprophyte also influence the incidence of these infectious or parasitic organisms. Microorganisms investigated for control include a range of nematode-trapping species, such as *Arthrobotrys* and *Dactylaria* spp. (Davies 2005; Singh et al. 2007) and infectious microbes, broadly divided into those that are facultative pathogens, such as the fungi *Paecilomyces lilacinus* and *Pochonia chlamydosporia* (formerly *Verticillium chlamydosporium*) (Kiewnick and Sikora 2006; Pérez-Rodriguez et al. 2007), obligate parasites, such as the fungi *Hirsutella rhossiliensis* and *H. minnessotensis* (Chen and Liu 2005; Mennan et al. 2006), and bacterial antagonists, such as *Pasteuria penetrans* and *Pseudomonas oryzihabitans* (Vagelas et al. 2007).

Nematode-trapping ('predatory') fungi from various taxonomic groups have been widely studied for control of root-knot and cyst nematodes. Numerous surveys have been done to document associations with important plant-parasitic species of nematodes and to determine their prevalence in different crops and soils. These organisms have a diverse range of trapping devices, which include adhesive hyphae, adhesive branches and knobs, and more complex multi-dimensional networks and constricting rings (Davies 2005). Formation of capture organs can be stimulated by the presence of nematodes, but the fungi generally respond to a limited host spectrum in both trap formation and initiation of a trapping response (Davies 2005). Nematophagous species also produce compounds that attract some nematodes while being repellent to others; this is related to the developmental and nutritional status of the fungus, and the nematode species (Davies 2005). All of these factors influence the efficacy of these fungi and, together with difficulties associated with the large-scale production and preservation of infectious inoculum for field-scale application, have restricted their development into reliable biological control agents for plant-parasitic nematodes.

As with insect pathogenic microbes, many nematophagous fungi and bacteria produce toxins and enzymes that play a key role in nematode infection and mortality, which are often primary virulence factors. Other microbes produce compounds that are directly toxic or show antagonistic effects against nematodes. Although the broad-spectrum, proteinaceous endotoxin produced by *B. thuringiensis* is better known for its insecticidal activity, some Bt proteins are also toxic to free-living and parasitic nematodes (Crickmore 2005; Tian et al. 2007). Their potential utility against plant-parasitic species has not been fully investigated. Rhizosphere-colonizing bacteria and fungi can reduce populations of plant-parasitic nematodes by regulating nematode behaviour, interfering with plant-nematode recognition, or by direct antagonism through the production of toxins, enzymes and other metabolites (Tian et al. 2007). Several *Bacillus* spp. are pathogenic (e.g., *B. subtilis*), while metabolites produced by rhizosphere-competent microbes, e.g., *Pseudomonas* spp., have antagonistic effects such as inhibiting egg hatch, and reducing survival and mobility of infective second-stage juveniles. Suppressive effects have been obtained with culture filtrates of the rhizo-competent bacterium, *Burkholderia cepacia*, and *Trichoderma* fungi against root-knot nematode, providing additional evidence that extracellular factors produced by these organisms are responsible for the observed results (Meyer et al. 2000; Sharon et al. 2001). Although *Trichoderma* spp. are more commonly used as biocontrol agents for soil-borne diseases, these findings indicate potential for a multi-functional role in plant bioprotection. The promise shown by several rhizobacteria has led to the development of commercial bionematicides based on *Bacillus* spp. and *B. cepacia* for control of root-knot and other nematodes on vegetables (Tian et al. 2007). Nematode suppression may also be related to the development of systemic resistance induced by the presence of the bacteria in the rhizosphere. Endophytic bacteria and fungi can also induce resistance, and secondary metabolites secreted by the microbes are thought to play an important role in this phenomenon (Siddiqui and Shaukat 2003; Siddiqui et al. 2005; Dababat and Sikora 2007; Tesfamariam et al. 2009). Interestingly, applications of entomopathogenic nematodes have also reduced plant-parasitic nematodes; the cause of this effect has been attributed to allelochemicals produced by the nematodes or their associated symbiotic bacteria (Somasekhar et al. 2002).

The greatest progress towards development of innundative controls has been achieved with fungi and bacteria that parasitize female nematodes or infect eggs.

Various *Pasteuria* spp. are obligate parasites of a wide range of economically important plant-parasitic nematodes (Davies 2005; Trudgill et al. 2000). High incidence of *P. penetrans* has been associated with soils that are suppressive to root-knot species and control has been achieved in some soils when amended with the bacterium (Trudgill et al. 2000). However, commercialization of the bacterium as a bionematicides has been hindered by the inability to culture the organism outside a nematode host and its host specificity (Davies 2005; Davies et al. 2001). Recent advances in *in vitro* production techniques, combined with research to understand the mechanisms behind endospore attachment and specificity may allow these hurdles to be overcome in the future (Davies 2005; Davies and Opperman 2006).

Both *P. lilacinus* and *P. chlamydosporia* (*V. chlamydosporium*) parasitize nematode eggs and have been widely studied for use against root-knot and cyst nematodes. In spite of early concerns around potential human-health risks associated with *P. lilacinus*, commercial products based on the fungus are now registered for use on a variety of field and greenhouse crops in several countries. The fungus is readily mass-produced on grains, but the high application rates required with early products, combined with inconsistency of efficacy, have inhibited their wider adoption (Kiewnick 2004). This prompted investigations into factors affecting host specificity and surveys for strains with broader host spectra (Kiewnick 2004; Freitas et al. 1995). *Pochonia chlamydosporia* is frequently present in nematode-suppressive soils; it colonizes the rhizosphere of some plants and proliferates in egg masses produced by root-knot nematodes. The fungus lends itself well to development as a myconematicide; optimal growth conditions have been defined on a variety of substrates and the organism produces resistant chlamydospores that are easily handled and applied for nematode suppression (Pérez-Rodriguez et al. 2007). [Sustainable strategies have been developed for the management of root-knot nematodes in organic vegetable production systems by integrating applications of *P. chlamydosporia* with the cultivation of plants that are poor nematode hosts, soil infestations before planting a susceptible crop in the following season (Kerry and Hidalgo-Diaz 2004)]. The strain used in these studies has subsequently been registered as a bionematicide in Cuba, and similar development programmes have been followed in several other countries (Pérez-Rodriguez et al. 2007).

Hirsutella rhossiliensis is an efficient natural pathogen of nematodes and has been recovered from many plant-parasitic species. The fungus has effectively suppressed nematode populations in greenhouse and small-plot studies, but field-scale trials have provided variable results (Chen and Liu 2005; Jaffee 2000). Addition of organic amendments to soil generally led to a reduction in nematode parasitism rates by *H. rhossiliensis* in spite of the fact that there was an increase in bacterivorous and fungivorous nematodes in the treated soils, demonstrating the importance of host species in fungal dynamics (Jaffee et al. 1994). This observation was borne out by trials to evaluate effects of crop sequence (nematode-susceptible versus non-susceptible) and tillage on nematode parasitism by the fungus, which demonstrated that observed effects could largely be attributed to the density-dependent relationship between the fungus and its host, whereas tillage practices appeared to have only a minor effect (Chen and Liu 2007).

Hirsutella minnesotensis also has a wide nematode host range and shows promise as a biocontrol agent for the soybean cyst and root-knot nematodes (Chen and Liu 2005; Mennan et al. 2006, 2007).

Microbial Plant Pathogens

Soil-borne diseases can cause major economic losses on a wide range of fruit, vegetable, arable, pasture and forest crops. Fungi are the main causal agents with important pathogenic species found within several genera including *Sclerotinia*, *Sclerotium*, *Fusarium*, *Pythium*, *Rhizoctonia*, *Phytophthora*, *Verticillium* and *Gaeumannomyces*. Selected bacterial genera such as *Agrobacterium*, *Erwinia* and *Pseudomonas* also include plant pathogenic species. Soil-borne diseases are often difficult to control because of the pathogens' ability to produce persistent survival structures that allows populations to build up over time. There are few disease resistant cultivars available to growers and control has relied on the use of soil-based pesticides. However, issues with loss of efficacy due to fungicide resistance or enhanced microbial degradation and health and environmental concerns leading to use restriction and/or deregistration has provided the impetus for the development of biological control alternatives (Martin 2003). The last decade has seen an increase in the number of commercial products brought to the market and these are now being integrated into mainstream production practices (Fravel 2005; Stewart 2001).

In contrast to the predominantly parasitism-based methods of biocontrol of insect pests and plant parasitic nematodes, biocontrol of microbial plant pathogens can be achieved by a number of different strategies, both direct and indirect. The direct approach involves the introduction of specific microbes into the soil where they interfere with pathogen survival, growth and/or plant infection. There are four direct mechanisms of biological control of plant diseases; parasitism when the biocontrol agent feeds directly on or inside the pathogen; antibiosis, which is the inhibition of the pathogen by a toxic metabolite produced by the antagonist; competition with the pathogen for limited resources; and hypovirulence, which is the use of less virulent strains of the pathogen to control the virulent strain (Chernin and Chet 2002). Indirect methods include induced resistance where the biological control agent activates physical and/or chemical self-defence responses within the host plant against a particular pathogen. All biological control agents utilize one or more of these general indirect or direct mechanisms and the most effective biological control agents often use two or three different mechanisms. In this section, we will highlight successful examples of microbes that utilise these different biocontrol strategies.

Parasitism

A parasite is an organism that lives either in or on its host, the latter serving as both the habitat and the substrate. In this relationship, the parasite obtains nutrients and energy from the host and the host is usually exploited to such an extent that its metabolism is adversely affected resulting in inhibition of growth and development, and in some circumstances, death. Parasitism of one microorganism upon another is common within the soil environment and the interaction can range from non-specific to highly host-specific.

In 1932, Weindling (1932) reported that *Trichoderma lignorum* could parasitize a number of soil-borne fungi in culture and suggested controlling certain pathogenic fungi by augmenting soil with an abundance of this parasite. This promulgated the idea of using mycoparasites for biological control. Two broad types of mycoparasitic interactions are recognised; namely biotrophic mycoparasitism, which is an obligate relationship whereby the parasite obtains nutrients from living cells with little or no apparent harm to the host, and necotrophic mycoparasitism, where the parasite destroys the host before, during or after invasion and utilises nutrients from the dying/dead host. Commercial exploitation of mycoparasites for biocontrol of soil-borne pathogens has principally been restricted to the latter category (Cortes-Pentagos et al. 2007). The most intensely studied of these latter mycoparasites include *Trichoderma* spp., *Gliocladium virens*, *Coniothyrium minitans, Pythium oligandrum, Sporidesmium sclerotivorum, Talaromyces flavus* and *Laetisaria arvalis*. All of these fungi have been reported to have potential for use in biocontrol against specific pathogens under certain conditions (Adams 1990). The broad aspects of mycoparasitism have been reviewed recently by Lartley (Lartey and Conway 2004). Here we cite a number of specific examples to illustrate the range of parasitic interactions that have been investigated and highlight those with the most commercial potential.

One of the most studied mycoparasites is the soil-borne saprophytic fungus, *Trichoderma*. This genus contains a number of species that act as mycoparasites against a range of economically important soil-borne plant pathogens and have been successfully used in the field to control plant disease. The sequential events involved in mycoparasitism by *Trichoderma* are described by Chernin and Chet (2002). *Trichoderma* can detect its host pathogen from a distance and the mycoparasite alters its growth pattern to grow towards the host in a positive chemotropic fashion (possibly related to a chemical gradient). Recognition is brought about by interactions between complementary molecules present at the surface of both the host and the parasite. Such recognition processes have been shown to include lectin-carbohydrate interactions. Lectins are sugar-binding proteins/glycoproteins that agglutinate cells. Elad et al. (1983a) demonstrated the role of lectins in the host-mycoparasite interaction between *T. harzianum* and *Rhizoctonia solani*, a soil-borne pathogen that causes damaging root, stem-base and tuber rots of a wide range of crop species. Barak et al. (1986) further investigated the role of lectins in the *Trichoderma-Rhizoctonia* interaction and demonstrated the presence of L-fucosyl residues on the *Trichoderma* cell-wall surface that could serve as receptors for the *Rhizoctonia* agglutinin. Using a biomimetic system based on nylon fibres designed to mimic pathogen hyphae, Inbar and Chet (1992) provided direct evidence for the involvement of lectins in mycoparasitism. Following recognition, *Trichoderma* hyphae attach to the host pathogen via hook- or appressorium-like structures and grow along or coil around the pathogen hyphae. Elad et al. (1983b), using *T. harzianum* and *R. solani/S. rolfsii* model systems, revealed lysed sites and penetration holes in the hyphae of the host fungus caused by the antagonist attaching and coiling around it. These observations suggested the *de novo* synthesis of lytic enzymes produced during the early stages of interaction with the pathogen and their secretion at the interaction sites resulting in degradation of the cell wall. This production of lytic enzymes has been shown to be a crucial property of mycoparasitic fungi. The role of lytic enzymes, in particular, chitinolytic enzymes, in fungal mycoparasitism and biocontrol activity are discussed in several contemporary reviews (Lorito 1998; Manocha and Govindsamy 1998; Steyaert et al. 2003).

Another successful example of hyphal parasitism occurs between the mycoparasite *Pythium oligandrum* and pathogenic *Pythium* spp. (e.g., *ultimum*). With this mycoparasitic fungus, contact occurs at random with no apparent directed growth of the mycoparasite towards the host. Parasitism involves lysis of the host hyphae and granulation of the cytoplasm. This can often occur very rapidly within seconds of penetration of the host by the mycoparasite. The hyphae of the mycoparasite then grow throughout the mycelium of the host (Laing and Deacon 1991). *P. oligandrum* has shown biocontrol activity against a number of important soil-borne plant pathogens in the field. For example, oospore preparations of *P. oligandrum* applied as a seed treatment have reduced the incidence of *P. ultimum* damping-off of sugarbeet (Veseley and Hejdanek 1984) and several commercial products have been developed. For example, Polyversum marketed by Biopreparaty Ltd, Czech Republic can be used as a seed treatment or soil spray to control clubroot of brassica caused by *Plasmodiophora brassicae*, *Sclerotinia* root rot and *Leptosphaeria* black leg of canola and damping-off diseases of vegetables (www.polyversum.cz).

Some fungal pathogens survive for many years in soil by forming black resting structures called sclerotia. These are long lived (2–8 years) and very difficult to eradicate once they have infested a soil, thus making them suitable targets for biocontrol using mycoparasites that are able to attack the sclerotia. A large number of sclerotial mycoparasites have been reported as potential biocontrol agents. For example, *Trichoderma* spp. and *G. virens* against pathogens such as *Sclerotium rolfsii*, *Sclerotium cepivorum*, *Sclerotinia minor*, *Sclerotinia sclerotiorum* and *R. solani; C. minitans* against sclerotia of *Sclerotinia* and *Sclerotium* spp. and *S. sclerotivorum* against mainly *Sclerotinia* spp. (Chet 1990). All of these organisms have been investigated as potential biocontrol agents with good success achieved in some circumstances (Jeffries and Young 1994). For example, *Coniothyrium* can destructively parasitize the sclerotia of *S. sclerotiorum*, *S. minor*, *S. trifoliorum*, *S. cepivorum* and *B. cinerea* (Jeffries and Young 1994). *Coniothyrium* was able to parasitize and kill sclerotia of *S. sclerotiorum* produced in the soil, on the root surface, inside the root and within the stem base of sunflower plants resulting in significant reductions in pathogen inoculum levels (Jeffries and Young 1994). Its use as a biocontrol agent for control of *S. sclerotiorum* has been widely investigated (Whipps et al. 2004; Gerlagh et al. 2003). Two products have been commercialised – Contans® (Prophyta GmbH, Germany), and Koni® (Hungary). In both cases, the biocontrol agent can be applied directly to soil containing sclerotia or as a spray on diseased crops. Contans WG, a water dispersable granule containing 1×10^9 conidia g^{-1} substrate is used as a soil application to infect *S. sclerotiorum* and prevent apothecial production (Luth 2001). Soil application is made several months before growing the crop, with conidia mixed into the upper 5 cm soil layer (Whipps and Gerlagh 1993).

From this extensive work on sclerotial mycoparasites, a number of key attributes have been identified that determine how effectively sclerotial numbers are reduced in the soil, including; the parasite's intrinsic parasitic ability, its growth potential in soil, activity over a range of environmental conditions and survival and reproduction potential. The biggest obstacles to utilising mycoparasites for practical disease control are the large quantity of the biocontrol agent necessary to achieve biocontrol when applied directly to soil in the field, and the biocontrol agent's ability to reproduce on the host and redistribute itself within the soil. It is clear that in order to achieve successful biocontrol with a mycoparasite detailed knowledge of the density and distribution of the pathogen in the soil and the efficiency of parasitism is required.

Antibiosis/Toxins

Antibiosis is antagonism mediated by specific or non-specific metabolites of microbial origin, by lytic agents, enzymes, volatile compounds or other toxic substances. Antibiotics are low molecular weight compounds produced by microbes, which are active at low concentrations and which inhibit the growth or metabolic activities of other microbes. Most are produced by soil inhabiting microbes (Fravel 1988). An extensive range of soil microbes have been reported to produce antibiotics that have a role in biocontrol including the bacterial genera *Agrobacterium, Bacillus* and *Pseudomonas*, the actinomycete genus, *Streptomyces*, and fungal genera including *Trichoderma, Gliocladium* and *Penicillium.* A selection of examples is given below to illustrate this diversity.

Bacteriocins are a subclass of antibiotics produced by bacteria with specificity towards bacterial strains closely related to the producer. The best-known example of this is with control of crown gall caused by *Agrobacterium tumefaciens* by the related *A. radiobacter* (Kerr 1980). The production of a bacteriocin, agrocin 84, by *A. radiobacter* is primarily responsible for the observed biocontrol of crown gall. Production of agrocin 84 is controlled genetically on a plasmid and sensitivity to agrocin 84 is determined genetically on the Ti (tumour inducing) plasmid in the target bacterium (Murphy and Roberts 1979). A mutant of *A. radiobacter*, which no longer produced agrocin 84 (because the plasmid had been removed), could not control crown gall (Das et al. 1978). The product NoGall™ containing 10^6 cells g^{-1} peat of *A. radiobacter* strain K1026 is marketed world-wide by Becker Underwood Pty Ltd (Australia). The product is applied as an aqueous solution to seeds, seedlings and cuttings before planting to protect the wound sites from infection by the crown gall pathogen (www.beckerunderwood.com).

Bacillus is a ubiquitous soil organism, and various species and strains are reported to produce a large variety of antifungal compounds, such as cepacin and pyrrolnitrin. *Bacillus cepacia* is an effective biocontrol agent for *Pythium*-induced damping-off and *Aphanomyces* root rot of peas (King and Parke 1993) and *R. solani* root rot of Poinsettia (Cartwright and Benson MD (February 1994). *Bacillus subtilis* isolate RB14-C produces the lipopeptide antibiotic iturin A. When the bacterium was introduced into the soil it was able to control *R. solani* on tomato (Szczech and Shoda 2006) and various other fungal pathogens (Hiraoka et al. 1992). Shortly after application to the soil, RB14-C was shown to produce iturin A and this was believed to play a major role in disease suppression. A mutant of RB14 RΔ1, deficient in the production of iturin showed only low protection of tomato plants compared with the parental strain (Asaka and Shoda 1996).

Fluorescent pseudomonads are known to produce a wide range of different antibiotics such as phenazines (Thomashow and Weller 1988), 2, 4-diacetylphloroglucinol (Fenton et al. 1992), pyrrolnitrin (Hill et al. 1994), pyoluteorin (Kraus and Loper 1995) and their involvement in plant disease biocontrol has also been well documented. For example, production of an unusual lipopeptide antibiotic (AFC-BC11) was shown to be responsible for the ability of *B. cepacia* BC11 to effectively control damping-off disease of cotton caused by the fungal pathogen *R. solani*. Howie and Suslow (1991) used an antifungal minus mutant of *P. fluorescens* to show that the antibiotic was of primary importance in the suppression of *P. ultimum* on cotton.

Strains of *P. fluorescens* are suppressive to *G. graminis* var. *tritici*, the causal agent of take-all of wheat, and production of a range of antibiotics has been implicated in this biocontrol (Hoffland et al. 1996). For example, bacterization of spring or winter wheat seeds with *P. fluorescens* 2-79 and *P. chlororaphis* 30-84 isolated from wheat grown in take-all suppressive soils, results in significant suppression of take-all (e.g., 2-79 increased yields by >10% in commercial scale tests) (Weller 1988). Both strains produced phenazine-1-carboxylic acid (PCA) and 2-79 also produced anthranilic acid (Weller 2007). Single site Tn5 insertions were used to make mutants defective in the production of the phenazine antibiotic. All of these mutants were less suppressive to take-all in greenhouse tests (Schippers et al. 1985).

Pyrrolnitrin is an active metabolite produced by a range of strains of *Pseudomonas* spp. that is active against *Rhizoctonia*, *Fusarium* spp. and other plant pathogen fungi. Ligon et al. (2000) produced *P. fluorescens* BL915 mutants that did not produce pyrrolnitrin to demonstrate the important role of pyrrolnitrin in the biocontrol activity shown by this strain. The author also genetically modified *P. fluorescens* strain BL915 to produce more pyrrolnitrin. The modified strain was compared to the wild-type in a field trial in 1997 on cotton for control of *R. solani* infections. Cotton seed was planted in the furrow and inoculated with *R. solani* by spreading wheat bran inoculant over the seed. Prior to closure of the furrow, the seeds were treated with a lyophilised preparation of the biocontrol agents. The wild-type strain gave 50% control compared to the chemical, while the genetically modified strains gave increased levels of control equivalent to those achieved by chemical (quintozene) treatments (Ligon et al. 2000).

The antimicrobial compounds produced by *Trichoderma* and *Gliocladium* constitute a diverse group with respect to structure and function and the group contains both volatile and non-volatile compounds (reviewed by Sivasithamparam and Ghisalberti (1998)). Different metabolites exhibiting antibiotic activity in *Trichoderma* can be classified into two main types: Low molecular weight volatile metabolites such as pyrones and butenoliches, volatile terpenes and the isocyane metabolites and high molecular weight polar metabolites like peptaibols (Cardoza et al. 2005). Gliovirin, a diketopiperazine antibiotic produced by *G. virens* is highly toxic to *P. ultimum*, but is inactive against other fungi associated with cotton seedling disease. A UV induced mutant of *G. virens* deficient for gliovirin production did not inhibit *P. ultimum* in culture and did not protect cotton seedlings from damping-off in *P. ultimum* infested soil indicating an active role for gliovirin in the control of the pathogen (Howell and Stipanovic 1983). The antibiotic, gliotoxin, isolated in 1936 (Weindling and Emerson 1936), was first associated with the biocontrol of *R. solani* on seed potatoes by *G. virens* (Aluko and Hering 1970). Since then there have been numerous reports associating gliotoxin production with biocontrol (Lumsden et al. 1992).

The pyrone, 6-pentyl-2H-pyran-2-one, is the representative metabolite common to the *Trichoderma* genus. This compound is a flavouring agent responsible for the coconut aroma associated with this fungus. This metabolite was used in plate tests against *R. solani* and *F. oxysporum* f. sp. *lycopersici* in which the addition of 0.3 mg mL^{-1} of pyrone to agar medium caused a 69.6% growth reduction in *R. solani* and a 31.7% reduction in *F. oxysporum* after 2 days (Claydon et al. 1987). When used in spore germination tests, 0.45 mg mL^{-1} was found to completely inhibit the germination of *Fusarium* spores (Claydon et al. 1987). Its antibiotic activity against a number of plant pathogens has been demonstrated (Ghisalberti et al. 1990).

Competition

Competition occurs when two or more organisms demand more of the same resource than is available. Competition between a biocontrol agent and a pathogen may lead to control, if the biocontrol agent's growth results in reduction of the pathogen population or inoculum production. Nutrients and space availability are two of the main limitations in the colonization of a particular ecological niche and so are often the main resources being competed for. Microbes with a high growth rate and optimal nutrient utilisation have clear advantages for survival. Many of the microbes utilised as biocontrol agents fulfil these characteristics (Cortes-Pentagos et al. 2007). Nitrogen, carbon and iron competition have also been associated with pathogen suppression by a number of biocontrol agents. A number of examples illustrating specific resource competition interactions leading to disease control are given below.

Space and Nutrients

Trichoderma and *Gliocladium* are common soil-borne saprophytic fungi. Their rapid growth, prolific conidiation and range of variation in substrate utilization make them very efficient saprophytes and their success as biocontrol agents is often linked to success in nutrient competition. They are biologically adapted to aggressive colonization of available nutrient bases and are able to persist as quiescent chlamydospores and conidia when nutrients are lacking. Optimum biocontrol is most likely if *Trichoderma* can rapidly colonize the site or nutrient base before the pathogen can establish. Rhizosphere competence is an important attribute that will facilitate a biocontrol agent's ability to compete for nutrients especially when the biocontrol agent is applied as a seed treatment (Ozbay and Newman 2004). Rhizosphere competent *Trichoderma* spp. have been shown to effectively utilise root exudates and cellulose substrates on or near the root (Cotes et al. 1996) leading to disease control, for example, in the control of *F. oxysporum* in the rhizosphere of cotton and melon by *T. harzarium* T-35 (Sivan and Chet 1989).

Idriella bolleyii is common on roots of cereals and grasses and has been shown to be able to control take-all disease (Kirk and Deacon 1987) and other root rot pathogens such as the eyespot fungus, *Tapezia yallundae* (Reinecke and Fokkema 1981). When applied as a seed treatment, *I. bolleyii* will colonize wheat roots and produce infective spores on inoculated plants. These spores are carried in percolating water and spread down the root system. The biocontrol agent controls the cereal root pathogen by competitive exclusion from naturally senescing host tissues (Allan et al. 1992). In glasshouse experiments, *I. bolleyii* significantly reduced infection of wheat roots by the take-all fungus *G. graminis* var. *tritici* when inocula were dispersed in the soil at ratios of 10:1 (Ib:Ggt) or more (Lascaris and Deacon 1991). *Idriella bolleyii* has several attributes that make it an effective biocontrol agent. It sporulates readily in liquid culture and retains viability as a seed coat making inoculum production cost efficient, it grows readily on root and stem base tissues and is a normal resident colonizer of the root and stem basal tissues of cereals in the field and it has a reasonably wide spectrum of biocontrol activity (Lascaris and Deacon 1991).

Iron Competition

Conditions of low iron in the soil are inhibitory to plant pathogens. Iron is commonly present in the soil in an insoluble form, but some bacterial and fungal species have developed a system for iron uptake. This involves the production of siderophores, which are peptides able to bind iron and transport it inside the cell. Iron chelated with these siderophores is unavailable to plant pathogens so their activity is thereby reduced (Cortes-Pentagos et al. 2007). *Pseudomonas* spp. are the best prokaryotic system studied in relation to siderophore synthesis. They secrete molecules such as pyoverdines or pseudobactins that bind ferric iron with high affinity (Raymond et al. 1984). Iron competition, mediated by siderophores, is the main mechanism by which *Pseudomonas putida* WCS 358 controls *Fusarium* pathogenesis in carnation and radish (Weisbeek and Gerrits 1999). Similarly, Loper (1988) showed that siderophore production by *P. fluorescens* was an important mechanism of *Pythium* inhibition during seedling emergence in cotton. The role of siderophores in plant disease expression has been demonstrated using *Pseudomonas* mutants lacking the ability to produce siderophores (Buysens et al. 1996).

Non-pathogenic and Hypovirulent Strains

Preliminary exposure of plant tissues to non-pathogens, or avirulent strains of a pathogen often results in increased resistance to subsequent challenge by the virulent strain. For example, Nel et al. (2006) reported the potential of non-pathogenic *F. oxysporum* and other biological control organisms for suppressing *Fusarium* wilt of banana. Glasshouse evaluations showed two non-pathogenic *F. oxysporum* isolates CAV255 and CAV 241 reduced *Fusarium* wilt incidence by 87.4% and 75%, respectively. Similarly, Shishido et al. (2005) reported effective biological control of *Fusarium* wilt of tomato by non-pathogenic *F. oxysporum* Fo-B2 in different environments. Inoculation of Fo-B2 onto tomato roots significantly reduced the severity of disease and the biocontrol agent was most effective when it colonized vascular tissues intensively. A key observation was that indigenous soil microbes were a primary factor negatively influencing the efficiency of Fo-B2. Therefore, early establishment of the antagonist in a non-competitive environment prior to out-planting could improve the efficacy of biocontrol. The soil-borne fungus *F. oxysporum* f. sp. *radicis-lycopersici* causes tomato foot and root rot (TFRR), which can be controlled by the addition of the non-pathogenic fungus *F. oxysporum* Fo47 to the soil. To examine biological interactions between the two fungi, they were labelled using different autofluorescent proteins as markers and visualised using confocal laser scanning microscopy (Bolwerk et al. 2005). This study revealed that there needed to be at least 50-fold more *F. oxysporum* Fo47 inoculum propagules than *F. oxysporum* f. sp. *radicis lycopersici* to give biocontrol. The study also showed that Fo47 hyphae attached to the root earlier than the pathogen and it was postulated that *F. oxysporum* Fo47 controls TFRR by germinating more quickly and being a better nutrient competitor on the rhizoplane than the pathogen.

Hypovirulence refers to the reduced virulence of selected isolates within a population of a plant pathogen. Hypovirulence has been associated with the presence of double stranded RNA (dsRNA) characteristic of fungal viruses (Nuss and Koltin 1990). However, other factors such as mitochondrial mutations, nuclear mutations and encapsidated fungal viruses have been associated with hypovirulent isolates. The potential in utilising hypovirulent isolates of fungal pathogens in a biocontrol strategy resides in the ability to transfer hypovirulence from hypovirulent isolates to virulent isolates and thereby reduce the mean disease severity of the population through overall reductions in virulence, growth, sporulation and/or survival (Harris 2000). Most work on hypovirulence has been conducted on *Rhizoctonia* spp.

The genus *Rhizoctonia* includes some of the most aggressive soil-borne plant pathogens, for example, the polynucleate *R. solani* that attacks solanaceous plants. Many *Rhizoctonia* isolates are avirulent or hypovirulent on plants, even isolates belonging to the same *R. solani* anastomosis group (AG) show a continuum between high virulence and avirulence (Herr 1995). Several hypovirulent strains have been shown to have biocontrol potential against virulent *R. solani* (Cardinale et al. 2006). Similarly, hypovirulent isolates of *S. sclerotivorum*, *S. minor* and *S. homeocarpa* have been evaluated for their role in reducing virulence in populations of these pathogens (Boland 1992). Transmission of the hypovirulent phenotype and dsRNA occurs through hyphal anastomosis and was shown to occur in culture and on lettuce tissue (Melzer and Boland 1996). With *S. sclerotivorum*, transmission did not occur consistently between isolates from the same or different mycelial compatibility groups and so this could represent a significant barrier to the use of hypovirulence as a biocontrol strategy. However, the number of mycelial compatibility groups in *S. minor* is relatively small and so transmissible hypovirulence may have more promise as a control strategy for this fungus. Mycelial suspensions of a hypovirulent isolate of *S. minor* applied to leaf lesions initiated by virulent isolates on lettuce suppressed lesion expansion by up to 100% and significantly reduced the development of sclerotia on diseased tissues (Melzer and Boland 1996). The earlier in lesion development the hypovirulent isolate was applied the more it suppressed lesion development by the virulent isolate.

Induced Resistance

Numerous biocontrol agents have been shown to be able to induce host plant defences. Induced resistance has been investigated more fully for control of foliar pathogens, but there are several examples of successful control of root pathogens including root parasitic nematodes (reviewed by Agrawal et al. (1999)). Examples of biocontrol agents shown to induce systemic host plant resistance include *Paenibacillus polymyxa* (Timmusk and Wagner 1999), *Pseudomonas* (Meera et al. 1994), rhizobacteria (Harman et al. 2004), streptomycetes (Lehr et al. 2008) and *Trichoderma* spp. (Harman et al. 2004). The soil isolate *Streptomyces* GB 4-24 prevented the development of *Heterobasidion* root rot in Norway spruce seedlings (Schrey and Tarkka 2008). Treatment of plant roots with GB 44-2 led to a reduction in the penetration ability of the pathogen. Whereas inner root cortex layers and vascular tissues of plants were fully colonised in plants inoculated with *Heterobasidion* only, pre-treatment with GB 4-2 resulted in inner root tissues virtually free of the pathogen (Lehr et al. 2008). Reduced root colonisation was accompanied by distinct changes in root anatomy, e.g., cell wall appositions, lignified walls in a phenomenon termed ‘priming’ by Conrath et al. (2002).

The nature of systemic acquired resistance (SAR) was shown using split root systems to demonstrate that treatment of one section of a root system with *Pseudomonas* bacteria induced resistance to soil-borne pathogens in the untreated roots (Chen et al. 1999). *Pseudomonas* may induce SAR directly by synthesizing

SAR signalling intermediates, such as salicylic acid, or by causing necrosis of host cells through the action of toxins, enzymes or elicitation of the hypersensitive reaction (HR). Preston (2004) showed that bacterization of carnation roots with *P. fluorescens* WCS417 reduced wilting caused by *F. oxysporum* f. sp. *dianthii*. Upon pathogen infection, higher amounts of the anthranilate derived phytoalexins dianthalexin (2-phenyl-7-hydroxy-1,3-benzoxazin-4H-one) and a group of diantramides accumulated in plants that had received a root pretreatment with the biocontrol strain (Van Peer and Schippers 1992).

Trichoderma spp. have the capacity to induce resistance to a range of diseases caused by various classes of plant pathogen (fungi, bacteria, viruses) in a variety of plants (Harman et al. 2004; Vinale et al. 2008). *Trichoderma* strains can produce three different classes of compounds that can induce resistance in plants; proteins with enzymatic functions e.g. xylanase (Anderson et al. 1993); homologues of proteins encoded by the avirulence (Avr) genes (de Wit et al. 2002); oligosaccharides/low molecular weight compounds released from cell walls by the activity of *Trichoderma* enzymes (Kubicek et al. 2001). Pre-inoculation of cotton plants with *T. virens* G6, G6-5, G-11 protected against attack by *R. solani*, as a result of the induction of fungitoxic terpenoid phytoalexins (Howell et al. 2000). Induction of phytoalexin synthesis in roots of cotton also protected the plants from vascular wilt pathogens that enter through the root. Similarly, seed treatment of cotton with *T. virens* resulted in colonisation of the developing root system, which led to suppression of symptom development when the plants were inoculated with *Verticillium* and *Fusarium* wilt pathogens (Hanson 2000; Zhang et al. 1996).

Constraints to Successful Biological Control

Abiotic Factors

Superficially, the soil appears an attractive environment in which to target microbial control efforts, protected from environmental extremes of temperature and moisture. Yet soil is a complex, heterogeneous environment, compartmentalised into nutrient-rich competitive (e.g., rhizosphere) and harsh environs, and microbial persistence and activity is influenced by a range of interacting environmental factors. Tolerance to abiotic fluctuations is, therefore, a prerequisite for the successful development of ecologically competent biocontrol agents for use in the field. Yet in most cases these conditions are not adequately defined. Pertinent examples of such interactions from the literature are discussed below.

Soil Type

Contrasting results have been obtained in studies investigating effects of soil type on the natural incidence, persistence and efficacy of biological control agents.

For example, Queseda-Moraga et al. (2007) observed a correlation between high soil organic matter and clay content and the incidence of entomopathogenic fungi, particularly *B. bassiana*. This was attributed to enhanced adsorption of conidia to soil particles which reduces potential biodegradation or leaching from the upper layers of the soil profile, inhibition of germination by organic acids and other fungistatic compounds produced in these soils, or greater incidence of potential arthropod hosts in soils with high organic content. However, Vänninen et al. (2000) found that persistence of *M. anisopliae* was poorer in peat than clay soils; Kessler et al. (2004) showed that *B. brongniartii* declined more quickly in soils with a high organic content when insect hosts were absent; while Milner et al. (2003) indicated that soil type had negligible effects on persistence. The ability of a biocontrol agent to perform in different soil types can differ significantly even between closely related biocontrol agents. For example, Larkin and Fravel (2002) observed variations in the efficacy of non-pathogenic *F. oxysporum* (CS-20) and *F. solani* (CS-1) used to control *Fusarium* wilt of tomato in different soil types. Isolate CS-20 effectively reduced disease in four different field soils varying in texture (sandy to clay) and organic matter content (0–3.2%), while isolate CS-1 reduced disease in sandy and loamy soil (49–60%) but was not effective in a heavy clay soil.

Soil Temperature

Temperature has a direct impact on persistence and efficacy. Although persistence is generally extended in cooler soils, most soil-borne fungal biocontrol agent strains are mesophilic so low temperatures in winter may constrain biocontrol activity. Also, these effects may be compounded by soil moisture content (Jaronski 2007). Soil temperature must be appropriate for insect infection to occur. While temperature optima for different isolates will vary, fungal development will generally occur between 15°C and 30°C (Jaronski 2007). At temperatures outside of this range, the infection process can be slowed or totally inhibited. For example, Glare et al. (1994) identified low soil temperature as a primary limiting factor to the effectiveness of *M. anisopliae* against New Zealand grass grub larvae, advocating selection of isolates with activity at cooler temperatures to overcome this limitation, similar to the approach taken in the development of a fungal biocontrol agent for cockchafer in Tasmania (Rath et al. 1992). Knudsen et al. (1991) reported that mycoparasitism of sclerotia of *S. sclerotiorum* by *T. harzianum* was reduced in soil at 15°C compared to 25°C. Other studies on the effects of temperature on spore germination and germtube growth, mycelial growth, competitive saprophytic ability and antibiotic production support this. For example, gliotoxin production by *T. virens* was low at 15°C increasing to a maximum at 25–30°C, thus identifying a potential adverse effect on biocontrol activity considering the low temperature of many soils at planting time (Lumsden et al. 1992). However, some studies have contradicted this general trend. For example, degradation of sclerotia of *S. cepivorum* by *T. viride* was optimal at 10°C with decreasing parasitism as temperatures increased (Clarkson et al. 2004). *Trichoderma viride* is a species group often associated with cold tolerance (Kredics et al. 2003), so there is clearly an opportunity to select for low temperature tolerant strains. Isolates with higher or lower thermal optima have been identified, and there does appear to be a link between the geographic origin of an isolate and its performance at lower temperatures (Fernandes et al. 2008); while more thermotolerant isolates have been recovered from agricultural soils which are exposed to elevated temperatures (Meyling and Eilenberg 2007).

Osmotolerance

Many biocontrol fungi have a low osmotolerance level and do not proliferate in dry soils with spore germination, conidial survival, germtube growth and mycelial growth all inhibited at low water potential. However, there is also an inverse relationship between persistence and soil moisture; where longevity declines as moisture levels increase, and the rate of decline increases as temperatures rise (Jaronski 2007). In trials examining effects of two irrigation schemes on conidial persistence in turfgrass, Thompson et al. (2006) showed that survival was greater at higher irrigation levels. Irrigation (and rainfall) only influence soil moisture levels in the short term and enhanced survival in this study was thought to be due to movement of the conidia into the soil profile where they were protected from higher temperatures and damaging UV. Similar observations were made by Milner et al. (2003), where rainfall appeared to have minor effects on persistence of *M. anisopliae* in different sugarcane soils in Australia. Soil moisture also appears to have a significant impact on efficacy. However, if a 'standard' soil treatment is applied, insect infection is commonly higher at low rather than high soil moisture levels (Jaronski 2007). Water activity (a_w) and pH were demonstrated to be the most important environmental parameters affecting the biocontrol activity of mycoparasitic *Trichoderma* spp. (Kredics et al. 2004). Optimal conditions for mycelial growth were defined as a_w 0.997 and pH 4. Kredics et al. (2004) also investigated optimal (a_w) and pH conditions for key extracellular enzymes known to play a role in mycoparasitism identifying the potential for the development of xenotolerant mutants with improved biocontrol performance in low a_w soils. Further to this, Clarkson et al. (2004) demonstrated a clear relationship between the efficacy of *T. viride* in degrading *S. cepivorum* sclerotia and soil water potential. Sclerotia were degraded at water potential levels as low as –4.03 MPa, however, degradation was most efficient at water potential levels over –0.022 MPa.

Carbon and Nitrogen

Mineral nutrition is essential for microbial growth and sporulation and can stimulate the production of secondary metabolites. Therefore, any nutrient limitation may constrain a biocontrol agent's competitiveness in the soil environment and ultimately its biocontrol activity. A soil sandwich assay was used to determine the influence of nitrogen on the saprophytic growth through soil of an isolate of *T. koningii*. A key result showed that nitrogen added as ammonium sulphate (NH_4^+ N) increased the saprophytic growth of *T. koningii* whereas nitrate (NO_3^- N) suppressed growth (Wakelin et al. 1999). This supports the findings of Danielson and Davey (1973) who reported that *Trichoderma* spp. grow best when supplied with ammonium-nitrate compared with nitrite-nitrate on artificial culture medium. Brian et al. (1946) showed that gliotoxin-levels produced by *T. virens* was optimum on medium containing glucose and phenylalamine at C-to-N ratios of 18:1, 31:1 or 42:1. Similarly, Howell and Stipanovic (1984) showed that viridiol production by *T. virens* was better on a rice substrate than on other substrates with lower C-to-N ratio, while Jones and Hancock (1987) showed that high C-to-N ratios enhanced and low C-to-N ratios suppressed viridiol production.

Soil pH

Soil pH may be influenced by a variety of agricultural practices, e.g., application of fertilizers, organic content, etc., which will affect the diversity and make-up of microbial populations. These, in turn, may exert a fungistatic effect on biocontrol agents, which will impact on their incidence and efficacy. Thus, while pH may not directly influence efficacy *per se*, its indirect effects via the soil biota may be more significant. The occurrence of *B. bassiana* in cultivated and natural soils in Spain was correlated directly with soil pH (Quesada-Moraga et al. 2007). The fungus was recovered with greater frequency from soils at pH 8–8.5, and both *B. bassiana* and *M. anisopliae* were infrequently recovered from soils with pH values >8.5. *Metarhizium anisopliae* was isolated over a greater pH range and was the predominant species isolated from soils with pH <7; *M. anisopliae* appears to be more tolerant of acidic conditions, which could explain these findings (Padmavathi et al. 2003). Low soil pH (<7) has long been known to favour *Trichoderma*-mediated suppression of a range of soil-borne pathogens. Harman and Taylor (1990) showed that acidification of the soil and seed environment can improve the competitiveness and biocontrol activity of introduced *Trichoderma* strains. Acidic conditions are known to enhance conidia production and germination, mycelial growth, and production and activation of antimicrobial compounds such as antibiotics and lytic enzymes. For example, Weindling (1941) demonstrated that the production of gliotoxin was higher at pH 3.5 than at 6.0, with this loss of activity at higher pH also demonstrated for viridin (Brian and McGowan 1945). Alkaline soils tend to evolve more ammonia than acidic soils, which reduces germination of *Trichoderma* conidia and has been suggested to reduce the biocontrol activity of *T. hamatum* (Papavizas 1985). Cellulase production by *T. koningii*, which is associated with competitive saprophytic ability and rhizosphere competence (Cotes et al. 1996), was negatively correlated with soil pH (Widden et al. 1988). Thus, soil pH will be a major constraint to effective biocontrol by *Trichoderma* spp., particularly in more alkaline soils and there is a need to screen for *Trichoderma* strains with tolerance to more alkaline conditions.

The effect of soil pH on biocontrol efficacy of *Pseudomonas* spp. has also been described (Ownley et al. 1992). *Pseudomonas chlororaphis* PCL1391 produces the secondary metabolite phenazine-1-carboxamide, which is an antifungal metabolite required for biocontrol activity of the strain. Identification of environmental/nutritional conditions that influence production of the antibiotic showed that pH decreasing from 7 to 6 and temperature from 21°C to 16°C could reduce production significantly (Tjeerd van Rij et al. 2004). Similarly, biocontrol by *P. fluorescens* 2-79 against take-all of wheat caused by *G. graminis* var. *tritici* was shown to increase with increasing pH (Ownley et al. 2003). Slininger and Shea-Wilbur (1995) reported a direct effect of pH on phenazine-1-carboxylic acid (PCA) accumulation by *P. fluorescens* strain 2-79. PCA production was optimal at pH 6–7 and declined at lower and higher pHs.

Interactions Between Soil Abiotic Factors

Interactions between soil factors may be as important in determining the outcome of biocontrol than any direct effect of one soil factor on the biocontrol agent. For example, *P. fluorescens* 2-79 RN_{10} protects wheat against take-all disease caused by *G. graminis* var. *tritici*, however, the level of protection in the field varies from site to site. Ownley et al. (2003) evaluated the relative importance of 28 soil properties on take-all suppression. Bacterized seed were planted in ten soils representative of the wheat-growing region in the Pacific Northwest. Biocontrol was positively correlated with ammonium-nitrogen, % sand, soil pH, sodium, sulphate-sulphur and zinc. In contrast, biocontrol was negatively correlated with cation exchange capacity (CEC), exchangeable acidity, iron, manganese,% clay,% organic matter, % silt, total carbon and total nitrogen. Six key soil properties that accounted for the variance in biocontrol activity were identified – ammonium-nitrogen, cation exchange capacity, iron,% silt, soil pH and zinc. Using this information, the researchers were able to identify specific crop management practices that could enhance biocontrol, for example, the performance of 2-79 RN_{10} was improved by amending a soil low in zinc with 50 μg of zinc-EDTA g^{-1} soil. A similar study was conducted by Duffy et al. (1997) on *T. koningii*, used to control take-all of wheat in Australia. Previous work had shown that soil treatment with *T. koningii* reduces the saprophytic growth of the pathogen and increases survival of wheat seedlings by ~50% and grain yield by 10% (Rovira et al. 1992). However, control was inconsistent between sites and seasons prompting the need to understand the influence of soil properties on the biocontrol activity of the introduced organism. Duffy et al. (1997) evaluated suppression of take-all by *T. koningii* in eight silt loam soils from Pacific North West, USA and the influence of 21 abiotic soil parameters on biocontrol activity. Biocontrol was positively correlated with iron, nitrate-nitrogen, boron, copper, soluble magnesium and percent clay and negatively correlated with soil pH and available phosphorus. Principal component factor analysis using these eight variables resulted in a three-component solution that accounted for 95% of the variation in disease rating. The authors proposed that this information could be used to help select sites where *T. koningii* will perform optimally.

Biotic Factors

There is a large body of literature on the influence of abiotic factors on biocontrol microbes in the soil environment, but surprisingly little is reported on the influence of biotic factors. This is probably due to the difficulty in identifying specific biotic influences. Soil microbial populations are relatively stable, so it would be naive to think that introducing a single organism into the soil as an augmentative treatment would always necessarily result in successful establishment. Clearly, there will be situations where a member or members of the soil microbe population will act in an inhibitory manner towards the biocontrol agent and constrain its ability to establish.

We can identify three principle biotic components that have a major influence on microbe persistence and efficacy. These are soil microorganisms, plants and invertebrates and examples of their effect on several biocontrol agents are discussed below. Generally speaking, entomopathogens are considered weak saprophytes in the competitive soil environment and inoculum levels will decline in the absence of an arthropod host. Metabolites produced by other soil microbes can adversely affect germination and growth, or be directly toxic, leading to reduced infectivity or multiplication; consequently, survival and efficacy of biocontrol agents is commonly observed to be greater in sterilized versus non-sterilized soils (Jaronski 2007). Even so, in natural soils conidia from entomopathogenic fungi will infect a susceptible host when they contact the insect cuticle; *Metarhizium* and *Beauveria* will germinate, grow and conidiate when applied to soil on nutrient granules or as mycelial aggregates; and amendment of soil with nutrients can overcome (apparent) fungistasis (e.g., Jaronski 2007; Brownbridge 2006). Hubbard et al. (1983) reported that *Trichoderma* biocontrol agents may decline in the soil or fail to proliferate in the plant rhizosphere due to the presence of antagonistic or competing microorganisms at the rhizosphere or rhizoplane level. Bae and Knudsen (2005) showed that higher levels of microbial soil biomass reduced the biocontrol activity of *T. harzianum* and postulated that this was due to a soil fungistatic effect brought about by the presence of *Pseudomonas* spp. and the production of antibiotic compounds. Indeed, *Pseudomonas* spp. were shown to compete with the biocontrol agents for iron in the rhizosphere and produce toxic metabolites that were inhibitory to the *Trichoderma*. Multiple logistic regression analysis used to determine the significance of associations between nine abiotic factors, total populations of fungi, bacteria and actinomycetes and the presence of 42 individual fungi over a 12-month period revealed that *T. harzarium* was positively associated with the total bacterial population and the occurrences of *Aspergillus ustus*, *A. ramarii*, *Penicillium citrinum*, *P. chrysogenum* and *P. griseoroseum*, but negatively correlated with the occurrences of *A. flavipes* and *P. miczynskii* (Eastburn and Butler 1988a, b). This suggests that biotic interactions in the soil occur at both the species and subspecies level. Inhibition of *Trichoderma* spp. by certain mycorrhizal fungi has also been reported. For example, the ectomycorrhizal fungus *Laccaria laccata* inhibited the germination of *T. virens* and *T. harzarium* conidia. Mantle hyphae of the ectomycorrhizal fungus coiled around the spores and occasionally caused breaks in their walls (Zadworny et al. 2004). Reduced growth of *T. harzarium* and *T. virens* as well as cytoplasm aggregation of their hyphae was observed in dual culture with *L. laccata* (Zadworny et al. 2004). These observations have not been validated in the soil environment so limited conclusions can be drawn on the significance of this inhibition to biocontrol activity.

Crop plant species and tillage practices affect the incidence and persistence of fungi. Some entomopathogens (*M. anisopliae*) are more commonly associated with agricultural (tilled) soils than natural habitats, although fungal prevalence and diversity is normally greater in undisturbed soils. Plant root exudates contain many nutrients that will support the development of microbial populations in the rhizosphere; *in vitro* tests demonstrated that carbohydrates and nitrogen compounds stimulate germination and growth of *M. anisopliae* conidia, while organic acids may inhibit germination (Li and Holdom 1995). This observation pre-empted later work showing that some *M. anisopliae* isolates are in fact rhizosphere-competent, a trait that enhances persistence in the root zone. *Beauveria bassiana* also shows specific plant adaptations in its existence as an endophyte; not all isolates can function as endophytes, but this capacity has been demonstrated in a diverse range of plant species (Vega et al. 2008).

Invertebrates have many effects on biocontrol agent levels in soil. Some, such as Collembola and earthworms, ingest conidia and play a role in their distribution within, and removal from, the soil profile. Insect hosts are critical to the long-term survival of entomopathogenic fungi. Access to and successful infection of a host is the only way in which fungi can significantly multiply within the soil; fungal loadings over time may thus be closely correlated with the presence of susceptible insect populations in both natural and agricultural soils (Meyling and Eilenberg 2007). Use of insecticides may contribute to the decline of fungal populations by reducing the availability of suitable hosts, as opposed to their having direct negative effects on fungal viability (Mietkiewski et al. 1997).

Fungi and arthropods have evolved complex relationships, and some soil-dwelling species show adaptive behavioural responses that prevent their coming in to contact with fungal inoculum. An avoidance response has been observed in mole crickets to conidia of both *M. anisopliae* and *B. bassiana*, which may lead to inconsistent performance of these fungi in the field (Thompson and Brandenburg 2005). However, there appears to be variation in the level of response to different isolates (Thompson and Brandenburg 2005). Insects may also be attracted to fungi. Engler and Gold (2004) showed that termites were attracted to mycelial preparations and volatile extracts of *M. anisopliae*, while Japanese beetle females preferentially oviposited in soils treated with mycelia (Villani et al. 1994). This recruitment effect was also seen with black vine weevil larvae that responded positively to *M. anisopliae*-treated media (Kepler and Bruck 2006). Such behavioural responses should be taken into consideration when selecting appropriate strains for insect pest management and may be useful in the development of more effective biocontrol strategies. Understanding the influence of soil properties on the pest and disease suppressive activities of introduced microbes should facilitate tailoring biocontrol agents for use in fields where biocontrol activity is maximised (Cook 1993). Further to this, the identification of soil factors that influence biocontrol will provide a biological basis for improved integration of biologicals with cultural practices that manipulate soil properties with an aim towards improved pest and disease control (Pierson and Weller 1994).

Production and Formulation

Fungal Inoculum

Product formulation is critical to the successful delivery and establishment of an effective population of a biocontrol agent. For soil pest control, fungi have generally been applied by direct incorporation of conidia, mycelial pellets, or inert or nutrient-based granules containing fungal propagules (conidia or mycelia). Suitable *in vitro* mass production systems are needed to provide sufficient inoculum for large-scale applications at a competitive price. The inoculum produced must be virulent, stable in storage, and environmentally competent after application. In the majority of cases, though, research emphasis has been placed on optimizing biomass production, with the assumption that control could be achieved if sufficient inoculum could be produced cheaply enough and applied at a high enough rate to the soil. However, soil is a complex environment and fungal activity against soil-inhabiting pests and pathogens is affected by many biotic and abiotic factors; environmental integrity is critical to performance, and maintenance of bioactivity must be a primary consideration when developing production media (Jaronski 2007; Brownbridge 2006). Formulation can enhance characteristics or render fungal preparations easier to apply, but their performance is ultimately reliant upon inclusion of robust biological material that is 'fit for purpose'. The production method selected will depend upon the nature of the inoculum required and isolates may have different growth characteristics on different production media.

Solid substrates have been widely used to produce aerial conidia of beneficial fungi. Temperature, pH, aeration and substrate components all influence conidial yield, viability, stability and virulence. Although these parameters are harder to regulate in a solid-substrate system, this remains the predominant method used for commercial products. This is due, in part, to the flexibility of a system that lends itself to the 'cottage-industry' scale used in many parts of the world, while solid-state fermentation bioreactors yielding 3×10^{13} conidia/kg of substrate have also been developed (Wraight et al. 2001). Small cereal grains are commonly used to produce fungi owing to the relatively high surface area-to-volume ratio of the substrate (which allows better aeration and access to nutrients) and their ready availability. Production varies on different grains, however, and virulence may be similarly affected (Wraight et al. 2001). Grains have been used to mass-produce conidia, chlamydospores or mycelial granules for soil incorporation (Jaronski 2007; Pérez-Rodriguez et al. 2007; McLean et al. 2005).

Large-scale liquid fermentation systems are successfully used for agriculturally important bacteria (e.g., *B. thuringiensis*, *S. entomophila*). In submerged culture, fungi generally produce vegetative propagules such as mycelia or yeast-like blastospores and culture conditions and media composition have a primary influence on the type and amount of inoculum produced. Production systems have largely been designed with high yield as a primary goal, but again the relative infectivity of the resulting biomass, its ecological competence and stability are key factors that must be considered during process development. Conditions and media can be manipulated to impart specific traits on the resulting biomass, including enhanced stability during drying and in storage, and infectivity. For example, conidia of *M. anisopliae* generated under nutritive stress (MM + 3 g L^{-1} lactose; MML) had increased virulence and increased germination speed than conidia produced under optimal nutritive conditions (e.g., potato dextrose agar + 1 g L^{-1} yeast extract; PDYA). MML medium produced conidia with greater adhesion capacity and that were more hydrophobic. There was also evidence that conidia production on MML were more reactive than PDYA produced conidia and enhanced binding efficiency to the surface of insect cuticles (Rangel et al. 2008). Similarly, studies by Jin et al. (1991) with *T. harzarium* demonstrated that by manipulating the growth conditions with polyethylene glycol (PEG), the trehalose content of the conidia could be significantly increased and this enabled the inocula to survive dessication better than unmodified, control conidia. Jaronski and Jackson (2008) recently described methods to induce production of microsclerotia by *M. anisopliae* in liquid media. The aggregates were readily air-dried, stable at room temperature, showed superior efficacy against sugarbeet root maggot in soil assays compared with conventional corn-grit granules, sporulated profusely in non-sterile soils and were active at low soil moisture levels. Such production/formulation techniques may overcome some of the biotic and abiotic constraints to fungal efficacy and result in increased opportunities to utilize these biocontrol agents against soil pests and pathogens.

Bacterial Inoculum

Bacterial agents are usually produced by liquid or solid state fermentation on low-cost carbohydrate or protein substrates. The method of production is determined by the characteristics of the specific organism with its economic viability dependent on the cost per unit of active ingredient applied in the field. The cost of production per unit can be reduced if the quality of the cells harvested is maximised during the fermentation. This can be achieved by optimising the harvest time for high yields of competent cells, maximising the cell density within the fermenter and by raising the scale of production. Visnovsky et al. (2008) showed that the most robust *S. entomophila* cells were produced after the cell culture entered the stationary phase of fermentation and that bacterial cells comprised 30–50% of the fermenter volume by the end of fermentation.

Media composition is also critical to the production of high yields of competent cells at harvest. Media can be optimised in the laboratory at a small scale then tested through scale up to high volumes. Liquid fermentation lends itself well to up-scaling to very large volumes. For production of *B. thuringiensis*, Valent Biosciences utilise fermenters of 50 m^3 volume utilising starch-based carbohydrates from the maize belt of the United States as a cheap and successful substrate (Georgis and Gaugler 1991). Where efficacy is determined by a toxin, increasing the production of the toxin can raise efficacy and modified media compositions have been used to develop high toxin cultures of *B. thuringiensis* (Marrone 1994).

Longevity and the maintenance of competency during storage are also critical to the development of a commercial product. Unfortunately, large-scale production creates a specific problem in that the large volume of cells may be vulnerable to factors in their surrounding environment. For non-spore forming bacteria, maintaining viability in the period between production and application poses even greater problems. These microbes lack a resistant stage and are susceptible to environmental factors such as temperature, toxins and desiccation, which usually lead to a short shelf life. Microbe producers have addressed this issue through mail order of sensitive microorganisms to reduce storage time, but the handling conditions after arrival cannot be guaranteed. Alternatively, the cell mass can be stabilised by cooling. However, low temperature storage is expensive and organisms can be sensitive to temperature fluctuations during distribution and application. Thus, other methods are usually required for long-term storage. For production of Bt toxins, the cell paste is concentrated by centrifugation and toxins subsequently stabilised by spray-drying or freeze-drying. The resultant dry powder containing spores and toxins can be stored for several years without loss in activity. An unprocessed fermenter broth containing a high-density, cell culture of *S. entomophila* was originally marketed as Invade®. This product could be stored for several months under refrigeration without deterioration (Johnson et al. 2001). Since then, a dry substrate coating system that allows cell survival at ambient storage temperatures has been developed for bacterial formulation and has been applied to a wide range of Enterobacteriaceae and other non-spore forming bacteria (Swaminathan et al. 2008). A second-generation *S. entomophila* product has been developed using Bioshield™ cell broth, which is incorporated into a biopolymer gel and coated onto a zeolite granule (Townsend et al. 2004). The resultant product can be stored at ambient temperatures without cell deterioration for 6–9 months at 25°C prior to application. Furthermore, the granules can be applied using standard farm equipment rather than the modified machinery used for application of Invade™. Similar products have been developed with Gram-positive bacteria, *Lactobacillus, Bifidobacterium* spp., *Pantoea* and *Pseudomonas* and other species, which can also be stabilised using these technologies, but with a lower water activity of the final product (Swaminathan et al. 2008).

Advances in formulation technologies now permit stabilization of environmentally sensitive microbes, and have applications to a diverse variety of beneficial organisms. Formulations can improve the handling characteristics and safety of a microorganism (e.g., by eliminating spore dust during preparation of a spray mixture), enhance stability pre- and post-application, improve persistence, promote efficacy, and facilitate easy delivery to the target pest/pathogen. Effective formulation is integral to the wider utility of biopesticides in agricultural production systems, and microbes can fail if formulated poorly (Jaffee et al. 1996). Formulations may be tailored to suit the environment in which the microbe will be used, the delivery system envisaged, and the nature of the inoculum being used. Like production systems, they must be rationally developed to ensure retention of key characteristics that are critical to microbial efficacy.

The Way Forward

Characteristics of Successful Biocontrol Agents

To be successful, a microbial control agent must provide the desired level of control of the target pest or disease. Some microbes that have been commercialised after providing consistent results in large-scale trials over a number of years or have been widely used in commercial operations are listed in Table 1. It is useful to examine this list to identify common biological factors that could assist in selection and accelerated development of new biocontrol agents.

For soil-dwelling insect pest species, a wide range of microbial agents (virus, bacteria, fungi, protozoa and nematodes) have been developed (Table 1). Interestingly no one agent has been developed as a general control for several soil dwelling species. This single product/single target effect appears to result from the intrinsic specificity of the relationship between microbial pathogens and their soil dwelling pest hosts. This contrasts with arboreal habitats where generalist strains of pathogen, such as *B. thuringiensis* var. *kurstaki*, have been used to control a very wide range of species (Navon 1993). Intrinsic specificity is demonstrated by plasmid bearing strains of the bacteria *Serratia* spp. which are only known to infect one species, the New Zealand grass grub (Jackson 2003). Similarly, *B. popilliae*, the cause of milky disease, is only known as a pathogen of the Scarabaeidae with a high degree of host species/bacterial strain specificity. While fungal products, based on *Beauveria* and *Metarhizium*, are listed several times in Table 1, the strains used are highly specific. For example, *B. brongniartii* strain 96 used against the scarab pest *Hoplochelus marginalis* in Reunion was isolated from that host in its original habitat of Madagascar and has little effect on other insects (Couteaudier et al. 1996). Other strains of fungi used for soil insect control, such as *M. anisopliae* DATF001 for pasture cockchafer control in Australia or *B. brongniartii* for May beetle control in Europe, also appear to be highly specific to the target pest as they provide effective control with little impact on non-target species (Traugott et al. 2005; Rath et al. 1995). While entomopathogenic nematodes appear to be less specific, the use of the mole cricket specific *S. scapterisci* (Parkman and Smart 1996) has proven more successful and consistent than the generalist nematode *Steinernema carpocapsae* in control of mole cricket and other turf pests (Georgis and Gaugler 1991).

Table 1 A selection of microbial control agents and products used for control of soil dwelling pests and diseases

Pest control agent	Pest targets	Product name	Reference
Oryctes virus	*Oryctes rhinoceros*	–	(Bedford 1981; Ramle et al. 2005)
Bacillus popilliae	*Popillia japonica*	Doom	(Klein 1992)
Serratia entomophila	*Costelytra zealandica*	Invade/Bioshield	(Jackson et al. 1992; Jackson 2007)
Bacillus thuringiensis var. *buibui*	*Popillia japonica, Anomala* spp.	–	(Alm et al. 1997)
Beauveria bassiana	*Cornitermes cumulans*	Boverial	(Alves et al. 1995)
Beauveria bassiana	*Heterotermes tenuis*	Termitrap	(Almeida et al. 1997; Almeida and Alves 1996)
Beauveria bassiana	*Cosmopolites sordidus*	–	(Badilla and Alves 1991)
Beauveria brongniartii	*Melolontha melolontha*	Engerlingspilz	(Keller 1992)
Beauveria brongniartii	*Hoplochelus marginalis*	Betel	(Vercambre et al. 1994)
Metarhizium anisopliae	*Adoryphorus couloni*	BiogreenChaferGuard®;	(Rath et al. 1995; Rath 2002)
Metarhizium anisopliae	*Cornitermes cumulans*	Metarril	(Alves et al. 1995)
Steinernema scapterisci	*Scapteriscus* spp.	–	(Parkman and Smart 1996)
Heterorhabditis bacteriophora	*Phyllopertha horticola*	Nema-green	(Sulistyanto and Ehlers 1996) www.e-nema.de
Heterorhabditis spp.	*Otiorhynchus* spp.	Dickmaulrüssler nematoden	Product literature, Andermatt Biocontrol AG
Heterorhabditis megidis	*Otiorhynchus* spp.	Larvanem	www.koppert.com
Thelohania solenopsae	*Solenopsis invicta*	–	(Oi and Williams 2002)
Nematode control agent	Nematode targets	Product name	Reference
Paecilomyces lilicanus	*Meloidogyne, Globodera* spp	Bionemat	www.biotech-int.com
Bacillus firmus	*Meloidogyne* spp	Bionem WP	(Blanchinsky et al. 2007)
Microbial control agent	Microbial targets	Product name	Reference
Agrobacterium radiobacter	*Agrobacterium tumefaciens*	NoGall™	(Ryder and Jones 1990), www.beckerunderwood.com
Bacillus subtilis	*Rhizoctonia, Fusarium* spp.	Kodiak®	(Brannen and Kenney 1998)

Bacillus subtilis	*Fusarium, Verticillium, Pythium* spp	Biosubtilin	www.biotech-int.com
Pseudomonas fluorescens	*Sclerotinia, Rhizoctonia, Pythium* spp.	Biomonas	www.biotech-int.com
Streptomyces griseoviridis	*Fusarium, Phytophtora, Pythium, Rhizoctonia* spp.	Mycostop	www.verdera.fi
Coniothyrium minitans	*Sclerotinia* spp.	Contans®	www.prophyta.de
Pythium oligandrum	*Sclerotinia, Pythium, Alternaria* spp.	Polyversum	www.polyversum.cz
Trichoderma atroviride	*Sclerotium cepivorum*	Tenet®	(Stewart and McLean 2007), www.agrimm.co.nz
T. harzianum T-22	*Pythium, Rhizoctonia, Fusarium* spp.	Rootshield®, PlantShield®	(Harman and Bjorkman 1998), www.bioworksinc.com
T. virens	*Pythium, Rhizoctonia* spp.	SoilGard®	(Lumsden et al. 1996; Lumsden and Knauss 2007)

In contrast to the soil pest situation, there is a much broader range of specificities exhibited by microbes targeted for plant disease control. This is likely to be a reflection of the broader mechanisms of action exhibited by these agents (e.g., parasitism, competition, antibiosis, induced resistance). For example, there is an extremely high degree of specificity and a single target focus exhibited by *A. radiobacter* due to the specific activity of the antibiotic produced and there is a relatively high degree of specificity where sclerotial parasitism occurs in *Sporidesmium* and *Coniothyrium*. In contrast, there are several biocontrol fungal and bacterial species that have a wider range of host targets covering several different taxonomic groups of soil-borne plant pathogens. This can be evident at the individual strain level, for example, *T. harzianum* T-22 is widely utilised for control of a range of soil-borne pathogens (e.g., *Rhizoctonia*, *Pythium* and *Fusarium* spp.) in numerous glasshouse and broad-acre cropping situations (Harman 2000).

The primary factors in the selection of biocontrol microbes are intrinsic pathogenicity or suppressive ability, but environmental competence – the ability of an applied microbe to persist and multiply in the pest's environment – is also essential (Jackson and O'Callaghan 1997). The soil comprises a microbial reservoir, where it is estimated that a Gram of fertile soil may contain as many as 10^5–10^8 bacteria, 10^6–10^7 actinomycetes and 10^5–10^6 fungal colony forming units (Metting 1993), and with as many as 3,000 species represented in a fertile pasture soil (Ovreas and Torsvik 1998). Therefore, applied microbes must be able to establish and survive in this competitive environment to be successful in pest or disease control. Some species, such as the spore-forming *B. popilliae*, are highly robust and can survive for many years in the soil before ingestion and germination in the specific conditions provided by the insect gut (Klein and Jackson 1992). Another spore former, *M. anisopliae* DAT F001 can survive in the soil for several years after application with persistence aided through recycling in the host population (Rath and Bullard 1997). Non-spore forming microbes, such as the bacterium *S. entomophila*, primarily persist by recycling through the host population. The nests of social insect are particularly challenging environments, where applied microbes must be able to survive both the stringent conditions within the nest and deterrent chemicals produced by the target insects. Testing from a wide variety of isolates has proven essential for selection of isolates that can survive these conditions and provide effective control of ants and termites (Almeida et al. 1997; Alves and Pereria 1998). Furthermore, lack of a repellent effect is a key trait when introducing a pathogen into the nest of social insects; this is important for acceptance of contaminated bait if this method of delivery is used, but also to promote development of an epizootic within a nest. Differences in repellency of fungal conidia have been previously observed and may be a critical factor in the selection of appropriate isolates for control of social insects (Wright et al. 2005). Persistence in soil of microbial agents used for plant disease control is highly variable depending on the system under investigation. Those microbes effective as sclerotial parasites often have good persistence because the host pathogen target acts as the survival conduit. Fravel et al. (1986) showed that *Talaromyces flavus*, a mycoparasite of *Rhizoctonia solani* and *Sclerotinia sclerotiorum* was able to persist in the soil on parasitized sclerotia from one season to the next. Similarly, a single application of *Sporidesmium sclerotivorum* to a field at

1,000 macroconidia g^{-1} soil provided significant control of lettuce drop for three consecutive crops over a 2 year period, indicating that the mycoparasite could be added to the field, establish, proliferate and persist (Adams and Ayers 1982). Although, biocontrol strains of *Trichoderma* and *Gliocladium* are common soil saprophytes able to survive on organic debris, they are not highly competitive in the soil environment with persistence often declining over a 3–6 month period. To utilize these agents effectively in biocontrol requires the addition of a nutrient base to sustain the population levels.

Identification and Selection of Desired Traits

Insect Pests

As outlined above, specificity and persistence are key traits of a successful biocontrol agent for soil pests. The importance of specificity in pest biocontrol suggests that targeted discovery will be more useful than high throughput screening from random sources in the selection of microbes. A review of the organisms developed into successful pest control products (Table 1) confirms that most isolates used in commercial products were taken from diseased insects in declining populations. Once a genetic mechanism is identified determining a desired trait, the strain selection process and quality control of developed products can be progressed to a higher level. Discovery that genes encoding pathogenicity were located on a specific plasmid (Glare et al. 1993) explained differences in pathogenicity between phenotypically similar strains and also loss of pathogenicity in extreme conditions. Identification of the gene sequence controlling pathogenicity allows pathogenic strain differences to be determined (Hurst et al. 2000). Knowledge of the genetic determinants allows quality control systems to be developed for monitoring stability in production and persistence in the field (O'Callaghan and Jackson 1993).

Insect-pathogenic fungi have generally been derived from the target pest directly, or from its environment. However, geographically exotic isolates or those from other hosts may be equally effective (Charnley and Collins 2007), and selection may simply be based on the prevailing regulations in force in the country where the pathogen will be used. The pathway to registration may be easier if a product contains a 'local' isolate (Chandler et al. 2008). Virulence, environmental competence and biological fitness all need to be considered in the selection of strains for further evaluation. Selection of strains, which can germinate, grow and infect the target pest under defined environmental parameters, will provide strains that have the desired pre-requisite characteristics for field use. Their suitability for mass-production and their genetic stability are additional important selection criteria.

Searches for fungi with specific traits have also been carried out. For example, surveys carried out to recover cold-tolerant isolates led to the discovery of *M. anisopliae* DATF001 and its development as a biocontrol agent for *A. couloni* (Rath et al. 1995). Differences in germination and growth of *M. anisopliae* and *B. bassiana* isolates at both high and low temperatures have been demonstrated with implications for strain selection and application in insect control strategies (Fernandes et al. 2008). Baiting of soil samples taken underneath pecan trees that had previously been sprayed with a regime of fungicides yielded fungal isolates with enhanced resistance to these chemicals (Shapiro-Ilan et al. 2002). Thus, selection pressures exerted on the natural population of entomopathogens probably results in the evolution of resistant strains and targeted sampling could yield strains with superior chemical compatibility (Mietkiewski et al. 1997).

Microbial Pathogens

For many years the standard laboratory dual culture test was used to identify and select microbes with antagonism towards plant pathogens. However, it has become apparent that isolate selections made using this method do not correlate well with results and efficacy in the field. The main exception to this is with sclerotial parasites where Henis et al. (1984) found a good correlation between sclerotial parasitism in agar plates and biocontrol in glasshouse and field tests. Thus, there is no general reliable laboratory test for biocontrol activity due to variability in plant pathogens, cropping systems, soils and temperatures. Selection assays are usually developed for specific mechanisms of bioactivity that are deemed likely to bring about biocontrol in any particular system. For example, many researchers have identified rhizosphere competency as a key biological attribute for microbes targeted to control root pathogens and so selection assays have been developed to identify this trait. Other researchers have targeted the selection of plant endophytic microbes on the basis that they would have greater induced resistance capability (Coombs et al. 2004) and assays comparing the carbon utilisation profiles of test strains relative to the target pathogen have been developed to select for microbes with high soil competitiveness (Marin et al. 1998).

Ecological Data

There has been considerable investment in the development of microbial biological control agents for inundative use, yet ways of preserving and enhancing the ecosystem services provided by these organisms, and developing a greater understanding of their fundamental ecology in soil have received scant attention. Research on factors affecting the performance of microbial biocontrol agents in soil is sparse. This is, in part, due to the complexity of the soil environment and the intricate interactions between different environmental and biological factors that can confound observations around cause and effect. While *in vitro* testing can provide valuable insights into microbial responses to specific inputs, they rarely yield data that can be directly extrapolated to predict field responses. More effort must be invested to evaluate effects of agricultural practices (e.g., Mietkiewski et al. 1997) on persistence and (particularly) efficacy under field conditions.

Production of good ecological data has also been impeded by a historic lack of tools to examine and quantify microbial populations in soil. Traditionally, studies investigating interactions with other soil microorganisms have relied on time-consuming isolation and plating techniques. Similarly, risk assessments have tended to focus on macroorganisms, while monitoring of interactions with other microbes has been limited and biased by our inability to culture all soil microorganisms. However, new tools and increasingly powerful molecular methods are becoming available to examine microbial communities in soil, and may be applied to the study of microbial biocontrol agents. Use of, for example, nuclear ITS and EF1-alpha sequences have enabled isolates of entomopathogenic fungi to be differentiated and phylogenetic relationships within species to be determined, enabling links to geographic and host origins to be defined. The ability to transform microbes to express beta-glucuronidase (GUS) and the green fluorescent protein (GFP) allows transformants to be observed in situ, and key ecological questions related to growth, sporulation, parasitism, and root colonisation etc. to be addressed (Bae and Knudsen 2005; Lorang et al. 2001). A variety of other molecular techniques such as RFLP, T-RFLP, AFLP and strain-specific microsatellite markers have been used as diagnostic tools allowing microbes to be tracked in the environment (Dodd et al. 2004; Enkerli et al. 2005; Inglis et al. 2008). Advances in the use of PCR techniques provide highly specific methods of monitoring microbial populations in 'real time' and in a quantitative manner, in soils, insects and *in planta* (Ownley et al. 2004; Wang et al. 2004; Entz et al. 2005; Schwarzenbach et al. 2007a; Meyling et al. 2009). Use of quantitative PCR with automated ribosomal intergenic spacer analysis (ARISA) allow soil microbial communities to be profiled and responses to specific events to be monitored; these techniques are set to be increasingly applied to the study of microbial biocontrol agents to assess their fate and impacts on microbial community structure and ultimately plant health (Schwarzenbach et al. 2007a, b; Enkerli et al. 2008).

Our ability to capitalize on the potential shown by microbes to regulate plant pest and pathogen populations will be greatly increased by improved insights into their ecology and behaviour in soil, their response to different environmental factors and farming practices, and interactions with other members of the soil biota (Jaronski 2007; Meyling and Eilenberg 2007). This knowledge can then be applied to the selection of robust, environmentally-competent strains; development of production systems that provide inoculum that is 'fit for purpose'; and to improve formulation and delivery techniques, etc., such that constraints to the effective utilization of these microbial control agents can be overcome.

Integrated Control

For the most part, this review has focused on the utilization of microbial biocontrol agents to control either soil-borne pests or diseases. In any given system, though, crops are likely to be affected by a suite of pests, requiring the concurrent use of more than one control agent. It is important, therefore, to ensure compatibility among the different control strategies so that their efficacy is not impaired in an integrated crop management programme.

The benefit of combining different biocontrol agents has been reported, for example, Alabouvette et al. (1996) demonstrated a synergistic effect in controlling *Fusarium oxysporum* f.sp. *radicis-lycopersici* by combining a fluorescent *Pseudomonas* sp. with a non-pathogenic *F. oxysporum*. The non-pathogenic *F. oxysporum* competes for carbon sources while the bacterial antagonist produces a siderophore competing for iron (Lemenceau et al. 1993). Effective disease control was also demonstrated by Domenech et al. (2006) for a combination of several plant growth promoting rhizobacteria. They reported enhanced biocontrol of *Fusarium* wilt and *Rhizoctonia* damping-off on tomato and pepper by the product LS213 that contained mixed inocula of *Bacillus subtilis* GB03 and *B. amyloliquefaciens* IN937a. A combined application of *M. anisopliae* with *S. entomophila* increased grass grub mortality in pot trials, suggesting synergies between biocontrol agents for pest and disease control could be explored (Jackson and Chinn 1993). However, a range of mortality responses were observed when entomopathogenic nematodes were combined with various beneficial fungi or bacteria against pecan weevil, thus, demonstrating the need to consider these interactions on a case-by-case basis rather than simply assuming that all biocontrol agents are compatible or interact positively (Shapiro-Ilan et al. 2004).

More usually, microbial biocontrol agents have been combined with a range of other crop management and pest management practices. Incorporation of antagonists following disinfestation of the soil with fumigants has been shown to be effective in controlling *R. solani* on carrots (Strashnow et al. 1985), and soil solarization proved a good control strategy when combined with *Gliocladium virens* against southern blight of tomatoes (Ristaino et al. 1991). Minuto et al. (2004) demonstrated that the combination of soil solarisation and *Streptomyces griseoviridis* was effective against *Fusarium* and *Verticillium* wilts and corky root rot, increasing the range of pathogens controlled with respect to the single treatments. Similarly, Stevens et al. (2003) showed that integration of soil solarisation with the biological control agent *T. virens* and the chemical fungicide PCNB significantly reduced southern blight and root knot disease on tomato compared to individual applications. The use of biocontrol agents with fertilization has also been shown to provide enhanced disease control. Hoynes et al. (1999) reported that soil treatment with ammonium sulphate followed by *Gliocladium virens* G1-3 resulted in greater reduction in viability of sclerotia of *S. rolfsii* and a higher bean seed germination than that achieved with each individual component. Howell (2007) reported optimum control of pre- and post-emergence damping-off of cotton seeds with a seed treatment of chloroneb plus *Trichoderma*. *Trichoderma* controlled the pre-emergence damping-off caused by *Pythium* spp. and the fungicide controlled post-emergence damping-off caused by *Rhizoctonia solani*. Similarly, *T. virens* in combination with metalaxyl as a seed treatment on cotton was shown to provide increased control of seedling diseases (Howell et al. 1997).

Similar synergistic interactions have been reported with insect pathogens, for example, *M. anisopliae* was applied to horticultural growing media or soil with sub-lethal doses of insecticides (imidacloprid or fipronil) to control black vine weevil (Shah et al. 2007) and citrus root weevil (Quintela and McCoy 1998) and co-application of *M. anisopliae* with neem seed cake enhanced fungal efficacy against black vine weevil (Shah et al. 2008). Improved efficacy of the biocontrol agents was attributed to effects of the insecticides on insect behaviour, which promoted acquisition of infectious material or reduced mechanical removal of conidia from the cuticle. Such an approach for insect pest management in soil serves to reduce pesticide use while improving opportunities to utilize fungi cost-effectively in this environment.

Conclusions

Microbial biocontrol agents have a demonstrated capacity to regulate pests and diseases, and possess numerous beneficial traits that favour their use in integrated crop management systems. The challenge lies in capitalizing on this potential, and developing products and strategies that exploit their unique characteristics. Presently, products containing insecticidal bacteria, primarily *Bacillus thuringiensis* (for foliar-feeding insects), dominate the biopesticides market. The bacterium has lent itself well to commercialization; industrial-scale liquid fermenters provide a raw material that is stable, readily-formulated and efficacious, for distribution and sale into to a global agricultural market. Few other microbial biocontrol agents (bacteria and fungi) have similar characteristics, which has been a limiting factor to their wider use. Products based on these microorganisms largely occupy niche markets, often within individual countries or geographically-linked regions. To increase utilization of these microbial biocontrol agents, several technical and regulatory challenges need to be overcome which, by-and-large, are common to all of these organisms. Fundamentally, we must recognize the strengths and weaknesses of microbial controls, and not simply try to replace chemical insecticides with biologically-derived materials.

In general, there is no shortage of excellent candidate organisms, some of which are already commercialized. To ensure their broader uptake in agriculture, more efficient mass production, formulation and delivery systems must be devised simply to supply a larger market; more testing under field conditions is required to identify effects of biotic and abiotic factors (and interactions between them) on efficacy and persistence, and potential limitations to the use of these biocontrol agents in certain crops or locations; and there has to be greater investment in the optimization of use practices. Advances in our understanding of infection processes, combined with the availability of new molecular tools which aid our ability to monitor the fate of these organisms in the environment and quantify effects of environmental factors on efficacy and persistence, continue to provide insights that will support the rational development of these technologies. Outreach and demonstration programmes that promote understanding of what growers can (or cannot) expect from microbial control agents, coupled with appropriate training on their use according to best practices, will further enhance their successful implementation.

The regulatory environment is creating conditions whereby biocontrol strategies are becoming the only available option for pest management. An increased interface between agricultural and urban environments and 'Greenbelt Legislation' is often preventing or strictly limiting the use of conventional crop protection materials.

Also, consumer awareness and demand is driving the move to implement more sustainable pest and disease management techniques. By ensuring that quality products are available and that farmers are equipped to apply them, microbial biocontrol agents can become important components of integrated crop production systems.

References

Adams PB (1990) The potential of mycoparasites for biological control of plant diseases. Annu Rev Phytopathol 28:59–72

Adams PB, Ayers WA (1982) Biological control of *Sclerotinia* lettuce drop in the field by *Sporidesmium sclerotivorum*. Phytopathology 72:485–488

Agrawal AA, Tuzan S, Bent E (1999) Induced plant defences against pathogens and herbivores; Biochemistry, ecology and agriculture. APS Press, St Paul, MN

Alabouvette C, Lemanceau P, Steinberg C (1996) Biological control of fusarium wilts: opportunities for developing a commercial product. In: Hall R (ed) Principles and practice of managing soilborne plant pathogens. APS Press, St Paul, MN, pp 192–212

Aldahmani JH, Abbasi PA, Sahin F, Hoitink HAJ, Miller SA (2005) Reduction of bacterial leaf spot severity on radish, lettuce, and tomato plants grown in compost-amended potting mixes Can J Plant Pathol 27:186–193

Allan RH, Thorpe CJ, Deacon JW (1992) Differential tropism to living and dead cereal root hairs by the biocontrol fungus *Idriella bolleyi*. Physiol Mol Plant Pathol 41:217–226

Alm SR, Villani MG, Yeh T, Shutter R (1997) *Bacillus thuringiensis* serovar. *japonensis* strain Buibui for control of Japanese and oriental beetle larvae (Coleoptera: Scarabaeidae). Appl Entomol Zool 32:477–484

Almeida JEM, Alves SB (1996) Mortality of *Heterotermes tenuis* (Hagen) attracted to traps treated with *Beauveria bassiana* (Bals.) Vuill. and imidacloprid. An Soc Entomol Bras 25:507–512

Almeida JEM, Alves SB, Pereira RM (1997) Selection of *Beauveria* spp. isolates for control of the termite *Heterotermes tenuis* (Hagen, 1858). J Appl Entomol 121:539–543

Aluko MO, Hering TF (1970) The mechanisms associated with the antagonistic relationship between *Corticium solani* and *Gliocladium virens*. T Brit Mycol Soc 55:173–179

Alves SB, Pereria RM (1998) Produção de fungos entomopatogênicos. In: Alves SB (ed) Controle microbiano de insetos, 2nd edn. FEALQ, Piracicaba, Brazil, pp 845–869

Alves SB, Almeida JEM, Moino Junior A, Stimac JL, Pereira RM (1995) Use of *Metarhizium anisopliae* and *Beauveria bassiana* for control of *Cornitermes cumulans* (Kollar, 1832) in pastures. Ecossistema 20:50–57

Amir H, Alabouvette C (1993) Involvement of soil abiotic factors in the mechanisms of soil suppressiveness to *Fusarium* wilts. Soil Biol Biochem 25:157–164

Anderson JD, Bailey BA, Taylor R, Sharon A, Avni A, Mattoo AK, Fuchs Y (1993) Fungal xylanase elicits ethylene biosynthesis and other defense responses in tobacco. In: Pech JC, Latche A, Balague C (eds) Cellular and molecular aspects of the plant hormone ethylene. Kluwer, Dordrecht, The Netherlands, pp 197–204

Ansari MA, Shah FA, Whittaker M, Prasad M, Butt TM (2007) Control of western flower thrips (*Frankliniella occidentalis*) pupae with *Metarhizium anisopliae* in peat and peat alternative growing media. Biol Control 40:293–297

Ansari MA, Brownbridge M, Shah FA, Butt TM (2008) Efficacy of entomopathogenic fungi against soil-dwelling life stages of western flower thrips, *Frankliniella occidentalis*, in plant-growing media. Entomol Exp Appl 127:80–87

Asaka O, Shoda M (1996) Biocontrol of *Rhizoctonia solani* damping-off of tomato with *Bacillus subtilis* RB14. Appl Environ Microbiol 62:4081–4085

Asher MJC, Shipton PJ (1981) Biology and control of take-all. Academic, London, p 538
Badilla F, Alves SB (1991) Control del picudo de la caña de azúcar *Sphenophorus levis* Vaurie (Col.: Curculionidae) con *Beauveria bassiana* y *Beauveria brongniartii* en condiciones de laboratorio y campo 20–21: 34–38
Bae YS, Knudsen GR (2005) Soil microbial biomass influence on growth and biocontrol efficacy of *Trichoderma harzianum*. Biol Control 32:236–242
Barak R, Elad Y, Chet I (1986) The properties of L-fructose-binding agglutinin associated with the cell wall of *Rhizoctonia solani*. Arch Microbiol 144:346–349
Bedford GO (1981) Control of the rhinoceros beetle by baculovirus. In: Burges HD (ed) Microbial control of pests and plant diseases. Academic, London, pp 418–426
Blanchinsky D, Antonov J, Bercovitz A, Elad B, Feldman K, Husid A, Lazore M, Marcov N, Shamai I, Keren-Zur M (2007) Bionem WP: a unique tool for nematode control. Biological control of fungi and bacterial plant pathogens. IOBC/wprs Bull 30:23–26
Boland GJ (1992) Hypovirulence and double-stranded RNA in *Sclerotinia sclerotiorum*. Can J Plant Pathol 14:10–17
Bolwerk A, Lagopodi AL, Lugtenberg BJJ, Bloemberg GV (2005) Visualization of interactions between a pathogenic and a beneficial *Fusarium* strain during biocontrol of tomato foot and root rot. Mol Plant Microbe Interact 18:710–721
Boulter JI, Boland GJ, Trevors JT (2002) Assessment of compost for suppression of *Fusarium* Patch (*Microdochium nivale*) and *Typhula* Blight (*Typhula ishikariensis*) snow molds of turfgrass. Biol Control 25:162–172
Brannen PM, Kenney DS (1998) KodiakReg.-a successful biological-control product for suppression of soil-borne plant pathogens of cotton. J Ind Microbiol Biotechnol 19: 169–171
Brian PW, McGowan JC (1945) Viridin: a highly fungistatic substance produced by *Trichoderma viride*. Nature 156:144
Brian PW, Curtis PJ, Hemming HG, McGowan JC (1946) The production of viridin by pigment-forming strains of *Trichoderma viride*. Ann Appl Biol 33:190–200
Brownbridge M (2006) Entomopathogenic fungi: status and considerations for their development and use in integrated pest management. Recent Res Dev Entomol 5:27–58
Brownbridge M, Glare T (2007) Impact of entomopathogenic fungi on soil-dwelling invertebrates. In: Ekesi S, Maniania NK (eds) Use of entomopathogenic fungi in biological pest management. Research Signpost, Trivandrum, India, pp 295–312
Brownbridge M, Nelson TL, Hackell DL, Eden TM, Wilson DJ, Willoughby BE, Glare TR (2006) Field application of biopolymer-coated *Beauveria bassiana* F418 for clover root weevil (*Sitona lepidus*) control in Waikato and Manawatu. NZ Plant Protection 59: 304–311
Bruck DJ (2005) Ecology of *Metarhizium anisopliae* in soilless potting media and the rhizosphere: implications for pest management. Biol Control 32:155–163
Bruck DJ (2006) Effect of potting media components on the infectivity of *Metarhizium anisopliae* against black vine weevil (Coleoptera: Curculionidae). J Environ Hortic 24:91–94
Brust GE (1994) Natural enemies in straw-mulch reduce Colorado beetle populations and damage in potato. Biol Control 4:163–169
Bush LP, Fannin FF, Siegel MR, Dahlman DL, Burton HR (1993) Chemistry, occurrence and biological effects of saturated pyrrolizidine alkaloids associated with endophyte-grass interactions. Agric Ecosyst Environ 44:81–102
Butt TM, Brownbridge M (1997) Fungal pathogens of thrips. In: Lewis T (ed) Thrips as crop pests. CAB International, Wallingford, UK, pp 399–434
Buysens S, Heungens K, Poppe J, Hofte M (1996) Involvement of Pyochelin and Pyoverdin in Suppression of *Pythium*-Induced Damping-Off of tomato by *Pseudomonas aeruginosa* 7NSK2. Appl Environ Microbiol 62:865–871
Campbell LG, Eide JD, Smith LJ, Smith GA (2000) Control of sugarbeet root maggot with the fungus *Metarhizium anisopliae*. J Sugar Beet Res 37:57–69

Campbell LG, Boetel MA, Jonason NB, Jaronski ST, Smith LJ (2006) Grower-adoptable formulations of the entomopathogenic fungus *Metarhizium anisopliae* (Ascomycota: Hypocreales) for sugarbeet root maggot (Diptera: Ulidiidae) management. Environ Entomol 35:986–991

Cardinale F, Ferraris L, Valentino D, Tamietti G (2006) Induction of systemic resistance by a hypovirulent *Rhizoctonia solani* isolate in tomato. Physiol Mol Plant Pathol 69:160–171

Cardoza RE, Hermosa MR, Vizcaino JA, Sanz L, Monte E, Gutierrez S (2005) Secondary metabolites produced by *Trichoderma* and their importance in the biocontrol process. In: Ekesi S, Maniania NK (eds) Microorganisms for industrial enzymes and biocontrol. Research Signpost, Trivandrum, India, pp 207–228

Cartwright DK, Benson MD (February 1994) *Pseudomonas cepacia* strain 5.5B and method of controlling *Rhizoctonia solani* therewith. US patent 5,288,633

Chandler D, Davidson G (2005) Evaluation of the entomopathogenic fungus *Metarhizium anisopliae* against soil-dwelling stages of cabbage maggot (Diptera: Anthomyiidae) in glasshouse and field experiments and effect of fungicides on fungal activity. J Econ Entomol 98:1856–1862

Chandler D, Davidson G, Grant WP, Greaves J, Tatchell GM (2008) Microbial biopesticides for integrated crop management: an assessment of environmental and regulatory sustainability. Trends Food Sci Tech 19:275–283

Charnley AK (2003) Fungal pathogens of insects: cuticle degrading enzymes and toxins. Adv Bot Res 40:241–321

Charnley AK, Collins SA (2007) Entomopathogenic fungi and their role in pest control. In: Kubicek CP, Druzhinina IS (eds) Environmental and microbial relationships, 2nd edn, The Mycota IV. Springer, Berlin, Germany, pp 159–187

Chen SY, Liu X (2005) Control of the soybean cyst nematode by the fungi *Hirsutella rhossiliensis* and *Hirsutella minnesotensis* in greenhouse studies. Biol Control 32:208–219

Chen S, Liu S (2007) Effects of tillage and crop sequence on parasitism of *Heterodera glycines* juveniles by *Hirsutella* spp. and on juvenile population density. Nematropica 37:93–106

Chen C, Belanger RR, Benhamou N, Paulitz TC (1999) Role of salicylic acid in systemic resistance induced by *Pseudomonas* spp. against *Pythium aphanidermatum* in cucumber roots. Eur J Plant Pathol 105:477–486

Chernin L, Chet I (2002) Microbial enzymes in the biocontrol of plant pathogens and pests. In: Burns RG, Dick RP (eds) Enzymes in the environment: activity, ecology and applications. Marcel Dekker, New York, pp 171–226

Chet I (1990) Mycoparasitism - Recognition, physiology and ecology. In: Barker RR, Dunn PE (eds) New directions in biological control: alternatives for suppressing, agricultural pests and diseases. Alan Liss, New York, pp 725–733

Chng SF (2009) Microbial factors associated with the natural suppression of take-all in wheat in New Zealand. PhD. Thesis, Lincoln University, Lincoln, New Zealand

Clarkson JP, Mead A, Payne T, Whipps JM (2004) Effect of environmental factors and *Sclerotium cepivorum* isolate on sclerotial degradation and biological control of white rot by *Trichoderma*. Plant Pathol 53:353–362

Claydon N, Allan M, Hanson JR, Avent AG (1987) Antifungal alkyl pyrones of *Trichoderma harzianum*. T Brit Mycol Soc 88:503–513

Clement SL, Elberson LR, Bosque-Perez NA, Schotzko DJ (2005) Detrimental and neutral effects of wild barley-*Neotyphodium* fungal endophyte associations on insect survival. Entomol Exp Appl 114:119–125

Conrath U, Pieterse CMJ, Mauch-Mani B (2002) Priming in plant-pathogen interactions. Trends Plant Sci 7:210–216

Cook RJ (1993) Making greater use of introduced microorganisms for biological control of plant pathogens. Annu Rev Phytopathol 31:53–80

Cook RJ (2003) Take-all of wheat. Physiol Mol Plant Pathol 62:73–86

Cook RJ (2007) Management of resident plant growth-promoting rhizobacteria with the cropping system: a review of experience in the US Pacific Northwest. Eur J Plant Pathol 119:255–264

Cook RJ, Baker KF (1983) Introduction of antagonists for biological control. In: Cook RJ, Baker KF (eds) The nature and practice of biological control of plant pathogens. APS Press, St Paul, MN, pp 281–311

Coombs JT, Michelsen PP, Franco CMM (2004) Evaluation of endophytic actinobacteria as antagonists of *Gaeumannomyces graminis* var. *tritici* in wheat. Biol Control 29:359–366

Cortes-Pentagos C, Olmedo-Monfil V, Herrera-Estrella A (2007) The nature of fungal mycoparasitic biocontrol agents. In: Chincholkar SB, Mukerji KG (eds) Biological control of plant diseases. The Haworth Press, New York, pp 327–353

Costa JLS, Menge JA, Casale WL (1996) Investigation of some of the mechanisms by which bioenhanced mulches can suppress *Phytophthora* root rot of avocados. Microbial Res 151:183–192

Cotes AM, Lepoivre P, Semal J (1996) Correlation between hydrolytic enzyme activities measured in bean seedlings after *Trichoderma koningii* treatment combined with pregermination and the protective effect against *Pythium splendens*. Eur J Plant Pathol 102:497–506

Couteaudier Y, Nueveglise C, Viaud M, Riba G (1996) The fungus *Beauveria* for microbial control of soil pests: molecular ecology and strain improvement. In: The 3rd international workshop on microbial control of soil dwelling pests, Lincoln, New Zealand pp 31–37

Craft CM, Nelson EB (1996) Microbial properties of composts that suppress damping-off and root rot of creeping bentgrass caused by *Pythium graminicola*. Appl Environ Microbiol 62: 1550–1557

Crickmore N (2005) Using worms to better understand how *Bacillus thuringiensis* kills insects. Trends Microbiol 13:347–350

Dababat AEA, Sikora RA (2007) Induced resistance by the mutualistic endophyte, *Fusarium oxysporum* strain 162, toward *Meloidogyne incognita* on tomato. Biocontrol Sci Technol 17:969–975

Danielson RM, Davey C (1973) Carbon and nitrogen nutrition of *Trichoderma*. Soil Biol Biochem 5:505–515

Das BC, Basu MM, Chaterjee GC (1978) Studies on the mode of action of agrocin 84. J Antibiot 31:490–492

Davidson G, Chandler D (2005) Laboratory evaluation of entomopathogenic fungi against larvae and adults of onion maggot (Diptera: Anthomyiidae). J Econ Entomol 98:1848–1855

Davies KG (2005) Interactions between nematodes and microorganisms: bridging ecological and molecular approaches. Adv Appl Microbiol 57:53–78

Davies KG, Opperman CH (2006) A potential role for collagen in the attachment of *Pasteuria penetrans* to nematode cuticle. Bull OILB/SROP 29:11–15

Davies KG, Fargette M, Balla G, Daudi A, Duponnois R, Gowen SR, Mateille T, Phillips MS, Sawadogo A, Trivino C, Vouyoukalou E, Trudgill DL (2001) Cuticle heterogeneity as exhibited by Pasteuria spore attachment is not linked to the phylogeny of parthenogenetic root-knot nematodes (*Meloidogyne* spp.). Parasitology 122:111–120

de Faria MR, Wraight SP (2007) Mycoinsecticides and mycoacaricides: a comprehensive list with worldwide coverage and international classification of formulation types. Biol Control 43:237–256

de Wit PJGM, Brandwagt BF, van der Burg HA, Cai X, van der Hoorn RAL, de Jong CF, van 't Klooster J, de Kock MJD, Kruijt M, Lindhout WH, Luderer R, Takken FLW, Westerink N, Vervoort JJM, Joosten MHAJ (2002) The molecular basis of co-evolution between *Cladosporium fulvum* and tomato. Antonie Leeuwenhoek 81:409–412

Dodd SL, Hill RA, Stewart A (2004) A duplex-PCR bioassay to detect a *Trichoderma virens* biocontrol isolate in non-sterile soil. Soil Biol Biochem 36:1955–1965

Domenech J, Reddy MS, Kloepper JW, Ramos B, Gutierrez-Manero J (2006) Combined application of the biological product LS213 with *Bacillus*, *Pseudomonas* or *Chryseobacterium* for growth promotion and biological control of soil-borne diseases in pepper and tomato. Biocontrol 51:245–258

Dong LQ, Zhang KQ (2006) Microbial control of plant-parasitic nematodes: a five-party interaction. Plant Soil 288:31–45

Duffy BK, Ownley BH, Weller DM (1997) Soil chemical and physical properties associated with suppression of take-all of wheat by *Trichoderma koningii*. Phytopathology 87:1118–1124

Eastburn DM, Butler EE (1988a) Microhabitat characterization of *Trichoderma harzianum* in natural soil: evaluation of factors affecting population density. Soil Biol Biochem 20:541–545

Eastburn DM, Butler EE (1988b) Microhabitat characterization of *Trichoderma harzianum* in natural soil: evaluation of factors affecting distribution. Soil Biol Biochem 20:547–553

Ekesi S, Maniania NK, Mohamed SA, Lux SA (2005) Effect of soil application of different formulations of *Metarhizium anisopliae* on African tephritid fruit flies and their associated endoparasitoids. Biol Control 35:83–91

Ekesi S, Dimbi S, Maniania NK (2007) The role of entomopathogenic fungi in the integrated management of fruit flies (Diptera: Tephritidae) with emphasis on species occurring in Africa. In: Ekesi S, Maniania NK (eds) Use of entomopathogenic fungi in biological pest management. Research Signpost, Trivandrum, India, pp 239–274

Elad Y, Barak R, Chet I (1983a) Possible role of lectins in mycoparasitism. J Bacteriol 154:1431–1435

Elad Y, Chet I, Boyle P, Henis Y (1983b) Parasitism of *Trichoderma* spp. on *Rhizoctonia solani* and *Sclerotium rolfsii* – SEM studies and fluorescence microscopy. Phytopathology 73:85–88

Engler KM, Gold RE (2004) Effects of multiple generations of *Metarhizium anisopliae* on subterranean termites feeding and mortality. Sociobiology 44:211–240

Enkerli J, Kolliker R, Keller S, Widmer F (2005) Isolation and characterization of microsatellite markers from the entomopathogenic fungus *Metarhizium anisopliae*. Mol Ecol Notes 5:384–386

Enkerli J, Schwarzenbach K, Widmer F (2008) Assessing potential effects of the *Beauveria brongniartii* biological control agent on fungal community structures in soil microcosms. In: Proceedings of the 41st annual meeting of the society for invertebrate pathology [Abstracts], Warwick University, UK, p 81

Entz SC, Johnson DL, Kawchuk LM (2005) Development of a PCR-based diagnostic assay for the specific detection of the entomopathogenic fungus *Metarhizium anisopliae* var. *acridum*. Mycol Res 109:1302–1312

Fenton AM, Stephens PM, Crowley J, O'Callaghan M, O'Gara F (1992) Exploitation of gene(s) involved in 2, 4-diacetylphloroglucinol biosynthesis to confer a new biocontrol capability to a Pseudomonas strain. Environ Microbiol 58:3873–3878

Fernandes EKK, Rangel DEN, Moraes AML, Bittencourt VREP, Roberts DW (2008) Cold activity of *Beauveria* and *Metarhizium*, and thermotolerance of *Beauveria*. J Invertebr Pathol 98:69–78

Fleming WE (1968) Biological control of the Japanese beetle. USDA Technical Bulletin No 1383. United States Department of Agriculture, Washington, DC, p 78

Fravel DR (1988) Role of antibiosis in the biocontrol of plant diseases. Annu Rev Phytopathol 26:75–91

Fravel DR (2005) Commercialization and implementation of biocontrol. Annu Rev Phytopathol 43:377–359

Fravel DR, Davis JR, Sorensen LH (1986) Effect of *Talaromyces flavus* and metham on *Verticillium* wilt incidence and potato yield, 1984–1985. Biol Cult Tests 1:7

Freitas LG, Ferraz S, Muchovej JJ (1995) Effectiveness of different isolates of *Paecilomyces lilacinus* and an isolate of *Cylindrocarpon destructans* on the control of *Meloidogyne javanica*. Nematropica 25:109–115

Georgis R, Gaugler R (1991) Predictability in biological control using entomopathogenic nematodes. J Econ Entomol 84:713–720

Gerard PJ (2001) Dependence of *Sitona lepidus* (Coleoptera: Curculionidae) larvae on abundance of white clover *Rhizobium* nodules. Bull Entomol Res 91:149–152

Gerlagh M, Goossen-van de Geijn HM, Hoogland AE, Vereijken PFG (2003) Quantitative aspects of infection of *Sclerotinia sclerotiorum* sclerotia by *Coniothyrium minitans* – timing of application, concentration and quality of conidial suspension of the mycoparasite. Eur J Plant Pathol 109:489–502

Ghisalberti EL, Narbey MJ, Dewan MM, Sivasithamparam K (1990) Variability among strains of *Trichoderma harzianum* in their ability to reduce take-all and to produce pyrones. Plant Soil 121:287–291
Gimsing AL, Kirkegaard JA (2006) Glucosinolate and isothiocyanate concentration in soil following incorporation of *Brassica* biofumigants. Soil Biol Biochem 38:2255–2264
Glare TR, Corbett GE, Sadler TJ (1993) Association of a large plasmid with amber disease of the New Zealand grass grub, *Costelytra zealandica*, caused by *Serratia entomophila* and *Serratia proteamaculans*. J Invertebr Pathol 62:165–170
Glare TR, Townsend RJ, Young SD (1994) Temperature limitations on field effectiveness of *Metarhizium anisopliae* against *Costelytra zealandica* (White) (Coleoptera: Scarabidae) in Canterbury. N Z Plant Protection 47:266–270
Green SJ, Inbar E, Michel FC, Hadar Y, Minz D (2006) Succession of bacterial communities during early plant development: transition from seed to root and effect of compost amendment. Appl Environ Microbiol 72:3975–3983
Hanson LE (2000) Red of Vertical wilt symptoms in cotton following seed treatment with *T. virens*. J Cotton Sci 4:224–231
Harman GE (2000) Myths and dogmas of biocontrol: changes in perceptions derived from research on *Trichoderma harzianum* T-22. Plant Disease 84:377–393
Harman GE, Bjorkman T (1998) Potential and existing uses of *Trichoderma* and *Gliocladium* for plant disease control and plant growth enhancement. In: Kubicek CP, Harman GE (eds) *Trichoderma* and *Gliocladium*: enzymes biological control and commercial applications, vol 2. Taylor & Francis, London, pp 229–265
Harman GE, Taylor AG (1990) Development of an effective biological seed treatment system. In: Hornby D (ed) Biological control of soil-borne plant pathogens. CABI, Wallingford, UK, pp 415–426
Harman GE, Howell CR, Viterbo A, Chet I, Lorito M (2004) *Trichoderma* species – opportunistic, avirulent plant symbionts. Nat Rev Microbiol 2:43–56
Harris AR (2000) Solid formulations of binucleate *Rhizoctonia* isolates suppress *Rhizoctonia solani* and *Pythium ultimum* in potting medium. Microbiol Res 154:333–337
Henis Y, Lewis JA, Papavizas GC (1984) Interactions between *Sclerotium rolfsii* and Trichoderma spp relationship between antagonism and disease control. Soil Biol Biochem 16:391–395
Herr LJ (1995) Biological control of *Rhizoctonia solani* by binucleate *Rhizoctonia* spp. and hypovirulent *R. solani* agents. Crop Prot 14:179–186
Hill DS, Stein JI, Torkewitz NR, Morse AM, Howell CR, Pachlatko JP, Becker JO, Ligon JM (1994) Cloning of genes involved in the synthesis of pyrrolnitrin from *Pseudomonas fluorescens* and role of pyrrolnitrin synthesis in biological control of plant disease. Appl Environ Microbiol 60:78–85
Hill RA, Eden MA, Cutler HG, Elmer PAG, Reglinski T, Parker SR (1999) Practical natural solutions for plant disease control. In: Biologically active natural products: agrichemicals. CRC Press, St Paul, MN, p 210
Hiraoka H, Asaka O, Ano T, Shoda M (1992) Characterization of *Bacillus subtilis* RB14, coproducer of peptide antibiotics iturin and surfactin. J Gen Appl Microbiol 38:635–640
Hoffland E, Hakulinen J, van Pelt JA (1996) Comparison of systemic resistance induced by avirulent and nonpathogenic *Pseudomonas* species. Phytopathology 86:757–762
Hoitink HAJ, Boehm MJ (1999) Biocontrol within the context of soil microbial communities: a substrate-dependent phenomenon. Annu Rev Phytopathol 37:427–446
Hoitink HAJ, Grebus ME (1994) Status of biological control of plant diseases with composts. Compost Sci Util 2:6–12
Hoitink HAJ, Madden LV, Boehm MJ (1996a) Relationships among organic matter decomposition level, microbial species diversity, and soilborne disease severity. In: Society TAP (ed) Principles and practises of managing soilborne plant pathogens. R. Hall, St. Paul, MN
Hoitink HAJ, Stone AG, Grebus ME (1996b) Suppression of plant diseases by composts. In: The science of composting: part 1. Blackie Academic & Professional, Glasgow, UK, pp 373–381
Hoper H, Steinberg C, Alabouvette C (1995) Involvement of clay type and pH in the mechanisms of soil suppressiveness to *Fusarium* wilt of flax. Soil Biol Biochem 27:955–967

Hornby D, Bateman GL, Gutteridge RJ, Ward E, Yarham DJ (1998) Take-all disease of cereals: a regional perspective. CAB International, Wallingford, UK, p 384

Howell CR (2007) Effect of seed quality and combination fungicide-*Trichoderma* spp. seed treatments on pre- and postemergence damping-off in cotton. Phytopathology 97:66–71

Howell CR, Stipanovic RD (1983) Gliovirin, a new antibiotic from *Gliocladium virens*, and its role in the biological control of *Pythium ultimum*. Can J Microbiol 29:321–324

Howell CR, Stipanovic RD (1984) Phytotoxicity to crop plants and herbicidal effects on weeds of viridiol produced by *Gliocladium virens*. Phytopathology 74:1346–1349

Howell CR, DeVay JE, Garber RH, Batson WE (1997) Field control of cotton seedling diseases with *Trichoderma virens* in combination with fungicide seed treatments. J Cotton Sci 1:15–20

Howell CR, Hanson LE, Stipanovic RD, Puckhaber LS (2000) Induction of terpenoid synthesis in cotton roots and control of *Rhizoctonia solani* by seed treatment with *Trichoderma virens*. Phytopathology 90:248–252

Howie WJ, Suslow T (1991) Role of antibiotic synthesis in the inhibition of *Pythium ultimum* in the cotton spermosphere and rhizosphere by *Pseudomonas fluorescens*. Mol Plant Microbe Interact 4:393–399

Hoynes CD, Lewis JA, Lumsden RD, Bean GA (1999) Biological control agents in combination with fertilization or fumigation to reduce sclerotial viability of *Sclerotium rolfsii* and disease of snap beans in the greenhouse. J Phytopathol 147:175–182

Hubbard JP, Harman GE, Hadar Y (1983) Effect of soilborne *Pseudomonas* species on the biocontrol agent *Trichoderma harzianum* on pea seeds. Phytopathology 73:655–659

Hurst MRH, Glare TR, Jackson TA, Robson CW (2000) Plasmid-located pathogenicity determinants of *Serratia entomophila*, the causal agent of amber disease of grass grub, show similarity to the insecticidal toxins of *Photorhabdus luminescens*. J Bacteriol 182:5127–5138

Hurst MRH, Young SD, Nelson TL, Jackson TA, Becher A, Glare TR (2007) Species identification and host range testing of a new entomopathogenic member of the Enterobacteriaceae. In: Proceedings of the 40th annual meeting of SIP, Quebec City, Canada

Inbar I, Chet I (1992) Biomics of fungal-cell recognition by the use of lectin-coated nylon fibres. J Bacteriol 174:1055–1059

Inglis GD, Duke GM, Goettel MS, Kabaluk JT (2008) Genetic diversity of *Metarhizium anisopliae* var. *anisopliae* in southwestern British Columbia. J Invertebr Pathol 98:101–113

Jackson TA (2003) Environmental safety of inundative application of a naturally occurring biocontrol agent, *Serratia entomophila*. In: Hokkanken HMT, Hajek AE (eds) Environmental impacts of microbial insecticides: need and methods for risk assessment. Kluwer, Dordrecht, The Netherlands, pp 169–176

Jackson TA (2007) A novel bacterium for control of grass grub. In: Vincent C, Goettel MS, Lazarovits G (eds) Biological control: a global perspective. CABI Publishing, Wallingford, UK, pp 160–168

Jackson TA, Chinn WG (1993) The effect of *Metarhizium anisopliae* formulations, and their combination with *Serratia entomophila*, on grass grub larvae. N Z Plant Protection 46:206–209

Jackson TA, Glare T (eds) (1992) Use of pathogens in scarab pest management. Intercept, Andover, UK, p 298

Jackson TA, Klein MG (2006) Scarabs as pests: a continuing problem. Coleopt Bull 60:102–119

Jackson TA, O'Callaghan M (1997) Environmental competence – an essential characteristic of successful microbial control agents for soil-dwelling pests. In: Proceedings of the 3rd Brisbane workshop on soil invertebrates. Bureau of sugar experimental stations, Brisbane, Australia, pp 74–77

Jackson TA, Pearson JF, O'Callaghan M, Mahanty HK, Willocks M (1992) Pathogen to product – development of *Serratia entomophila* (Enterobacteriaceae) as a commercial biological control agent for the New Zealand grass grub (*Costelytra zealandica*). In: Jackson TA, Glare TR (eds) Use of pathogens in scarab pest management. Intercept, Andover, UK, pp 191–198

Jackson TA, Boucias DG, Thaler JO (2001) Pathobiology of amber disease, caused by *Serratia* spp., in the New Zealand grass grub, *Costelytra zealandica*. J Invertebr Pathol 78:232–243

Jaffee BA (2000) Augmentation of soil with the nematophagous fungi *Hirsutella rhossiliensis* and *Arthrobotrys haptotyla*. Phytopathology 90:498–504

Jaffee BA, Ferris H, Stapleton JJ, Norton MVK, Muldoon AE (1994) Parasitism of nematodes by the fungus *Hirsutella rhossiliensis* as affected by certain organic amendments. J Nematol 26:152–161

Jaffee BA, Muldoon AE, Westerdahl BB (1996) Failure of a mycelial formulation of the nematophagous fungus *Hirsutella rhossiliensis* to suppress the nematode *Heterodera schachtii*. Biol Control 6:340–346

Jaronski ST (2007) Soil ecology of the entomopathogenic Ascomycetes: a critical examination of what we (think) we know. In: Ekesi S, Maniania NK (eds) Use of entomopathogenic fungi in biological pest management. Research Signpost, Trivandrum, India, pp 91–143

Jaronski ST, Jackson MA (2008) Efficacy of *Metarhizium anisopliae* microsclerotial granules. Biocontrol Sci Technol 18:849–863

Jeffries P, Young TWK (1994) Interfungal parasitic relationships. CAB International, Wallingford, UK

Jiang Z-Q, Guo Y-H, Li S-M, Qi H-Y, Guo J-H (2006) Evaluation of biocontrol efficiency of different *Bacillus* preparations and field application methods against *Phytophthora* blight of bell pepper. Biol Control 36:216–223

Jin X, Harman GA, Taylor AG (1991) Conidial biomass and desiccation tolerance of *Trichoderma harzarium* produced at different water potentials. Biol Control 1:237–243

Johnson VW, Pearson JF, Jackson TA (2001) Formulation of *Serratia entomophila* for biological control of grass grub. N Z Plant Protection 54:125–127

Jones RW, Hancock JG (1987) Conversion of viridin to viridiol by viridin-producing fungi. Can J Microbiol 33:963–966

Kannangara T, Utkhede RS, Bactawar B (2004) Compost effect on greenhouse cucumbers and suppression of *Fusarium oxysporum*. Compost Sci Util 12:308–313

Keller S (1992) The *Beauveria-Melolontha* project: experiences with regard to locust and grasshopper control. In: Lomer CJ, Prior C (eds) Biological control of locusts and grasshoppers. CAB International, Wallingford, UK, pp 279–286

Keller S (2000) Use of *Beauveria brongniartii* in Switzerland and its acceptance by farmers. Bull OILB/SROP 23:67–71

Keller S, Kessler P, Schweizer C (2003) Distribution of insect pathogenic soil fungi in Switzerland with special reference to *Beauveria brongniartii* and *Metharhizium anisopliae*. Biocontrol 48:307–319

Kepler RM, Bruck DJ (2006) Examination of the interaction between the black vine weevil (Coleoptera: Curculionidae) and an entomopathogenic fungus reveals a new tritrophic interaction. Environ Entomol 35:1021–1029

Kerr A (1980) Biological control of crown gall through production of agrocin 84. Plant Dis 64:25–30

Kerry B, Hidalgo-Diaz L (2004) Application of *Pochonia chlamydosporia* in the integrated control of root-knot nematodes on organically grown vegetable crops in Cuba. Bull OILB/SROP 27:123–126

Kessler P, Enkerli J, Schweizer C, Keller S (2004) Survival of *Beauveria brongniartii* in the soil after application as a biocontrol agent against the European cockchafer *Melolontha melolontha*. Biocontrol 49:563–581

Kiewnick S (2004) Biological control of plant parasitic nematodes with *Paecilomyces lilacinus*, strain 251. Bull OILB/SROP 27:133–136

Kiewnick S, Sikora RA (2006) Biological control of the root-knot nematode *Meloidogyne incognita* by *Paecilomyces lilacinus* strain 251. Biol Control 38:179–187

King EB, Parke JL (1993) Biocontrol of *Aphanomyces* root rot and *Pythium* damping-off by *Pseudomonas cepacia* AMMD on four pea cultivars. Plant Dis 77:1185–1188

Kirk JJ, Deacon JW (1987) Control of the take-all fungus by *Microdochium bolleyi*, and interactions involving *M. bolleyi*, *Phialophora graminicola* and *Periconia macrospinosa* on cereal roots. Plant Soil 98:231–237

Kirkland BH, Eisa A, Keyhani NO (2005) Oxalic acid as a fungal acaricidal virulence factor. J Med Entomol 42:346–351

Klein MG (1992) Use of *Bacillus popilliae* in scarab control. In: Jackson TA, Glare TR (eds) Use of pathogens in scarab pest management. Intercept, Andover, UK, pp 179–189

Klein MG, Jackson TA (1992) Bacterial diseases of scarabs. In: Jackson TJ, Glare T (eds) Use of pathogens in scarab pest management. Intercept, Andover, UK, pp 43–61

Knudsen GR, Eschen DJ, Dandurand LM, Bin L (1991) Potential for control of *Sclerotinia sclerotiorum* through colonization of sclerotia by *Trichoderma harzianum*. Plant Dis 75:466–470

Kraus J, Loper JE (1995) Characterization of a genomic region required for production of the antibiotic pyoluteorin by the biological control agent *Pseudomonas fluorescens* Pf-5. Appl Environ Microbiol 61:849–854

Kredics L, Antal Z, Manczinger L, Szekeres A, Kevei F, Nagy E (2003) Influence of environmental parameters on *Trichoderma* strains with biocontrol potential. Food Technol Biotechnol 41:37–42

Kredics L, Manczinger L, Antal Z, Pénzes Z, Szekeres A, Kevei F, Nagy E (2004) In vitro water activity and pH dependence of mycelial growth and extracellular enzyme activities of *Trichoderma* strains with biocontrol potential. J Appl Microbiol 96:491–498

Kubicek CP, Mach RL, Peterbauer CK, Lorito M (2001) *Trichoderma*: from genes to biocontrol. J Plant Pathol 83:11–23

Laing SAK, Deacon JW (1991) Video microscopical comparison of mycoparasitism by *Pythium oligandrum, P. nunn* and an unnamed *Pythium* species. Mycol Res 95:469–479

Larkin RP, Fravel DR (2002) Effects of varying environmental conditions on biological control of *Fusarium* wilt of tomato by nonpathogenic *Fusarium* spp. Phytopathology 92:1160–1166

Larkin RP, Hopkins DL, Martin FN (1996) Suppression of *Fusarium* wilt of watermelon by nonpathogenic *Fusarium oxysporum* and other microorganisms recovered from a disease-suppressive soil. Phytopathology 86:812–819

Lartey RT, Conway KE (2004) Novel considerations in biological control of plant pathogens: microbial interactions. In: Lartey RT, Caesar AJ (eds) Emerging concepts in plant health management. Research Signpost, Trivandrum, India, pp 141–157

Lascaris D, Deacon JW (1991) Colonization of wheat roots from seed-applied spores of *Idriella* (*Microdochium*) *bolleyi*: a biocontrol agent of take-all. Biocontrol Sci Technol 1:229–240

Lazarovits G (2001) Management of soil-borne plant pathogens with organic soil amendments: a disease control strategy salvaged from the past. Can J Plant Pathol 23:1–7

Lehr NA, Schrey SD, Hampp R, Tarkka MT (2008) Root inoculation with a forest soil streptomycete leads to locally and systemically increased resistance against phytopathogens in Norway spruce. New Phytol 177:965–976

Lemenceau P, Bakker PAHM, De Kogel WJ, Alabouvette C, Schippers B (1993) Antagonistic effect of nonpathogenic *Fusarium oxysporum* strain Fo47 and pseudobactin 358 upon pathogenic *Fusarium oxysporum* f. sp. *dianthi*. Appl Environ Microbiol 59:74–82

Li DP, Holdom DG (1995) Effects of nutrients on colony formation, growth, and sporulation of *Metarhizium anisopliae* (Deuteromycotina: Hyphomycetes). J Invertebr Pathol 65:253–260

Ligon JM, Hill DS, Hammer PE, Torkewitz NR, Hofmann D, Kempf HJ, van Pee, KH (2000) Natural products with antifungal activity from *Pseudomonas* biocontrol bacteria. Pest Manage Sci 56:688–695

Litterick AM, Harrier LA, Wallace P, Watson CA, Wood M (2004) The role of uncomposted materials, composts, manures and compost extracts in reducing pest and disease incidence and severity in sustainable temperate agricultural and horticultural crop production – A review. Crit Rev Plant Sci 23:453–479

Lodha S (1995) Soil solarization, summer irrigation and amendments for the control of *Fusarium oxysporum* f.sp. *cumini* and *Macrophomina phaseolina* in arid soils. Crop Prot 14:215–219

Loper JE (1988) Role of fluorescent siderophore production in biological control of *Pythium ultimum* by a *Pseudomonas fluorescens* strain. Phytopathology 78:166–172

Lorang JM, Tuori RP, Martinez JP, Sawyer TL, Redman RS, Rollins JA, Wolpert TJ, Johnson KB, Rodriguez RJ, Dickman MB, Ciuffetti LM (2001) Green fluorescent protein is lighting up fungal biology. Appl Environ Microbiol 67:1987–1994

Lorito M (1998) Chitnolytic enzymes and their genes. In: Harman GE, Kubicek CP (eds) *Trichoderma* and *Gliocladium*. Taylor & Francis, London, pp 73–99

Lumsden RD, Knauss JF (2007) Commercial development of Gliocladium virens for damping-off disease. In: Vincent C, Goettel MS, Lazarovits G (eds) Biological control: a global perspective. CAB International, Kew, UK, pp 203–210

Lumsden RD, Locke JC, Adkins ST, Walter JF, Ridout CJ (1992) Isolation and localisation of the antibiotic gliotoxin produced by *Gliocladium virens* from alginate prill in soil and soilless media. Phytopathology 82:230–235
Lumsden RD, Lewis JA, Fravel D (1995) Formulation and delivery of biocontrol agents for use against soilborne plant pathogens. In: Hall FR, Barry JW (eds) Biorational pest control agents: formulation and delivery. American Chemical Society, Washington, DC, pp 166–182
Lumsden RD, Walter JF, Baker CP (1996) Development of *Gliocladium virens* for damping-off control. Can J Plant Pathol 18:463–468
Luth P (2001) The control of *Sclerotinia* spp. and *Sclerotium cepivorum* with the biological fungicide Contrans(R)WG – experiences from field trials and commercial use. In: Proceedings of the XI international *Sclerotinia* workshop, Central Science Laboratory, York, England, pp 37–38
MacSpadden Gardener BB, Weller DM (2001) Changes in populations of rhizosphere bacteria associated with Take-all disease of wheat. Appl Environ Microbiol 67:4414–4425
Maniania NK, Ekesi S, Songa JM (2002) Managing termites in maize with the entomopathogenic fungus *Metarhizium anisopliae*. Insect Sci Appl 22:41–46
Manocha MS, Govindsamy V (1998) Chitinolytic enzymes of fungi and their involvement in biocontrol of plant pathogens. In: Boland GJ, Kuykendall LD (eds) Plant-microbe interactions and biological control. Marcel Dekker, New York, pp 309–327
Marin S, Sanchis V, Ramos AJ, Vinas I, Magan N (1998) Environmental factors, *in vitro* interactions, and niche overlap between *Fusarium moniliforme, F. proliferatum*, and *F. graminearum, Aspergillus* and *Penicillium* species from maize grain. Mycol Res 102:831–837
Marrone PG (1994) Present and future use of *Bacillus thuringiensis* in integrated pest management systems: an industrial perspective. Biocontrol Sci Technol 4:517–526
Martin FN (2003) Development of alternative strategies for management of soilborne pathogens currently controlled with methyl bromide. Annu Rev Phytopathol 41:325–350
McKellar ME, Nelson EB (2003) Compost-induced suppression of *Pythium* damping-off is mediated by fatty-acid-metabolizing seed-colonizing microbial communities. Appl Environ Microbiol 69:452–460
McLean KL, Swaminathan J, Frampton CM, Hunt JS, Ridgway HJ, Stewart A (2005) Effect of formulation on the rhizosphere competence and biocontrol ability of *Trichoderma atroviride* C52. Plant Pathol 54:212–218
Meera MS, Shivanna MB, Kageyama K, Hyakumachi M (1994) Plant growth promoting fungi from zoysiagrass rhizosphere as potential inducers of systemic resistance in cucumbers. Phytopathology 84:1399–1406
Melzer MS, Boland GJ (1996) Transmissible hypovirulence in *Sclerotinia minor*. Can J Plant Pathol 18:19–28
Mennan S, Chen SY, Melakeberhan H (2006) Suppression of *Meloidogyne hapla* populations by *Hirsutella minnesotensis*. Biocontrol Sci Technol 16:181–193
Mennan S, Chen SY, Melakeberhan H (2007) Effects of *Hirsutella minnesotensis* and N-Viro Soil on populations of *Meloidogyne hapla*. Biocontrol Sci Technol 17:233–246
Metting FB (1993) Structure and physiological ecology of soil microbial communities. In: Metting FB (ed) Soil microbial ecology. Marcel Dekker, New York, pp 3–25
Meyer SLF, Massoud SI, Chitwood DJ, Roberts DP (2000) Evaluation of *Trichoderma virens* and *Burkholderia cepacia* for antagonistic activity against root-knot nematode, *Meloidogyne incognita*. Nematology 2:871–879
Meyling NV, Eilenberg J (2006) Occurrence and distribution of soil-borne entomopathogenic fungi within a single organic agroecosystem. Agric Ecosyst Environ 113:336–341
Meyling NV, Eilenberg J (2007) Ecology of the entomopathogenic fungi *Beauveria bassiana* and *Metarhizium anisopliae* in temperate agroecosystems: potential for conservation biological control. Biol Control 43:145–155
Meyling NV, Lübeck M, Buckley EP, Eilenberg J, Rehner SA (2009) Community composition, host range and genetic structure of the fungal entomopathogen *Beauveria* in adjoining agricultural and seminatural habitats. Mol Ecol 18:1282–1293

Mietkiewski RT, Pell JK, Clark SJ (1997) Influence of pesticide use on the natural occurrence of entomopathogenic fungi in arable soils in the UK: field and laboratory comparisons. Biocontrol Sci Technol 7:565–575

Milner RJ, Staples JA, Lutton GG (1998) The selection of an isolate of the hyphomycete fungus, *Metarhizium anisopliae*, for control of termites in Australia. Biol Control 11:240–247

Milner RJ, Samson P, Morton R (2003) Persistence of conidia of *Metarhizium anisopliae* in sugarcane fields: effect of isolate and formulation on persistence over 3.5 years. Biocontrol Sci Technol 13:507–516

Minuto A, Spadaro D, Garibaldi A, Gullino ML (2004) Control of soilborne pathogens of tomato using a commercial formulate of *Streptomyces griseoviridis* and solarization. Crop Prot 25:468–475

Moorhouse ER, Gillespie AT, Charnley AK (1992a) Effect of potting media on the control of *Otiorhynchus sulcatus* larvae on outdoor strawberry plants using the entomogenous fungus *Metarhizium anisopliae*. Biol Control 2:238–243

Moorhouse ER, Charnley AK, Gillespie AT (1992b) A review of the biology and control of the vine weevil, *Otiorhynchus sulcatus* (Coleoptera: Curculionidae). Ann Appl Biol 121:431–454

Moorhouse ER, Easterbrook MA, Gillespie AT, Charnley AK (1993a) Control of *Otiorhynchus sulcatus* (Fabricius) (Coleoptera: Curculionidae) larvae on a range of hardy ornamental nursery stock species using the entomogenous fungus *Metarhizium anisopliae*. Biocontrol Sci Technol 3:63–72

Moorhouse ER, Easterbrook MA, Gillespie AT, Charnley AK (1993b) The development of *Otiorhynchus sulcatus* (Fabricius) (Coleoptera: Curculionidae) larvae on a range of ornamental pot plant species and the potential for control using *Metarhizium anisopliae*. J Hortic Sci 68:627–635

Moorhouse ER, Gillespie AT, Charnley AK (1993c) Application of *Metarhizium anisopliae* (Metsch.) Sor. conidia to control *Otiorhynchus sulcatus* (F.) (Coleoptera: Curculionidae) larvae on glasshouse pot plants. Ann Appl Biol 122:623–636

Murphy PJ, Roberts WP (1979) A basis for agrocin 84 sensitivity in *Agrobacterium radiobacter*. J Gen Microbiol 114:207–213

Murphy JA, Sun S, Betts LL (1993) Endophyte-enhanced resistance to billbug (Coleoptera: Curculionidae), sod webworm (Lepidoptera: Pyralidae), and white grub (Coleoptera: Scarabaeidae) in tall fescue. Environ Entomol 22:699–703

Nakasaki K, Hiraoka H, Nagata H (1998) A new operation for producing disease-suppressive compost from grass clippings. Appl Environ Microbiol 64:4015–4020

Navon A (1993) Control of lepidopteran pests with *Bacillus thuringiensis*. In: Entwistle PF, Cory JS, Bailey MJ et al (eds) *Bacillus thuringiensis*, an environmental biopesticide: theory and practice. Wiley, New York, pp 125–146

Nel B, Steinberg C, Labuschagne N, Viljoen A (2006) The potential of nonpathogenic *Fusarium oxysporum* and other biological control organisms for suppressing *Fusarium* wilt of banana. Plant Pathol 55:217–223

Noble R, Coventry E (2005) Suppression of soil-borne plant diseases with composts: a review. Biocontrol Sci Technol 15:3–20

Nuss DL, Koltin Y (1990) Significance of dsRNA genetic elements in plant pathogenic fungi. Annu Rev Phytopathol 28:37–58

O'Callaghan M, Brownbridge M (2009) Environmental impacts of microbial control agents used for control of invasive pests. In: Hajek AE, Glare T, O'Callaghan M (eds) Use of microbes for control and eradication of invasive arthropods. Springer Science and Business Media BV, The Netherlands, pp 305–327

O'Callaghan M, Jackson TA (1993) Isolation and enumeration of *Serratia entomophila* – a bacterial pathogen of the New Zealand grass grub, *Costelytra zealandica*. J Appl Bacteriol 75:307–314

Oerke EC, Dehne H-W, Schonbeck F, Weber A (eds) (1994) Crop production and crop protection: estimated losses in major food and cash crops. Elsevier, Amsterdam, The Netherlands, pp. 808

Oi DH, Williams DF (2002) Impact of *Thelohania solenopsae* (Microsporidia: Thelohaniidae) on polygyne colonies of red imported fire ants (Hymenoptera: Formicidae). J Econ Entomol 95:558–562

Olivain C, Alabouvette C (1999) Process of tomato root colonization by a pathogenic strain of *Fusarium oxysporum* f.sp. *lycopersici* in comparison with a non-pathogenic strain. New Phytol 141:497–510

Ovreas L, Torsvik V (1998) Microbial diversity and community structure in two different agricultural soil communities. Microb Ecol 36:303–315

Ownley BH, Weller DM, Thomashow LS (1992) Influence of *in situ* and *in vitro* pH on suppression of *Gaeumannomyces graminis* var. *tritici* by *Pseudomonas fluorescens* 2-79. Phytopathology 82:178–184

Ownley BH, Duffy BK, Weller DM (2003) Identification and manipulation of soil properties to improve the biological control performance of phenazine-producing *Pseudomonas fluorescens*. Appl Environ Microbiol 69:3333–3343

Ownley BH, Pereira RM, Klingeman WE, Quigley NB, Leckie BM (2004) *Beauveria bassiana*, a dual purpose biocontrol organism, with activity against insect pests and plant pathogens. In: Ekesi S, Maniania NK (eds) Emerging concepts in plant health management 2004. Research Signpost, Trivandrum, India, pp 255–269

Ozbay N, Newman SE (2004) Biological control with *Trichoderma* spp. with emphasis on *T. harzianum*. Pak J Biol Sci 7:478–484

Padmavathi J, Devi KU, Rao CUM (2003) The optimum and tolerance pH range is correlated to colonial morphology in isolates of the entomopathogenic fungus *Beauveria bassiana* – a potential biopesticide. World J Microbiol Biotechnol 19:469–477

Papavizas GC (1985) *Trichoderma* and *Gliocladium*: biology, ecology and potential for biocontrol. Annu Rev Phytopathol 23:23–54

Parkman JP, Smart GC (1996) Entomopathogenic nematodes, a case study: introduction of *Steinernema scapterisci* in Florida. Biocontrol Sci Technol 6:413–419

Pérez-Rodriguez I, Doronteo-Mendoza A, Franco-Navarro F, Santiago-Santiago V, Montero-Pineda A (2007) Isolates of *Pochonia chlamydoporia* var. *chlamydosporia* from Mexico as potential biological control agents of *Nacobbus aberrans*. Nematropica 37:127–134

Pharand B, Carisse O, Benhamou N (2002) Cytological aspects of compost-mediated induced resistance against *Fusarium* crown and root rot in tomato. Phytopathology 92:424–438

Pierson EA, Weller DM (1994) Use of mixtures of fluorescent pseudomonads to suppress take-all and improve the growth of wheat. Phytopathology 84:940–947

Preston GM (2004) Plant perceptions of plant growth-promoting *Pseudomonas*. Philos Trans R Soc Lond B Biol Sci 359:907–918

Quesada-Moraga E, Carrasco-Diaz JA, Santiago-Alvarez C (2006) Insecticidal and antifeedant activities of proteins secreted by entomopathogenic fungi against *Spodoptera littoralis* (Lep., Noctuidae). J Appl Entomol 130:442–452

Quesada-Moraga E, Navas-Cortes JA, Maranhao EAA, Ortiz-Urquiza A, Santiago-Alvarez C (2007) Factors affecting the occurrence and distribution of entomopathogenic fungi in natural and cultivated soils. Mycol Res 111:947–966

Quintela ED, McCoy CW (1998) Synergistic effect of imidacloprid and two entomopathogenic fungi on the behavior and survival of larvae of *Diaprepes abbreviatus* (Coleoptera: Curculionidae) in soil. J Econ Entomol 91:110–122

Ramle M, Wahid MB, Norman K, Glare TR, Jackson TA (2005) The incidence and use of *Oryctes* virus for control of rhinoceros beetle in oil palm plantations in Malaysia. J Invertebr Pathol 89:85–90

Rangel DEN, Alston DG, Roberts DW (2008) Effects of physical and nutritional stress conditions during mycelial growth on conidial germination speed, adhesion to host cuticle and virulence of *Metarhizium anisopliae*, an entomopathogenic fungus. Mycol Res 112:1355–1361

Rath AC (2002) Ecology of entomopathogenic fungi in field soils. In: Proceedings of the meeting of the society for invertebrate pathology, Iguassu Falls, Brazil, pp 65–71

Rath AC, Bullard GK (1997) Persistence of *Metarhizium anisopliae* DAT F-001 in pasture soils for 7.5 years – implications for sustainable soil-pest management. In: Allsopp PG, Rogers DJ, Robertson LN (eds) Soil invertebrates in 1997. Bureau of Sugar Experimental Stations, Brisbane, Australia, pp 78–80

Rath AC, Koen TB, Yip HY (1992) The influence of abiotic factors on the distribution and abundance of *Metarhizium anisopliae* in Tasmanian pasture soils. Mycol Res 96:378–384

Rath AC, Worledge D, Koen TB, Rowe BA (1995) Long-term field efficacy of the entomogenous fungus *Metarhizium anisopliae* against the subterranean scarab, *Adoryphorus couloni*. Biocontrol Sci Technol 5:439–451

Raymond RN, Muller G, Matzanke F (1984) Complexation of iron by sideophores. A review of their solution and structural chemistry and biological function. Top Curr Chem 123:49–102

Redmond CT, Potter DA (1995) Lack of efficacy of *in vitro* and putatively *in vitro-* produced *Bacillus popilliae* against field populations of Japanese beetle (Coleoptera: Scarabaeidae) grubs in Kentucky. J Econ Entomol 88:846–854

Reinecke P, Fokkema NJ (1981) An evaluation of methods of screening fungi from the haulm base of cereals for antagonism to *Pseudocercosporella herpotrichoides* in wheat. T Brit Mycol Soc 77:343–350

Ristaino JB, Perry KB, Lumsden RD (1991) Effect of solarization and *Gliocladium virens* on sclerotia of *Sclerotium rolfsii*, soil microbiota, and the incidence of southern blight of tomato. Phytopathology 81:1117–1124

Rovira AD, Ryder MH, Harris AR (1992) Biological control of root diseases with pseudomonads. In: Tjamos ES, Papavizas GC, Cook RJ (eds) Biological control of plant diseases – Progress and challenges for the future. Plenum, New York, pp 175–184

Ryder MH, Jones DA (1990) Biological control of crown gall. In: Hornby D (ed) Biological control of soilborne plant pathogens. CAB International, Wallingford, UK, pp 45–63

Schippers B, Geels FP, Hoekstra O, Lamers JG, Maenhout CAAA, Scholte K (1985) Yield depressions in narrow rotations caused by unknown microbial factors and their suppression by selected pseudomonads. In: Parker CA, Rovira AD, Moore KJ et al (eds) Ecology and management of soilborne plant pathogens. The American Phytopathological Society, St. Paul, MN, pp 127–130

Schrey SD, Tarkka MT (2008) Friends and foes: streptomycetes as modulators of plant disease and symbiosis. Antonie Leeuwenhoek 94:11–19

Schwarzenbach K, Widmer F, Enkerli J (2007a) Cultivation-independent analysis of fungal genotypes in soil by using simple sequence repeat markers. Appl Environ Microbiol 73:6519–6525

Schwarzenbach K, Enkerli J, Widmer F (2007b) Objective criteria to assess representativity of soil fungal community profiles. J Microbiol Meth 68:358–366

Sequeira L (1962) Influence of organic amendments on survival of *Fusarium oxysporum* f. *cubense* in the soil. Phytopathology 52:976–982

Serra-Wittling C, Houot S, Alabouvette C (1996) Increased soil suppressiveness to *Fusarium* wilt of flax after addition of municipal soild waste compost. Soil Biol Biochem 28:1207–1214

Shah FA, Ansari MA, Prasad M, Butt TM (2007) Evaluation of black vine weevil (*Otiorhynchus sulcatus*) control strategies using *Metarhizium anisopliae* with sublethal doses of insecticides in disparate horticultural growing media. Biol Control 40:246–252

Shah FA, Gaffney M, Ansari MA, Prasad M, Butt TM (2008) Neem seed cake enhances the efficacy of the insect pathogenic fungus *Metarhizium* anisopliae for the control of black vine weevil, *Otiorhynuchs sulcatus* (Coleoptera: Curculionidae). Biol Control 44:111–115

Shapiro-Ilan DI, Reilly CC, Hotchkiss MW, Wood BW (2002) The potential for enhanced fungicide resistance in *Beauveria bassiana* through strain discovery and artificial selection. J Invertebr Pathol 81:86–93

Shapiro-Ilan DI, Jackson MA, Reilly CC, Hotchkiss MW (2004) Effects of combining and entomopathogenic fungi or bacterium with entomopathogenic nematodes on mortality of *Curculio caryae* (Coleoptera: Curculionidae). Biol Control 30:119–126

Sharon E, Bar-Eyal M, Chet I, Herrera-Estrella A, Kleifeld O, Spiegel Y (2001) Biological control of the root-knot nematode *Meloidogyne javanica* by *Trichoderma harzianum*. Phytopathology 91:687–693

Shishido M, Miwa C, Usami T, Amemiya Y, Johnson KB (2005) Biological control efficiency of *Fusarium* wilt of tomato by nonpathogenic *Fusarium oxysporum* Fo-B2 in different environments. Phytopathology 95:1072–1080

Siddiqui IA, Shaukat SS (2003) Suppression of root-knot disease by *Pseudomonas fluorescens* CHA0 in tomato: importance of bacterial secondary metabolite, 2, 4-diacetylpholoroglucinol. Soil Biol Biochem 35:1615–1623

Siddiqui IA, Haas D, Heeb S (2005) Extracellular protease of *Pseudomonas fluorescens* CHA0, a biocontrol factor with activity against the root-knot nematode *Meloidogyne incognita*. Appl Environ Microbiol 71:5646–5649

Singh KP, Jaiswal RK, Kumar N, Kumar D (2007) Nematophagous fungi associated with root galls of rice caused by *Meloidogyne graminicola* and its control by *Arthrobotrys dactyloides* and *Dactylaria brochopaga*. J Phytopathol 155:193–197

Sivan A, Chet I (1989) The possible role of competition between *Trichoderma harzianum* and *Fusarium oxysporum* on rhizosphere colonization. Phytopathology 79:198–203

Sivasithamparam K, Ghisalberti EL (1998) Secondary metabolism in *Trichoderma* and *Gliocladium*. In: Kubicek CP, Harman GE, Ondik KL (eds) *Trichoderma* and *Gliocladium*: basic biololgy, taxonomy and genetics. CRC Press, Boca Raton, FL, pp 139–191

Slininger PJ, Shea-Wilbur MA (1995) Liquid-culture pH, temperature, and carbon (not nitrogen) source regulate phenazine productivity of the take-all biocontrol agent *Pseudomonas fluorescens* 2–79. J Appl Microbiol Biotechnol 43:194–800

Somasekhar N, Grewal PS, EABd N, Stinner BR (2002) Non-target effects of entomopathogenic nematodes on the soil nematode community. J Appl Ecol 39:735–744

Stevens C, Khan VA, Rodriguez-Kabana R, Ploper LD, Backman PA, Collins DJ, Brown JE, Wilson MA, Igwegbe ECK (2003) Integration of soil solarization with chemical, biological and cultural control for the management of soilborne diseases of vegetables. Plant Soil 253:493–506

Stewart A (2001) Commercial biocontrol- reality or fantasy. Australas Plant Pathol 30:127–131

Stewart A, McLean KL (2007) Biological control of onion white rot. In: Mukerji KG, Chincholkar SB (eds) Biological control of plant disease. The Haworth Press, New York, pp 123–149

Steyaert JM, Ridgway HJ, Elad Y, Stewart A (2003) Genetic basis of mycoparasitism: a mechanism of biological control by species of *Trichoderma*. N Z J Crop Hortic Sci 31:281–291

Strashnow Y, Elad Y, Sivan A, Chet I (1985) Integrated control of *Rhizoctonia solani* by methyl bromide and *Trichoderma harzianum*. Plant Pathol 34:146–151

Strasser H, Vey A, Butt TM (2000) Are there any risks in using entomopathogenic fungi for pest control, with particular reference to the bioactive metabolites of *Metarhizium*, *Tolypocladium* and *Beauveria* species? Biocontrol Sci Technol 10:717–735

Sulistyanto D, Ehlers RU (1996) Efficacy of the entomopathogenic nematodes *Heterorhabditis megidis* and *Heterorhabditis bacteriophora* for the control of grubs (*Phyllopertha horticola* and *Aphodius contaminatus*) in golf course turf. Biocontrol Sci Technol 6:247–250

Sun JZ, Fuxa JR, Henderson G (2003) Effects of virulence, sporulation, and temperature on *Metarhizium anisopliae* and *Beauveria bassiana* laboratory transmission in *Coptotermes formosanus*. J Invertebr Pathol 84:38–46

Swaminathan J, Bunt CR, Jackson TA (2008) Coating technology to enhance the shelf life of probiotic bacteria on cereals. In: Proceedings of the IPA world congress, Beverley Hills Hilton, Beverley Hills, CA

Szczech M, Shoda M (2006) The effect of mode of application of *Bacillus subtilis* RB14-C on its efficacy as a biocontrol agent against *Rhizoctonia solani*. J Phytopathol 154:370–377

Tesfamariam M, Kiewnick S, Sikora R (2009) Endophytic bacteria from Ethiopian coffee plants and their potential to antagonise *Meloidogyne incognita*. Nematology 11:117–127

Thomashow LS, Weller DM (1988) Role of a phenazine antibiotic from *Pseudomonas fluorescens* 2-97 in biological control of *Gaeumannomyces graminis* var. *tritici*. J Bacteriol 170:3499–3508

Thompson SR, Brandenburg RL (2005) Tunneling responses of mole crickets (Orthoptera: Gryllotalpidae) to the entomopathogenic fungus, *Beauveria bassiana*. Environ Entomol 34:140–147

Thompson SR, Brandenburg RL, Arends JJ (2006) Impact of moisture and UV degradation on *Beauveria bassiana* (Balsamo) Vuillemin conidial viability in turfgrass. Biol Control 39:401–407

Tian B, Yang J, Zhang K (2007) Bacteria used in the biological control of plant-parasitic nematodes: populations, mechanisms of action, and future prospects. FEMS Microbiol Ecol 61:197–213

Timmusk S, Wagner EGH (1999) The plant-growth-promoting *Rhizobacterium Paenibacillus polymyxa* Induces Changes in *Arabidopsis thaliana* Gene Expression: a Possible Connection Between Biotic and Abiotic Stress Responses. Mol Plant Microbe Interact 12:951–959

Tjeerd van Rij E, Wesselink M, Chin-A-Woeng TFC, Bloemberg GV, Lugtenberg BJJ (2004) Influence of environmental conditions on the production of phenazine-1-carboxamide by *Pseudomonas chlororaphis* PCL1391. Mol Plant Microbe Interact 17:557–566

Toussoun TA, Menzinger W, Smith RS Jr (1969) Role of conifer litter in ecology of *Fusarium*: stimulation of germination in soil. Phytopathology 59:1396–1399

Townsend RJ, Glare TR, Willoughby BE (1995) The fungi *Beauveria* spp. cause epizootics in grass grub populations in Waikato. NZ Plant Protection 48:237–241

Townsend RJ, Ferguson CM, Proffitt JR, Slay MWA, Swaminathan J, Day S, Gerard E, O'Callaghan M, Johnson VW, Jackson TA (2004) Establishment of *Serratia entomophila* after application of a new formulation for grass grub control. NZ Plant Protection 57:310–313

Toyota K, Ritz K, Young IM (1996) Microbiological factors affecting the colonisation of soil aggregates by *Fusarium oxysporum* f. sp. *raphani*. Soil Biol Biochem 28:1513–1521

Traugott M, Weissteiner S, Strasser H (2005) Effects of the entomopathogenic fungus *Beauveria brongniartii* on the non-target predator *Poecilus versicolor* (Coleoptera: Carabidae). Biol Control 33:107–112

Trillas MI, Casanova E, Cotxarrera L, Ordovas J, Borrero C, Aviles M (2006) Composts from agricultural waste and the *Trichoderma asperellum* strain T-34 suppress *Rhizoctonia solani* in [illegible] Biol Control 39:32–38

Trillas-Gay MI, Hoitink HAJ, Madden LV (1986) Nature of suppression of [illegible] radish in a container medium amended with composted hardwood bark. Plant Dis 70: 1023–1027

Trudgill DL, Blok VC, Bala G, Daudi A, Davies KG, Gowen SR, Fargette M, Madulu JD, Mateille T, Mwageni W, Netscher C, Phillips MS, Sawadogo A, Trivino CG, Voyoukallou E (2000) The importance of tropical root-knot nematodes (*Meloidogyne* spp.) and factors affecting the utility of *Pasteuria penetrans* as a biocontrol agent. Nematology 2:823–845

Vagelas IK, Pembroke B, Gowen SR, Davies KG (2007) The control of root-knot nematodes (*Meloidogyne* spp.) by *Pseudomonas oryzihabitans* and its immunological detection on tomato roots. Nematology 9:363–370

Van Driesche RG, Bellows TSJ (1996) Biological control. Chapman & Hall, New York, p 539

Van Peer R, Schippers B (1992) Lipopolysacchamdes of plant growth promoting *Pseudomanas* sp. strain WCC5417r induce resistance in carnations to *Fusarium* wilt. Neth J Plant Pathol 98:129–139

Vanninen I, Tyni-Juslin J, Hokkanen H (2000) Persistence of augmented *Metarhizium anisopliae* and *Beauveria bassiana* in Finnish agricultural soils. Biocontrol 45:201–222

Vega FE, Posada F, Aime MC, Pava-Ripoll M, Infante F, Rehner SA (2008) Entomopathogenic fungal endophytes. Biol Control 46:72–82

Vercambre B, Goebel O, Riba G, Marchal M, Neuvéglise C, Ferron P (1994) Success in microbial control of a soil pest, *Hoplochelus marginalis*, in Reunion Island; choice of a suitable fungus. In: Proceedings of the VIth international colloquium on invertebrate pathology and microbial control, SIP, Montpellier, France, pp 283–288

Veseley D, Hejdanek S (1984) Microbial relations of *Pythium oligandrum* and problems in the use of this organism for the biological control of damping-off in sugar beet. Zbl Mikrobiol 139:257–265

Vey A, Hoagland RE, Butt TM (2001) Toxic metabolites of fungal biocontrol agents. In: Butt TM (ed) Fungi as biocontrol agents: progress, problems and potential. CABI Publishing, Wallingford, UK, pp 311–346

Villani MG, Krueger SR, Schroeder PC, Consolie F, Consolie NH, Preston-Wilsey LM, Roberts DW (1994) Soil application effects of *Metarhizium anisopliae* on Japanese beetle (Coleoptera: Scarabaeidae) behavior and survival in turfgrass microcosms. Environ Entomol 23:502–513

Vinale F, Krishnapillai S, Ghisalberti EL, Marra R, Woo SL, Lorito M (2008) *Trichoderma*-plant-pathogen interactions. Soil Biol Biochem 40:1–10

Vincent C, Goettel MS, Lazarovits G (2007) Biological control: a global perspective. CAB International, Wallingford, UK, p 440

Visnovsky GA, Smalley DJ, O'Callaghan M, Jackson TA (2008) Influence of culture medium composition, dissolved oxygen concentration and harvesting time on the production of Serratia entomophila, a microbial control agent of the New Zealand grass grub. Biocontrol Sci Technol 18:87–100

Wakelin SA, Sivasithamparam K, Cole ALJ, Skipp RA (1999) Saprophytic growth in soil of a strain of *Trichoderma koningii*. N Z J Agric Res 42:337–345

Wang C, St Leger RJ (2007) A scorpion neurotoxin increases the potency of a fungal insecticide. Nat Biotechnol 25:1455–1456

Wang C, Fan M, Li Z, Butt TM (2004) Molecular monitoring and evaluation of the application of the insect-pathogenic fungus *Beauveria bassiana* in southeast China. J Appl Microbiol 96:861–870

Weindling R (1932) *Trichoderma lignorum* as a parasite of other soils fungi. Phytopathology 11:273–292

Weindling R (1941) Experimental consideration of the mold toxins of *Gliocladium* and *Trichoderma*. Phytopathology 31:991–1003

Weindling R, Emerson OH (1936) The isolation of a toxic substance from the culture filtrate of *Trichoderma*. Phytopathology 26:1068–1070

Weisbeek PJ, Gerrits H (1999) Iron and biocontrol. In: Stacey G, Keen NT (eds) Plant-microbe interactions. APS Press, St Paul, MN, pp 217–250

Weller DM (1988) Biological control of soil-borne plant pathogens in the rhizosphere with bacteria. Annu Rev Phytopathol 26:379–407

Weller DM (2007) Pseudomonas biological control agents of soil-borne pathogens: looking back over 30 years. Phytopathology 97:250–256

Weller DM, Raaijmakers JM, MacSpadden Gardener BB, Thomashow LS (2002) Microbial populations responsible for specific soil suppressiveness to plant pathogens. Annu Rev Phytopathol 40:309–348

Whipps JM (1997) Interactions between fungi and plant pathogens in soil and the rhizosphere. In: Gange AC, Brown VK (eds) Multitrophic interactions in terrestrial systems. Blackwell Science, Oxford, pp 47–65

Whipps JW, Gerlagh M (1993) Biology of *Coniothyrium minitans* and its potential for use in disease biocontrol. Mycol Res 96:897–907

Whipps J, Bennett A, Challen M, Hill R, Jiang D, Jones E, McQuilken M, Rinzema A, Rogers C, Stewart A, Tomprefa M (2004) Recent developments in inoculums production and application, ecology and pathogenicity in the biocontrol agent *Coniothyrium minitans*. Management of Plant Diseases and Arthropod pests by BCAs. IOBC/wprs Bull 27:281–284

Widden P, Cunningham J, Breil B (1988) Decomposition of cotton by *Trichoderma* species: influence of temperature, soil type, and nitrogen levels. Can J Microbiol 35:469–473

Wraight SP, Jackson MA, SLd K (2001) Production, stabilization and formulation of fungal biocontrol agents. In: Butt TM, Jackson C, Magan N (eds) Fungi as biocontrol agents: progress, problems and potential. CABI Publishing, Wallingford, UK, pp 253–287

Wright MS, Osbrink WLA, Lax AR (2002) Transfer of entomopathogenic fungi among Formosan subterranean termites and subsequent mortality. J Appl Entomol 126:20–23

Wright MS, Raina AK, Lax AR (2005) A strain of the fungus *Metarhizium anisopliae* for s controlling subterranean termites. J Econ Entomol 98:1451–1458

Zadworny M, Werner A, Idzikowska K (2004) Behaviour of the hyphae of *Laccaria laccata* in the presence of *Trichoderma harzianum in vitro*. Mycorrhiza 14:401–405

Zhang J-X, Howell CR, Starr JL (1996) Suppression of *Fusarium* colonization of cotton roots and *Fusarium* wilt by seed treatments with *Gliocladium virens* and *Bacillus subtilis*. Biocontrol Sci Technol 6:175–187

Zimmermann G (2007a) Review on safety of the entomopathogenic fungi *Beauveria bassiana* and *Beauveria brongniartii*. Biocontrol Sci Technol 17:553–596

Zimmermann G (2007b) Review on safety of the entomopathogenic fungus *Metarhizium anisopliae*. Biocontrol Sci Technol 17:879–920

Chapter 10
How Will Climate Change Impact Soil Microbial Communities?

Teri C. Balser, Jessica L.M. Gutknecht, and Chao Liang

Introduction

More than a century ago Svante Arrhenius predicted that continued combustion of fossil fuels would lead to a doubling of carbon dioxide in the atmosphere and associated climate warming (Arrhenius 1896). Despite this warning, we are now faced with the predicted doubling of atmospheric carbon dioxide and global temperature increase of 1.3°C by the end of this century if no policy changes are made (Cubasch et al. 2001). Furthermore, not only are we faced with rising global temperature but also shifting weather patterns, ocean acidification, and the potential loss of many species on earth (Intergovernmental Panel on Climate Change (IPCC) 2001). These factors will all have a marked impact on land use, land cover, soil quality, and productivity.

Climate change will have direct and indirect impacts on terrestrial ecosystems, both above- and belowground (Fig. 1). Aboveground, the effects of global change will be largely direct: elevated atmospheric carbon dioxide as well as changes in temperature, precipitation, and nitrogen availability will all result in changes to the abundance of plant species and altered land cover in unmanaged ecosystems (Tylianakis et al. 2008). In managed systems, changes in seasonal climate and precipitation will influence our choice of crop species, and the production of and stresses on those chosen species, which will in turn influence management decisions for irrigation, fertilization, and pathogen dynamics (Dixon 2009). Indirectly, land use, plant species composition, and plant productivity all feedback to further alter plant communities via changes in the belowground microbial community.

T.C. Balser(✉) and C. Liang
Department of Soil Science, University of Wisconsin-Madison, 1525 Observatory Drive, Madison, WI 53706, USA
e-mail: tcbalser@wisc.edu; chaoliang@wisc.edu

J.L.M. Gutknecht
Department of Soil Ecology, Helmholtz-Centre for Environmental Research-UFZ, Theodor-Lieser-Str, 4, D06120 Halle (Saale), Germany
e-mail: Jessica.gutknecht@ufz.de

G.R. Dixon and E.L. Tilston (eds.), *Soil Microbiology and Sustainable Crop Production*, DOI 10.1007/978-90-481-9479-7_10, © Springer Science+Business Media B.V. 2010

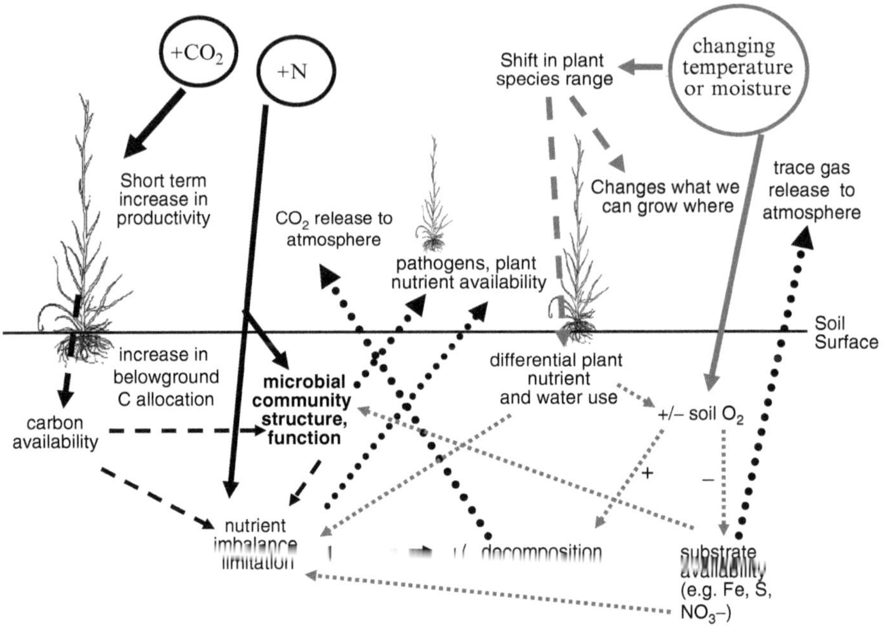

Fig. 1 Schematic of direct and indirect effects of global change drivers an below grand communities, including potential Feedbacks. Direct effects: elevated CO_2 or N deposition; Indirect effects: elevated CO_2 or N deposition; Direct effects: altered temperature or precipitation; Indirect effects: altered temperature or precipitation; Feedbacks to atmosphere or plant growth

Belowground, the response to climate change is more complex. Plant responses affect the type and amount of carbon entering the soil system as well as the physical architecture of the plant root zone. This has an indirect impact on microbial community composition and biomass. Water, temperature, and nitrogen all have direct effects on the microbial community as the organisms respond to temperature or drought stress along with altered resource availability. Microbial responses to climatic changes, whether direct or indirect, will feedback to influence plants via nutrient availability or pathogen productivity (Fig. 1). Thus a critical step in our understanding of ecosystem response to climate change must be to increase our understanding of the microbial community response.

However, too often researchers study ecosystem response to climate change without explicitly including the microbial communities. Nevertheless, through their diverse metabolic activities soil microbial communities are the major drivers of soil nutrient cycling and their importance in mediating climate change and ecosystem functioning should not be underestimated (Balser et al. 2001). Microbial metabolic activities can be altered by many ecosystem-scale factors such as climate change, nitrogen deposition, elevated carbon dioxide, or disturbance (Dhillion et al. 1996; Ajwa et al. 1999; Mayr et al. 1999). Shifts in microbial activity can in turn lead to changes in decomposition, nitrogen mineralization, organic carbon storage, and other ecosystem processes (Carreiro et al. 2000; Sinsabaugh et al. 2002; Henry et al. 2005a; Sowerby et al. 2005). Predicting microbial metabolic responses to these global changes and disturbance factors is thus essential for an understanding about how global change will affect soil and ecosystem function. For example, nitrogen deposition is likely to decrease mycorrhizal fungal biomass while increasing bacterial and saprotrophic fungal biomass (Treseder 2004; Rinnan et al. 2007) and has the potential to increase carbon cycling by increasing the activity of microbial enzymes related to carbon cycling (Henry et al. 2005a; Gutknecht et al. 2010). Elevated carbon dioxide may mitigate the effect of nitrogen deposition, but results have been highly varied (Treseder 2004, 2008). The response of microbial communities to climate change, and interactions between climate change and other global change factors, may be highly dependent on specific ecosystems and historical adaptation of the community (Rinnan et al. 2007; Balser et al. 2005; Fraterrigo et al. 2005, 2006).

In this chapter we will discuss the impact of climate change on belowground microbial community structure and function.

Microbial Response to Climate Change

As a preface to a more detailed discussion of microbial response to specific global change drivers later in the text, here we give an overview of microbial community climate change research. Until recently, the focus of most climate change research has been on how ecosystems respond to elevated atmospheric carbon dioxide. While it is relatively well established that elevated carbon dioxide changes belowground carbon allocation by plants (root biomass and root respiration), there is less consensus on the response of microbial communities (Zak et al. 2000a). Root symbionts such as mycorrhizal fungi and nitrogen fixing bacteria may (O'Neill 1994; Treseder and Allen 2000) or may not (Treseder and Allen 2000) increase in biomass under elevated carbon dioxide. Microbial total biomass and community structure also have shown highly varied responses to elevated carbon dioxide (Zak et al. 1993, 2000b; Diaz et al. 1993; Hungate et al. 1996; Lussenhop et al. 1998; Bruce et al. 2000; Montealegre et al. 2002), even between similar soils or between plant species (Hungate et al. 1996; Niklaus 1998). Over longer time spans especially, elevated carbon dioxide may have little direct effect on soil microbial communities (Niklaus et al. 2003). For instance, after 6 years of elevated carbon dioxide in a temperate grassland ecosystem, Niklaus et al. (2003) reported no major effect on microbial biomass, microbial community composition, or microbial carbon and nitrogen despite an increase in plant productivity and shift in soil aggregation (although there were small effects on microbial variables that changed each year of the study).

In addition to the effort to understand the microbial response to elevated carbon dioxide, other major drivers of global change, namely nitrogen deposition, elevated temperature, and altered precipitation have been the focus of much research. Nitrogen addition to unmanaged as well as managed ecosystems is consistently associated with a decrease in mycorrhizal fungal biomass (Balser et al. 2001;

Treseder 2004; Klironomos et al. 1997) as well as shifts in microbial community composition and decomposition rates (Balser et al. 2001; Carreiro et al. 2000; Henriksen and Breland 1999). Increased precipitation and hence increased soil moisture may result in increased microbial predation by mesofauna (Clarholm 1985; Kuikman et al. 1991; Taylor et al. 2004). It is also thought that the osmotic potential of the soil solution (soil salinity effects on water movement and availability) may play a more important role than absolute water content (Stark and Firestone 1995; Zak et al. 1999). Finally, increased temperature may increase microbial metabolic activity in temperate ecosystems (Zak et al. 1999; Contin et al. 2000; Vinolas et al. 2001), but not as much in northern ecosystems (Allison and Treseder 2008). Temperature increase is also predicted to shift carbon use toward old carbon or toward new carbon pools (MacDonald et al. 1995; Zogg et al. 1997; Andrews et al. 2000) or just generally increase soil respiration (Lin et al. 1999; Niinisto et al. 2004).

Experiments focused on manipulating single factors alone (such as those with elevated carbon dioxide, or altered moisture or temperature) offer valuable insight on fundamental responses, but there is a need for studies which examine the realistic future of global change: that of multiple, simultaneously interacting factors (Intergovernmental Panel on Climate Change (IPCC) 2007). Multiple factor, long-term studies may allow for a more mechanistic, predictive understanding of how microbial communities will respond to future global changes.

Those studies that are focused on interacting factors tend to focus on two general groups; (1) interactions between microbial resources (such as carbon and nitrogen), and (2) interactions between climate factors and resources (for instance, temperature with nitrogen addition). Many investigations have focused on the possible interaction between elevated carbon dioxide and nitrogen deposition, with less attention on how elevated carbon dioxide or nitrogen deposition may interact with climate change factors (temperature or precipitation).

Elevated carbon dioxide or nitrogen deposition may push a system toward nitrogen or carbon limitation (by increasing available soil carbon or nitrogen, respectively), so when both are elevated, the responses may mitigate each other (Hu et al. 1999; Lee et al. 2003; Schaeffer et al. 2003). For instance, elevated carbon dioxide lessens the positive effect of nitrogen addition on soil respiration and decomposition of litter (Lutze et al. 2000). Elevated carbon dioxide may also slow a decline in the abundance of saprotrophic fungi seen with nitrogen addition alone (Řezáčová et al. 2005). At the same time that effects of nitrogen deposition may be altered by elevated carbon dioxide, the effects of elevated carbon dioxide on plants and microorganisms may be dependent on soil nitrogen and fertility levels (thus, creating a cycle of nutrient limitation; Zak et al. 2000a; Lee et al. 2003; Tate and Ross 1997; Martin-Olmedo et al. 2002). Elevated nitrogen has been shown to slow the increase in carbon storage seen with elevated carbon dioxide alone (Tate and Ross 1997; van Groenigen et al. 2002), perhaps by increasing carbon cycling rates. The response of arbuscular mycorrhizal fungi (AMF) to elevated carbon dioxide may also depend on soil nutrient levels (Rillig and Field 2003) and determine whether AMF biomass becomes a carbon sink or decomposes organic matter and contributes to carbon cycling (Treseder and Allen 2000).

Climate change factors (e.g. elevated moisture or temperature) also interact with elevated carbon dioxide and nitrogen addition. For instance, increased moisture can increase the apparent effect of nitrogen addition by increasing the movement of nitrogen to the plant (Henry et al. 2005a). Or water may act in concert with nitrogen addition to increase decomposition of plant tissues (Henry et al. 2005b). Increased moisture, or alleviation of water stress, can also alter the lignification of plant cell walls (Henry et al. 2005b), increase grassland productivity, or impact soil carbon (Tate and Ross 1997). Soil moisture coupled with elevated carbon dioxide also decreases abundance of ammonium oxidizing bacteria, potentially altering the soil nitrogen cycle (Horz et al. 2004). Warming with elevated carbon dioxide may act additively to increase soil respiration (Niinisto et al. 2004; van Veen et al. 1991; Körner and Arnone 1992; Peterjohn et al. 1993; Johnson et al. 1994; Nakayama et al. 1994; Pajari 1995; Vose et al. 1995; Hungate et al. 1997). Although there have been few reports on the interactions between elevated temperature and moisture, there is evidence to suggest that together they may lead to shifts in the structure of methane oxidizing bacterial communities (Horz et al. 2005). In sum, while there are reported differences, it is difficult to synthesize the varied patterns seen among temperature or soil moisture and elevated carbon dioxide or nitrogen. There is also little information about how climate factors will interact with elevated carbon dioxide and/or nitrogen to affect microbial biomass or community structure (Pendall et al. 2004). More research is needed to generate a synthetic explanation and build usable conceptual models.

It is apparent from the research outlined above that there is no straightforward or clear model or prediction of how microbial communities will respond to future global climatic changes. In fact, several recent reviews have commented on our need for better general understanding of soil communities in order to understand soil feedbacks to future global change (Zak et al. 2000a; Pendall et al. 2004; Foley and Ramankutty 2004). In the next section of this chapter we will review microbial response to specific climate perturbations in more detail.

Response to Specific Changes in Climate

Impact of Increasing Temperature

As climate warms, microbial populations must acclimate or die. Many studies have shown an increase in microbial biomass in short-term experiments but over the long-term under elevated temperature biomass is more likely to decrease. This is because the efficiency of microbial growth changes at higher temperatures (Schimel et al. 2007; Hyvonen et al. 2005). Rather than biomass being directly correlated with decomposition, the reverse becomes true as organisms utilize labile carbon for energy production rather than biomass production (Contin et al. 2000; Zogg et al. 1997; Schimel et al. 2007). A specific instance is that higher temperatures alter cell membrane fluidity and permeability, requiring membrane lipid re-synthesis

(Petersen and Klug 1994). The high energetic cost of this stress response is one mechanism by which carbon can be utilized for energy instead of biomass. When the microbial energy demand exceeds the limit of labile carbon pools, biomass cannot be maintained and may decline at higher temperatures (Balser 2000; Balser and Firestone 2005). If however, the microbial community can access necessary labile carbon, then increased temperature will result in a shift in carbon allocation from growth to acclimation with a concomitant decrease in growth efficiency (i.e. increase in respiration per unit biomass; Schimel et al. 2007). In this case microbial biomass may be maintained rather than decline.

In one of the few long-term studies including elevated temperature that explicitly focused on the microbial community, Gutknecht (2007) considered 8 years of field-based data. They found that ambient inter-annual and seasonal variations in the microbial community were greater than effects of elevated temperature or moisture treatments. Overall there were few significant responses of microbial biomass or community structure to climatic treatments (elevated precipitation or temperature), with responses to elevated temperature being stronger than elevated precipitation. Importantly, however, these treatments did impact community process responses. Exposure to 8 years of elevated temperature or elevated precipitation in the field affected community responses to a short-term increase in substrate availability. In detail, Gutknecht (2007) reported a microbial response to elevated temperature that was similar to that obtained with nitrogen addition. Mycorrhizal abundance decreased, but other bacterial and general fungal microbial indicators increased in relative abundance. Elevated precipitation was related to lower relative abundance of a mycorrhizal indicator, but this trend was only statistically significant in one of 6 years of this study. While there were few significant effects from climate manipulation alone, temperature interacted to modify the effects of elevated carbon dioxide or nitrogen addition alone. Increased temperature may directly enhance arbuscular mycorrhizal (AM) colonization and development (Pendall et al. 2004; Fitter et al. 2000; Gavito et al. 2003), or may have the opposite effect. These changes in mycorrhizal fungal biomass may be important as mycorrhizas play such an important role in plant nutrition and plant, community, and ecosystem responses to global change.

Model systems can also be valuable for looking in detail at microbial responses to elevated temperature. Bardgett and Shine (1999), in a model ecosystem study, explored how a microbial community responded to elevated temperature over three plant generations. Microbial analyses revealed that biomass increased significantly, but only during the first plant generation. By the last plant generation, total microbial biomass was in decline. The initial increase in microbial biomass was likely due to fast-growing bacteria that initially responded to elevated temperature, while slower growing microbes such as fungi and actinomycetes were unaffected. Similarly, the relative abundance of Gram-positive and Gram-negative bacteria may increase with temperature, perhaps due to a shift in available substrates (Zogg et al. 1997), while fungal and actinomycete biomarkers may decline at higher temperatures (Waldrop and Firestone 2004). These studies together illustrate the importance of understanding how different microorganisms respond to elevated temperature, and how these varied responses impact on the timing and duration of the overall community response to elevated temperature or other potentially to other global changes.

Finally, there is a growing body of work indicating that the size and quality of the carbon pool accessed by microorganisms changes with varying temperature (MacDonald et al. 1995; Zogg et al. 1997; Andrews et al. 2000; Tison and Pope 1980; Linkins et al. 1984; Ellert and Bettany 1992; Nicolardot et al. 1994). Models that define decomposition or respiration rate coefficients (k) as a function of temperature assume constant substrate pool size and uniform substrate preference (Ellert and Bettany 1992). However, MacDonald et al. (1995) found that the pool size of carbon substrate available to the microbial community can vary substantially with temperature. In addition, several researchers have found that not only does the size of the carbon pool accessed change with temperature, but microbial use of specific substrates also changes (Zogg et al. 1997; Andrews et al. 2000; Balser 2000; Waldrop and Firestone 2004; Nicolardot et al. 1994). Further, some data suggest that the response is not always consistent with simple kinetics. In the microbial community of a Californian annual grassland for instance, substrate-utilization varied substantially with incubation temperature (Balser and Wixon 2009). Strikingly, despite predictions that polymeric carbon will be more readily degraded at higher temperatures, we have seen the opposite effect. At cooler incubation temperatures, polymers were preferentially degraded (Balser and Wixon 2009). The significance of this for soil carbon dynamics remains to be explored.

However, the issue of whether soil will act as a net carbon source or sink in response to climate warming remains a matter of intense interest in global change policy and research communities (Cox et al. 2000; Shaver et al. 2000; Rustad et al. 2001; Knorr et al. 2005). Because global soil organic carbon concentration is greater than twice that of the atmosphere (Post et al. 1982; Schimel 1995; Schlesinger 1996), even small changes in flux can have a significant impact on atmospheric carbon dioxide (Kirschbaum 2000; Rustad et al. 2000; Schlesinger and Andrews 2000).

In particular, the sensitivity of recalcitrant ('older') carbon to temperature is a critical parameter for predicting the role of soil as a feedback agent in climate warming (Knorr et al. 2005; Giardina and Ryan 2000; Luo et al. 2001; Fang et al. 2005). While there is a reasonable level of agreement that younger (labile) carbon will generally display a predictable pattern of response to temperature (e.g. it has a Q_{10} of approximately 2.4 and increasing rate of mass loss as temperature rises), the dynamics of older carbon largely remain a mystery (Kirschbaum 2000; Agren and Bosatta 2002). The sensitivity of recalcitrant carbon to rising temperature has been predicted to increase (Knorr et al. 2005; Fierer et al. 2005), decrease (Agren and Wetterstedt 2007; Wagai et al. 2008), or remain invariant (Giardina and Ryan 2000; Luo et al. 2001). This variability is likely due to the web of interacting factors that influence carbon stability in soil (Davidson et al. 2000). As litter transforms to 'soil organic matter', and then ages to stable (humic) forms it becomes increasingly chemically altered (Balser 2005) and associated with soil minerals (Sollins et al. 1996). Further, litter of differing chemical quality will vary in its transformation, as will the availability of organisms to degrade it (Balser 2005). As a result, temperature sensitivity of older carbon is not a simple function of enzyme response, but instead is the product of a complex suite of interactions among the varying temperature responses of competing processes such as activation energy (Bosatta and Agren 1999), altered substrate diffusion (Mikan et al. 2002), mineral adsorption or occlusion (Wagai et al. 2008; Thornley and Cannell 2001), historical carbon input and land use (Davidson et al. 2000), and acclimation of the decomposer community (Balser and Firestone 2005; Waldrop and Firestone 2004; Agren and Bosatta 2002). To date, studies including or focusing on more than one of these factors at a time are rare, and consequently, results of existing studies often appear idiosyncratic or surprising (Davidson et al. 2000; Mack et al. 2004).

Impact of Elevated Carbon Dioxide

To date, the majority of studies investigating ecosystem carbon storage response to increasing atmospheric carbon dioxide have been focused on the role of aboveground tissue chemistry and biomass production (Pan et al. 1998; Mooney et al. 1999). However, initial predictions that increases in plant biomass and decreases in above-ground litter quality in response to elevated carbon dioxide concentrations would lead to increased soil carbon storage have not been borne out (Norby and Cotrufo 1998; Schlesinger and Lichter 2001). In fact, it has become clear in recent years that more often than not, aboveground biomass remains unchanged, and plant tissue chemistry at senescence is indistinguishable from that grown at ambient carbon dioxide concentrations (O'Neill 1994; Curtis et al. 1989; Franck et al. 1997). However, there is a growing body of evidence indicating that carbon dioxide enrichment coupled with increased plant nitrogen requirements leads consistently to enhanced carbon allocated belowground (O'Neill 1994; Niinisto et al. 2004; Cardon et al. 2001). Although responses vary across systems, increased carbon flow to belowground pools appears to be driven by mechanisms for nitrogen acquisition such as:

1. Increased allocation to root structural tissues (Zak et al. 2000a; Owensby 1993; Kampichler et al. 1998)
2. Accelerated root turnover (Norby 1994; Pregitzer et al. 2000)
3. Rhizodeposition (Cardon et al. 2001; Hungate et al. 1999)
4. Mycorrhizal development (O'Neill 1994; Treseder and Allen 2000)
5. Nitrogen-fixation (Hungate et al. 1999; Montealegre et al. 2000)

Few studies to date have addressed, however, the relative importance and long-term consequences of different mechanisms by which increases in belowground carbon impact microbial communities, and thus carbon dynamics in soil.

Increasing atmospheric carbon dioxide can affect plant responses in several ways: plant growth can increase or decrease (Poorter 1993; Joel et al. 2001), nutrient usage and allocation can change (Curtis et al. 1989; Cotrufo et al. 1998), and above- and belowground patterns of biomass can be altered (Zak et al. 2000a; Hungate et al. 1997). These responses to elevated carbon dioxide affect the competitive abilities of plant species, and in turn, can alter plant community composition (Zangerl and Bazzaz 1984). Within plant communities containing exotic, invasive species, the growth and competitive dynamics of plants may change with rising carbon dioxide to favour the exotic species or the native species. Fast-growing C_3 species, such as many exotic, invasive species, tend to respond most strongly to elevated carbon dioxide (Poorter 1993). Therefore, there is concern whether or not invasive species will become more aggressive in their growth and establishment under enriched atmospheric carbon dioxide. Hungate et al. (1996) showed that the impact of invasive grasses on the cycling of nitrogen in Californian grassland was greater under elevated carbon dioxide than under ambient levels. The invasive grasses showed increased plant nitrogen pools and $^{15}NH_4^+$ (ammonium) uptake under elevated carbon dioxide while the native species exhibited smaller increases or decreases. A study by Dukes (2002) showed that growth and competition of *Centaurea* was enhanced under elevated carbon dioxide. Similarly, Smith et al. (2000) found that both the native desert annuals and the exotic, invasive grass (*Bromus tectorum*) showed greater total aboveground biomass and individual plant biomass under elevated carbon dioxide, but the total density of the native annuals decreased while the density of *Bromus* increased. Additionally, *Bromus* exhibited a three-fold increase in seed rain, while the whole-plot seed rain of the native annuals did not increase significantly. These studies illustrate that changes in the growth dynamics of invasive and native species under elevated carbon dioxide may lead to ecosystem-level changes in primary productivity, which will in turn impact on soil microbial and carbon dynamics as plants senesce.

Impact of Changing Soil Moisture

The soil moisture relationship to microbial community is more highly variable and complicated than that of temperature, and less studied (Lavigne et al. 2004; Saiz et al. 2007). It is easy to understand intuitively that soil water content will vary negatively with temperature, and has been shown to be the case (Luo et al. 2001; Davidson et al. 1998). However, the relationship is challenging to describe empirically. There is no consensus as to an equation describing soil respiration and moisture (Emmett et al. 2004), or moisture and temperature (Lavigne et al. 2004). As with other factors, interactions may be seen as a key reason that temperature relationships are not clear. The time scales and spatial scales of water change are different from those of temperature change. Moisture changes may come in the form of wet-dry cycles, drought, flooding, or smaller shifts. These different changes have different community structural and functional impacts, and are conditioned by a community's native regime.

There are several mechanisms or physical processes affecting microbial communities that vary with moisture content (Rodrigo et al. 1997). Precipitation is generally agreed to constrain decomposition at its extremes of dry (water stress) and wet (anoxia). Although a general interaction between oxygen concentrations and soil moisture is intuitively obvious, soil moisture effects are not limited to anoxia. For example, considering aerobic respiration at low moisture contents, substrate diffusion has been shown to be the main regulating process at low water contents in clay soils (Schjonning et al. 2003).

Despite logical mechanisms by which microbial communities may be altered by changes in soil moisture, these effects have so far only been observed in some studies, in dry summers (Hui and Luo 2004; Rey and Jarvis 2006). The impact of experimental drought has been likewise inconsistent (Emmett et al. 2004). Water is a critical factor in global change, nevertheless. Wetter soils such as peats and wetlands constitute large carbon sinks. Heterotrophic respiration overall is commonly seen as negatively related to water content above the soil moisture optima, typically considered 60–80% of water holding capacity. Waterlogged soils are so successful at creating physical and chemical barriers to aerobic respiration that a future carbon sequestering management strategy might involve saturating soils (Sylvia et al. 2005). Other evidence considering the interacting factor of elevated carbon dioxide suggests that anaerobic wetland conditions may be less of a carbon sink with global climate change (Wolf et al. 2007).

The adaptation of a microbial community to a local precipitation regime can make generalizing moisture response figures as dangerous as stating a universal temperature optimum. Response to conditions such as flooding depends on the life history of the community. Communities more tolerant of flooding, or drying and rewetting, have different responses to these disturbances (Fierer et al. 2003; Mentzer et al. 2006). Fierer et al. (2003) provide an example of community adaptation and process response to moisture levels. Drying-wetting regimes were found to significantly impact on bacterial community composition in oak woodland soils, which are less frequently exposed to moisture stress, but not in grassland soils. Size and function of litter decomposers can also be strongly affected by their moisture stress history (Schimel et al. 1999). Thus, the historical adaptation of the microbial community may determine how an ecosystem responds to changing moisture regimes.

Drought, irrigation, flooding, and re-wetting or other pulse events all can be expected to have different responses compared with modest water stress or modest increases in soil moisture (Schimel et al. 2007). Microbial activity or biomass may increase during wet seasons or after ‘wet-up’ (the period when soils regain moisture after the dry season; Fierer et al. 2003; Waldrop and Firestone 2006). Rewetting after drying also has unique process implications, and several studies address it (Reichstein et al. 2005; Wu and Brookes 2005). For example, dissolved organic carbon release is enhanced when dry soils are rewetted (Marschner and Bredow 2002), but these effects may be community-specific (Reichstein et al. 2005). Increasing soil moisture may also alter microbial activity but not biomass. For instance one study (Waldrop and Firestone 2004) reported that increased soil water did not affect the size of the available substrate pool, the utilization of differing ages of soil carbon, or microbial community composition. However, increased soil moisture did decrease hydrolytic enzymatic activity and reduced peroxidase enzyme activity, associated with complex carbon degradation, by 42%.

It is likely in at least some cases, however, that soil moisture has an influence even at levels other than the extremes. Considering the importance of soil microsites, water films, osmotic stress tolerances, ion concentrations, and the differential water retention of different pore size, it is clear that the effect of moisture is structurally and biologically complex. Any environmental conditions, including smaller moisture changes that limit or accelerate the diffusion of carbon dioxide from soils or the surface layer, can create non-equilibrium conditions that impact carbon dioxide efflux rates (Hanson et al. 2000). Studies have shown that moderate drying can significantly affect decomposition (Reichstein et al. 2005). Even modest water stress can cause a 25–50% decline in soil respiration (Lavigne et al. 2004).

A last consideration is that certain factors may interact with soil moisture regimes to either affect microbial community structure, or activity, or both simultaneously. A study by Chena et al. (2008) considered two near-optimal moisture levels (60% and 80% of field capacity) and two different plant covers (with differing biomass). The community structure (assessed via phospholipid fatty acid analysis) was more affected by plant species, but community-level physiological profiles were affected by both plant species and soil moisture. Both physiological and acclimation mechanisms occurred, with differing factors contributing to each mechanism.

Changing Agricultural Management Regimes and Soil Microbial Communities

Some of the most critical impacts of climate change on microbial communities will not be through direct changes to the soil environment, but through those that occur indirectly via changes in land management and vegetation cover as humans respond to a changing climate.

It is well-established that conversion of native ecosystems to agricultural uses can strongly affect microbial community structure, composition and diversity. For example, conversions of tropical forest to plantations (Waldrop et al. 2000) have been found to engender distinct soil microbial community structures, and agricultural intensification has been reported to decrease microbial diversity (Steenwerth et al. 2005). Additionally, the type of land management practices in agroecosystems also affects microbial community structure and function through a variety of different mechanisms. Numerous studies have documented changes in microbial community structure as a result of physical disturbance, especially tillage (Frey et al. 1999; Guggenberger et al. 1999). Tillage represents a severe disturbance to fungi by severing hyphal connections and no-till systems favor fungi over bacterial community components (Minoshima et al. 2007; Kennedy and Schillinger 2006). Conversions to agriculture and cultivation practices also alter microbial communities through changes to temperature, soil moisture (through irrigation and alteration of soil structure), and other physical parameters.

Land-use changes also alter soil microbial community structure through alterations in carbon availability and quality, pH (Cookson et al. 2007), nutrient availability, or other chemical parameters. For example, fungal-to-bacterial ratios are commonly measured as indicators of microbial community structure, and the relative proportions of fungi are increased by no-till practices, crop rotations, and use of cover crops (Six et al. 2006). Studies in both agroecosystems and natural systems report that nitrogen additions decrease the relative abundance of fungi to bacteria (Bardgett and Shine 1999; Bradley et al. 2006). Seghers et al. (2004) found that nitrogenous fertilizers decreased populations of methanotrophs in the bulk soil microbial community, as well as members of the root endophytic community. They also found differential effects of organic fertilizers versus inorganic, consistent with other studies. For example, Wander et al. (1995) reported that manure-amended plots contained a less diverse population of microorganisms than cover cropped soil, but that the microbial biomass was more metabolically active (Wander et al. 1995). Ulrich et al. (2008) found that manure applications led to an increase in the population densities of cellulytic bacteria within the soil microbial community.

In addition to physical disturbance effects, alterations in vegetation, including alterations in plant diversity and species-specific plant traits, can in turn cause alterations in aboveground litter quantity and quality, and belowground root dynamics. However, alterations in vegetation tend to be idiosyncratic effects of particular plant species or particular plant functional traits, and difficult to draw more generalized patterns from (Porazinska et al. 2003). The effects of plant litter quality and quantity, in particular, are limiting factors to microorganisms, and thus species-specific differences in plant litter can especially affect microbial community structure and function (Wardle 2002).

Additionally, effects may persist for many years and decades after a given land use has stopped (Steenwerth et al. 2003). For example, Fraterrigo et al. (2006) found long-term alterations to microbial community structure in forest stands that had been cultivated but not logged. Fungal markers, especially, were lower in previously cultivated sites, suggesting that fungi may need more time to recover from agriculture (Fraterrigo et al. 2006). Likewise, Spiegelberger et al. (2006) found, 70 years after agricultural abandonment, changes to the microbial community due to lasting changes in pH due to former agricultural liming. Evidence is accumulating that the history of land-use can leave contingency effects in the soil microbial community, ultimately influencing successional dynamics of future plant communities (Kardol et al. 2007) and hence providing a mechanism by which changes due to agricultural management practices may persist far into the future.

Linking the Structure and Function of Microbial Communities in Changing Landscapes

As conversion of native ecosystems to arable land is likely to continue into the future, future agricultural management practices are likely to alter the essential functions microbial communities provide to plant productivity, including providing structural stability to soils, nutrient (especially limiting nutrients such as nitrogen and phosphorus) mineralization and transformation, and disease suppression (Barea et al. 2002; Peterson et al. 2002; Kennedy and Papendick 1995). In addition to effects on plant productivity, land-use shifts are likely to affect numerous other processes that the microbial community carries out in agroecosystems, including decomposition, carbon cycling, soil aggregation and stability, plant productivity, and greenhouse gas emissions. For example, Bossuyt et al. (2002) found that elimination of soil fungal populations decreased macroaggregate formation in the soil. The impact of climatic change on soil ecosystem services and products is likely to be profound and demands immediate and detailed studies not least because of its relevance to food security.

Studies that link community structure and functional measurements of land-use change effects are relatively rare but increasing. For example, the study of Waldrop et al. (2000) on conversion of tropical forest to plantation linked measurements to community structure to enzyme activities, and found that land conversion to plantations increased the activities of key enzymes involved in the carbon cycle, such as phenol oxidase and peroxidase (lignin degradation) and cellobiohydrolase (cellulose degradation). Thus changes to the overall physiology of the soil microbial community provide a mechanism whereby land-use can affect ecosystem functions.

It is well established that land-use changes alter mineralization rates of key limiting nutrients and thus affect nutrient availability to plants. Recent studies have found that alterations to microbial community structure can often explain differences in mineralization rates in different altered ecosystems (Fraterrigo et al. 2006; Hawkes et al. 2005). Shifts in nitrogen processes important to agricultural productivity, such as nitrogen mineralization rates (Balser et al. 2005; Fraterrigo et al. 2005) or phosphorus cycling rates (Cleveland et al. 2003), can thus be altered with shifts in microbial community composition.

Land-use caused shifts in microbial mediated processes ultimately feed back to plants themselves, impacting agricultural productivity, although the mechanisms linking aboveground and belowground systems may be complex and context-dependent (Wardle et al. 2004). Additionally, recent research has suggested that particular land uses (including cultivation) may cause changes in microbial community structure and function that may persist many years after cultivation has ceased (Fraterrigo et al. 2006; Dupouey et al. 2002). These biotic legacies in the soil may influence future plant communities and environmental conditions at the site, and so microbial functions may reflect past land uses as much as contemporary ones. Through biotic legacies left by previous vegetation that has altered the soil microbial community, changes to soil microbial physiology and function induced by land-use changes can ultimately feed back to subsequent vegetation that is planted, succeeds, or is restored. More work is needed to elucidate how interacting biotic and abiotic factors can influence microbial community structure and function, over short and long time periods of land-use changes.

Caveats and Conclusions

Climate Modulators and Native Regimes

Temperature and water are both environmental factors that are important for microbial growth. They are considered to be environmental "modulators" (that influence organism activity) as opposed to "resources" that are used by organisms to grow and reproduce (Chapin 2003). Modulator changes affect the whole community, but often ultimately cause change by their impacts on resources. For example, a reduction in water potential (modulator) may act to limit substrate availability (resource) via diffusion, improving the success of hyphal strategists. Energy must be used to adjust processes when a modulator is not optimal (Balser et al. 2002). Microbial responses to temperature and moisture stress can occur via several mechanisms, including physiological changes such as lipid membrane changes, substrate preference changes, or the formation of dormant structures. Regarding such stresses, Schimel et al. (2007) conclude that even when microbial community response to stress is limited, the physiological costs imposed on soil microbes are large enough that they may cause large shifts in the allocation and fate of carbon and nitrogen.

Response to temperature and moisture change is dependent on native regimes. This importance of native regimes makes theoretical sense. From a microbial physiology perspective, a stressed microbial community is more likely to show changes than one operating within accustomed conditions. Prevailing conditions at a given location select for communities with particular adaptations to temperature (Dalias et al. 2001), and to its variability. If an environment rarely changes, there is a disadvantage to maintaining sufficient genetic material to have a broad functional range (Balser et al. 2002).

A community may be more sensitive to temperature and moisture changes if the new condition is outside of its normal climate range. This has been demonstrated experimentally (Balser and Firestone 2005; Wolf et al. 2007). For example, Waldrop and Firestone (2006) demonstrated that a forest microbial community perturbed beyond its environmental range due to transplant was more sensitive to climate changes than a transplanted grassland microbial community with a broader native range. In this case, both community structure and biomass changed. This is consistent with a study by Balser and Firestone (2005) showing that response to soil transplant depended on the deviation from the community's ambient regime.

Temperature optima do not necessarily relate to average local temperature (Lipson 2007), but the community may instead be driven by substrate availability resulting from plant phenologies (Yuste et al. 2004). Optima changes with high temperature incubations may also be more related to substrate limitation, and may not increase as might be expected (Dalias et al. 2001).

Importance of Study Context and Length

Even among studies that examine multiple interacting global change factors, few have studied microbial responses for more than one growing season. Variation caused by seasonal oscillations in weather over time significantly affects microbial community structure and function (Wolf et al. 2007; Ebersberger et al. 2003; Jin and Evans 2007), as well as changing responses to global change manipulations (Ebersberger et al. 2003). Year to year climate variation may also change the intensity of treatment effects (Saiya-Cork et al. 2002). Neglect of seasonal trends can lead to misinterpretation of results (Boerner et al. 2005), and in much the same way neglect of inter-annual variation could lead to misinterpretation or over-generalization of results that may have been particular to one growing season. Multiple year studies are more able to take into consideration seasonal changes in weather, perhaps giving a broader or more realistic perspective of how multiple, interacting factors alter soil communities.

A critical reason to study microbial communities across long-term global change manipulation is to understand how chronic, multiple long-term stresses, such as the stress of elevated temperature or shifting nutrient dynamics, will alter the microbial community. It's possible, for instance, that cumulative stress could result in large community shifts depending on external 'tipping points', such as the extreme wet or dry years that may occur more often in a globally changing climate (Intergovernmental Panel on Climate Change (IPCC) 2007). Multiple stresses simulated in global change experiments may interact to impact above and belowground communities in ways distinct from single factor responses (Henry et al. 2005a, b; Flannigan et al. 2006; Kandeler et al. 1998). For example, nitrogen deposition from urban and agricultural areas may result in a shift from mycorrhizal to bacterially dominated communities (Aber et al. 1998) that may be altered further by elevated carbon dioxide or climate factors (Treseder and Allen 2000). Despite the abundance of work on single factors there is still relatively little known about how microbial communities will respond to multiple simultaneous stresses represented by global change treatments (Pendall et al. 2004; Kandeler et al. 1998).

This added interpretive strength and understanding of temporal aspects of soil community response to global change is also important from a standpoint of its consequences. What does it mean that the community acts in one way or another under global change treatments? How will ecosystem function change as a result of changes in small scale, microbial processes? Should soil management decisions change in wet versus dry years? These questions can only be answered from understanding the microbial community on a longer time-scale than has traditionally been studied. For instance, several studies have shown that it takes from 2 to several years for microbial activity and nutrient cycling to recover after burning (Smithwick et al. 2005; Zhang et al. 2005; Yong-Mei et al. 2005; Boerner et al. 2006; Turner et al. 2007). Over this time frame, or until final recovery, any one time-point may give a very different picture of response to disturbance that may not accurately reflect long-term patterns of recovery and thus, poor interpretations for proper landscape management.

Conclusions

Climate change will impact microbial community structure and activities both directly, through alteration of the soil chemical and physical environment, and indirectly through changes in land use and cover. We have most often studied climate change impacts by isolating the various change drivers and manipulating them independently. However, in reality changes in the soil habitat will occur in concert and may produce unexpected results as factors like temperature interact with changes in water or nitrogen availability and plant species cover. We need a better understanding of the nature of climate change interactions from studies that include multiple factors. It is especially critical to note that the specific response of a community to a given environmental change will not be dependent only on the severity of the disturbance, but also on the history of the affected soil and soil community. Evidence is accumulating that rather than being infinitely plastic in their response to and recovery from environmental perturbations, microbial communities are uniquely adapted to their native climatic regime and vegetation cover. They will thus respond uniquely to climate and land use change across ecosystem types or climate regimes. Further, it is increasingly being shown that land use practices can create legacy effects that persist for decades.

Soil microorganisms are essential components in the response of agricultural ecosystems to climate change through their capacity to cycle nutrients and process soil carbon. In order to fully understand and manage the impacts of climate change on soil communities we must be sure to include assessments of their composition and biomass in our studies, and design studies over the longer-term. Short-term studies are inadequate to capture the impact of climate change on soil microbial dynamics, and are inadequate. Finally, we must consider how our land use choices impact microbial communities into the future.

Acknowledgments The authors would like to acknowledge the members of the Balser laboratory at the University of Wisconsin-Madison, especially Dr. Harry Read and Kevin Budsberg, for their untiring support and assistance. We would also like to thank Dr. Chris Field, Dr. Nona Chiariello, and supporting staff at the Jasper Ridge Global Change Experiment where much of our own work has taken place. Finally we thank the book editors for their valuable comments in strengthening this chapter.

References

Aber J, McDowell W, Nadelhoffer K, Magill A, Berntson G, Kamakea M, McNulty S, Currie W, Rustad L, Fernandez I (1998) Nitrogen saturation in temperate forest ecosystems: hypothesis revisited. Bioscience 48:921–934

Agren GI, Bosatta E (2002) Reconciling differences in predictions of temperature response of soil organic matter. Soil Biol Biochem 34:129–132

Agren GI, Wetterstedt JAM (2007) What determines the temperature response of soil organic matter decomposition? Soil Biol Biochem 39:1794–1798

Ajwa H, Dell CJ, Rice CW (1999) Changes in enzyme activities and microbial biomass of tallgrass prairie soil as related to burning and nitrogen fertilization. Soil Biol Biochem 31:769–777

Allison S, Treseder KK (2008) Warming and drying suppress microbial activity and carbon cycling in boreal forest soils. Glob Change Biol 14:2898–2909

Andrews JA, Matamala R, Westover KM, Schlesinger WH (2000) Temperature effects on the diversity of soil heterotrophs and the $\partial^{13}C$ of soil-respired CO_2. Soil Biol Biochem 32: 699–706

Arrhenius S (1896) On the influence of carbonic acid in the air upon the temperature of the ground. Phil Mag J Sci 41:237–276

Balser TC (2000) Linking soil microbial communities and ecosystem functioning. Doctoral dissertation, University of California, Berkeley, CA

Balser TC, Kinzig A, Firestone MK (2001) Linking soil microbial communities and ecosystem functioning. In: Kinzig A, Pacala S, Tilman D (eds) The functional consequences of biodiversity: empirical progress and theoretical extensions. Princeton University Press, Princeton, NJ, pp 265–294

Balser TC, Kinzig AP, Firestone MK (2002) Linking soil microbial communities and ecosystem functioning. In: Kinzig AP, Pacala SW, Tilman D (eds) The functional consequences of biodiversity: empirical process and theoretical extensions. Princeton University Press, Princeton, NJ

Balser TC (2005) Humification. In: Hillel D et al (eds) Encyclopedia of soils in the environment. Elsevier, Oxford, UK, pp 195–208

Balser T, Treseder K, Ekenler M (2005) Using lipid analysis to estimate AM and saprotrophic fungal abundances along a soil chronosequence. Soil Biol Biochem 37:601–604

Balser TC, Firestone MK (2005) Linking microbial community composition and soil processes in a California annual grassland and mixed-conifer forest. Biogeochemistry 73:395–415

Balser TC, Wixon D (2009) Investigating biological control over soil carbon temperature sensitivity. Glob Change Biol. doi:10.1111/j.1365-2486.2009.01946.x

Bardgett RD, Shine A (1999) Linkages between plant litter diversity, soil microbial biomass and ecosystem function in temperate grasslands. Soil Biol Biochem 31:317–321

Barea JM, Azcon R, Azcon-Aguilar C (2002) Mycorrhizosphere interactions to improve plant fitness and soil quality. Antonie Leeuwenhoek 81:343–351

Boerner R, Brinkman JA, Smith A (2005) Seasonal variations in enzyme activity and organic carbon in soil of a burn and unburned hardwood forest. Soil Biol Biochem 37:1419–1426

Boerner R, Waldrop TA, Shelburne VB (2006) Wildfire mitigation strategies affect soil enzyme activity and soil organic carbon in loblolly pine (*Pinus taeda*) forests. Can J For Res 36:3148–3154

Bosatta E, Agren GI (1999) Soil organic matter quality interpreted thermodynamically. Soil Biol Biochem 31:1889–1891

Bossuyt H, Six J, Hendrix PF (2002) Aggregate-protected carbon in no-tillage and conventional tillage agroecosystems using Carbon-14 labeled plant residue. Soil Sci Soc Am J 66: 1965–1973

Bradley K, Drijber RA, Knops J (2006) Increased N availability in grassland soils modifies their microbial communities and decreases the abundance of arbuscular mycorrhizal fungi. Soil Biol Biochem 38:1583–1595

Bruce KD, Jones TH, Bezemer TM, Thompson LJ, Ritchie DA (2000) The effect of elevated atmospheric carbon dioxide levels on soil bacterial communities. Glob Change Biol 6:427–434

Cardon ZG, Hungate BA, Cambardella CA, Chapin FS III, Field CB, Holland EA, Mooney HA (2001) Contrasting effects of elevated CO_2 on old and new soil carbon pools. Soil Biol Biochem 33:365–373

Carreiro MM, Sinsabaugh RL, Repert DA, Parkhurst DF (2000) Microbial enzyme shifts explain litter decay responses to simulated nitrogen deposition. Ecology 81:2359–2365

Chapin FS (2003) Effects of plant traits on ecosystem and regional processes: a conceptual framework for predicting the consequences of global change. Ann Bot 91:455–463

Chena M, Zhu Y, Sua Y, Chena B, Fu B, Marschner P (2007) Effects of soil moisture and plant interactions on the soil microbial community structure. Eur J Soil Biol 43:31–38

Clarholm M (1985) Possible roles for roots, bacteria, protozoa and fungi in supplying nitrogen to plants. In: Fitter A et al (eds) Ecological interactions in soil. Blackwell Scientific, Oxford, UK, pp 355–365

Cleveland CC, Townsend AR, Schmidt SK, Constance BC (2003) Soil microbial dynamics and biogeochemistry in tropical forests and pastures, southwestern Costa Rica. Ecol Appl 13:314–326

Contin M, Corcimaru S, Nobili MD, Brookes PC (2000) Temperature changes and the ATP concentration of the soil microbial biomass. Soil Biol Biochem 32:1219–1225

Cookson WR, Osman M, Marschner P, Abaye DA, Clark I, Murphy DV, Stockdale EA, Watson CA (2007) Controls on soil nitrogen cycling and microbial community composition across land use and incubation temperature. Soil Biol Biochem 39:744–756

Cotrufo MF, Ineson P, Scott A (1998) Elevated CO_2 reduces the nitrogen concentration of plant tissues. Glob Change Biol 4:43–54

Cox PM, Betts RA, Jones CD, Spall SA, Totterdell IJ (2000) Acceleration of global warming due to carbon cycle feedbacks in a coupled climate model. Nature 408:184–187

Cubasch U, Meehl GA, Boer GJ, Stouffer RJ, Dix M, Noda A, Senior CA, Raper S, Yap KS (2001) Projections of future climate change. In: Houghton JT, Ding Y, Griggs DJ, Noguer M, van der Linden PJ, Dai C, Maskell K, Johnson CA (eds) Climate change 2001: the scientific basis. Cambridge University Press, New York, pp 525–582

Curtis P, Drake B, Whigham D (1989) Nitrogen and carbon dynamics in C3 and C4 estuarine marsh plants grown under elevated carbon dioxide in situ. Oecologia 78:297–301

Dalias P, Anderson JM, Bottner P, Couteaux M (2001) Temperature responses of carbon mineralization in conifer forest soils from different regional climates incubated under standard laboratory conditions. Glob Change Biol 7:181–192

Davidson EA, Belk E, Boone BD (1998) Soil water content and temperature as independent or confounded factors controlling soil respiration in a temperate mixed hardwood forest. Glob Change Biol 4:217–227

Davidson EA, Trumbore SE, Amundson R (2000) Biogeochemistry: soil warming and organic carbon content. Nature 408:789–790

Dhillion S, Roy J, Abrams M (1996) Assessing the impact of elevated CO_2 on soil microbial activity in a Mediterranean model ecosystem. Plant Soil 187:333–342

Diaz S, Grime JP, Harris J, McPherson E (1993) Evidence of a feedback mechanism limiting plant response to elevated carbon dioxide. Nature 364:616–617

Dixon GR (2009) The impact of climate and global change on crop production. In: Letcher TM (ed) Climate change: observed impacts on planet earth. Elsevier, Oxford, UK/Amsterdam, Netherlands, pp 307–324

Dukes JS (2002) Comparison of the effect of elevated CO_2 on an invasive species (*Centaurea solstitialis*) in monoculture and community settings. Plant Ecol 160:225–234

Dupouey JL, Dambrine E, Lafftie JD, Moares C (2002) Irreversible impact of past land use on forest soils and biodiversity. Ecology 83:2978–2984

Ebersberger D, Niklaus PA, Kandeler E (2003) Long term CO_2 enrichment stimulates N-mineralisation and enzyme activities in calcareous grassland. Soil Biol Biochem 35:965–972

Ellert BH, Bettany JR (1992) Temperature-dependence of net nitrogen and sulfur mineralization. Soil Sci Soc Am J 56:1133–1141

Emmett BA, Beier C, Estiarte M, Tietema A, Kristensen HL, Williams D, Penuelas J, Schmidt I, Sowerby A (2004) The response of soil processes to climate change: results from manipulation studies of shrublands across an environmental gradient. Ecosystems 7:625–637

Fang CM, Smith P, Moncrieff JB, Smith JU (2005) Similar response of labile and resistant soil organic matter pools to changes in temperature. Nature 433:57–59

Fierer N, Schimel JP, Holden PA (2003) Influence of drying-rewetting frequency on soil bacterial community structure. Microb Ecol 45:63–71

Fierer N, Craine J, McLauchlan K, Schimel JP (2005) Litter quality and the temperature sensitivity of decomposition. Ecology 86:320–326

Fitter AH, Heinemeyer A, Staddon PL (2000) The impact of elevated CO_2 and global climate change on arbuscular mycorrhizas: a mycocentric approach. New Phytol 147:179–187

Flannigan M, Amiro BD, Logan KA, Stocks BJ, Wotton BM (2006) Forest fires and climate change in 21st century. Mitig Adapt Strateg Glob Change 11:847–859

Foley JA, Ramankutty N (2004) A primer on the terrestrial carbon cycle: what we don't know but should. In: Field CB, Raupach MR (eds) The global carbon cycle: integrating humans, climate, and the natural world. Island Press, Washington, DC, pp 279–294

Franck VM, Hungate BA, Chapin FS III, Field CB (1997) Decomposition of litter produced under elevated CO_2: dependence on plant species and nutrient supply. Biogeochemistry 36:223–237

Fraterrigo JM, Turner MG, Pearson SM, Dixon P (2005) Effects of past land use on spatial heterogeneity of soil nutrients in southern Appalachian forests. Ecol Monogr 75:215–230

Fraterrigo JM, Balser TC, Turner MG (2006) Microbial community variation and its relationship with nitrogen mineralization in historically altered forests. Ecology 87:570–579

Frey SD, Elliott ET, Paustian K (1999) Bacterial and fungal abundance and biomass in conventional and no-tillage agroecosystems along two climatic gradients. Soil Biol Biochem 31:573–585

Gavito ME, Schweiger P, Jakobsen I (2003) P uptake by arbuscular mycorrhizal hyphae: effect of soil temperature and atmospheric CO_2 enrichment. Glob Change Biol 9:106–116

Giardina CP, Ryan MG (2000) Evidence that decomposition rates of organic carbon in mineral soil do not vary with temperature. Nature 404:858–861

Guggenberger G, Frey SD, Six J, Paustian K, Elliott ET (1999) Bacterial and fungal cell-wall residues in conventional and no-tillage agroecosystems. Soil Sci Soc Am J 63:1188–1198

Gutknecht JLM (2007) Exploring long-term microbial responses to simulated global change. Doctoral dissertation, University of Wisconsin, Madison, WI

Gutknecht JLM, Henry HA, Balser TC (2010) Inter-annual variation in soil extra-cellular enzyme activity in response to simulated global change and fire disturbance. Pedobiologia 53:283–293

Hanson PJ, Edwards NT, Garten CT, Andrews JA (2000) Separating root and soil microbial contributions to soil respiration: a review of methods and observations. Biogeochemistry 48: 115–146

Hawkes CV, Wren IF, Herman DJ, Firestone MK (2005) Plant invasion alters nitrogen cycling by modifying the soil nitrifying community. Ecol Lett 8:976–985

Henriksen T, Breland T (1999) Nitrogen availability effects on carbon mineralization, fungal and bacterial growth, and enzyme activities during decomposition of wheat straw in soil. Soil Biol Biochem 31:1121–1134

Henry HA, Juarez JD, Field CB, Vitousek PM (2005a) Interactive effects of elevated CO2, N deposition and climate change on extracellular enzyme activity and soil density fractionation in a California annual grassland. Glob Change Biol 11:1808–1815

Henry HA, Clelad EE, Field CB, Vitousek PM (2005b) Interactive effects of CO_2, N deposition, and climate change on plant litter quality in a California annual grassland. Oecologia 142:465–473

Horz HP, Barbrook A, Field CB, Bohannan BJM (2004) Ammonia-oxidizing bacteria respond to multifactorial global change. Proc Natl Acad Sci USA 101:15136–15141

Horz HP, Rich V, Avrahami S, Bohannan BJM (2005) Methane-oxidizing bacteria in a California upland grassland soil: diversity and response to simulated global change. Appl Environ Microbiol 71:2642–2652

Hu S, Firestone MK, Chapin FSC (1999) Soil microbial feedbacks to atmospheric CO_2 enrichment. Trends Ecol Evol 14:433–437

Hui DF, Luo Y (2004) Evaluation of soil CO_2 production and transport in Duke Forest using a process-based modeling approach. Glob Biogeochem Cy 18:GB4029

Hungate BA, Canadell J, Chapin FS (1996) Plant species mediate changes in soil microbial N in response to elevated CO_2. Ecology 77:2505–2515

Hungate BA, Lund CP, Pearson HL, Chapin FS (1997) Elevated CO_2 and nutrient addition alter soil N cycling and N trace gas fluxes with early season wet-up in a California annual grassland. Biogeochemistry 37:89–109

Hungate BA, Dijkstra P, Johnson DW, Hinkle CR, Drake BG (1999) Elevated CO_2 increases nitrogen fixation and decreases soil nitrogen mineralization in Florida scrub oak. Glob Change Biol 5:781–789

Hyvonen R, Agren GI, Dalias P (2005) Analysing temperature response of decomposition of organic matter. Glob Change Biol 11:770–778

Intergovernmental Panel on Climate Change (IPCC) (2001) Climate change 2001: synthesis report. Cambridge University Press, New York/Cambridge, p 398

Intergovernmental Panel on Climate Change (IPCC) (2007) Contribution of working group I to the fourth assessment report of the IPCC, the physical science basis. Cambridge University Press, Cambridge, UK/New York

Jin VL, Evans RD (2007) Elevated CO_2 increases microbial carbon substrate use and nitrogen cycling in Mojave Desert soils. Glob Change Biol 13:452–465

Joel G, Chapin FS, Chiariello NR, Thayer SS, Field CB (2001) Species-specific responses of plant communities to altered carbon and nutrient availability. Glob Change Biol 7:435–450

Johnson DW, Geisinger DR, Walker RF, Newman J, Vose JM, Elbot K, Ball JT (1994) Soil pCO_2, soil respiration, and root activity in CO_2-fumigated and nitrogen fertilized ponderosa pine. Plant Soil 165:129–138

Kampichler C, Kandeler E, Bardgett RD, Jones TH, Thompson LJ (1998) Impact of elevated atmospheric CO_2 concentration on soil microbial biomass and activity in a complex, weedy field model ecosystem. Glob Change Biol 4:335–346

Kandeler E, Tscherko D, Bardgett RD, Hobbs PJ, Kamplichler C, Jones TH (1998) The response of soil microorganisms and roots to elevated CO_2 and temperature in a terrestrial model ecosystem. Plant Soil 202:251–262

Kardol P, Cornips NJ, Van Kempen MML, Bakx-Schotman JMT, Van Der Putten WH (2007) Microbe-mediated plant–soil feedback causes historical contingency effects in plant community assembly. Ecol Monogr 77:147–162

Kennedy AC, Papendick RI (1995) Microbial characteristics of soil quality. J Soil Water Conserv 50:243–248

Kennedy AC, Schillinger WF (2006) Soil quality and water intake in conventional-till vs. no-till paired farms in Washington's Palouse Region. Soil Sci Soc Am J 70:940–949

Kirschbaum MUF (2000) Will changes in soil organic carbon act as a positive or negative feedback on global warming? Biogeochemistry 48:21–51

Klironomos JN, Rillig MC, Allen MF, Zak DR, Kubiske M, Pregitzer KS (1997) Soil fungal-arthropod responses to Populus tremuloides grown under enriched atmospheric CO_2 under field conditions. Glob Change Biol 3:473–478

Knorr W, Prentice IC, House JI, Holland EA (2005) Long-term sensitivity of soil carbon turnover to warming. Nature 433:298–301

Körner C, Arnone JA (1992) Responses to elevated carbon dioxide in artificial tropical ecosystems. Science 257:1672–1675

Kuikman PJ, Jansen AG, van Veen JA (1991) ^{15}N-nitrogen mineralization from bacteria by protozoan grazing at different soil moisture regimes. Soil Biol Biochem 23:193–200

Lavigne MB, Foster RJ, Goodine G (2004) Seasonal and annual changes in soil respiration in relation to soil temperature, water potential and trenching. Tree Physiol 24:415–424

Lee TD, Tjoelker MG, Russelle MP, Reich PB (2003) Contrasting response of an N-fixing and non N-fixing forb to elevated CO_2: dependence on soil N supply. Plant Soil 255:475–486

Lin G, Ehleringer JR, Rygiewicz PT, Johnson MG, Tingey D (1999) Elevated CO_2 and temperature impacts on different components of soil CO_2 efflux in Douglas-fir terracosms. Glob Change Biol 5:157–168

Linkins AE, Melillo JM, Sinsabaugh RL (1984) Factors affecting cellulase activity in terrestrial and aquatic ecosystems. In: Klug MJ, Reddy CA (eds) Current perspectives in soil microbiology. American Society for Microbiology press, Washington, DC, pp 572–670

Lipson DA (2007) Relationships between temperature responses and bacterial community structure along seasonal and altitudinal gradients. FEMS Microbiol Ecol 59:418–427

Luo Y, Wan S, Hui D, Wallace L (2001) Acclimatization of soil respiration to warming in a tall grass prairie. Nature 413:622–625

Lussenhop J, Treonis A, Curtis PS, Teeri JA, Vogel CS (1998) Response of soil biota to elevated atmospheric CO_2 in poplar model systems. Oecologia 113:247–251

Lutze JL, Gifford RM, Adams HN (2000) Litter quality and decomposition in *Danthonia richardsonii* swards in response to CO_2 and nitrogen supply over four years growth. Glob Change Biol 6:13–24

MacDonald NW, Zak DR, Pregitzer KS (1995) Temperature effects on kinetics of microbial respiration and net nitrogen and sulfur mineralization. Soil Sci Soc Am J 59:233–240

Mack MC, Schuur EAG, Bret-Harte MS, Shaver GR, Chapin FS III (2004) Ecosystem carbon storage in arctic tundra reduced by long-termnutrient fertilization. Nature 431:440–443

Marschner B, Bredow A (2002) Temperature effects on release and ecologically relevant properties of dissolved organic carbon in sterilised and biologically active soil samples. Soil Biol Biochem 34:459–466

Martin-Olmedo P, Rees RM, Grace J (2002) The influence of plants grown under elevated CO_2 and N fertilization on soil nitrogen dynamics. Glob Change Biol 8:643–657

Mayr C, Miller M, Insam H (1999) Elevated CO_2 alters community-level physiological profiles and enzyme activities in alpine grassland. J Microbiol Meth 36:35–43

Mentzer J, Goodman RM, Balser TC (2006) Microbial response over time to hydrologic and fertilization treatments in a simulated wet prairie. Plant Soil 284:85–100

Mikan C, Schimel J, Doyle A (2002) Temperature controls of microbial respiration above and below freezing in Arctic tundra soils. Soil Biol Biochem 34:1785–1795

Minoshima H, Jackson LE, Cavagnaro TR, Sanchez-Moreno S, Ferris H, Temple SR, Goyal S, Mitchell JP (2007) Soil food webs and carbon dynamics in response to conservation tillage in California. Soil Sci Soc Am J 71:952–963

Montealegre CM, Kessel CV, Blumenthal JM, Hur HG, Hartwig UA, Sadowsky MJ (2000) Elevated atmospheric CO_2 alters microbial population structure in a pasture ecosystem. Glob Change Biol 6:475–482

Montealegre CM, Kessel CV, Russelle MP, Sadowsky MJ (2002) Changes in microbial activity and composition in a pasture ecosystem exposed to elevated atmospheric carbon dioxide. Plant Soil 243:197–207

Mooney HA, Canadell J, Chapin SF III, Ehleringer JH, Körner C, McMurtrie RE, Parton WJ, Pitelka LF, Schulze ED (1999) Ecosystem physiology responses to global change. In: Walker B et al (eds) The terrestrial biosphere and global change, vol 4. Cambridge University Press, Cambridge, UK, pp 141–189

Nakayama PS, Huluka G, Kimball BA, Lewin KF, Nagy J, Hendrey GR (1994) Soil carbon dioxide fluxes in natural and CO_2 enriched systems. Agr Forest Meteorol 70:131–140

Nicolardot B, Fauvet G, Cheneby D (1994) Carbon and nitrogen cycling through soil microbial biomass at various temperatures. Soil Biol Biochem 26:253–261

Niinisto SM, Silvola J, Kellomaki S (2004) Soil CO_2 efflux in a boreal pine forest under atmospheric CO_2 enrichment and air warming. Glob Change Biol 10:1–14

Niklaus PA (1998) Effects of elevated atmospheric CO_2 on soil microbiota in calcerous grassland. Glob Change Biol 4:451–458

Niklaus PA, Alphei J, Ebersberger D, Kampichler C, Kandeler E, Tscherko D (2003) Six years of *in situ* CO_2 enrichment evoke changes in soil structure and soil biota of nutrient-poor grasslands. Glob Change Biol 9:585–600

Norby R (1994) Issues and perspectives for investigating root responses to elevated atmospheric carbon dioxide. Plant Soil 165:9–20

Norby RJ, Cotrufo MF (1998) A question of litter quality. Nature 396:17–18

O'Neill EG (1994) Responses of soil biota to elevated atmospheric carbon dioxide. Plant Soil 165:55–65

Owensby CE (1993) Potential impacts of elevated CO_2 and above- and belowground litter quality of a tallgrass prairie. Water Air Soil Pollut 70:413–424

Pajari B (1995) Soil respiration in a poor upland site of scots pine stand subjected to elevated temperature and atmospheric carbon concentration. Plant Soil 169:563–570

Pan Y, Melillo JM, McGuire AD, Kicklighter DW, Pitelka LF, Hibbard K, Pierce LL, Running S, Ojima D, Parton WJ, Schimel DS, Members OV (1998) Modeled responses of terrestrial ecosystems to elevated atmospheric CO_2: a comparison of simulations by the biogeochemistry models of the Vegetation/Ecosystem Modeling and Analysis Project (VEMAP). Oecologia 114:389–404

Pendall E, Bridgham S, Hanson PJ, Hungate B, Kicklighter DW, Johnson DW, Law BE, Luo YQ, Megonigal JP, Olsrud M, Ryan MG, Wan SQ (2004) Below-ground process responses to elevated CO_2 and temperature: a discussion of observations, measurement methods, and models. New Phytol 162:311–322

Peterjohn WT, Melillo JM, Bowles FP, Steudler PA (1993) Soil warming and trace gas fluxes- experimental design and preliminary flux rates. Oecologia 93:18–24

Petersen SO, Klug MJ (1994) Effects of sieving, storage, and incubation-temperature on the phospholipid fatty-acid profile of a soil microbial community. Appl Environ Microbiol 60:2421–2430

Peterson SO, Pamela S, Frohne, Kennedy AC (2002) Dynamics of a Soil Microbial Community under Spring Wheat. Soil Sci Soc Am J 66:826–833

Poorter H (1993) Interspecific variation in the growth-response of plants to an elevated ambient CO_2 concentration. Vegetatio 104:77–97

Porazinska DL, Bardgett RD, Blaauw MB, Hunt HW, Parsons AN, Seastedt TR, Wall DH (2003) Relationships at the aboveground-belowground interface: plants, soil biota, and soil processes. Ecol Monogr 73:377–395

Post WM, Emanuel WR, Zinke PJ, Stangenberger AG (1982) Soil carbon pools and life zones. Nature 298:156–159

Pregitzer KS, Zak DR, Maziasz J, DeForest J, Curtis PS, Lussenhop J (2000) Interactive effects of atmospheric CO_2 and soil-N availability on fine roots of *Populus tremuloides*. Ecol Appl 10:18–33

Reichstein M, Subke JA, Angeli AC, Tenhunen JD (2005) Does the temperature sensitivity of decomposition of soil organic matter depend upon water content, soil horizon, or incubation time? Glob Change Biol 11:1754–1767

Rey A, Jarvis P (2006) Modelling the effect of temperature on carbon mineralization rates across a network of European forest sites (FORCAST). Glob Change Biol 12:1894–1908

Řezáčová V, Blum H, Hršelová H, Gamper H, Gryndler M (2005) Saprobic microfungi under *Lolium perenne* and *Trifolium repens* at different fertilization intensities and elevated atmospheric CO_2 concentration. Glob Change Biol 11:224–230

Rillig MC, Field CB (2003) Arbuscular mycorrhizae respond to plants exposed to elevated CO_2 as a function of soil depth. Plant Soil 254:383–391

Rinnan R, Michelsen A, Baath E, Jonasson S (2007) Fifteen years of climate change manipulations alter soil microbial communities in a subarctic heath ecosystem. Glob Change Biol 13:28–39

Rodrigo A, Recous S, Neel C, Mary B (1997) Modelling temperature and moisture effects on C-N transformations in soils: comparison of nine models. Ecol Model 102:325–339

Rustad LE, Huntington TG, Boone RD (2000) Controls on soil respiration: implications for climate change. Biogeochemistry 48:1–6

Rustad LE, Campbell JL, Marion GM, Norby RJ, Mitchell MJ, Hartley AE, Cornelissen JHC, Gurevitch J, GCTE-NEWS (2001) A meta-analysis of the response of soil respiration, net nitrogen mineralization, and aboveground plant growth to experimental ecosystem warming. Oecologia 126:543–562

Saiya-Cork K, Sinsabaugh RL, Zak DR (2002) The effects of long term nitrogen deposition on extracellular enzyme activity in an Acer saccharum forest soil. Soil Biol Biochem 34: 1309–1315

Saiz G, Black K, Reidy B, Lopez S, Farrell EP (2007) Assessment of soil CO_2 efflux and its components using a process-based model in a young temperate forest site. Geoderma 139: 79–89

Schaeffer SM, Billings SA, Evans RD (2003) Responses of nitrogen dynamics in an intact Mojave Desert ecosystem to manipulations in soil carbon and nitrogen availability. Oecologia 134: 547–553

Schimel DS (1995) Terrestrial ecosystems and the carbon cycle. Glob Change Biol 1:77–91

Schimel JP, Gulledge JM, Clein-Curley JS, Lindstrom JE, Braddock JF (1999) Moisture effects on microbial activity and community structure in decomposing birch litter in the Alaskan taiga. Soil Biol Biochem 31:831–838

Schimel J, Balser TC, Wallenstein M (2007) Microbial stress-response physiology and its implications for ecosystem function. Ecology 88:1386–1394

Schjonning P, Thomsen IK, Moldrup P, Christensen BT (2003) Linking soil microbial activity to water- and air-phase contents and diffusivities. Soil Sci Soc Am J 67:156–165

Schlesinger WH (1996) Biogeochemistry; an analysis of global change, 2nd edn. Academic Press, San Diego, CA

Schlesinger WH, Andrews JA (2000) Soil respiration and the global carbon cycle. Biogeochemistry 48:7–20

Schlesinger WH, Lichter J (2001) Limited carbon storage potential in soil and litter of experimental forest plots under increased atmospheric CO_2. Nature 411:466–469

Seghers D, Wittebolle L, Top EM, Verstraete W, Siciliano SD (2004) Impact of agricultural practices on the *Zea mays*, L. Endophytic community. Appl Environ Microbiol 70: 1475–1482

Shaver GR, Canadell J, Chapin FS, Gurevitch J, Harte J, Henry G, Ineson I, Jonasson S, Melillo J, Pitelka L, Rustad L (2000) Global warming and terrestrial ecosystems: a conceptual framework for analysis. Bioscience 50:871–882

Sinsabaugh RL, Careirro MM, Repert DA (2002) Allocation of extracellular enzymatic activity in relation to litter decomposition, N deposition, and mass loss. Biogeochemistry 60:1–24

Six J, Frey SD, Thiet RK, Batten KM (2006) Bacterial and fungal contributions to carbon sequestration in agroecosystems. Soil Sci Soc Am J 70:555–569

Smith SD, Huxman TE, Zitzer SF, Charlet TN, Housman DC, Coleman JS, Fenstermaker LK, Seemann JR, Nowak RS (2000) Elevated CO_2 increases productivity and invasive species success in an arid ecosystem. Nature 408:79–82

Smithwick EH, Turner MG, Metzger KL, Balser TC (2005) Variation in NH^{4+} mineralization and microbial communities with stand age in lodgepole pine (Pinus contorta) forests, Yellowstone National Park (USA). Soil Biol Biochem 37:1546–1559

Sollins P, Homann P, Caldwell BA (1996) Stabilization and destabilization of soil organic matter: mechanisms and controls. Geoderma 74:65–105

Sowerby A, Emmett B, Beier C, Tietema A, Peneulas J, Estiarte M, van Meeteren MJM, Hughes S, Freeman C (2005) Microbial community changes in heathland soil communities along a geographical gradient: interaction with climate change manipulations. Soil Biol Biochem 37:1805–1813

Spiegelberger T, Hegg O, Diethart M, Hedlund K, Schaffner U (2006) Long-term effects of short-term perturbation in a subalpine grassland. Ecology 87:1939–1944

Stark JM, Firestone MK (1995) Mechanisms for soil moisture effects on activity of nitrifying bacteria. Appl Environ Microbiol 61:218–221

Steenwerth KL, Jackson LE, Calderon FJ, Stromberg MR, Scow KM (2003) Soil microbial community composition and land use history in cultivated and grassland ecosystems of coastal California. Soil Biol Biochem 35:489–500

Steenwerth KL, Jackson LE, Calderon FJ, Scow KM, Rolston DE (2005) Response of microbial community composition and activity in agricultural and grassland soils after a simulated rainfall. Soil Biol Biochem 37:2249–2262

Sylvia DM, Fuhrmann JJ, Hartel PG, Zuberer DA (2005) Principles and applications of soil microbiology. Pearson Education, Upper Saddle River, NJ

Tate KR, Ross DJ (1997) Elevated CO_2 and moisture effects on soil carbon storage and cycling in temperate grasslands. Glob Change Biol 3:225–235

Taylor AR, Schroter D, Pflug A, Wolters V (2004) Response of different decomposer communities to the manipulation of moisture availability: potential effects of changing precipitation patterns. Glob Change Biol 10:1–12

Thornley JHM, Cannell MGR (2001) Soil carbon storage response to temperature: an hypothesis. Ann Bot 87:591–598

Tison DL, Pope DH (1980) Effect of temperature on mineralization by heterotrophic bacteria. Appl Environ Microbiol 39:584–587

Treseder KK, Allen MF (2000) Mycorrhizal fungi have a potential role in soil carbon storage under elevated CO_2 and nitrogen deposition. New Phytol 147:189–200

Treseder K (2004) A meta-analysis of mycorrhizal responses to nitrogen, phosphorus, and atmospheric CO_2 in field studies. New Phytol 164:347–355

Treseder KK (2008) Nitrogen additions and microbial biomass: a meta-analysis of ecosystem studies. Ecol Lett 11:1111–1120

Turner MG, Smithwick EAH, Metzger KL, Tinker DB, Romme WH (2007) Inorganic nitrogen availability after sever stand-replacing fire in the greater Yellowstone ecosystem. Proc Natl Acad Sci USA 104:4782–4789

Tylianakis JM, Didham RK, Bascompte J, Wardle DA (2008) Global change and species interactions in terrestrial ecosystems. Ecol Lett 11:1351–1363

Ulrich A, Klimke GK, Wirth S (2008) Diversity and activity of cellulose-decomposing bacteria, isolated from a sandy and a loamy soil after long-term manure application. Microb Ecol. 55:512–522

van Groenigen KJ, Harris D, Horwath WR, Hartwig UA, Kessel CV (2002) Linking sequestration of ^{13}C and ^{15}N in aggregates in a pasture soil following 8 years of elevated atmospheric CO_2. Glob Change Biol 8:1094–1108

van Veen K, Liljeroth E, Lekkerkerk J (1991) Carbon fluxes in plant-soil systems at elevated atmospheric carbon dioxide levels. Ecol Appl 1:175–181

Vinolas LC, Vallejo VR, Jones DL (2001) Control of amino acid mineralization and microbial metabolism by temperature. Soil Biol Biochem 33:1137–1140

Vose JM, Elliot KJ, Johnson DW, Walker RF, Johnson MG, Tingey DT (1995) Effects of elevated CO_2 and N fertilization on soil respiration from ponderosa pine (*Pinus ponderosa*) in open-top chambers. Can J For Res 25:1243–1251

Wagai R, Mayer LM, Kitayama K, Knicker H (2008) Climate and parent material controls on organic matter storage in surface soils: a three-pool, density-separation approach. Geoderma 147:23–33

Waldrop M, Balser T, Firestone MK (2000) Linkages between microbial community composition and enzyme activities in soil. Soil Biol Biochem 32:1837–1846

Waldrop MP, Firestone MK (2004) Altered utilization patterns of young and old soil C by microorganisms caused by temperature shifts and N additions. Biogeochemistry 67:235–248

Waldrop M, Firestone MK (2006) Response of microbial community composition and function to soil climate change. Microb Ecol 52:716–724

Wander MN, Hedrick DS, Kaufman D, Traina SJ, Stinner BR, Kehrmeyer SR, White DC (1995) The functional significance of the microbial biomass in organic and conventionally managed soils. Plant Soil 170:87–97

Wardle DA (2002) Communities and ecosystems: linking the aboveground and belowground components. Princeton University Press, Princeton, NJ

Wardle DA, Bardgett RD, Klironomos JN, Setala H, van der Putten WH, Wall DH (2004) Ecological linkages between aboveground and belowground biota. Science 304:1629

Wolf AA, Drake BG, Erickson JE, Megonigal JP (2007) An oxygen-mediated positive feedback between elevated carbon dioxide and soil organic matter decomposition in a simulated anaerobic wetland. Glob Change Biol 13:2036–2044

Wu J, Brookes PC (2005) The proportional mineralisation of microbial biomass and organic matter caused by air-drying and rewetting of a grassland soil. Soil Biol Biochem 37:507–515

Yong-Mei Z, Ning W, Guo-Yi Z, Wei-Kai B (2005) Changes in enzyme activities of spruce (*Picea balfouriana*) forest soil as related to burning in the eastern Qinghai-Tibetan Plateau. Appl Soil Ecol 30:215–225

Yuste JC, Janssens IA, Carrara A, Ceulemans R (2004) Annual Q(10) of soil respiration reflects plant phenological patterns as well as temperature sensitivity. Glob Change Biol 10:161–169

Zak DR, Pregitzer KS, Curtis PS, Teeri JA, Fogel R, Randlett DL (1993) Elevated atmospheric CO_2 and feedback between carbon and nitrogen cycles. Plant Soil 151:105–117

Zak DR, Holmes WE, MacDonald NW, Pregitzer KS (1999) Soil temperature, matric potential, and the kinetics of microbial respiration and nitrogen mineralization. Soil Sci Soc Am J 63:575–584

Zak DR, Pregitzer KS, King JS, Holmes WE (2000a) Elevated atmospheric CO_2, fine roots and the response of soil microorganisms: a review and hypothesis. New Phytol 147:201–222

Zak DR, Pregitzer KS, Curtis PS, Holmes WE (2000b) Atmospheric CO_2 and the composition and function of soil microbial communities. Ecol Appl 10:47–59

Zangerl AR, Bazzaz FA (1984) The response of plants to elevated CO_2: competitive interactions among annual plants under varying light and nutrients. Oecologia 62:412–417

Zhang YM, Wu N, Zhou GY, Bao WK (2005) Changes in enzyme activities of spruce (*Picea balfouriana*) forest soil as related to burning in the eastern Qinghai-Tibetan Plateau. Appl Soil Ecol 30:215–225

Zogg GP, Zak DR, Ringelberg DB, MacDonald NW, Pregitzer KS, White DC (1997) Compositional and functional shifts in microbial communities due to soil warming. Soil Sci Soc Am J 61:475–481

Chapter 11
Evaluating the Economic and Social Impact of Soil Microbes

Harpinder S. Sandhu, Vadakattu V.S.R. Gupta, and Stephen D. Wratten

Introduction

Developments in agriculture have been very successful in producing sufficient amounts of food to meet the growing demand by human population in the twentieth century. This was possible due to breakthroughs in the development of high yielding cultivars and the combined efforts of agricultural research and public policy to support high input-based farming in different parts of the world. The 'Green Revolution' yielded desired outcomes due to agricultural intensification, accomplished through the use of input-responsive plant cultivars, increased chemical use in the form of fertilisers and chemical based plant protection measures. These modern technologies are practised in agro-ecosystems which are the largest managed ecosystems on earth and comprise 1.5 billion hectares of cropping area and about 3.5 billion hectares of pasture land area (FAO 2009).

Modern cropping and livestock farming practices have made agriculture a major driver of land use change (Vitousek et al. 1997; Goldewijk and Ramankutty 2004; UNEP 2005) which resulted in both gains and losses in several ecosystem services (ES) (Heywood 1995; Costanza et al. 1997; Daily 1997; Tilman et al. 2001). Ecosystem services are the benefits people obtain either directly or indirectly from functioning ecological systems (Daily 1997; MA 2003; Reid et al. 2005). They include products such as food, fuel, and fibre; regulating services such as climate and water regulation and flood control; and nonmaterial assets such as recreational or aesthetic benefits (de Groot et al. 2002). These ES support life on earth through

H.S. Sandhu (✉)
CSIRO Sustainable Ecosystems, PMB No. 2, Glen Osmond, South Australia 5064, Australia
e-mail: Harpinder.Sandhu@csiro.au

V.V.S.R. Gupta
CSIRO Entomology, PMB No. 2, Glen Osmond, South Australia 5064, Australia
e-mail: gupta.vadakattu@csiro.au

S.D. Wratten
Bio-Protection Research Centre, Lincoln University, P.O. Box 84, Lincoln 7647, New Zealand
e-mail: steve.wratten@lincoln.ac.nz

G.R. Dixon and E.L. Tilston (eds.), *Soil Microbiology and Sustainable Crop Production*, DOI 10.1007/978-90-481-9479-7_11, a wide range of processes and functions (Myers 1996; Daily 1997; Daily et al. 1997) which have been demonstrated to be of very high economic value in natural (Costanza et al. 1997) and agro-ecosystems (Porter et al. 2009). Current trends in degradation of agro-ecosystems threaten to alter radically not only the capabilities to produce food and fibre but also the delivery of ES by agro-ecosystems (Pretty 2002). The key challenge, therefore, is to meet the food demands of a growing population by enhancing productivity and maintaining ES (UN 1992).

Agricultural intensification has also resulted in monocultures. The remarkable scientific and technological achievements of the Green Revolution resulted in reduced plant biodiversity in some regions and promoted planned diversity of crop systems (Matson et al. 1997). Reduction in crop diversity (above ground plant diversity) can result in a variety of effects on belowground (soil biota) diversity and abundance of soil biota (Hooper et al. 2000), which in turn can affect soil function or ES. Higher diversity of plant species can result in higher belowground biodiversity through a broader range of crop residue types (resource heterogeneity) and diverse root exudates. Thus, to maintain agricultural productivity and high resource use efficiency, it is essential to maintain and even enhance the capacity of soils to provide adequate ES via functions performed by soil biota. These ES provided by soil, support crop production by providing growth media for seeds, aeration, plant support, nutrients (timely supply of nutrients and biological N inputs through fixation), water and accumulation of carbon (Brady 1990; Daily et al. 1997; de Groot et al. 2002). These bio-physical processes mediate necessary soil functions to support crop production. Biological components of soils include macro-, meso- and micro-fauna as well as micro-flora (Nannipieri et al. 2003). They carry out ecosystem functions in the form of supplying nutrients, biological control, organic matter turn over, maintaining soil structure and support the provision of ES by soil (Brussaard et al. 1997; Gupta et al. 2010). These ES are the result of complex interactions between biotic (living) and abiotic (chemical and physical) components of soil ecosystems through the universal driving forces of matter and energy (de Groot et al. 2002). To capture the benefits of soil biological activity, a better understanding of the linkages between soil biota and ecosystem function is required. Detailed discussions on the roles of different soil biota groups are in previous chapters of this Book.

An increase by 50% of the world human population, expected by 2050, will double the food demand thereby placing enormous pressures on agro-ecosystems and their components especially on soils. This could have severe social and economic impacts. Agricultural sustainability and productivity in coming decades requires not only water and crop management for more efficient use of resources but also the optimal use and management of soil fertility and physical properties to improve soil health, which relies on soil biota to maintain biological processes and biodiversity.

In the following sections, we discuss the social and economic impacts of soil microbes in agriculture using specific case studies and examples from Australia and other parts of the world.

Social and Economic Impacts of Soil Microbes

Soil microbes deliver substantial social and economic incentives annually to the global economy and to society in general by supporting various ecosystem functions and processes (Fig. 1). Soils are a reservoir of a large proportion of Earth's biodiversity and are considered as one of the last frontiers of unknown microbial diversity (Wall and Virginia 2000). Biological processes in the soil are crucial in the effective functioning and overall health of both the terrestrial (soil) and aquatic ecosystems. There is a two-way relationship between the soil microbes and agricultural activities as human activities are increasingly intervening, both positively and negatively, in the microbial diversity and the biological processes they mediate both through direct effects and indirectly by affecting the soil habitat.

Economic Impacts

Global economic growth is projected to remain in the 1.7% range while growth in the developing-countries is expected to average 2.1% in 2009 (World Bank 2009). This growth can change the demand side of the world food equation. High income growth in developing and least developed countries may readily lead to increased food consumption. The estimated world population by the end of the twenty-first

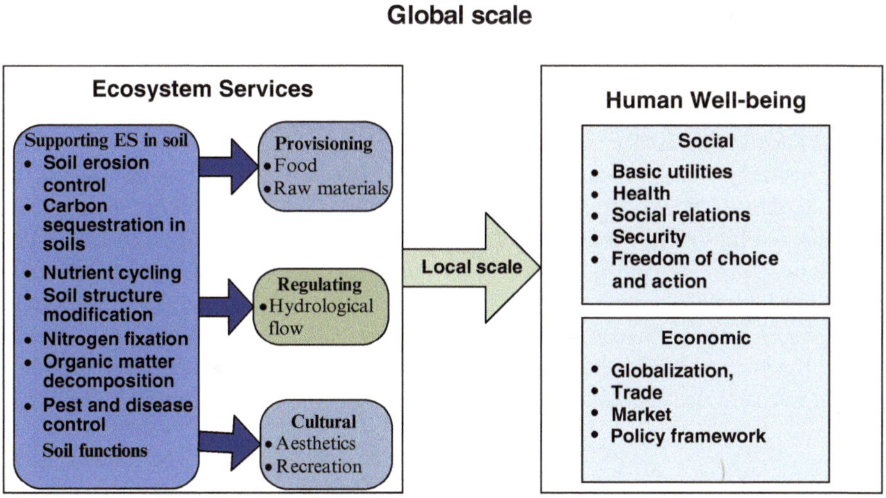

Fig. 1 Linkages between ecosystem functions provided by soil microbes and human well-being at local and global scale (Modified from Millennium Assessment, Reid et al. 2005)

century is approx. ten billion and to feed this population will require considerable efforts and measures to improve and maintain soil health (Lal 2009) apart from other measures such as improved resource use efficiency and improved formulation of public policies.

Soils sustain global ecosystems as plants and animals both need soil or its products for their survival. Terrestrial ecosystems provide both essential ES as well as food supply for humans (FAO 2007). The living populations in soil (soil microbes) are responsible for carrying out 90% of the soil processes (Coleman et al. 2004). Thus soil biota provides major inputs to the global agricultural economy through the functions they perform that lead to the provision of ES. These ES provided by soil organisms have been demonstrated to be of very high economic value (US$1.5 trillion/year globally) (Pimentel et al. 1997). An example is biological nitrogen fixation (BNF) in agriculture by legume-rhizobia symbiosis. The global estimates of biological nitrogen fixation into agriculture range between 33 and 46 million tonnes of nitrogen each year (Herridge et al. 2008), which approximately equates (at current fertilizer prices) to $50–70 billion annually. But the importance of legume and fertilizer nitrogen sources varies with the region and is influenced by a range of socio-economic factors including patterns of land use, farming traditions and population density. The use of nitrogenous fertilizers has greatly increased global food production. The benefits of using these fertilisers are numerous, but nitrogen fixing legumes are driven by solar energy and differentiated from the fertiliser nitrogen which requires non-renewable fossil fuels (Peoples et al. 2008). Nitrogen fertilizers also have high external costs (Pretty 2005) largely associated with losses of some of the nitrogen applied, high energy costs, contributions to net carbon dioxide emissions. Recent increases with the costs of nitrogenous fertilizers also highlight the need for this element to be supplied to crops in the most efficient ways possible. Inefficient use of fertilizer inputs and concomitantly increasing costs can threaten the economic sustainability of food production practices that solely depend upon external inputs. However, legumes offer to restore the functional biodiversity in soils and to yield a more sustainable agriculture (Altieri 1999).

An economic assessment of the global microbial inoculant industry (Phillips 2004) embraces three types of technology: nitrogen fixing rhizobial bacteria, phosphate-solubilising *Penicillium* fungi and insecticidal inoculants. These technologies can provide significant economic and social gains, e.g. legumes contribute 20% of the global food proteins (Montanez 2000). Legumes appear to fix on an average 277 kg N ha^{-1} $year^{-1}$ (FAO 1984). At this rate, total inoculants' area of 43 million hectares will fix about 12 million tons of nitrogen, equal to about 13% of current nitrogen use globally each year (Phillips 2004) which is worth US$7.8 billion annually at the price of nitrogen (US$0.65 kg^{-1} N; 2004). Recent increases in the cost of nitrogen fertilizer (US$ > 1.00 kg^{-1} N) would only increase the importance of nitrogen inputs through biological fixation. In addition, there is a growing push to utilize the antibiotic potential of soil microorganisms (bacteria, actinomycetes and fungi) as biocontrol agents to reduce the impact of plant pathogens on agricultural production (Pal and McSpadden-Gardener 2006).

Nutrient cycling (Brady and Weil 2004) is one of the most important functions provided by soil microbes in soil. Sandhu and colleagues (Sandhu et al. 2008) investigated the economic value of nutrient cycling carried out by soil microbes and the role of land management practices in the maintenance and enhancement of this ES (nutrient cycling) in agricultural land. They quantified the economic value of this ES at the field level based on an experimental approach. The study sites included 29 arable fields, distributed over the Canterbury Plains in New Zealand and comprised 14 organic and 15 conventional fields. The economic value of this ES carried out by soil microbes was in the range of US$25.60 to 425.50 $ha^{-1}year^{-1}$ (mean US$160.65 $ha^{-1}year^{-1}$) in organic fields and US$30.00–348.00 $ha^{-1}year^{-1}$ (mean US$142. $ha^{-1}year^{-1}$) in conventional ones (Sandhu et al. 2008).

Microorganisms play important roles in the degradation of the diverse range of agrochemicals and the bioremediation of both soil and water environments. There is an extensive effort to utilise biological processes for the removal of contaminants and take advantage of the amazing catabolic diversity and adaptability of microorganisms to degrade toxic contaminants into non-toxic forms (Singh and Ward 2004; Diaz 2008).

Social Impacts

For centuries many communities have survived by harvesting resources from nature. In spite of the unprecedented increase in agricultural production during the 'Green Revolution', the majority of small farmers in developing countries are too poor to purchase seeds and inputs and increase their farm productivity. Inadequate access to appropriate technical support with which to harness benefits from the new technology is a key bottleneck inhibiting improved productivity. Food insecurity has been identified as a key issue by United Nations in its Millennium Development Goals (MDGs) to be achieved by 2015 (UN 2005) which contains social, economic and environmental objectives. The vast majority of the world's poor farmers reside in developing countries in South Asia and the African continent. The success of the Green Revolution by managing soil constraints has not been realised by these small-scale farmers. The United Nations led Millennium Assessment (Reid et al. 2005) concluded that natural and modified ecosystems and the services they provide are declining. It also pointed out the consequences for global stability if that rate of decline continues. The livelihood of large populations in developing and least developed countries depend on healthy natural and agro-ecosystems. There is greater need to address land degradation and soil health so that agriculture can sustain growing food demand and increasing human population.

Human livelihoods and social and economic well-being are inextricably linked to our dependence on soil health (see Box 1). Therefore, environmental and social sustainability can be achieved by maintaining soil. Soil health and fertility is critical in improving impoverished agricultural communities, which constitute approximately 2.4 billion people worldwide (Lal 2009).

Box 1 Soil Health and Social and Economic Well-being

Haiti is one of the least-developed countries. It ranks very low in the United Nations Human Development Index 2006 (UNDP 2006). Agriculture provides work to about two-thirds of its population but this makes up only one-third of the GDP of low income countries. Per capita food production has dropped 30% from 1991 to 2002. There are several constraints. Farmers are unable to afford prime agricultural land, irrigation or inputs and increase farm productivity.

Sustainable Organic Integrated Livelihoods (SOIL) is a non-profit organization which is dedicated to work with communities to increase agricultural production and to provide livelihoods on farm. For this, SOIL dedicated its efforts to protect soil resources and transform wastes into resources. Composting toilets are built in rural communities to get much needed organic matter and fertility back into fields. Composting of such organic materials is performed by soil macro- and micro-fauna. Soil microbes like fungi and bacteria decompose the composted material resulting in the release nutrients for plant use. Incorporating organic matter can also lead to improvements in the soil structure and soil aggregation. Managing soil health by using this technology has potential to improve farm production and household income.

SOIL thus promotes integrated approaches that utilize biological activities to tackle the socio-economic issues and help achieve sustainability. SOIL is promoting close collaboration between local communities and international academics and activists and channels resources to build the soil health, strengthen and empower grassroots communities (http://www.oursoil.org).

Soils had profound social impacts in ancient civilizations (Marris 2006). In the Amazon, *terra preta* is a soil rich in vital minerals such as: calcium, manganese, phosphorus and zinc, which are scarce in most tropical soils. *Terra preta* is blacker than the blackest coffee and is very deep soil extending from the surface down as much as 2 m. In the Amazon region it is found where humans inhabited the area and it is known to be artificially made by mankind between 450 BC and AD 950 to support agriculture production. Unlike ordinary tropical soils, *terra preta* remains fertile after centuries of exposure to tropical sun and rain. It has been suggested that the charcoal component of the *terra preta* provides an important habitat for the microbial populations hence improving its soil biological fertility.

The use of a variety of pesticides (e.g. insecticides, herbicides, fungicides) has become an integral part of modern farming systems with the main aim of removing constraints caused by insects and plant pathogens and competition from weeds to achieve higher productivity and resource efficiency. Microorganisms play an important role in the degradation of all types of pesticides in soil and water. Chemical pollution of land and water resources has the potential to threaten environmental sustainability and can result in harmful effects on human health.

In addition, there are a number of examples of the contamination of natural environments by the pollutants transported from nearby agricultural fields (Young et al. 1989; Manz et al. 2001). Although the benefits from judicious use of pesticides have been demonstrated to provide economic benefits, long-term exposure of farmers to pesticides may cause increased health costs (Pingali and Roger 1995) and associated social impacts to their quality of life. Pingali and Roger (1995) identified the potential for social gains from reducing pesticide use in Philippine rice production. They suggested that productivity losses from reduced pest control can be offset by the gains from improved farmer health.

Moderate and extreme poverty occur in many least developed and developing countries, in which more than 25% of the population lives on less than US$2 per day. The most important reason for declining economic growth of these countries is their poor food productivity. Most of the poverty occurs in rural areas where populations are increasing while food production is stagnant or declining in many cases. Resource poor farmers cannot afford expensive inputs. Therefore, crop yields are adversely affected by extractive farming. Technologies such as biological nitrogen fixation (BNF), no-till husbandry, mulching with residues and cover cropping and integrated nutrient management that enhance the functions performed by soil microorganisms can play a crucial roles in reducing environmental pollution, raise yields and alleviate poverty. Biofertilisers consisting of beneficial microorganisms can increase the growth and yield of plants (Mulongoy et al. 1992). Use of biofertilisers helps in reducing off-site environmental pollution and avoiding non-target interactions.

Soil erosion by wind is a natural process but its occurrence and severity has increased along with the expansion of agriculture. The major affects of soil loss can occur through dust storms affecting human health, reducing visibility, diminishing air quality and disrupting electricity supplies (Leys 2003). Williams and Young (1999) estimated the off-site impacts of dust storms in South Australia to be in the range of $2–15 million annually, largely due to impacts on human health (asthma and general respiratory problems). There has been significant decrease in the annual frequency of dust storms in Australia since the 1970s (Hamblin 2001) largely due to the improved control of rabbits, the spread of woody weeds such as the invasive *Acacia nilotica* and the adoption of conservation tillage. Some farming practices impair soil structure through tillage and that increases the likelihood of dust storms. Soil microbes play crucial roles in maintaining soil structure through their involvement in aggregate formation and subsequent stabilization and in turn helping in decreasing soil erosion that contributes towards reduction in dust storms.

Ecosystem Functions

Soil organisms contribute to a wide range of services essential for the sustainable functioning of all ecosystems. These functions include carbon and nutrient (nitrogen, phosphorus and sulphur) cycling, turnover of soil organic matter, soil carbon sequestration and reducing greenhouse gas emissions; modifying soil physical structure and water regimes, enhancing the amount and efficiency of nutrient acquisition by the vegetation, and enhancing plant health (Wall and Virginia 2000; Brussaard et al. 2007). In addition, the diversity and complexity of soil microbes is also attributed to the suppression of plant diseases (Hornby 1983; Gupta and Neate 1999). Soils not only harbour the majority of plant pathogens, they are the sources of several human pathogens (Ajello 1956). Soil organisms are grouped into different functional and trophic groups in order to illustrate linkages between various groups of biota and specific ecosystem processes (Lavelle and Spain 2001; Coleman et al. 2004). The soil biota mediated services are not only essential to the functioning of natural ecosystems but constitute an important resource for the sustainable management of agricultural systems (Altieri 1999; Roper and Gupta 2007).

The importance of ecosystem functions carried out by soil microbes in low-input and organic systems is generally accepted however, their importance in high input agro-ecosystems may be neglected because external inputs of fertilizers and pesticides are expected to replace the products of biological processes (Barrios 2007). Optimum functioning of biological processes is necessary, however, for the efficient use of human inputs and reducing or eliminating any off the [illegible]

Recent studies have proposed an ES framework to examine the linkages between biodiversity, ecosystem functions and services (Reid et al. 2005). According to the Millennium Ecosystem Assessment (2003), ES can be classified into four categories depending on their functions – provisioning, supporting, regulating and cultural services. Soils contribute to all four categories of ES (Table 1). The ecosystem functions carried by soil microorganisms contribute to the provision of ES that leads to improved economic and ecological sustainability of agro-ecosystems. Consequently, it has potential to improve the existing socioeconomic conditions of the farming community

Table 1 Relationship between soil functions and socio-economic impacts

	Ecosystem functions	Social impacts	Economic impacts
Regulating Functions			
1	Maintenance of soil structure	High	High
2	Regulation of soil hydrological processes	Low	Medium
3	Gas exchanges and carbon sequestration	High	High
4	Soil detoxification	Medium	Low
Supporting functions			
5	Nutrient cycling	Medium	High
6	Decomposition of organic matter	Medium	High
7	Suppression of pests, parasites and diseases	Medium	High
8	Symbiotic and asymbiotic relationships with plants and their roots	Low	High
9	Plant growth control (positive and negative)	Low	Medium
Provisioning functions			
10	Sources of food and medicines	Low	Medium
Cultural functions			
11	Aesthetics/spiritual	Low	Low

(Altieri 1995). Linkages between ecosystem functions performed by soil microbes and their social and economic impacts are summarised in Table 1. Two regulating functions have high social and economic impacts through maintenance of soil structure and by regulating gas exchanges and carbon sequestration. Whereas, three supporting functions (nutrient cycling, decomposition of organic matter and suppression of pests, parasites and pathogens) have medium social impacts but high economic impacts as food production depends on these functions that are performed by soil microorganisms. Symbiotic relationships have low social but high economic impacts. Soil microbes have low social and economic impacts for provisioning and cultural functions.

Supporting Functions

The cycling of nutrients is a critical ecosystem function carried out by soil microbes. Soil microbes facilitate biological nitrogen fixation (BNF) (e.g. *Rhizobium* spp.) and phosphorus uptake through arbuscular mycorrhizal fungi (AMF) (Smith and Read 1997). Nitrogen fixation through the legume-*Rhizobium* symbiosis dominates the global inputs of biologically fixed nitrogen. It is estimated that ~50 million tonnes of nitrogen is harvested globally each year in food crops and microbial functions play an important role in making available of both biologically fixed and fertilizer nitrogen inputs. Soil microorganisms also perform the decomposition of organic matter and its transformation and play essential roles in biogeochemical cycles of essential plant nutrients, e.g. nitrogen, phosphorus and sulphur (Coleman et al. 2004). Microorganisms play an important role in the degradation of all types of pesticides, i.e. insecticides, herbicides and fungicides, in soil and water. Soil microbial functions support the production of provisioning goods and services in agriculture (Porter et al. 2009) by providing essential activities in soil and resulting in food and fibre production along with other essential ES.

Regulating Functions

Soil biota regulate essential ecological processes through bio-geochemical cycles and other biospheric processes. Soil organisms help in aggregate formation and the maintenance of soil structure. Micro-aggregates, which are stabilized by persistent organic binding agents of microbial origin, are bound together by temporary (roots and fungal hyphae) and transient (plant and microbial polysaccharides) binding agents forming macro-aggregates (Tisdall and Oades 1982).

Soil organisms have profound influences on global crop losses due to soil-borne pests and pathogens, as described elsewhere in this Book. For example, soil-borne diseases are one of the major bottlenecks to achieving maximum yield potential in rainfed and irrigated crops all over the World. Brennan and Murray (Murray and Brennan 2009) estimated that yield losses resulting from the four major soil-borne diseases of wheat crop in Australia would cost more than US$325 million annually. In undisturbed natural ecosystems with high above-ground diversity, soil-borne plant disease epidemics are rare when compared with disturbed agricultural systems (Cook and Baker 1983; Schisler and Linderman 1984). Poor nutrition can accentuate the effects resulting from pathogen and pest incidences resulting in productivity and economic losses. In general, there is a strong relationship between soil biota, fertility and plant health (Altieri and Nichols 2003), and the diversity of microbial communities is one of the critical factors contributing to the proper functioning of soil suppressiveness (Gupta and Neate 1999; Garbeva et al. 2004).

Case Studies

Assessing the impacts of soil microbes is difficult as there are so many ways by which their actions can have direct or indirect influence on societies or economies. These effects are the product of complex interactions between microbial processes, physical and chemical components of soils and are influenced by environmental and management practices. This study includes two case studies which elaborate the role of soil in general and soil microbes in particular and their socio-economic impacts through their functions.

The first case study that we discuss below, deals with the adoption of new technology that enhances soil microbial population and/or the processes which they mediate yielding multiple benefits and also having social and economic effects. A recent study by Llewellyn and colleagues (Llewellyn et al. 2009) examined trends in the adoption of no-till husbandry across southern cropping regions in Australia using socio-economic data. It identified opportunities for the research, its adoption through extension and changing policies to achieve improved farming systems with advantages for both private and public good.

The second case study elaborates the role of soil microbes in improving soil health in sub-Saharan Africa, through the discussion of various individual projects.

Adoption of No-Till and Conservation Farming Practices in Australia

Conservation agriculture and no-till husbandry practices enhance soil biological activity and modify the diversity and population structure of soil microflora and fauna by reducing soil disturbance and improving the levels of biologically available carbon (energy source) (Roper and Gupta 1995; Kladivko 2001). Long-term adoption of these practices could lead to the removal of biological constraints to plant growth and improve the overall soil quality. Thus soil biota plays a crucial role in the successful adoption of no-till husbandry and conservation farming (Stubbs et al. 2004). No-till farming also reduces fuel inputs in particular during non-cropping season. According to one estimate (Derpsch 2005), no-tillage and conservation farming practices cover 95 million hectare worldwide. In Australia, it covers about 9 million hectares spread over the main cropping districts. Data in this study (Llewellyn et al. 2009) were collected from 1,172 primary cropping decision-makers in each household and only farms cropping greater than 200 ha in a 'normal' season were included. The results show that no-tillage seeding practices have now been adopted by the majority of grain growers across all regions. The proportion of growers using at least some no-tillage is now peaking at levels of around 90% in many regions. In regions with relatively low adoption 5 years ago, there have been very rapid increases in adoption, particularly in the period 2003–2006.

In general, relatively few growers appear to have made the decision not to adopt no-tillage in the near future and continue cropping without adopting it at all. The proportion of growers using no-tillage is expected to exceed 80% in a majority of regions by 2013. Substantial net gains in no-tillage adoption have been observed and further net increases are expected in the future, except where no-tillage adoption is already reaching a plateau as it approaches full adoption.

For no-tillage practices to be widespread in the landscape there needs to be both high adoption rates and extensive use by the farmers. The results show that it remains common for no-tillage adopters to still use some forms of cultivation. However, the extent of no-tillage use by adopters is typically high. The average percentage of crop sown using no-tillage by adopters exceeds 70% in all regions.

Because a majority of growers in most districts still use some tillage, it should be expected that economic and agri-environmental factors may cause seasonal shifts in extent of tillage use. Just as falls in the price of glyphosate herbicide led to increased no-tillage adoption (D'Emden and Llewellyn 2006), growers in this study indicated that recent rises in the price of this chemical have led to their increased use of tillage in many regions. In regions such as those in Western Australia where it is more common that 100% of crop area is sown without tillage, the rising price of glyphosate has had little or no reported influence on tillage use. No-tillage adopters are more than twice as likely to use a paid cropping consultant. This suggests that availability of high quality cropping advisory support may be an important factor in encouraging further adoption and/or maintaining its extensive use. Use of no-tillage husbandry is also strongly associated with larger farm sizes.

This study shows the remarkable diffusion of no-tillage across Australian cropping regions, confirming that it is highly adoptable and that extensive use has so far been sustained across a wide range of agro-ecosystems. The last 5 years have seen many regions with previously lower no-tillage adoption rapidly adapt to levels similar to early-adopting regions. It needs to be recognised however, that based on current indicators, over the next 5 years, and possibly at peak adoption, there will still be several regions with a combination of a relatively lower no-tillage adoption levels and lower extent of use than the average, which stands at 70%. To achieve the same reductions in cultivation use and erosion risk in these regions requires new and innovative approaches which encourage adoption through research, development and extension of soil conservation practices.

Long-term adoption of no-tillage practices by farmers in Australia has the potential for significantly improving soil physical, chemical and biological conditions in particular when combined with retention of crop residues (Fig. 2a and b). This often results in increases in microbial population and diversity (Stubbs et al. 2004). This can lead to large improvements in productivity, profitability and sustainability in agriculture (Williams and Young 1999) that has potential to enhance the socio-economic conditions of the rural communities.

Fig. 2 (**a**) Extensive cultivation with no stubble protection at Loxton, South Australia. (**b**) Intensive cropping with zero-tillage and stubble retention at Waikerie, South Australia Soil type at both sites is Arenic Calcisol and a Mediterranean climate

Soil Improvement in Sub-Saharan Africa

Agricultural land in sub-Saharan Africa is considered to be severely degraded (FAO 1994; GEF 2003). Increasing agricultural productivity to feed growing population in these regions is possible only if substantial efforts are made to restore and maintain the productivity of their soils (Lal 2009). Various technologies are available to restore and maintain the biophysical structure of soils (conservation tillage and no-tillage farming, controlled grazing, mulching with natural materials, leaving crop residue *in situ*, application of manure and biosolids, incorporation of cover crops in the rotation cycle), improve the efficiency of water and nutrient use (integrated nutrient management, biological nitrogen fixation) and manipulate the rhizosphere to enhance the soil and plant performance (NRC 2008). Another useful technology is the incorporation of grain legumes into rotations to improve nutritional components in the food and also indirect benefits through nitrogen fixation.

Soil degradation is common in Sub-Saharan Africa due to the nutrient deficiencies in soils (Henao and Baanante 2006) which affects 95 million hectares of arable land. This has serious implications on the social and economic conditions of the farmers. Large investments and efforts are required for soil restoration to make them productive again (IFDC 2006).

Social, political, and cultural factors have deep influences on the practices in Sub-Sahara that affect soil quality. Soil quality can be restored with established management practices by increasing carbon content, enhancing water infiltration, reducing erosion, creating a nutrient surplus and encouraging beneficial organisms. Research has shown that these management practices will increase crop productivity and improve soil health (Lal 1987, 2006; Kapkiyai et al. 1999; Sanchez 2002; Wani et al. 2003; Singh et al. 2005).

There have been several barriers to the adoption of new technologies including weak government institutions and extension, inadequate infrastructure, and resource-poor agricultural systems.

In spite of these barriers, African Highland Initiative (AHI; http://www.african-highlands.org) is working to improve livelihoods and reverse natural resource degradation in the densely settled highlands of eastern and central Africa. The AHI is an eco-regional programme of the Consultative Group for International Agricultural Research (CGIAR) and a network of the Association for Strengthening Agricultural Research in Eastern and Central Africa (ASARECA) hosted by the World Agroforestry Centre (ICRAF). The AHI's targeted beneficiaries and partners in this work include national and international research organizations and networks, development organizations, local government, civil society organizations, service providers, policy makers, community-based organizations, and farmers.

In one of its studies, this initiative helped farmers of Tanzania to restore and improve soil nutrient content by using native shrubs (Wickama and Mowo 2001). Farmers in the Tanzanian village of Kwalei use very little mineral fertiliser to improve their crop yields as it is expensive. In this region, farmers used leaves of certain local

Table 2 Inventory of useful shrubs identified by farmers to improve soil fertility in Tanzania (Wickama and Mowo 2001)

Local name in Kisambaa	Botanical name	Family
Tughutu	*Vernonia subligera* (O.Hoffn.)	Compositae
Mhasha	*Vernonia amyridiantha* (Hook, J.)	Compositae
Mshai	*Albizia schiniperiana*	Mimosaceae
Mkuyu	*Ficus vallis-choudae* (Del.)	Moraceae
Sopolwa	*Kalanchoe crinata* (Andrew) Haw.	Crassulaceae
Tundashozi	*Justicia glabra* (Roxb.)	Acanthaceae
Boho	*Bothriocline tementosa* (S. Moore) M. Gilbert	Compositae

shrubs as green manure, these seemed to improve soil fertility when ploughed into the fields. Researchers and extension workers did not know their taxonomic description or nutrient value, and it was agreed that an inventory should be made of this important local source of nutrients. Farmers identified seven shrubs that seemed to improve soil nutrient content (Table 2). Their observations were confirmed by soil sampling and mineralisation studies. Participatory research methods helped to identify the important shrubs and shared local knowledge has helped to increase soil health.

In another study, Amede and Kirkby (2004) used socio-economic criteria to develop guidelines for the integration of legumes into multiple cropping systems of East African Highlands. They conducted participatory research to evaluate the performance of six legume cover crops and two food crops in southern Ethiopian Highlands, to be used for soil fertility improvement.

These examples show that using local knowledge combined with the scientific and technical expertise it is possible to manage natural resources and to increase food production more beneficially in Sub-Saharan African countries.

Summary

Soil microbes have social, environmental and economic impacts through the functions they perform in soils. Improved management of soils and the functions performed by soil microbes are imperative to ensure food security and achieve MDGs. Modern agricultural technology has been very successful in reducing hunger and poverty in some developing countries in the last 50 years (Smil 2000; Tilman et al. 2002; McNeeley and Scherr 2003; Federico 2005); however with increasing new demand to food during the next 20–50 years and lack of additional land with which to expand agriculture, a new approach is needed in order to achieve the next generation of the Green Revolution. This involves better utilization of ES (which requires the services of soil microbes) and extension of technological developments both in terms of cultivars and farming systems to all parts of the world. The need may be greater in the regions that are dependant on low fertility soils and have poor economic status. In such environments there is an even greater need to utilize microbial services in support of higher productivity systems. The increasing world population also puts pressure on the health of both soil and water systems in terms of their abilities to provide the diverse range of ES. In addition there is a greater urgency to utilize the enormous and wide ranging capabilities of microorganisms to maintain or ameliorate and improve the health of soil and water systems for both social and economic needs of the society.

Therefore developing sustainable agriculture by incorporating traditional and modern methods of farming is the most appropriate choice for developing countries (UN 2008). Management of soils by maintaining and enhancing soil microbial functions will help to alleviate some of the barriers as explained in the case studies discussed above. The challenge in promoting economic and social benefits from biological processes to farming community is to allow the ecosystem functions in farming systems to operate optimally with a minimum of external inputs. The current and future challenge is to develop cost-effective, low-input eco-technologies, for their rapid implementation and uptake by end-users (Swaminathan 2001). Policy makers should aim for sustained economic growth within ecological constraints to eliminate hunger and poverty as agreed in the MDGs (UN 2008).

References

Ajello L (1956) Soil as natural reservoir for human pathogenic fungi. Science 123:876–879

Altieri MA (1995) Agroecology: the science of sustainable agriculture. Westview Press, Boulder, CO, p 433

Altieri MA (1999) The ecological role of biodiversity in agroecosystems. Agric Ecosyst Environ 74:19–31

Altieri MA, Nichols CI (2003) Soil fertility management and insect pests: harmonizing soil and plant health in agroecosystems. Soil Till Res 72:203–211

Amede T, Kirkby R (2004) Guidelines for integration of legumes into the farming systems of East African highlands. In: Bationo A (ed) Managing nutrient cycles to sustain soil fertility in Sub-Saharan Africa. Academic Science Publishers, Nairobi, Kenya, pp 43–64

Barrios E (2007) Soil biota, ecosystem services and land productivity. Ecol Econ 64:269–285

Brady NC (1990) The nature and properties of soils. Macmillan, New York

Brady NC, Weil RR (2004) Elements of the nature and properties of soils. Prentice Hall, Old Tappan, NJ

Brussaard L, Behan-Pelletier VM, Bignell DE, Brown VK, Didden WAM, Folgarait PJ, Fragoso C, Wall Freckman D, Gupta VVSR, Hattori T, Hawksworth D, Klopatek C, Lavelle P, Malloch D, Rusek J, Sbderstrom B, Tiedje JM, Virginia RA (1997) Biodiversity and ecosystem functioning in soil. Ambio 26:563–570

Brussaard L, de Ruiter PC, Brown GG (2007) Soil biodiversity and agricultural sustainability. Agric Ecosyst Environ 121:233–244

Coleman DC, Crossley DA, Hendrix PF (2004) Fundamentals of soil ecology, 2nd edn. Elsevier Academic, Amsterdam, The Netherlands

Cook RJ, Baker KF (1983) The nature and practice of biological control of plant disease. The American Phytopathological Society, St Paul, MN, p 539

Costanza R, d'Arge R, De Groot R, Farber S, Grasso M, Hannon B, Limburg K, Naeem S, O'Neill RV, Paruelo J, Raskin RG, Sutton P, van den Belt M (1997) The value of the world's ecosystem services and natural capital. Nature 387:253–260

D'Emden FH, Llewellyn RS (2006) No-till adoption decisions in southern Australian cropping and the role of weed management. Aust J Exp Agr 46:563–569

Daily GC (ed) (1997) Nature's services: societal dependence on natural ecosystems. Island Press, Washington, DC

Daily GC, Alexander S, Ehrlich PR, Goulder L, Lubchenco J, Matson PA, Mooney HA, Postel S, Schneider SH, Tilman D, Woodwell GM (1997) Ecosystem services: benefits supplied to human societies by natural ecosystems. Issues Ecol 2:18

de Groot RS, Wilson MA, Boumans RMJ (2002) A typology for the classification, description and valuation of ecosystem functions, goods and services. Ecol Econ 41:393–408

Derpsch R (2005) The extent of conservation agriculture adoption worldwide: implications and impact. In: Proceedings of the third world congress on conservation agriculture: linking production, livelihoods and conservation, Nairobi, Kenya

Diaz E (2008) Microbial biodegradation: genomics and molecular biology. Caister Academic Press/Horizon Scientific Press, Norfolk, UK, p 402

FAO (1984) Legume inoculation and their use. Food and Agriculture Organization of the United Nations, Rome

FAO (1994) Land degradation in South Asia: its severity, causes and effects upon the people. World Soil Resources Report 78. FAO, Rome

FAO (2007) The state of food and agriculture: paying farmers for environmental services, vol 38, FAO Agriculture Series 2007. Food and Agriculture Organisation of the United Nations, Rome

FAO (2009) FAOSTAT Division. http://faostat.fao.org

Federico G (2005) Feeding the world: an economic history of agriculture, 1800–2000. Princeton University Press, Princeton, NJ

Garbeva P, Veen JA, Elsas JD (2004) Microbial diversity in soil: selection of microbial populations by plant and soil type and implications for disease suppressiveness. Annu Rev Phytopathol 42:243–270

GEF (Global Environmental Facility) (2003) What kind of world? The challenge of land degradation. Global Environmental Facility, Washington, DC

Goldewijk KK, Ramankutty N (2004) Land cover change over the last three centuries due to human activities: the availability of new global data sets. GeoJournal 61:335–344

Gupta VVSR, Neate SM (1999) Root disease incidence-A simple phenomenon or a product of diverse microbial/biological interactions. In: Magarey RC (ed) Proceedings of the first Australasian soil borne disease symposium. BSES, Brisbane, Australia, pp 3–4

Gupta VVSR, Rovira AD, Roget DK (2010) Principles and management of soil biological factorsfor sustainable rainfed farming systems. In: Tow P, Cooper I, Partridge I, Birch C (eds) Rainfed farming systems. Springer Science and Business Media

Hamblin A (2001) Australia state of the environment report 2001 – land theme report. CSIRO Publishing on behalf of the Department of the Environment and Heritage, Canberra. Dust storm and accelerated erosion indices

Henao J, Baanante C (2006) Agricultural production and soil nutrient mining in Africa: implications for resource conservation and policy development. International Center for Soil Fertility & Agricultural Development (IFDC), Muscle Shoals, AL

Herridge DF, Peoples MB, Boddey RM (2008) Marschner review: global inputs of biological nitrogen fixation in agricultural systems. Plant Soil 311:1–18

Heywood VH (ed) (1995) United Nations environment program, global biodiversity assessment. Cambridge University Press, Cambridge

Hooper DU, Bignell DE, Brown VK, Brussaard L, Dangerfield JM, Wall DH, Wardle DA, Coleman DC, Giller KE, Lavelle P, Van der Putten WH, De Ruiter PC, Rusek J, Silver WL, Tiedje JM, Wolters V (2000) Interactions between aboveground and belowground biodiversity in terrestrial ecosystems: patterns, mechanisms, and feedbacks. Bioscience 50:1049–1061

Hornby D (1983) Suppressive soils. Ann Rev Phytopathol 21:65–85

IFDC (International Center for Soil Fertility & Agricultural Development) (2006) African soil exhaustion. Science 312:31

Kapkiyai JJ, Karanja NK, Qureshi JN, Smithson PC, Woomer PJ (1999) Soil organic matter and nutrient dynamics in a Kenyan Nitisol under a long-term fertilizer and organic input management. Soil Biol Biochem 31:1773–1782

Kladivko EJ (2001) Tillage systems and soil ecology. Soil Till Res 61:61–76

Lal R (1987) Managing soils of sub-Saharan Africa. Science 236:1069–1076

Lal R (2006) Managing soils for feeding a global population of 10 billion. J Sci Food Agric 86:2273–2284

Lal R (2009) Soils and food sufficiency. A review. Agron Sustain Dev 29:113–133

Lavelle P, Spain AV (2001) Soil ecology. Kluwer, Dordrecht, The Netherlands

Leys J (2003) Wind erosion, 2nd edn. Centre for Natural Resources, NSW Department of Infrastructure, Planning and Natural Resources (CNR 2003.069)

Llewellyn R, D'Emden F, Gobbett D (2009) Adoption of no-till and conservation farming practices in Australian grain growing regions: current status and trends. CSIRO report, Adelaide, Australia

MA (Millennium Ecosystem Assessment) (2003) Ecosystems and human well-being: a framework for assessment. Island Press, Washington, DC

Manz M, Wenzel KD, Dietze U, Schüürmann G (2001) Persistent organic pollutants in agricultural soils of central Germany. Sci Total Environ 27:187–198

Marris E (2006) Black is the new green. Nature 442:624–626

Matson PA, Parton WJ, Power AG, Swift MJ (1997) Agricultural intensification and ecosystem properties. Science 277:504–509

McNeeley JA, Scherr SJ (2003) Ecoagriculture. Island Press, Washington, DC

Montanez A (2000) Case study B2-Overview and case studies on biological nitrogen fixation: perspectives and limitations. Soil Biodiversity Portal, Land and Water Division, Food and Agriculture Organisation of the United Nations, Rome

Mulongoy K, Gianinazzi S, Roger PA, Dommergues Y (1992) Biofertilisers: agronomic and environmental impacts and economics. In: DaSilva EJ, Ratledge C, Sasson A (eds) Biotechnology: economic and social aspects: issues for developing countries. Cambridge University Press, London

Murray GM, Brennan JP (2009) The current and potential costs from diseases of wheat. Grain Research Development Corporation Report, Kingston, ACT, Australia, p 70. http://www.grdc.com.au/

Myers N (1996) Environmental services of biodiversity. Proc Natl Acad Sci USA 93:2764–2769

Nannipieri P, Ascher J, Ceccherini MT, Landi L, Pietramellara G, Renella G (2003) Microbial diversity and soil functions. Eur J Soil Sci 54:655–670

NRC (National Research Council) (2008) Emerging technologies to benefit farmers in sub-Saharan Africa and South Asia. Committee on a study of technologies to benefit farmers in Africa and South Asia, National Research Council. The National Academies Press, Washington, DC

Pal KK, McSpadden-Gardener B (2006) Biological control of plant pathogens. The Plant Health Instructor. doi:10.1094/PHI-A-2006-1117-02

Peoples MB, Hauggaard-Nielsen H, Jensen ES (2008) The potential environmental benefits and risks derived from legumes in rotation. In: Emerich DW, Krishnan HB (eds) Nitrogen fixation in crop production. Agronomy monograph 52, ASA, CSSA, SSSA, Madison, WI, USA

Phillips PWB (2004) An economic assessment of the global inoculant industry. Crop management. http://www.plantmanagementnetwork.org/sub/cm/review/2003/industry

Pimentel D, Wilson C, McCullum C, Huang R, Dwen P, Flack J, Tran Q, Saltman T, Cliff B (1997) Economic and environmental benefits of biodiversity. Bioscience 47:747–757

Pingali PL, Roger PA (eds) (1995) Impact of pesticides on farmer health and the rice environment. Kluwer/International Rice Research Institute, Norwell, MA/ Los Baños, Philippines

Porter J, Costanza R, Sandhu H, Sigsgaard L, Wratten S (2009) The value of producing food, energy and ecosystem services within an agro-ecosystem. Ambio 38:186–193

Pretty JN (2002) Agri-culture: reconnecting people, land and nature. Earthscan, London

Pretty JN (2005) The pesticide detox: towards a more sustainable agriculture. Earthscan, London

Reid WV, Mooney HA, Cropper A, Capistrano D, Carpenter SR, Chopra K, Dasgupta P, Dietz T, Duraiappah AK, Hassan R, Kasperson R, Leemans R, May RM, McMichael AJ, Pingali P, Samper C, Scholes R, Watson RT, Zakri AH, Shidong Z, Ash NJ, Bennett E, Kumar P, Lee MJ, Raudsepp-Hearne C, Simons H, Thonell J, Zurek NB (2005) Millennium ecosystem assessment synthesis report. Island Press, Washington, DC

Roper MM, Gupta VVSR (1995) Management practices and soil biota. Aust J Soil Res 33:321–339

Roper MM, Gupta VVSR (2007) The living soil – an agricultural perspective. Microbiol Aust 28:104–106

Sanchez PA (2002) Soil fertility and hunger in Africa. Science 295:2019–2020

Sandhu HS, Wratten SD, Cullen R, Case B (2008) The future of farming: the value of ecosystem services in conventional and organic arable land. An experimental approach. Ecol Econ 64:835–848

Schisler DA, Linderman RG (1984) Evidence from the involvement of the soil microbiota in the exclusion of *Fusarium* from coniferous forest soils. Can J Microbiol 30:142–150

Singh A, Ward OP (2004) Biodegradation and bioremediation. Springer, Heidelberg, p 204

Singh Y, Singh B, Timsina J (2005) Crop residue management for nutrient cycling and improving soil productivity in rice-based cropping systems in the tropics. Adv Agron 85:269–407

Smil V (2000) Feeding the world. MIT Press, Cambridge, MA

Smith SE, Read DJ (1997) Mycorrhizal symbiosis. Academic, San Diego, CA

Stubbs TL, Kennedy AC, Schillinger WF (2004) Soil ecosystem changes during the transition to no-till cropping. J Crop Improvement 11:105–135

Swaminathan MS (2001) Ecotechnology: meeting global and local challenges of food insecurity and poverty. Development 44:17–22

Tilman D, Fargione J, Wolff B, D'Antonio C, Dobson A, Howarth R, Schindler D, Schlesinger WH, Simberloff D, Swackhamer D (2001) Forecasting agriculturally driven global environmental change. Science 292:281–284

Tilman D, Cassman G, Matson PA, Naylor R, Polasky S (2002) Agricultural sustainability and intensive production practices. Nature 418:671–677

Tisdall JM, Oades JM (1982) Organic-matter and water-stable aggregates in soils. J Soil Sci 17:429–441

UN (1992) Promoting sustainable agriculture and rural development. United Nations Conference on Environment & Development. Rio de Janerio, Brazil, 3–14 June. Agenda 21 1992; 14.1-14.104. On WWW at http://www.un.org/esa/sustdev/agenda21.htm

UN (2005) The millennium development goals report. United Nations, New York

UN (2008) Organic agriculture and food security in Africa. United Nations Conference on Trade and Development-United Nations Environment Programme, Capacity-building task force on trade, environment and development. http://www.unep-unctad.org/cbtf/publications/UNCTAD_DITC_TED_2007_15.pdf

UNDP (2006) Human Development Report 2006. UNDP, New York. http://hdr.undp.org/en/media/HDR06-complete.pdf

UNEP (2005) One planet, many people: atlas of our changing environment. UN, New York

Vitousek PM, Mooney HA, Lubchenco J, Melillo JM (1997) Human domination of earth's ecosystems. Science 277:494–499

Wall DH, Virginia RA (2000) The world beneath our feet: soil biodiversity and ecosystem functioning. In: Raven P, Williams TA (eds) Nature and human society: the quest for a sustainable world. National Academy of Sciences Press, Washington, DC

Wani SP, Pathak P, Jangewd LS, Eswaran H, Singh P (2003) Improved management of Vertisols in the semi-arid tropics for increased productivity and soil carbon sequestration. Soil Use Manage 19:217–222

Wickama JM, Mowo JG (2001) Using local resources to improve soil fertility in Tanzania. Managing Africa's Soils No. 21. IIED. www.iied.org/drylands

Williams P, Young M (1999) Costing dust – how much does wind erosion cost the people of South Australia. CSIRO Land and Water, Policy and Economic Research Unit. www.clw.csiro.au/publications/consultancy/1999/costing_dust.pdf

World Bank (2009) Global economic prospects 2009: commodities at the crossroads. http://siteresources.worldbank.org/INTGEP2009/Resources/10363_WebPDF-w47.pdf

Young RA, Onstad CA, Bosch DD, Anderson WP (1989) AGNPS, a non-point source pollution model for evaluating agricultural watersheds. J Soil Water Conserv 44:168–173

Index

G

H

I